Theory of Rank Tests

This is a volume in
PROBABILITY AND MATHEMATICAL STATISTICS

Z. W. Birnbaum, founding editor
David Aldous, Y. L. Tong, series editors

A list of titles in this series appears at the end of this volume.

other subsections, sections and chapters. The same convention is used for theorems. The end of the proof is now designated by □.

As with the first edition, the bibliography placed at the end of the book is not intended to be complete though it has been substantially updated. Each chapter is also supplied with a paragraph presenting some problems and supplementary material that have been thoroughly updated. Certain passages of the text are set in a smaller font than the rest of the text. The reason is that they contain topics of minor importance, mathematical, computational and historical remarks etc.

Acknowledgements

Z. Šidák wishes to express his gratitude to the Grant Agency of the Academy of Sciences of the Czech Republic, since the preparation of this book was partially supported by its grant No. A119109. His most sincere thanks go to L. Šidáková, his daughter-in-law, for her painstaking work in typesetting almost the whole manuscript on computer, and to O. Šidák, his son, for preparing the final computer version for printing. Further thanks go also to B. Koubová, Y. Petrusová, M. Jarník for their technical help, and to K. Šidáková, his wife, for her patience during the years of preparation of this book.

P.K. Sen wishes to acknowledge help from Prof. Antonio Carlos Pedroso de Lima (Sao Paulo, Brazil) and Prof. Bahjat Qaqish (Chapel Hill) in matters relating to LaTex preparations and electronic transfers of the manuscript. He also wants to thank his wife Gauri Sen for her consistent encouragement during this project.

Z. Šidák and P.K. Sen would like to acknowledge the impact of the pioneering and on-going work in this field by researchers at the Department of Statistics and Probability of Charles University. P.K. Sen is especially grateful to his colleagues from the Charles University for the long-lasting collaboration that inspired him to undertake this project. They also thank the wife and daughters of the late Prof. J. Hájek for their permission to reprint parts of the first edition.

November 1998

Zbyněk Šidák
Mathematical Institute
Academy of Sciences
of the Czech Republic
Žitná 25, 115 67 Prague 1
The Czech Republic

Pranab K. Sen
Departments of Biostatistics
and Statistics
University of North Carolina
at Chapel Hill
Chapel Hill, NC 27599-7400
USA

Preface to the second edition

This treatise is the second edition, revised and substantially extended, of the book Theory of Rank Tests, by J. Hájek and Z. Šidák, published in 1967.

Sadly, J. Hájek died very untimely in 1974, at the age of only forty-eight. For many years, as the research area of the book continued to flourish, statisticians from all over the world and former colleagues of the late Professor Hájek were pressing Z. Šidák to publish an updated second edition. But without his friend and colleague, this task appeared to be almost too difficult. Finally, by happy chance, he asked Pranab Sen for active cooperation to bring this project to completion. This is how the new edition came to life and why the original two authors of the first edition have changed into three authors for this second edition. We offer this work as a tribute to the memory of Professor J. Hájek.

Substantial material has been taken from the first edition. But, after some forty years of active research, certain approaches had changed, new results needed to be incorporated or given a stronger emphasis. This is why the book went through a substantial revision.

The aim was two-fold: first, to refresh some well established methods (especially those related to contiguity), and, second, to develop many new topics that were not covered in the first edition. As a consequence, this edition might have lost its former compactness, but, on the other hand, it contains much more material of various kinds and of recent interest. A more detailed analysis of what is original and what is more recent is given in Chapter 1.

The organization and style of the book is changed slightly. The book now has ten chapters, each chapter being divided into sections and subsections. The style of referring to individual formulae and theorems was simplified. Equation (1) of Subsection 3.4.3 in Chapter 3 is as before simply referred by (1) throughout this subsection but now by (3.4.3.1) in any

This book is printed on acid-free paper.

ACADEMIC PRESS
525B Street
Suite 1900
San Diego, California 92101-4495, USA
http://www.apnet.com

ACADEMIC PRESS
24–28 Oval Road
LONDON
NW1 7DX, UK
http://www.hbuk.co.uk/ap/

A catalogue record for this book is available from the British Library

ISBN 0-12-642350-4

Printed in the United States of America

99 00 01 02 03 QW 9 8 7 6 5 4 3 2 1

THEORY OF RANK TESTS

Jaroslav Hájek
Zbyněk Šidák
Pranab K. Sen

(Second Edition)

ACADEMIC PRESS
San Diego London Boston New York
Sydney Tokyo Toronto

Preface to the first edition

This book is designed for specialists, teachers and advanced students in statistics. We assume the reader to be acquainted with the basic facts about the theory of testing hypotheses, measure theory, stochastic processes, and the central limit theorem. Our two basic reference books in these respects are Lehmann (1959) and Loève (1955). The main body of the present book is based on the ideas of LeCam and of one of the authors. Asymptotic methods developed here are likely to be useful also in other statistical problems. The overlapping of this book with the related books by Fraser (1957a), Siegel (1956) and Walsh (1962) is indeed small.

Striving for compactness and lucidity of the theory, we concentrated on contiguous alternatives and on problems concerning location and scale parameters. In this respect the results obtained are almost complete. The two most serious gaps still left are the absence of an effective method for the estimation of the type of a density, and the failure to carry out an adequate asymptotic treatment of the alternatives for the hypotheses of independence. Relatively little space has been given to the non-contiguous alternatives and to the famous Chernoff-Savage theorem, and no space at all to the interesting investigations on the possibility of employing rank tests for estimations problems, started by Hodges and Lehmann (1963), Lehmann (1963b).

The bibliography placed at the end of the book, though rather extensive, is naturally still far from being complete; we have tried to gather here only those publications that are more closely related to the topics treated in the book. A very complete bibliography has been compiled by Savage (1962).

The book has seven chapters, each chapter being divided into paragraphs, and each paragraph into sections. The style of referring to individual formulas and theorems is as usual: e.g. formula (1) of Section 4.3 in

Chapter II is throughout this section referred to simply as (1), in other sections and paragraphs of Chapter II as (4.3.1), in other chapters as (II.4.3.1); Theorem II.4.2 means the theorem presented in Section 4.2 of Chapter II, etc. If there are several theorems (lemmas, definitions, remarks, examples) in a section, they are distinguished by adding letters a, b, c, ...

Each chapter is supplied with a paragraph presenting some problems and complements. These paragraphs contain much additional material; we therefore recommend that every reader at least read them, even if he is not willing or has no time to solve the problems in detail.

For those who are interested in the elementary theory of rank tests only, or who wish to get acquainted with a survey of rank tests more for practical purposes, we suggest reading the first three or four chapters (maybe omitting Paragraph I.4 and Sections II.2.1 and II.2.2).

At the end of the book an index of frequently used symbols is given for the convenience of the reader, in addition to the subject index and the author index. The symbol Q.E.D. denotes the end of a proof.

We wish to express our thanks to Messrs. V. Dupač and M. Josífko for their critical examination of the manuscript, to Mr. M. Basch for revising the English, and to all who helped prepare the typewritten copies.

Prague, March 1965

J. Hájek
Z. Šidák

Contents

Chapter 1

Introduction and coverage

1.1 THE BACKGROUND

Our treatise of the theory of rank tests comprises a specialized and yet important sector of the general theory of *testing statistical hypotheses* with due attention to the dual rank-based R-estimation theory. The genesis of rank tests is in *nonparametric* or *distribution-free methods* that generally put much less emphasis on the specific forms of the underlying probability distributions. In this simple setup, the ranks are *maximal invariant* with respect to the group of strictly monotone transformations on the sample observations, and hence, they lead to rank tests that are simple, computationally attractive, and applicable even when only ranking data are available. In the current statistical literature, rank tests have also been labelled as a broader class of tests based on ranks of sample observations; for suitable *hypotheses of invariance* under appropriate *groups of transformations*, such rank tests may be genuinely *(exact) distribution-free* (EDF), while in more composite setups, they are either *conditionally distribution-free* (CDF), or *asymptotically distribution-free* (ADF). This feature makes it possible to prescribe rank based statistical inference procedures under relatively less stringent regularity assumptions than in a conventional parametric setup based on some specific distributional models. Nevertheless, the development of the theory of rank tests, particularly over the past 40 years, goes far beyond the traditional nonparametric interpretations; it will be seen in the sequel that rank tests have their natural appeal from a broader perspective incorporating scope for applicability, *global robustness* and *(asymptotic) efficiency* considerations all blended harmoniously. Yet it is worth noting that rank tests are closely allied to *permutation* or *randomization tests* that commonly arise in testing statistical hypotheses of invariance. The current treatise of the theory of rank tests includes a broad class of *semiparametric models* and is amenable to various practical applications as well.

What made the theory of rank tests a flourishing branch of statistical research is no doubt the success of rank tests in both theory and practice. The scenario is quite simple in traditional nonparametric models. The usual characteristics (namely, EDF, simplicity and computational flexibilities) may not, however, be fully tenable without an hypothesis of invariance. Nevertheless, the main thrust underlying the popularity of rank tests is their global robustness with usually moderate to little (and sometimes asymptotically negligible) *loss of power-efficiency* properties; this appraisal constitutes the main objective of this updated and revised version of the theory of rank tests. In this context, general nonparametric and semiparametric models pertaining to various *univariate* as well as *multivariate, single* as well as *multisample* problems, *semiparametric linear models*, and even some simple *sequential models*, are covered to depict the general structure and *performance characteristics* of rank tests.

The intricate relationship between the theory of statistical tests and the dual (*point* as well as *set/interval*) estimation theory have been fully exploited in the parametric case, and some of these relationships also hold for many semiparametric models. The recent text by Jurečková and Sen (1996) provides an up-to-date account of *robust statistical procedures* (theory and methodology) in *location-scale* and *regression models*, encompassing the so called M-, L-, and R-*estimation* procedures, along with their siblings. We find it quite appropriate to examine the duality of the theory of rank tests and the theory of R-estimators.

An *alignment* principle having its genesis in *linear statistical inference* methodology, as incorporated in rank based (typically non-linear) inference methodology, has opened the doors for a large class of rank test statistics and estimates. These are known as *aligned rank statistics.* It will be quite in line with our general objectives to emphasize R-estimates based on aligned rank statistics, in order to examine the effective role of the theory of rank tests in this prospective domain too.

Multivariate statistical analysis, once thought as invincible by nonparametrics, has already been annexed to this domain by the successful intervention of the theory of rank tests that has been developed at the cost of sacrificing the EDF property in favour of suitable CDF/ADF properties. With the initial lead by the Calcutta school in the early 1960s, *multivariate rank tests* (theory and methodology) acquired a solid foundation within a few years. The Prague school, under the pioneering leadership of the late Jaroslav Hájek, has made a significant contribution toward this development. A treatise of *multivariate nonparametrics*, covering the developments in the 1960s, is due to Puri and Sen (1971), although it has been presented in a somewhat different perspective. Again, significant developments have cropped up during the past 25 years, and they would be tied up with our current treatise of the theory of rank tests.

Intricate distribution-theoretical problems for rank statistics under general alternatives stood, for a while, in the way of developing the theory of rank tests for *general linear models.* A breakthrough in this direction is due to Hájek (1968), and following his lead, the Prague school has made significant contributions in this area also. Puri and Sen (1985) contains a comprehensive account of some of these developments up to the early 1980s. We find it quite appropriate to update and appraise the theory of rank tests in general linear models. Within this framework, in the context of *subhypothesis testing* problems, because of *nuisance parameters*, an hypothesis of invariance may not generally be appropriate here. *Aligned rank tests* have emerged as viable alternatives (see for example, Sen (1968b), Sen and Puri (1977), Adichie (1978), and others), and for these tests a theoretical foundation can be fully appraised by incorporating the so-called *uniform asymptotic linearity of rank statistics in location/regression parameters* results. A significant part of these developments took place in Prague, and are reported systematically in Jurečková and Sen (1996). In our unifying and updating task of the theory of rank tests, due emphasis will be placed on the profound impact of such asymptotic linearity results on the theory of (aligned) rank tests.

Asymptotic theory or *asymptotics* occupy a focal point in the developments of the theory of rank tests. These asymptotics are pertinent in the study of the distribution theory of rank statistics (under null as well as suitable alternative hypotheses), and more so, in the depiction of local and asymptotic *power* and *optimality* properties of rank tests. These asymptotics also crop up in the study of *asymptotic relative efficiency* (ARE) properties of rank tests. In this respect, a very useful tool (mostly developed by Hájek (1962) from LeCam's (1960) original but somewhat more abstract formulation), namely, the *contiguity of probability measures*, has reshaped the entire flavour of asymptotics in the theory of rank tests. This has indeed been the bread and butter of the general asymptotics presented in a systematic and unified manner in the original edition of the *Theory of Rank Tests.* It is of natural interest to contrast this contiguity based approach to some alternative ones, such as the general case treated in Hájek (1968), with special attention to the developments that have taken place during the past 30 years.

By construction, rank statistics are generally neither linear functions of the sample observations nor have they an *independent summands* structure. Nevertheless, an important property of a general class of rank statistics is their accessibility to the general *martingale* methodology under appropriate hypotheses of invariance, and this feature extends to general *contiguous alternatives* as well. This martingale approach to rank test theory, exploited fully (in a relatively more general *sequential* setup) in Sen (1981), may also be tied up with the general theory of rank tests. *Weak convergence* of probability measures or *invariance principles*, only partly introduced in the original text, will also be updated to facilitate the accessibility of this contiguity approach in a broader setup.

One of the open problems encountered in the early 1960s in the context of rank tests is the following: In order to make a rational choice from within a class of rank tests, all geared to the same hypotheses testing problem, we need to have a knowledge of the form of the underlying distribution or density functions that are generally unknown, though assumed to have finite *Fisher information* with respect to location or scale parameters. The characterization of *locally optimal rank* tests (even in an asymptotic setup) may invariably involve the so-called *Fisher score function* that depends on the logarithmic derivative of the unknown density function when the latter is assumed to be absolutely continuous. Thus, there is a genuine need to estimate the underlying (and assumed to be absolutely continuous) density function to facilitate the construction of such (asymptotically) optimal rank tests (against parametric alternatives). This problem, treated in an intuitive manner, in the very last chapter of the original text, requires an enormously large sample size in order to be suitable for practical adoption. This drawback has been eliminated to a great extent, for rank tests and allied R-estimates, by incorporating *adaptive rank statistics* based on suitable *orthonormal expansions* of the Fisher score function, along with robust estimation of the associated Fourier coefficients based on linear rank statistics; we refer to Hušková and Sen (1985, 1986) for details and for a related bibliography as well. This piece of development naturally places the formulation of the theory of *aligned adaptive rank tests* on a stronger footing.

In the contemplated updating task, attempts have been made to cover the entire field of developments on the theory of rank tests. During the past fifteen years or so there has been an increase of development on *semiparametric* models where rank tests often crop up in some way or other. For example, in the original formulation of the *proportional hazards model*, due to Cox (1972), the *log-rank statistic* provides the link with conventional nonparametrics. This model has led to a vigorous growth of statistical literature on semiparametrics, and in its complete generality such a semiparametric model, treated in Andersen et al. (1993), involves some (multivariate) *counting processes*, and the developed methodology rests on suitable martingale theory. A somewhat different approach to asymptotically optimal semiparametric procedures has been pursued by Bickel et al. (1993). We shall not, however, attempt to intrude into this specialized branch of the asymptotic theory of statistical inference, beyond an introduction to the relevance of semiparametrics to the theory of rank tests.

A synopsis of the basic organization of the present version of the theory of rank tests is provided in the next section.

1.2 ORGANIZATION OF THE PRESENT TREATISE

Our dual objective is to preserve, to the greatest possible extent, the flavour of the original treatise of the *Theory of Rank Tests*, and at the same time, to capture the highlights of developments that have taken place mostly during the past 35 years to update the coverage of the original text. As such, our delicate task is primarily geared to retaining the original presentation of the classical part as far as possible, along with a unified and integrated supplementation of new (and later) developments, either in the form of new chapters, or as new sections added to some existing ones. For reasons well understood, some parts of the original text were to be virtually replaced by more general, updated versions; there are, however, certain sections that have gone through an extensive updating and/or gross replacement. Moreover, though it might have been quite tempting to incorporate a complete coverage of recent developments in this vast field in our contemplated updating task, in view of their rather encyclopedic nature, and also due to the appearance of other contemporary texts and advanced monographs (most of which being primarily devoted to more specialized topics), we shall mainly focus on the basic relationship of these recent developments with the original text in a unified manner, so as to comprehend the basic role of the theory of rank tests with regard to the theoretical foundation of these later developments. In this vein, cross-references to their original sources, as well as to relevant texts and monographs, will be provided along with appropriate discussions on their interrelationship and diversity as well. Further, it is to be noted here that our goal is to present a treatise of the theory of rank tests on a sound statistical and probabilistic basis. In many contemporary developments, there has been a basic emphasis on applications in many other fields, notably in biomedical and clinical sciences, reliability and survival analysis and applied sciences in general. Even dealing with some of these application-oriented problems involving fruitful adoption of rank tests, we shall mainly confine ourselves to the basic theoretical perspectives so as to highlight their impact in the respective field of application. In this way, it may be more convenient for an application-oriented reader to comprehend the basic theoretical and methodological foundations of such applications. Finally, in spite of our primary emphasis on theoretical foundations, we shall refrain from sheer abstractions as far as possible, without, of course, sacrificing rigour and clarity of presentation.

Our Chapter 2 is adapted from Chapter I of the original text (except the former Section I.1.1 that has been updated to the current Section 1.1). It contains the theoretical basis of mathematical statistics that is particularly relevant to the theory of testing statistical hypotheses. The presentation of the original text for most of the sections has been preserved here, including the classical theorem on the distribution of a quadratic form in random vectors having normal distributions. A new inclusion in Chapter 2 is an introduction (in 2.1.3) to *martingales, reversed martingales* and *submartingales* that play an important role in the study of the properties of rank statistics in finite as well as large sample sizes.

Chapter 3 is a reinstatement of Chapter II of the original text, and it deals with the elementary (finite sample) theory of rank tests. The basic *hypotheses of invariance*, arising mainly in the context of testing for *randomness* (two or several sample models as well as regression models), *symmetry* of a distribution about a specified location (one-sample and paired-sample location models), *interchangeability* of the elements of a stochastic vector (randomized block designs models), and *bivariate independence*, are formulated; in the context of conventional *permutation (randomization) tests* and *invariant* tests, the genesis of rank tests is appraised. First and second order moments of linear rank statistics under such hypotheses of invariance are obtained in closed form. An important topic, dealt with in Section 3.4, is a general treatise of *locally most powerful rank* (LMPR) tests for such hypotheses of invariance against specific types of (parametric) alternatives in the single parameter case. A general discussion is appended on multiparameter cases as well. For the sake of completeness, Hoeffding's (1948a) *U-statistics* and related von Mises (1947) functionals are introduced in a new Section 3.5.

Chapter 4 retains the flavour of Chapter III of the original text. For testing statistical hypotheses of invariance (introduced in the preceding chapter), some selected rank statistics are considered and their basic finite sample properties are displayed. *Kolmogorov-Smirnov* type tests, originally presented for the *goodness of fit* problems, are also considered for the same hypotheses testing problems. New inclusions to the current chapter are the following: (1) Kolmogorov-Smirnov type tests for symmetry problem, and (2) a detailed treatment of rank tests under various types of *censoring*, with due emphasis on the classical *life testing* problems (see Section 4.9). Multivariate rank tests for suitable hypotheses of invariance that have their genesis in the theory of rank tests are introduced in Section 4.10. In that way, the passage from the univariate to the multivariate hypotheses testing problems has been fortified.

Chapter 5 of the present text, like Chapter IV of the original version, is devoted to the computation of exact distributions (under appropriate null hypotheses of invariance) of various rank test statistics; some recursive relations have been exploited in this context. The importance of the *exact distributions* and *null hypotheses* should not be overemphasized. In this respect, the current version differs somewhat from the original one. Namely, the exposition of computational problems for distributions of some of the test statistics was shortened a little bit. *Asymptotic expansions*, from theoretical perspectives, belong to the domain of asymptotics, with often too little (and mostly empirical) justifications for finite sample sizes. Hence, these are deferred to Subsection 8.4.1 in view of their relation to Hodges-Lehmann deficiency introduced in 8.4.2.

Back in the 1960s when a unified treatise of weak convergence of probability measures and some other related concepts was lacking, Chapter V of the original text had a special appeal of its own. The present Chapter 6 retains the salient features of that chapter, but is updated considerably in order to keep pace with the most significant developments that have taken place during the past 30 years or so. Nevertheless, the first two sections follow closely the original text. Section 6.3 has been recast, and to incorporate new developments a new section on *functional central limit theorems* and *weak invariance principles* for rank statistics has been added. In this context, martingale characterizations for various rank statistics, mostly adapted from Sen (1981), play a fundamental role. These invariance principles also paved the way for the development of *sequential nonparametrics*, and some of them are outlined in the form of exercises at the end of the chapter.

Chapter 7 of the current text is an updated version of the original Chapter VI, where the main thrust has been on the exploitation of the notion of contiguity of probability measures for the derivation of asymptotic distribution theory under (local) alternatives, in a simple and yet unified manner. These developments relate to a much bigger class of statistics that need not be of the rank type, and in view of this broader perspective, the first four sections in the present chapter are closely adopted from the original text. Nevertheless, to outline the significant role of contiguity based approaches in statistical inference, the original presentation has been elaborated a bit more here. In Section 7.1, Subsection 7.1.6 is added to outline the relation of the *Hellinger distance* to contiguity; Subsection 7.1.5 added there deals with the multiparameter case arising in linear models and multivariate problems. The original Section VI.5 has been recast completely. At that time, in the early 1960s, the case of *non-contiguous alternatives* referred mostly to the classical *Chernoff-Savage* (1958) type results. However, during the later 1960s, there was a significant research accomplishment on the asymptotic distribution theory of rank statistics under alternatives of various types, and it truly culminated with the most outstanding work of Hájek (1968). As such, in the present Section 7.5, due emphasis has been laid on the *projection method* for rank statistics developed in Hájek (1968).

Although Chapter 8 of the current text has its genesis in the original Chapter VII, in order to cope with the updating task, there has been a reorganization to a considerable extent. The first two Sections 8.1 and 8.2 closely follow the original version, though the former Subsections VII.1.5 and VII.1.6 on estimation of density and Fisher score function have been relegated to Section 8.5, where *adaptive rank tests* have been introduced in greater generality to cover these topics as well. Among other new additions in the present text, it is worth mentioning the developments of *Bahadur efficiency* and *Hodges-Lehmann deficiency* concepts (Sections 8.3 and 8.4). Asymptotic efficiency of rank tests for multiparameter hypotheses has also been discussed in this chapter.

The last two chapters, namely, Chapter 9 and 10, relate exclusively to some important developments that have taken place mostly during the past 30 years, and were not covered in the original text. These topics include *ranking after alignment* leading to *aligned rank tests* (see Section 10.1) that are of two types: CDF or ADF. The CDF procedures generally relate to block designs where the estimation of the nuisance parameters (for example, block effects) may still render interchangeability of the aligned observations (within each block) and thereby permits the formulation of CDF rank tests. The ADF rank tests, on the other hand, mainly refer to *subhypothesis testing* problems where in the presence of nuisance parameters, the process of alignment yields residuals that may no longer be interchangeable. The formulation of both CDF and ADF aligned rank tests has been considered here in a general mold. Secondly, rank test statistics in conjunction with suitable *alignment principles* lead to suitable point as well as interval (set) estimators of parameters in some conventional semi-parametric models. These developments were sparked in the early 1960s (Hodges and Lehmann (1963), and Sen (1963), see Section 9.1) but gained considerable momentum in the years to follow (Adichie (1967), Sen (1968a), and others). The diverse tracks in this novel area of *R-estimation theory* were later unified by an approach, now labelled as the *uniform asymptotic linearity of rank statistics in shift/regression parameters* (Jurečková (1969, 1971a,b), Koul (1969), Sen (1969), and others, see Section 9.2 and 9.3). This unification rests heavily on the theory of rank tests, and more so on related asymptotics. These developments in a more general setup of *robust statistical inference* (see Section 10.3), including L-, M- and R-estimators, have recently been presented in a unified manner in Jurečková and Sen (1996). Therefore, we shall mainly confine ourselves to the basic impact of the theory of rank tests on such asymptotic linearity results underlying R-estimation theory in a general setup. The concept of contiguity of probability measures, reported in Chapter 7, plays a fundamental role in this context too, and therefore that aspect is also highlighted. Another important development that has its genesis in the theory of rank tests relates to the *regression rank scores* (see Section 10.2). These scores extend the Hájek (1965) scores (also treated in the original text) to linear models, and were introduced by Gutenbrunner and Jurečková (1992) for robust estimation in linear models, and by now (see Chapter 6 of Jurečková and Sen (1996)) they pertain to robust testing as well. We shall comment briefly on certain asymptotic equivalence results for regression rank scores estimates and classical R-estimates in linear models, which also hold for allied tests.

Developments in the area of multivariate nonparametrics took place mostly in the 1960s and 1970s. Our updating mission would have remained somewhat incomplete if the theory of rank tests for some genuine multivariate problems were totally left out. In the first edition of the book these problems were intentionally omitted in order to keep the exposition concise and compact. However, in this second edition we have decided to add to the original text also some older ideas and developments (of course, not speaking of the new ones) to cover the field of nonparametrics as much as possible. Thus, a short survey of basic multivariate nonparametric tests is contained in the new Section 4.10. These tests are mainly based on the *rank permutational invariance principles* (Chaterjee and Sen (1964)), so that they are usually CDF. Moreover, there are some brief remarks on the multivariate tests spread at other places of the book, e.g. in Subsection 6.4.4 on invariance principles, or in Subsection 7.1.4 on contiguous alternatives, etc. The first book devoted to multivariate nonparametrics was Puri and Sen (1971); as for the rank procedures in linear models, the interested reader may find more thorough information in the monographs of Puri and Sen (1985), and Jurečková and Sen (1996). Of course, in this book we will provide cross-references for details, if it seems useful.

Finally, the *Bibliography* appearing in the original text has been updated in the current version. This has been done with the inclusion of the principal references pertaining to the new material imported in the current text, as well as updating the previous set with some other important ones that were not listed. It is almost an encyclopedic task to provide a complete list of all publications pertaining to the general field of the theory of rank tests, and is certainly beyond the scope of the current updating task. Therefore, the present Bibliography, appended at the end of the text, should by no means be regarded as an exhaustive one.

Chapter 2

Preliminaries

2.1 BASIC NOTATION

This section deals with the basic notation adopted in the formulation of a probability space and random elements defined on it. Statistics, σ-fields and λ-fields are also introduced in the same vein. The last subsection deals with the preliminaries of martingales, reversed martingales and submartingales which play a basic role in subsequent developments.

2.1.1 Probability space and observations. From a probabilistic point of view, a random experiment is represented by a *probability space* $(\Omega, \mathcal{A}, P)$, where Ω is the space of all possible outcomes of the experiment, $\mathcal{A}$ is a σ-field determining how finely the outcomes are distinguished, and $P(.)$ is a probability measure. In this setup, observations as well as other random variables are defined as $\mathcal{A}$-measurable functions.

From a statistical point of view, an experiment is described by an indexed set of *observations* $X = (X_1, \dots, X_N)$, each observation X_i taking its values x_i on the real line or on a proper subset of the real line. The set of observations X is determined by its (joint) distribution function

$$F(x_1, \dots, x_N) = P(X_1 \leq x_1, \dots, X_N \leq x_N), \qquad -\infty < x_1, \dots, x_N < \infty. \tag{1}$$

In this framework, it is not necessary to assume that the X_i are independent, though in many statistical model, they are assumed to be so. In the latter case, we have

$$F(x_1, \ldots, x_N) = \prod_{i=1}^{N} P(X_i \leq x_i) = \prod_{i=1}^{N} F_i(x_i).$$

If more convenient, the individual observations may be distinguished by indices other than $1, \ldots, N$, or by using several generic letters X, Y, Z, etc. For example, we may work with $(X_1, \ldots, X_m, Y_1, \ldots, Y_n)$, $(X_1, Y_1, \ldots, X_N, Y_N)$, $(X_{11}, \ldots, X_{nk})$, and so on.

2.1.2 Statistics, σ-fields and λ-fields. A set of observations X together with a distribution function $F(x)$ generates a probability space $(\mathsf{X}, \mathcal{A}, P)$ as follows: X is the space of all possible values $x = (x_1, \ldots, x_N)$ of X, i.e. it is an N-dimensional Euclidean space; $\mathcal{A}$ is the σ-field of Borel subsets of X, i.e., the smallest σ-field containing all events appearing on the right side of (2.1.1.1); finally, $P(\cdot)$ is the probability measure uniquely determined by (2.1.1.1). The set of observations X may be considered to be an identity map $X(x) = x$, $x \in \mathsf{X}$. In this book we shall use the term probability distribution instead of probability measure, in accordance with statistical usage.

Throughout the book we shall usually follow the convention of using capital letters X, Y, Z, $R, \ldots$ for random variables and the corresponding small letters x, y, z, $r, \ldots$ for their particular values. To every measurable function $s(x)$ on $(\mathsf{X}, \mathcal{A})$ there corresponds a random variable $s(X)$, denoted by S or by some other capital letter. The expectation of S will be denoted either by $\mathsf{E}S$ or by $\mathsf{E}s(X)$, and, if it exists, defined as the N-dimensional Lebesgue-Stieltjes integral

$$\mathsf{E}s(X) = \int s(x)\,\mathrm{d}P. \tag{1}$$

Here we do not indicate the dimensionality by multiple integral signs $\int \ldots \int$ if there is no danger of confusion. As a rule the probability distribution $P(\cdot)$ will be determined by a density p,

$$P(A) = \int_A p(x)\,\mathrm{d}x, \quad A \in \mathcal{A}. \tag{2}$$

Consequently the expectation may then be expressed as the Lebesgue integral:

$$\mathsf{E}s(X) = \int s(x)p(x)\,\mathrm{d}x. \tag{3}$$

However, even in such a case, we shall prefer the shorter notation (1). Similarly $\mathrm{d}Q$ will stand for $q\,\mathrm{d}x$ etc.

Occasionally we shall write $P(X \in A)$ for $P(A)$, to indicate that A is a subset of x-values. If $s = s(X, Y)$ is a function of two independent random variables, then $\mathsf{E}s(x, Y)$ denotes the integral over the y-values while x is fixed. This usage may be extended to functions of several independent random variables as well.

We do not share the opinion that, in statistics, the notion of a σ-field is solely a concession to mathematical formalism. On the contrary, it is a very useful notion and should belong to the basic apparatus. Making this statement, we have in mind various sub-σ-fields of the σ-field of Borel subsets as well as monotone sequences of σ-fields we shall make use of in the sequel.

The notion of a sub-σ-field is intimately connected with the notion of a statistic. Generally speaking, a statistic $s(x)$ is a measurable map of $(\mathcal{X}, \mathcal{A})$ to another measurable space $(\mathcal{S}, \mathcal{B})$. It is easily seen that the class $\{s^{-1}(B),\, B \in \mathcal{B}\}$ represents a sub-σ-field of $\mathcal{A}$, invariant with respect to some map of $\mathcal{X}$ onto itself, by partitions of $\mathcal{X}$, etc.

Let us recall that a σ-field $\mathcal{A}$ is a class of subsets (events) of $\mathcal{X}$ containing $\mathcal{X}$ and closed with respect to taking complements and countable unions. It is also convenient to introduce the notion of a λ-field: A class of subsets of $\mathcal{X}$ will be called a λ-field if it contains $\mathcal{X}$ and is closed under taking complements and countable unions of *disjoint* members. Thus the only difference in the two definitions lies in the word 'disjoint'. Obviously, every σ-field is also a λ-field, but the converse is not true.

Most important λ-fields are generated by classes of similar events for a system of probability distributions $P_\Theta(\cdot)$, $\Theta \in \boldsymbol{\Theta}$. Recall that an event A is called similar if its probability $P_\Theta(A)$ is independent of $\Theta \in \boldsymbol{\Theta}$. We shall see that permutation tests may be defined as tests whose critical regions belong to the λ-field of similar events.

The very fact that the similar events do not constitute a σ-field but only a λ-field is the source of difficulties encountered in the application of conditional arguments. Actually, the conditioning is always defined with respect to some σ-field contained in the λ-field of similar events, which is a process without end, since there generally is no such maximal σ-field.

The notion of a λ-field is also connected with another fundamental problem. We may ask how far it is justified to insist that a probability distribution be defined on a σ-field. Actually, the basic relation

$$P\Big(\bigcup_{i=1}^{\infty} A_i\Big) = \sum_{i=1}^{\infty} P(A_j), \quad (A_i \cap A_j = \emptyset,\ i \neq j)$$

makes sense on λ-fields as well. Moreover, we can imagine practical experiments in which the class of all events with which a numerical probability may reasonably be associated does not constitute a σ-field, but constitutes a λ-field (see Problem 4).

We think that there is no extramathematical reason for preferring σ-fields. However, the mathematico-technical reasons are very strong. They lie in the existence of conditional probabilities and expectations relative to σ-fields, which has no counterpart for λ-fields. Assuming a λ-field only, we could not even define the probability density of a probability distribution with respect to some dominating measure, and hence cannot derive the basic Neyman-Pearson lemma in the theory of hypothesis testing.

2.1.3 Martingales and related sequences. First, we extend the notion of random variables to *stochastic processes* that are collections of random variables. Let T be an *index set* and $\{X_t, t \in T\}$ be a collection of one-dimensional random variables, defined on a probability space $(\Omega, \mathcal{A}, P)$, such that X_t is $\mathcal{A}$-measurable for each $t \in T$. In the discrete case, T can be taken as the set of all non-negative integers, or a subset of it. Let $F = \{F_t, t \in T\}$ be a system of sub-σ-fields of $\mathcal{A}$, such that X_t is F_t-measurable, for each $t \in T$. Then we say that $X = \{X_t, t \in T\}$ is adapted to F, and is written as $\{X_t, F_t; t \in T\}$. Note that if F_t is non-decreasing (or non-increasing) then $F_t \subset F_{t'}$ (or $F_t \supset F_{t'}$), for every $t < t'$ $(\in T)$. Assume further that $\mathsf{E}|X_t| < \infty$ for all $t \in T$.

Now if F_t is non-decreasing and

$$\mathsf{E}\{X_{t'} \mid F_t\} = X_t \text{ a.e., for all } t' \geq t \ (\in T),$$

then $\{X_t, F_t; t \in T\}$ is termed a *martingale.* If, in the above definition, the '$=$' sign is replaced by '$\geq$' (or '$\leq$'), then $\{X_t, F_t; t \in T\}$ is termed a *sub-martingale* (or *super-martingale*). Similarly, if F_t is non-increasing and

$$\mathsf{E}\{X_{t'} \mid F_t\} = X_t \text{ a.e., for all } t' \leq t \ (\in T),$$

then $\{X_t, F_t; t \in T\}$ is a *reversed* or backward *martingale*, and replacing the '$=$' by '$\geq$' (or '$\leq$'), we obtain a *reversed sub-martingale* (or *super-martingale*).

Note that in the above definitions, T can be replaced by a linearly ordered index set Γ, and defining $\{F_\gamma, \gamma \in \Gamma\}$ in an analogous way, we have parallel results for $\{X_\gamma, F_\gamma; \gamma \in \Gamma\}$. Moreover, if $\{X_t, F_t; t \in T\}$ is a (reversed) martingale and $g(.)$ is convex (or concave), then $\{g(X_t), F_t;\ t \in T\}$ is a (reversed) sub- (or super-) martingale. In particular, for $g(z) = |z|^p$, for $p \geq 1$, we have $g(.)$ convex, and for $0 < p < 1$, $g(.)$ concave. Finally, a reversed martingale is *uniformly integrable.*

2.2 FAMILIES OF ONE-DIMENSIONAL DENSITIES

As a rule we shall assume that the observations $X_1, \ldots, X_N$ are independent, so that the N-dimensional densities p, q will be products of the one-dimensional densities f, g. Here we shall establish some auxiliary results concerning one-dimensional densities needed later.

2.2.1 Absolutely continuous densities. We say that a density $f(x)$ is absolutely continuous on finite intervals if, for every $C > 0$, $\varepsilon > 0$, there exists a $\delta = \delta(\varepsilon, C)$ such that for each finite set of disjoint intervals (a_k, b_k) the relation

$$\sum_{k=1}^{n} |b_k - a_k| < \delta \quad \text{and} \quad -C < a_k,\ b_k < C \tag{1}$$

implies

$$\sum_{k=1}^{n} |f(b_k) - f(a_k)| < \varepsilon. \tag{2}$$

It is a well-known theorem of calculus that $f(x)$ is absolutely continuous on finite intervals if and only if there exists a function $f'(x)$ integrable on finite intervals and such that

$$f(b) - f(a) = \int_a^b f'(x)\, \mathrm{d}x, \quad -\infty < a < b < \infty. \tag{3}$$

We further know that $f'(x)$ represents the derivative of $f(x)$ almost everywhere. However, the converse is not true, i.e. the existence of the derivative a.e. does not imply (3), unless $f(x)$ is absolutely continuous on finite intervals.

In what follows we shall drop the specification 'on finite intervals', and speak about absolutely continuous densities only. A useful criterion is given by the following.

Theorem 1 *If a finite derivative of $f(x)$ exists at all but a finite number of points, at which, however, the right-hand and left-hand derivatives exist, and is integrable over finite intervals, then $f(x)$ is absolutely continuous.*

2.2.2 Strongly unimodal densities. If $-\log f(x)$ is a convex function within some open interval (a, b) such that $-\infty \le a < b \le \infty$ and $\int_a^b f(x)\, \mathrm{d}x = 1$, the density $f(x)$ is called strongly unimodal. Such densities are absolutely continuous within (a, b) and

$$[-\log f(x)]' = -\frac{f'(x)}{f(x)}$$

is a non-decreasing function. In Table 1 a few typical examples of strongly unimodal densities are given.

Theorem 1 *A convolution of two strongly unimodal densities is again a strongly unimodal density.*

Proof. See Ibragimov (1956). □

Table 1

Density	$f(x)$ and domain, if different from $(-\infty, \infty)$	$-\dfrac{f'(x)}{f(x)}$	$\varphi(u, f)$	$1 + \varphi_1(u, f)$
Normal (1)	$(2\pi)^{-\frac{1}{2}} \exp(-\frac{1}{2}x^2)$	x	$\Phi^{-1}(u)$	$[\Phi^{-1}(u)]^2$
Double-exponential (2)	$\frac{1}{2}e^{-\lvert x\rvert}$	$\operatorname{sign} x$	$\operatorname{sign}(2u-1)$	$-\log(1 - \lvert 2u-1\rvert)$
Exponential (3)	e^{-x} for $x > 0$	1	not defined	$-\log(1-u)$
Logistic (4)	$e^{-x}(1+e^{-x})^{-2}$	$(1-e^{-x})(1+e^{-x})^{-1}$	$2u-1$	$(2u-1)\log[u/(1-u)]$
Uniform (5)	1 for $0 < x < 1$	0	not defined	not defined
Triangular (6)	$1 - \lvert x\rvert$ for $-1 < x < 1$	$(\operatorname{sign} x)(1-\lvert x\rvert)^{-1}$	$[\operatorname{sign}(2u-1)](1-\lvert 2u-1\rvert)^{-\frac{1}{2}}$	not interesting
No name (7)	$\exp(x - e^x)$	$e^x - 1$	$-1 - \log(1-u)$	not interesting
No name (8)	$\frac{1}{2}(1+\lvert x\rvert)^{-2}$	$2(\operatorname{sign} x)(1+\lvert x\rvert)^{-1}$	$2(1-\lvert 2u-1\rvert)\operatorname{sign}(2u-1)$	$2\lvert 2u-1\rvert$
No name (9)	1 for $\lvert x\rvert \le \frac{1}{4}$ $1/16x^2$ for $\lvert x\rvert > \frac{1}{4}$	0 for $\lvert x\rvert \le \frac{1}{4}$ $2/x$ for $\lvert x\rvert > \frac{1}{4}$	0 for $\lvert u-\frac{1}{2}\rvert \le \frac{1}{4}$ $16(1-\lvert 2u-1\rvert)\operatorname{sign}(2u-1)$ for $\lvert u-\frac{1}{2}\rvert > \frac{1}{4}$	$1 + \operatorname{sign}(\lvert 2u-1\rvert - \frac{1}{2})$

2.2.3 Densities with finite Fisher information. If $f(x)$ is absolutely continuous and

$$I(f) = \int_{-\infty}^{\infty} \left(\frac{f'(x)}{f(x)}\right)^2 f(x)\,\mathrm{d}x < \infty, \tag{1}$$

we say that f has finite Fisher information $I(f)$. If f is not absolutely continuous, we put $I(f) = \infty$.

Theorem 1 *If $G(y)$ is an arbitrary distribution function and*

$$h(x) = \int_{-\infty}^{\infty} f(x-y)\,\mathrm{d}G(y),$$

then

$$I(h) \leq I(f).$$

Proof. If $I(f) = \infty$, we have nothing to prove. If $I(f) < \infty$, then f is absolutely continuous and

$$\int_{-\infty}^{\infty} |f'(x)|\,\mathrm{d}x \leq \left\{\int_{-\infty}^{\infty} \left[\frac{f'(x)}{f(x)}\right]^2 f(x)\,\mathrm{d}x\right\}^{\frac{1}{2}} < \infty.$$

Hence

$$\int_{-\infty}^{\infty} \left[\int_{-\infty}^{\infty} |f'(x-y)|\,\mathrm{d}x\right] \mathrm{d}G(y) < \infty,$$

and therefore, by Fubini's theorem,

$$\begin{aligned}\int_{-\infty}^{x} \int_{-\infty}^{\infty} f'(t-y)\,\mathrm{d}G(y)\,\mathrm{d}t &= \int_{-\infty}^{\infty} \left[\int_{-\infty}^{x} f'(t-y)\,\mathrm{d}t\right] \mathrm{d}G(y) \\ &= \int_{-\infty}^{\infty} f(x-y)\,\mathrm{d}G(y) = h(x),\end{aligned}$$

so that $h(x)$ is absolutely continuous as well and

$$h'(x) = \int_{-\infty}^{\infty} f'(x-y)\,\mathrm{d}G(y).$$

Finally, according to the Schwartz inequality,

$$\frac{[h'(x)]^2}{h(x)} \leq \int_{-\infty}^{\infty} \frac{[f'(x-y)]^2}{f(x-y)}\,\mathrm{d}G(y),$$

from which the proof is easily concluded. □

Generally, Fisher's information is related to a parametric family $d(x, \Theta)$ of densities. Assume that Θ varies over an open interval and that $d(x, \Theta)$ is absolutely continuous in Θ. Put

$$\dot{d}(x, \Theta) = \frac{\partial}{\partial \Theta} d(x, \Theta)$$

and define Fisher's information as follows:

$$I(d, \Theta) = \int_{-\infty}^{\infty} \left[\frac{\dot{d}(x, \Theta)}{d(x, \theta)}\right]^2 d(x, \Theta)\, \mathrm{d}x.$$

Without additional assumptions, the finiteness of $I(f, \Theta)$ may hardly be a basis for deeper conclusions. However, in the particular cases where Θ is a location or a scale parameter, no further assumptions are necessary as we shall see.

If Θ is a location parameter, i.e. $d(x, \Theta) = f(x - \Theta)$, then $I(d, \Theta)$ coincides with $I(f)$ defined above. Let us define *the scale parameter family with centre* μ as follows:

$$d(x, \Theta) = e^{-\Theta} f[(x - \mu)e^{-\Theta}], \quad -\infty < \Theta < \infty. \tag{2}$$

For $\mu = 0$ this is equivalent to the traditional model

$$d(x, \sigma) = \frac{1}{\sigma} f\left(\frac{x}{\sigma}\right), \quad \sigma > 0,$$

with $\sigma = e^{\Theta}$. However, the model (2) will be more suitable for our needs. From (2) it follows that

$$\dot{d}(x, \Theta) = -d(x, \Theta) - (x - \mu)e^{-2\Theta} f'[(x - \mu)e^{-\Theta}]$$

where $f' = \mathrm{d}f/\mathrm{d}x$. Consequently, denoting the associated $I(d, \Theta)$ by $I_1(f)$, we obtain by easy computations that

$$I_1(f) = \int_{-\infty}^{\infty} \left[-1 - x\frac{f'(x)}{f(x)}\right]^2 f(x)\, \mathrm{d}x, \tag{3}$$

constantly in Θ and μ, $-\infty < \Theta, \mu < \infty$.

2.2.4 The φ-functions.

Put

$$F(x) = \int_{-\infty}^{x} f(y)\, \mathrm{d}y, \tag{1}$$

and let $F^{-1}(u)$ be the inverse of F, more precisely,

$$F^{-1}(u) = \inf\{x : F(x) \geq u\}. \tag{2}$$

In this book an important role is assigned to the functions

$$\varphi(u,f) = -\frac{f'(F^{-1}(u))}{f(F^{-1}(u))}, \quad 0 < u < 1. \tag{3}$$

This definition relates to the location model. Generally, for any model $\{d(x,\Theta)\}$ considered in the previous subsection, we put

$$\varphi(u,d,\Theta) = \frac{\dot{d}(D^{-1}(u,\Theta),\Theta)}{d(D^{-1}(u,\Theta),\Theta)}, \tag{4}$$

where $D^{-1}(u,\Theta)$ refers to

$$D(x,\Theta) = \int_{-\infty}^{x} d(x,\Theta)\,\mathrm{d}x.$$

In the case of a scale parameter family with arbitrary centre μ and arbitrary Θ the function (4) becomes

$$\varphi_1(u,f) = -1 - F^{-1}(u)\frac{f'(F^{-1}(u))}{f(F^{-1}(u))} \tag{5}$$

and will be denoted by $\varphi_1(u,f)$ throughout the book.

Lemma 1 *If f is absolutely continuous and if*

$$\int_{-\infty}^{\infty} |f'(x)|\,\mathrm{d}x < \infty, \tag{6}$$

then

$$\int_0^1 \varphi(u,f)\,\mathrm{d}u = 0. \tag{7}$$

Proof. From (6) it follows that

$$\lim_{b,b'\to\infty} |f(b') - f(b)| = \lim_{b,b'\to\infty} \int_b^{b'} |f'(x)|\,\mathrm{d}x = 0, \quad (b < b').$$

Consequently the limit $\lim_{b\to\infty} f(b)$ exists and equals 0, since $f(x)$ is integrable. Similarly $\lim_{a\to-\infty} f(a) = 0$. Consequently

$$\int_{-\infty}^{\infty} f'(x)\,\mathrm{d}x = \lim_{\substack{a\to-\infty\\ b\to\infty}} \int_a^b f'(x)\,\mathrm{d}x = \lim_{\substack{a\to-\infty\\ b\to\infty}} [f(b) - f(a)] = 0. \tag{8}$$

Finally, by the change of variables $u = F(x)$, we obtain (see Problem 6)

$$\int_0^1 \varphi(u,f)\,\mathrm{d}u = \int_{-\infty}^{\infty} -f'(x)\,\mathrm{d}x = 0. \tag{9}$$

□

By similar considerations we can prove

Lemma 2 *If f is absolutely continuous and if*

$$\int_{-\infty}^{\infty} |xf'(x)| \,\mathrm{d}x < \infty, \tag{10}$$

then

$$\int_0^1 \varphi_1(u, f) \,\mathrm{d}u = 0. \tag{11}$$

Lemma 3 *If $f(x)$ is strongly unimodal, then $\varphi(u, f)$ is non-decreasing, and conversely.*

Proof. Immediate. □

Lemma 4 *If $(f) < \infty$, then $\varphi(u, f)$ is square integrable and*

$$I(f) = \int_0^1 \varphi^2(u, f) \,\mathrm{d}u. \tag{12}$$

This holds generally for $I(d, \Theta)$ and $\varphi(u, d, \Theta)$; in particular, $I_1(f) < \infty$ implies

$$I_1(f) = \int_0^1 \varphi^2(u, f) \,\mathrm{d}u. \tag{13}$$

Proof. Immediate. □

Now denote

$$f_{\mu,\sigma}(x) = \frac{1}{\sigma} f\Big(\frac{x-\mu}{\sigma}\Big)$$

and

$$f_\sigma(x) = \frac{1}{\sigma} f\Big(\frac{x}{\sigma}\Big).$$

Lemma 5

$$\varphi(u, f_{\mu,\sigma}) = \frac{1}{\sigma}\varphi(u, f) \tag{14}$$

and

$$\varphi_1(u, f_\sigma) = \varphi_1(u, f). \tag{15}$$

Proof. Immediate. □

Lemma 6 *Assume that for a given $\varphi(u)$*

$$\int_t^1 \varphi(u) \,\mathrm{d}u > 0, \quad 0 < t < 1. \tag{16}$$

Then the distribution given by

$$F_c^{-1}(u) = \int_c^u \Big[\int_t^1 \varphi(v)\,\mathrm{d}v\Big]^{-1} \mathrm{d}t, \quad 0 < u < 1, \tag{17}$$

where $c \in (0,1)$, *is the only distribution such that* (3) *holds for* $\varphi(u,f) = \varphi(u)$ *and* $F(0) = c$.

Proof. Under (16), the function F_c^{-1} is strictly increasing and $F_c^{-1}(c) = 0$. Consequently its inverse $F(x)$ is uniquely determined and satisfies $F(0) = c$. Further,

$$\int_t^1 \varphi(u)\,\mathrm{d}u = \Big[\frac{\partial F_c^{-1}(t)}{\partial t}\Big]^{-1} = \frac{\partial F(x)}{\partial x}\Big|_{x=F^{-1}(t)} = f\big(F^{-1}(t)\big),$$

whence

$$\varphi(t) = -\frac{\partial}{\partial t} f\big(F^{-1}(t)\big) = -\frac{f'(F^{-1}(t))}{f(F^{-1}(t))}, \tag{18}$$

so that the distribution actually satisfies (3). On the other hand there is only one solution of the first order differential equation (18), which satisfies $F^{-1}(c) = 0$. The proof is thus finished. □

Similarly we could prove

Lemma 7 *If* $\varphi_1(u)$ *is a symmetric function,* $\varphi_1(u) = \varphi_1(1-u)$, *and such that*

$$\int_t^1 \varphi_1(u)\,\mathrm{d}u > 0, \quad \tfrac{1}{2} < t < 1,$$

then the symmetric distribution with

$$F_c^{-1}(u) = \exp\Big\{\int_c^u \Big[\int_t^1 \varphi_1(v)\,\mathrm{d}v\Big]^{-1} \mathrm{d}t\Big\}, \quad \tfrac{1}{2} < u < 1, \tag{19}$$

where $c \in (\frac{1}{2}, 1)$, *is the only symmetric solution of* (5) *for* $\varphi_1(u,f) = \varphi_1(u)$ *such that* $F(1) = c$.

Remark. If $\varphi_1(u)$ is non-decreasing, then (19) is applicable with $c \in (0,1)$ and $0 < u < 1$.

Examples of the functions $\varphi(u,f)$ and $\varphi_1(u,f)$ are given in Table 1.

2.2.5 Types of densities and the measurement of their distance. Two densities $f(x)$ and $g(x)$ are of the same type if for some real μ and $\sigma > 0$

$$g(x) = \frac{1}{\sigma} f\Big(\frac{x-\mu}{\sigma}\Big). \tag{1}$$

In other words, distributions differing only in location and scale are of the same type. If we say that the distribution of X is normal, logistic, etc., we often mean that it is of normal type, logistic type, etc., respectively.

Let us associate with each $f(x)$ such that $I(f) < \infty$ the function

$$\gamma(u, f) = [I(f)]^{-\frac{1}{2}} \varphi(u, f), \tag{2}$$

where $\varphi(u, f)$ is given by (2.2.4.3) and $I(f)$ by (2.2.3.1). Since (2.2.4.7) and (2.2.4.12) hold, we have

$$\int_0^1 \gamma(u, f)\,\mathrm{d}u = 0, \tag{3}$$

$$\int_0^1 [\gamma(u, f)]^2\,\mathrm{d}u = 1. \tag{4}$$

Moreover, $\gamma(u, f)$ is the same for all densities of the same type. Thus

$$d(f, g) = \Big\{ \int_0^1 [\gamma(u, f) - \gamma(u, g)]^2\,\mathrm{d}u \Big\}^{\frac{1}{2}} \tag{5}$$

may be used to measure the distance between the respective types to which f and g belong. If f is normal and

$$d(f, g) < 0.1, \tag{6}$$

we may say that g is approximately normal. As we shall see (Problem 14 and Subsection 8.2.1), (6) implies that, in testing problems concerning differences in location, the loss incurred by assuming normality, while g is true, does not exceed 1%, i.e. is less than the loss incurred by dropping one observation out of each 100.

More realistically, we say that a random variable X is approximately normal if

$$X = Y + X_\varepsilon,$$

where Y has density g such that (6) holds for normal f, and X_ε is negligibly small with probability negligibly different from 1.

The distance (5) is not universally applicable and should be modified in an obvious manner if differences between the distributions in various parts of the sample do not reduce to location differences.

2.3 TESTING HYPOTHESES

2.3.1 Statement of the problem. In statistics a hypothesis is represented by a family of probability distributions of the set of observations X. Since we shall permanently assume that $P(\cdot)$ is determined by a density $p(x)$, the hypothesis may preferably be represented by the corresponding family of densities. We shall, however, use both interpretations simultaneously, i.e. regard the hypothesis sometimes as a family of densities, and other times, if need be, as a family of corresponding probability distributions, without changing the notation. Thus we attach to expressions of the types '$\int \Psi p \, \mathrm{d}x$, $p \in H$' and '$\int \Psi \, \mathrm{d}P$, $P \in H$' the same meaning. Where convenient, we shall use the word 'hypothesis' also for the conjecture or statement that p belongs to a certain family.

We shall distinguish the *null hypothesis* $H = \{p\}$, or equivalently $H = \{P\}$, and *alternative hypothesis* $K = \{q\}$ or equivalently $K = \{Q\}$. The members of the null hypothesis will be denoted by p with or without affixes, and the members of the alternative hypothesis will be denoted by q with or without affixes. We shall also occasionally speak simply about the *hypothesis* and *alternative* instead of the null hypothesis and alternative hypothesis, respectively. If H or K contains only one member, it will be called *simple* and denoted by p or q, respectively. Otherwise it will be called *composite*.

Although formally equivalent, the two hypotheses induce different attitudes in the researcher's mind. The null hypothesis is usually based on the assumptions of equality, symmetry or independence and expresses a reserved position. The alternative hypothesis is based on the assumptions of the presence of differences, asymmetries or dependences, which the researcher is hoping to prove or support by the results of the experiment. There are a few basic null hypotheses in contrast to the abundance of conceivable alternative hypotheses. Thus the tendency to adhere to null hypotheses as long as it is reasonable is fully justified by the economy of thought and the limited capacity of the human brain. In theory, the difference between the two hypotheses shows up in that the distribution and other problems are much easier to deal with under the null hypothesis than under the alternative.

A *test* of the hypothesis H against the alternative K consists in deciding whether or not the hypothesis is true (or might be true). The decision is based on the observed value x of X in the following manner: The space X is divided into two disjoint parts, the *critical region* (region of rejection) A_K, and the *region of acceptance* A_H, $A_K \cup A_H = \mathsf{X}$, $A_K \in \mathcal{A}$; whenever x falls into A_K, the null hypothesis H is rejected, and in the contrary case $x \in A_H$ the null hypothesis H is accepted. The choice of A_K should be made before carrying out the experiment. When performing a test one may arrive at the correct decision, or one may commit one of the two errors:

rejecting H when it is true (the error of the first kind) or accepting it when it is false (the error of the second kind).

First of all, due to the special role of H, one wants to keep the probability of the error of the first kind low. Therefore, one chooses a number α, lying between 0 and 1, and imposes the condition

$$P(X \in A_K) \leq \alpha \quad \text{for all } P \in H. \tag{1}$$

The number α is called the *level of significance.*

In certain cases (it is the usual situation in rank tests), having prescribed α in advance, this upper bound of $P(X \in A_K)$ is attained for no $P \in H$. Thus it is convenient to introduce a special term for the number

$$\sup_{P \in H} P(X \in A_K); \tag{2}$$

it will be called the *size* of the test or critical region A_K.

As a rule, the test is based on a proper statistic $t(x)$, called the test statistic. The correspondence between A_K and $t(x)$ is usually one of the following three types:

$$[x \in A_K] \Longleftrightarrow [t(x) \geq c_u], \tag{3}$$

$$[x \in A_K] \Longleftrightarrow [t(x) \leq c_\ell], \tag{4}$$

$$[x \in A_K] \Longleftrightarrow [t(x) \geq c_u \text{ or } t(x) \leq c_\ell]. \tag{5}$$

In the first two cases we speak of *one-sided tests* based on t, in the last case of the *two-sided test* based on t. The numbers c_u and c_ℓ are called the upper critical value and the lower critical value, respectively. Admitting the infinite values $c_\ell = -\infty$ and $c_u = \infty$, the first two regions turn into a special case of the third one.

As a generalization we introduce the notion of a *randomized test.* A randomized test is defined by a measurable function $\Psi(x)$, such that $0 \leq \Psi(x) \leq 1$ for all x, and the hypothesis H is rejected with probability $\Psi(x)$ if $X = x$. The function $\Psi(x)$ is called the *critical function* of the test. Since the correspondence between randomized tests and critical functions is one-to-one, we shall occasionally use the abbreviated expression 'the test Ψ' instead of 'the test with the critical function Ψ'. The size of the test Ψ is defined as $\int \Psi \, \mathrm{d}P$, or as $\sup_{P \in H} \int \Psi \, \mathrm{d}P$ for a composite hypothesis. If we base a test on the test statistic $t(x)$, the critical function $\Psi(x)$ is usually defined so that

$$\Psi(x) = \begin{cases} 0, & \text{if } c_\ell < t(x) < c_u \\ 1, & \text{if } t(x) < c_\ell \text{ or } t(x) > c_u \end{cases} \tag{6}$$

and the intermediate values $0 < \Psi(x) < 1$, if any, are chosen only for x such that $t(x) = c_u$ or $t(x) = c_\ell$.

The theory is considerably simplified by introducing randomized tests. With randomized tests we are always able to ensure equality between the size and the significance level, and moreover the set of all possible tests becomes convex in the sense that for any two critical functions Ψ_1 and Ψ_2 of respective sizes $\leq \alpha_1$ and $\leq \alpha_2$, the function

$$\Psi(x) = \lambda\Psi_1(x) + (1-\lambda)\Psi_2(x) \tag{7}$$

is again a critical function of size not exceeding $\alpha = \lambda\alpha_1 + (1-\lambda)\alpha_2$ for any $0 \leq \lambda \leq 1$. Thus it is convenient to develop the theory for randomized tests even if we may feel reluctant to use the randomization in practice.

If dealing with a simple alternative q, we call

$$\int \Psi \, \mathrm{d}Q$$

the *power* of the respective test. Obviously, the power is the complement to 1 or the error of the second kind. When the alternative $K = \{q\}$ is composite, the power is defined by

$$\inf_{Q\in K} \int \Psi \, \mathrm{d}Q,$$

i.e. by the greatest lower bound of the powers for individual alternatives q from K. Frequently we are interested in a parametric set of simple alternatives $\{Q_\Delta\}$ indexed by a real parameter Δ, and then

$$\pi(\Delta) = \int \Psi \, \mathrm{d}Q_\Delta$$

is called the *power function* of Ψ. The main purpose of the theory of testing hypotheses is to provide tests with largest power for a given level of significance.

In the opinion of some statisticians, there is no need to prescribe a fixed significance level, and furthermore there is no reasonable rule for its choice. They do not regard testing hypotheses as a decision procedure leading to an irreversible decision, but rather as an intellectual procedure in the mind of a researcher, whose attitude to various hypotheses comes to be more or less changed on the basis of the experimental evidence. Such a point of view may lead to preferring the so-called *level actually attained* by a test statistic t to the fixed significance level. The level actually attained is defined as follows: Assuming the test is based on the right tail of the distribution of a test statistic $t(x)$, and that this distribution is the same for all $p \in H$, we put

$$\ell(x) = P\big(t(X) \geq t(x)\big)$$

and call the random variable $\ell(X)$ the level actually attained. The smaller

$\ell(x)$ we observe, the stronger evidence against the null hypothesis H the set of observations $X = x$ provides. The results presented in this book might be utilized with this approach to testing hypotheses, too, but we shall prefer the more clear-cut framework of the Neyman-Pearson theory.

In conclusion, let us mention that in proving optimum properties of rank tests no advantage is gained by considering randomized rank tests. Actually, the optimum properties are either of local character, and then the optimum tests are either non-randomized (for proper values of α) or they depend on the derivatives of second and higher orders, and, consequently, are too difficult to be dealt with; alternatively the optimum properties are of asymptotic character, and then we may ensure the asymptotic equality between the size and the significance level with non-randomized rank tests, too.

2.3.2 The envelope power function. Consider first a simple hypothesis p and a simple alternative q. Denote the set of critical functions of size not exceeding α by $M(\alpha, p)$, i.e.

$$\int \Psi \,\mathrm{d}P \leq \alpha \quad \text{if and only if} \quad \Psi \in M(\alpha, p),$$

where $\mathrm{d}P = p\,\mathrm{d}x = p(x_1, \ldots, x_N)\,\mathrm{d}x_1, \ldots, \mathrm{d}x_N$. We are now interested in choosing a Ψ from $M(\alpha, p)$ which maximizes the power. We shall call

$$\beta(\alpha, p, q) = \sup_{\Psi \in M(\alpha, p)} \int \Psi \,\mathrm{d}Q$$

the *envelope power function* for p against q at level α. It may be considered as a function of q while α and p are fixed, or as a function of α while p and q are fixed. Obviously, a test maximizes the probability of accepting q, when q is true, if it makes this probability equal to $\beta(\alpha, p, q)$.

If the hypothesis is composite, $H = \{p\}$, we shall denote by $M(\alpha, H)$ the set of critical functions such that

$$\sup_{P \in H} \int \Psi \,\mathrm{d}P \leq \alpha,$$

and call

$$\beta(\alpha, H, q) = \sup_{\Psi \in M(\alpha, H)} \int \Psi \,\mathrm{d}Q$$

the envelope power function for H against q at level α.

Finally if the alternative $K = \{q\}$ is composite, too, we call

$$\beta(\alpha, H, K) = \sup_{\Psi M(\alpha, H)} \inf_{Q \in K} \int \Psi \,\mathrm{d}Q$$

the envelope power function for H against K at level α. Obviously, the last definition embraces the previous ones as particular cases when K or H, or both, are simple, and is based on the maximin principle (the reverse form of the minimax principle).

Lemma 1 *The envelope power function considered as a function of α is concave.*

Proof. For any three levels $\alpha_1 < \alpha < \alpha_2$ we can find λ such that $\alpha = \lambda\alpha_1 + (1-\lambda)\alpha_2$, $0 < \lambda < 1$. Further let Ψ_1 and Ψ_2 be critical functions from $M(\alpha_1, H)$ and $M(\alpha_2, H)$ such that

$$\beta(\alpha_1, H, K) - \varepsilon \le \inf_{Q\in K} \int \Psi_1 \,\mathrm{d}Q$$

and

$$\beta(\alpha_2, H, K) - \varepsilon \le \inf_{Q\in K} \int \Psi_2 \,\mathrm{d}Q$$

for a given $\varepsilon > 0$. Then the critical function Ψ given by (2.3.1.7) is of size not exceeding α, and hence

$$\begin{aligned}\beta(\alpha, H, K) &\ge \inf_{Q\in K} \int \Psi \,\mathrm{d}Q \\ &\ge \lambda \inf_{Q\in K} \int \Psi_1 \,\mathrm{d}Q + (1-\lambda) \inf_{Q\in K} \int \Psi_2 \,\mathrm{d}Q \\ &\ge \lambda\beta(\alpha_1, H, K) + (1-\lambda)\beta(\alpha_2, H, K) - \varepsilon.\end{aligned}$$

This concludes the proof. □

2.3.3 Most powerful test. In testing H against q a test Ψ^0 is said to be *most powerful* at level α if $\Psi^0 \in M(\alpha, H)$ and if

$$\int \Psi^0 \,\mathrm{d}Q = \beta(\alpha, H, q). \tag{1}$$

When testing H against K, a test $\Psi^0 \in M(\alpha, H)$ will be called *uniformly most powerful* at level α if (1) holds for all $q \in K$.

In testing H against K, a test $\Psi^0 \in M(\alpha, H)$ will be called *maximin most powerful* at level α if

$$\inf_{Q\in K} \int \Psi^0 \,\mathrm{d}Q = \beta(\alpha, H, K).$$

Lemma 1 *If H and K are defined as families of densities, then there always exists a maximin most powerful test for every $\alpha \in (0,1)$.*

Proof. We shall now write explicitly $q\,\mathrm{d}x$ for $\mathrm{d}Q$. There exists a sequence $\{\Psi_n\}$ of critical functions from $M(\alpha, H)$ such that

$$\inf_{q\in K} \int \Psi_n q\,\mathrm{d}x > \beta(\alpha, H, K) - \frac{1}{n}, \quad n \geq 1. \tag{2}$$

Furthermore, there is a subsequence $\{j\} \subset \{n\}$ such that the subsequence $\{\Psi_j\}$ of critical functions converges to a critical function Ψ^0 in the sense that for every density p

$$\lim_j \int \Psi_j p\,\mathrm{d}x = \int \Psi^0 p\,\mathrm{d}x. \tag{3}$$

This proposition is proved in Lehmann (1959, p. 354). From (3) it follows immediately that Ψ^0 also belongs to $M(\alpha, H)$. Also, for every $q \in K$ we have, in view of (2),

$$\begin{aligned} \int \Psi^0 q\,\mathrm{d}x &= \lim_j \int \Psi_j q\,\mathrm{d}x \geq \lim_j \inf_{q\in K} \int \Psi_j q\,\mathrm{d}x \\ &= \beta(\alpha, H, K), \quad q \in K. \end{aligned}$$

Hence

$$\inf_{q\in K} \int \Psi^0 q\,\mathrm{d}x \geq \beta(\alpha, H, K),$$

where only the case of equality may occur. The proof is finished. □

In conclusion let us consider the still more general situation of testing H against $K = \bigcup_\sigma K_\sigma$, where the subalternatives K_σ are disjoint. The test will be called uniformly maximin most powerful at level α if it is maximin most powerful for all K_σ. This concept embraces the previous ones as particular cases.

2.3.4 The Neyman-Pearson lemma.

This basic lemma shows that the most powerful test for testing a simple hypothesis against a simple alternative may be found quite easily.

Lemma 1 *In testing p against q at level α the most powerful test may be found as follows:*

$$\Psi^0(x) = \begin{cases} 1, & \text{if } q(x) > k\,p(x), \\ 0, & \text{if } q(x) < k\,p(x), \end{cases} \tag{1}$$

where k and $\Psi^0(x)$ for x such that $q(x) = k\,p(x)$ should and can be defined so that

$$\int \Psi^0\,\mathrm{d}P = \alpha. \tag{2}$$

Proof. Observe that $\alpha(c) = P(q(X) > c\,p(X))$ is a non-increasing and right-continuous function of c such that $\alpha(0-0) = 1$ and $\alpha(\infty) = 0$. Therefore for each $\alpha \in (0,1)$ there exists a $k \geq 0$ such that

$$\alpha(k-0) \geq \alpha \geq \alpha(k). \tag{3}$$

If k is a continuity point, then (2) follows from (1) regardless of the choice of $\Psi^0(x)$ for x such that $q(x) = k\,p(x)$. If k is a point of discontinuity, it suffices to put

$$\Psi^0(x) = \frac{\alpha - \alpha(k)}{\alpha(k-0) - \alpha(k)} \tag{4}$$

for x such that $q(x) = k\,p(x)$.

Now for any other critical function Ψ, $0 \leq \Psi \leq 1$ and (1) imply that either $\operatorname{sign}(\Psi^0 - \Psi) = \operatorname{sign}(q - kp)$ or at least one of these expressions equals 0, so that for all x

$$[\Psi^0(x) - \Psi(x)][q(x) - k\,p(x)] \geq 0. \tag{5}$$

Consequently, if $\int \Psi\,\mathrm{d}P \leq \alpha$,

$$\int \Psi^0\,\mathrm{d}Q - \int \Psi\,\mathrm{d}Q \geq k \int (\Psi^0 - \Psi)\,\mathrm{d}P = k\Big(\alpha - \int \Psi\,\mathrm{d}P\Big) \geq 0, \tag{6}$$

which was to be proved. □

Remark 1. We also see that Ψ has the same power as Ψ^0 only if (5) equals 0 a.s., i.e. if Ψ satisfies (1) a.s. with respect to Lebesgue measure.

Remark 2. We have made no use of the fact that the measure, with respect to which the densities are defined, is the Lebesgue measure. And as a matter of fact the Neyman-Pearson lemma holds for densities defined with respect to any σ-finite measure.

2.3.5 Least favourable densities. If there is a $p^0 \in H$ such that the most powerful test Ψ^0 for p^0 against q belongs to $M(\alpha, H)$, we call p^0 a *least favourable density* in H for testing H against q at level α

Similarly, if there are $p^0 \in H$ and $q^0 \in K$ such that the most powerful test Ψ^0 for p^0 against q^0 belongs to $M(\alpha, H)$ and

$$\int \Psi^0\,\mathrm{d}Q \geq \int \Psi^0\,\mathrm{d}Q^0, \quad \text{for all } Q \in K, \tag{1}$$

we call (p^0, q^0) a *least favourable pair of densities* for H against K at level α. This terminology is justified by the following.

Lemma 1 *If* (p^0, q^0) *is a least favourable pair of densities, then* $\beta(\alpha, p^0, q^0) = \beta(\alpha, H, K)$, *further*

$$\beta(\alpha, p, q) \geq \beta(\alpha, p^0, q^0) \quad \text{for all } p \in H,\ q \in K \tag{2}$$

and Ψ^0 *above is the maximin most powerful test for* H *against* K.

Proof. Since $\Psi^0 \in M(\alpha, H)$, also $\Psi^0 \in M(\alpha, p)$ for all $p \in H$. Consequently, in view of (1),

$$\beta(\alpha, p^0, q^0) = \int \Psi^0 \, \mathrm{d}Q^0 \leq \int \Psi^0 \, \mathrm{d}Q \leq \beta(\alpha, p, q)$$

for every $p \in H$ and $q \in K$. The rest of the proof is obvious. □

Remark. If $\Psi \in M(\alpha, H)$, then also $\Psi \in M(\alpha, \bar{H})$, where $\bar{H}$ is the convex hull of H. Thus, the replacement of H by $\bar{H}$ in no way makes the testing problem more difficult. On the other hand, if there is no least favourable density in H, we may still hope to find it in $\bar{H}$.

2.4 AUXILIARY RESULTS FOR NORMAL SAMPLES

Most definitive results in the theory of hypothesis testing have been obtained under the assumption of normality. In this section we shall quote a few of them for later reference, since many problems in rank testing may, for large N, be approximated by problems concerning normal samples. We begin with a useful proposition concerning the sum of squares of normal random variables correlated in a special way.

2.4.1 χ^2-distribution for correlated normal random variables. Consider a normal random vector $(Z_1, \ldots, Z_k)$ with respective expectations $(\mu_1, \ldots, \mu_k)$ and the following covariances:

$$\begin{aligned} \mathsf{var}\, Z_j &= 1 - \kappa_j^2, \quad j = 1, \ldots, k, & (1)\\ \mathsf{cov}(Z_j, Z_g) &= -\kappa_j \kappa_g, \quad 1 \leq j \neq g \leq k, & (2) \end{aligned}$$

where $\kappa_1, \ldots, \kappa_k$ are non-negative numbers such that

$$\sum_{j=1}^{k} \kappa_j^2 = 1. \tag{3}$$

Theorem 1 *Under the above assumptions the statistic*

$$Q = Z_1^2 + \ldots + Z_k^2 \tag{4}$$

has a non-central χ^2-distribution with $k-1$ degrees of freedom and non-centrality parameter

$$\delta = \sum_{j=1}^{k} \mu_j^2 - \Big(\sum_{j=1}^{k} \mu_j \kappa_j\Big)^2. \tag{5}$$

Proof. Let $Y_1, \ldots, Y_k$ be independent normal random variables with respective expectations $\mu_1, \ldots, \mu_k$ and variances 1. Put

$$Y_0 = \sum_{j=1}^{k} \kappa_j Y_j. \tag{6}$$

Then, in view of (3),

$$Y_1^2 + \ldots + Y_k^2 - Y_0^2 \tag{7}$$

has a non-central χ^2-distribution with $k-1$ degrees of freedom and non-centrality parameter (5), and is independent of Y_0. Consequently, (7) has the same distribution conditionally for $Y_0 = 0$. However, for $Y_0 = 0$ the conditional expectations and covariances of $Y_1, \ldots, Y_k$ coincide with those assumed concerning $Z_1, \ldots, Z_k$. □

2.4.2 Most powerful tests.

Theorem 1 *Let Z be a normal random variable with parameters (μ, σ^2), and consider the problem of testing $\mu = 0$ against $\mu = b > 0$, with σ known.*

Then the critical region of the most powerful level α test is $\{Z > k_{1-\alpha}\sigma\}$, where $k_{1-\alpha}$ denotes the $(1-\alpha)$-quantile of the standardized normal distribution.

The power of the test equals

$$1 - \Phi(k_{1-\alpha} - b\sigma^{-1}), \tag{1}$$

where Φ is the standardized normal distribution function.

Theorem 2 *With the same random variable Z let us test $\mu = 0$ against $|\mu| = b > 0$. Then the critical region of the maximin most powerful level α test is $\{|Z| > k_{1-\frac{1}{2}\alpha}\sigma\}$.*

The power of the test equals

$$1 - \Phi(k_{1-\frac{1}{2}\alpha} - b\sigma^{-1}) + \Phi(k_{\frac{1}{2}\alpha} - b\sigma^{-1}). \tag{2}$$

The proof of this as well as of the following theorem is based on an invariance argument and the Hunt-Stein theorem. See Example 8 on page 338 of Lehmann (1959) with equality instead of inequality in (16).

Theorem 3 *Let* $(Z_1, \ldots, Z_k)$ *be the normal vector considered in Theorem 2.4.1.1 and let us test the hypothesis* $\mu_1 = \ldots = \mu_k = 0$ *against the alternative*

$$\sum_{j=1}^{k} \mu_j^2 = b^2 > 0 \quad \text{and} \quad \sum_{j=1}^{k} \kappa_j \mu_j = 0. \tag{3}$$

Then the maximin most powerful level α *test has the critical region*

$$\{Z_1^2 + \ldots + Z_k^2 \geq \chi^2_{1-\alpha,k-1}\}, \tag{4}$$

where $\chi^2_{1-\alpha,k-1}$ *is the* $(1-\alpha)$*-quantile of the* χ^2*-distribution with* $k-1$ *degrees of freedom.*

The power of the test equals

$$1 - F_{k-1}(\chi^2_{1-\alpha,k-1}, b^2), \tag{5}$$

where $F_{k-1}(\cdot, \delta)$ *denotes the distribution function of non-central* χ^2 *with* $k-1$ *degrees of freedom and non-centrality parameter* δ.

PROBLEMS AND COMPLEMENTS TO CHAPTER 2

Section 2.1

1. Let $H_0 = \{P\}$ be the system of two-dimensional distributions corresponding to the densities $p(x, y) = f(x)f(y)$, with f running through the set of all one-dimensional densities.

(i) Let B be a two-dimensional Borel set such that $x < y$ for all $(x, y) \in B$. Let $\bar{B}$ consist of all points (x, y) such that $x > y$ and $(y, x) \notin B$. Then the event $C = B \cup \bar{B}$ is similar for H_0, and $P(C) = \frac{1}{2}$ for $P \in H_0$.

(ii) Let $\mathcal{C}$ consist of the empty set $\emptyset$, of the whole plane X, and of all the events C considered in (i). Then an event D is similar for H_0 if and only if there exists $C \in \mathcal{C}$ such that the Lebesgue measure of the symmetric difference $C \div D$ equals 0.

(iii) Each $C \in \mathcal{C}$ is similar also under $H_0' = \{P\}$, H_0' being the system of distributions induced by distribution functions $H(x, y) = F(x)F(y)$, with F running through the set of all continuous one-dimensional functions. However, there are events of Lebesgue measure 0 which are not similar under H_0' (in contradistinction to H_0).

(iv) Let $C_1, \ldots, C_N$ be disjoint events which are similar for H_0 and have positive probability. Then $N \leq 2$.

(v) Let $\mathcal{A}$ be a sub-σ-field consisting of similar events, and let us regard events differing by a null set as equivalent. Then $\mathcal{A}$ cannot contain more than four elements, namely $\mathcal{A} = \{\emptyset, \mathsf{X}, C, \mathsf{X} - C\}$, where X denotes the whole plane and C is an event considered in (i).

Definitions. We say that $\mathcal{E}$ is a π-field of events if $E \in \mathcal{E}$ and $F \in \mathcal{E}$ entails $(E \cap F) \in \mathcal{E}$. We say that a σ-field is generated by $\mathcal{E}$ if it is the smallest σ-field containing $\mathcal{E}$.

2. Let $\mathcal{D}$ be a λ-field, $\mathcal{E}$ be a π-field, and $\mathcal{A}$ be the σ-field generated by $\mathcal{E}$. Then $\mathcal{D} \supset \mathcal{E}$ entails $\mathcal{D} \supset \mathcal{A}$.

3. Let $t(x)$ be a statistic such that the events $C_a = \{x : t(x) < a\}$ are similar for every real a, with respect to a certain system of probability distributions on $(\mathcal{X}, \mathcal{A})$. Then the events $C_B = \{x : t(x) \in B\}$, with B being a linear Borel subset, are similar as well.

4. Consider an experiment consisting in throwing a discus under conditions where neither the man performing the throws nor the time and place are kept fixed. Let the lengths of the throws, say x, be an integer-valued variable, denoting the number of centimetres. Consider two events: A occurring if x is odd, and B occurring if either x is odd and < 4500, or x is even and ≥ 4500.

Then we may reasonably assume that $P(A) = P(B) = \frac{1}{2}$, whereas the probability of $A \cap B$ will be completely indeterminate.

Section 2.2

5. Prove that $F(F^{-1}(u)) \geq u$ holds generally, and $F(F^{-1}(u)) = u$ for any continuous F, $0 < u < 1$. [Hint: (2.2.4.2) entails $F(F^{-1}(u) + \varepsilon) \geq u$ and $F(F^{-1}(u) - \varepsilon) < u$ for $\varepsilon > 0$.]

6. Prove that $F^{-1}(F(x)) \leq x$ with the inequality possible even for continuous F, but that generally

$$P\big(F^{-1}(F(X)) \neq X\big) = 0, \tag{1}$$

if X is distributed according to F. [Hint: $F^{-1}(F(x)) \leq x$ is entailed by (2.2.4.2) and by $F(x) \geq F(x)$. Further, the set of points x such that $F^{-1}(F(x)) < x$ is a union of disjoint intervals, each having probability 0.]

7. If X is a random variable with distribution function $F(x)$, then $P(F(X) \leq u) \leq u$. If F is continuous, then $P(F(X) \leq u) = u$, $0 < u < 1$, i.e. $U = F(X)$ is uniformly distributed.

8. Let f be continuous and $f(F^{-1}(t)) \neq 0$. Then $[\partial/\partial t]F^{-1}(t)$ exists and equals $[f(F^{-1}(t))]^{-1}$. Thus (2.2.4.18) should be read as follows: $\varphi(U) = -f'(F^{-1}(U))/f(F^{-1}(U))$ with probability 1.

Definition. A density is said to have monotone likelihood ratio (with respect to the location parameter) if

$$\frac{f(x-\Theta')}{f(x-\Theta)} \leq \frac{f(x'-\Theta')}{f(x'-\Theta)} \quad \text{for all } x < x',\ \Theta < \Theta'.$$

9. Show that f has monotone likelihood ratio if and only if it is strongly unimodal.

10. Show that $I(f_{\mu,\sigma}) = \sigma^{-2}I(f)$, with $f_{\mu,\sigma} = \sigma^{-1}f((x-\mu)/\sigma)$.

11. Let $f(x) = 0$ for $x < 0$. Put $\bar{f}(x) = e^x f(e^x)$, $-\infty < x < \infty$. Then $\varphi(u, \bar{f}) = \varphi_1(u, f)$ and $I(\bar{f}) = I_1(f)$. Cf. densities No 3 and No 7 in Table 1.

12. Let $f(x) = f(-x)$. Put $\tilde{f}(x) = 2e^x f(e^x)$. Then $\varphi(u, \tilde{f}) = \varphi_1(\frac{1}{2}u + \frac{1}{2}, f)$ and $I(\tilde{f}) = I_1(f)$. Cf. densities No 8 and No 4 in Table 1.

13. Let $f(x) = f(-x)$. Put $f^+(x) = 2f(x)$ for $x \geq 0$ and $f^+(x) = 0$ for $x < 0$. Then $\varphi_1(u, f^+) = \varphi_1(\frac{1}{2}u + \frac{1}{2}, f)$. Cf. densities No 2 and No 3 in Table 1.

14. Put $\varrho = \int_0^1 \gamma(u, f)\gamma(u, g)\,\mathrm{d}u$. Then $d(f, g) \geq (1 - \varrho^2)^{\frac{1}{2}}$.

Section 2.3

15. Let the random variable $L = \ell(X)$ be the level actually attained. Then, under H, generally L is stochastically at least as large as a uniform random variable on $(0, 1)$. Hence the size of the test which rejects H if and only if $L \leq \alpha$ is bounded by α; in other words, $P(L \leq \alpha) \leq \alpha$. [Theorem 8.3.1.3.] If X has a continuous distribution under H, then the distribution of $L = \ell(X)$ is, under H, exactly uniform on $(0, 1)$.

Definition. For randomized tests, L may depend additionally on a uniformly distributed random variable U independent of X, and we assume $P(\ell(X, U) \leq \alpha) \leq \alpha$ for $\alpha \in (0, 1)$, $P \in H$.

16. Prove that $\mathsf{E}_p(L) \geq \frac{1}{2}$ for $p \in H$ and $\mathsf{E}_q(L) \geq \int_0^1 [1 - \beta(\alpha, H, q)]\,\mathrm{d}\alpha$. Furthermore, $\mathsf{E}_p(L) = \frac{1}{2}$ if and only if $P(L \leq \alpha) = \alpha$, $\alpha \in (0, 1)$, and $\mathsf{E}_q(L) = \int_0^1 [1 - \beta(\alpha, H, q)]\,\mathrm{d}\alpha$ if and only if $\{L \leq \alpha\}$ is the critical region of the most powerful level α test for each $\alpha \in (0, 1)$. (Within the framework mentioned in Subsection 2.3.3 a test based on L may be called most powerful if $\mathsf{E}_q(L) =$ minimum.)

17. Let L_0 be a random variable depending on X and U and such that $0 \leq L_0 \leq 1$ and that $\mathsf{E}_p(L_0) \geq \frac{1}{2}$ for all $p \in H$. Then $\mathsf{E}_q(L_0) \geq 1 - \beta(\frac{1}{2}, H, q)$.

18. $1 - \beta(\frac{1}{2}, H, q) \leq \int_0^1 [1 - \beta(\alpha, H, q)]\,\mathrm{d}\alpha$.

Section 2.4

19. Proof of Theorem 2.4.2.2: If $\Psi(x) \in M(\alpha, p)$, then $\bar{\Psi}(x) = \frac{1}{2}[\Psi(x) + \Psi(-x)]$ also belongs to $M(\alpha, p)$ and $\min\{\int \Psi\,\mathrm{d}Q_b, \int \Psi\,\mathrm{d}Q_{-b}\} \leq \int \bar{\Psi}\,\mathrm{d}Q_b = \int \bar{\Psi}\,\mathrm{d}Q_{-b}$ where $Q_{\pm b}$ denote normal distributions $N(\pm b, \sigma^2)$. Thus for each

critical function $\Psi(x)$ there exists a critical function $\bar{\Psi} = \bar{\Psi}(X)$, the power of which is not smaller. Further, the distribution of Z is unique under both the hypothesis and the alternative, so that the most powerful test of the latter type may be found by means of the Neyman-Pearson lemma.

Chapter 3

Elementary theory of rank tests

3.1 RANKS AND ORDER STATISTICS. BASIC NULL HYPOTHESES

3.1.1 Ranks and order statistics. Let $o_i(x)$ denote the value of the i-th smallest coordinate in $x = (x_1, \dots, x_N)$, so that $o_1(x)$ denotes the minimal value and $o_N(x)$ denotes the maximal value. Obviously, $o_i(x)$ is well defined even if some coordinates coincide. Putting $x^{(i)} = o_i(x)$, we have

$$x^{(1)} \leq x^{(2)} \leq \ldots \leq x^{(N)}. \tag{1}$$

The statistic $X^{(i)} = o_i(X)$ will be called the *i-th order statistic*, and the vector of order statistics $(X^{(1)}, \dots, X^{(N)})$ will be briefly denoted by $X^{(\cdot)}$.

Now, for $x = (x_1, \dots, x_N)$ such that no two coordinates coincide, let $r_i(x)$ denote the number of $x's \leq x_i$, i.e. the rank of x_i in the sequence (1):

$$x_i = x^{(r_i)}, \quad 1 \leq i \leq N. \tag{2}$$

The statistic $R_i = r_i(X)$ will be called the *rank of* X_i, and R will stand for the vector of ranks $(R_1, \dots, R_N)$. The ranks are well defined only if the probability of coincidence of any pair of coordinates equals 0. This is, however, true if the distribution in $(\mathsf{X}, \mathcal{A})$ is determined by a density, as we shall constantly assume in the sequel. It will be convenient to have a special symbol for the space of all permutations $r = (r_1, \dots, r_N)$ of $(1, \dots, N)$. Let us denote it by R. Obviously R contains $N!$ points. Furthermore, let us denote by $\mathsf{X}^{(\cdot)}$ the subspace of X containing points $x^{(\cdot)} = (x^{(1)}, \dots, x^{(N)})$ such that $x^{(1)} < \ldots < x^{(N)}$. Let $\mathcal{A}^{(\cdot)}$ be the σ-field consisting of the Borel subsets of $\mathsf{X}^{(\cdot)}$. If $P(\cdot)$ is determined by a density, then $P(X^{(\cdot)} \in \mathsf{X}^{(\cdot)}) = 1$.

Note that X may be reconstructed from $X^{(\cdot)}$ and R. Actually, if $X^{(\cdot)} = x^{(\cdot)}$ and $R = r$, then $X = x$ with coordinates given by (2). Thus the pair $(X^{(i)}, R)$ is a sufficient statistic for any system of distributions determined by densities.

Theorem 1 *If X is governed by the density q, then $X^{(\cdot)}$ is governed by the density*

$$\bar{q}(x^{(1)}, \ldots, x^{(N)}) = \sum_{r \in \mathsf{R}} q(x^{(r_1)}, \ldots, x^{(r_N)}), \quad x^{(\cdot)} \in \mathcal{X}^{(\cdot)}. \tag{3}$$

Moreover,

$$Q(R = r \mid X^{(\cdot)} = x^{(\cdot)}) = \frac{q(x^{(r_1}, \ldots, x^{(r_N)})}{\bar{q}(x^{(1)}, \ldots, x^{(N)})}, \; r \in \mathsf{R}, \; x^{(\cdot)} \in \mathcal{X}^{(\cdot)}, \tag{4}$$

holds with Q being the probability distribution corresponding to q.

Proof. For any $A \in \mathcal{A}^{(\cdot)}$ it holds that

$$\begin{aligned} &\int \cdots \int_{X^{(\cdot)} \in A} q(x_1, \ldots, x_N) \, \mathrm{d}x_1 \ldots \mathrm{d}x_N \\ &= \sum_{r \in \mathsf{R}} \int \cdots \int_{R = r, X^{(\cdot)} \in A} q(x_1, \ldots, x_N) \, \mathrm{d}x_1 \ldots \mathrm{d}x_N \\ &= \sum_{r \in \mathsf{R}} \int \cdots \int_A q(x^{(r_1)}, \ldots, x^{(r_N)}) \, \mathrm{d}x^{(1)} \ldots \mathrm{d}x^{(N)} \\ &= \int \cdots \int_A \bar{q}(x^{(1)}, \ldots, x^{(N)}) \, \mathrm{d}x^{(1)} \ldots \mathrm{d}x^{(N)}, \end{aligned} \tag{5}$$

where we make use of the fact that the correspondence between $(x_1, \ldots, x_N)$ and $(x^{(1)}, \ldots, x^{(N)})$ is one-to-one and linear with the Jacobian $= 1$ on each subset $\{R = r\}$. Obviously (5) proves (3). By the same argument,

$$\begin{aligned} &Q(R = r, X^{(\cdot)} \in A) \\ &= \int \cdots \int_{R = r, X^{(\cdot)} \in A} q(x_1, \ldots, x_N) \, \mathrm{d}x_1 \ldots \mathrm{d}x_N \\ &= \int \cdots \int_A q(x^{(r_1)}, \ldots, x^{(r_N)}) \, \mathrm{d}x^{(1)} \ldots \mathrm{d}x^{(N)} \\ &= \int \cdots \int_A \frac{q(x^{(r_1)}, \ldots, x^{(r_N)})}{\bar{q}(x^{(1)}, \ldots, x^{(N)})} \bar{q}(x^{(1)}, \ldots, x^{(N)}) \, \mathrm{d}x^{(1)} \ldots \mathrm{d}x^{(N)}, \end{aligned} \tag{6}$$

where we make use of the fact that $\bar{q}(x^{(1)}, \ldots, x^{(N)}) = 0$ implies $q(x^{(r_1)}, \ldots, x^{(r_N)}) = 0$ for each $r \in \mathsf{R}$. Clearly, (6) justifies (4). □

3.1.2 Hypotheses H_0 and H_* (randomness). Let us denote by H_* the family of all N-dimensional densities which are symmetric with respect to permutations of arguments. Consequently,

$$p(x_{r_1}, \ldots, x_{r_N}) = p(x_1, \ldots, x_N), \quad r \in \mathcal{R}, \tag{1}$$

if and only if $p \in H_*$.

Further, let us denote by H_0 the system of densities such that

$$p = \prod_1^N f(x_i), \tag{2}$$

where $f(x)$ may be an arbitrary one-dimensional density. Obviously $H_0 \subset H_*$.

In more usual terminology, under H_* the distribution of $(X_1, \ldots, X_N)$ is assumed to be symmetric and determined by a density, and under H_0 the observations X_i are assumed to be independent and identically distributed according to some density f. From Theorem 3.1.1.1 we obtain immediately

Theorem 1 *If the distribution of X is governed by a density $p \in H_0$ (or $p \in H_*$), then the random vectors $R = (R_1, \ldots, R_N)$ and $X^{(\cdot)} = (X^{(1)}, \ldots, X^{(N)})$ are independent.*

$$P(R = r) = \frac{1}{N!}, \quad r \in \mathcal{R}, \tag{3}$$

holds, and the density of $X^{(\cdot)}$ equals

$$N!\,p(x^{(\cdot)}), \quad x^{(\cdot)} \in \mathcal{X}^{(\cdot)}. \tag{4}$$

Theorem 2 *For any $p \in H_0$ (or $p \in H_*$) and any statistic $t(x)$*

$$\mathsf{E}\{t(X_1, \ldots, X_N) \mid R = r\} = \mathsf{E}\{t(X^{(r_1)}, \ldots X^{(r_N)})\} \tag{5}$$

holds. In particular, if q is a density such that $p(x) = 0$ entails $q(x) = 0$, $x \in \mathcal{X}$, we have

$$Q(R = r) = \mathsf{E}\left\{\frac{q(X^{(r_1)}, \ldots, X^{(r_N)})}{p(X^{(r_1)}, \ldots, X^{(r_N)})}\right\}\frac{1}{N!}. \tag{6}$$

Proof. Since $X_i = X^{(r_i)}$, $1 \le i \le N$, under $R = r$, the independence of R and $X^{(\cdot)}$, proved in Theorem 1, implies

$$\begin{aligned} \mathsf{E}\{t(X) \mid R = r\} &= \mathsf{E}\{t(X^{(r_1)}, \ldots, X^{(r_N)}) \mid R = r\} \\ &= \mathsf{E}\{t(X^{(r_1)}, \ldots, X^{(r_N)})\}. \end{aligned}$$

□

In the following theorem we denote the order statistics by $X_N^{(i)}$ instead of $X^{(i)}$, in order to indicate the dependence on the respective sample size N.

Theorem 3 *Let* $X_N^{(1)} < \ldots < X_N^{(N)}$ *be the order statistics of a set of independent observations* $X_1, \ldots, X_N$*, each having the same distribution function* $F(x)$*. Then the distribution function of* $X_N^{(i)}$ *is*

$$F_N^{(i)}(x) = P(X_N^{(i)} \le x) = \sum_{k=i}^{N} \binom{N}{k} [F(x)]^k [1 - F(x)]^{N-k} \tag{7}$$

$$= \frac{N!}{(i-1)!(N-i)!} \int_0^{F(x)} u^{i-1}(1-u)^{N-i}\, du. \tag{8}$$

Therefore, if $F(x)$ *has the density* $f(x)$*, then the density of* $X_N^{(i)}$ *equals*

$$f_N^{(i)}(x) = N \binom{N-1}{i-1} [F(x)]^{i-1} [1 - F(x)]^{N-i} f(x), \tag{9}$$

all formulas holding for $-\infty < x < \infty$*,* $1 \le i \le N$*.*

Proof. $P(X_N^{(i)} \le x) = P(\sum_{j=1}^{N} I(X_j \le x) \ge i)$, where I is the indicator function of the event in parentheses. But $\sum_{j=1}^{N} I(X_j \le x)$ is the random variable having a binomial distribution with parameters N and $P(X_j \le x) = F(x)$, so that

$$P(X_N^{(i)} \le x) = \sum_{k=i}^{N} \binom{N}{k} [F(x)]^k [1 - F(x)]^{N-k},$$

which proves (7). Using now the incomplete Beta function ratio $I_z(a, b)$, its well-known relation to the tails of the binomial distribution and its definition, we get

$$P(X_N^{(i)} \le x) = I_{F(x)}(i, N - i + 1)$$

$$= \frac{N!}{(i-1)!(N-i)!} \int_0^{F(x)} u^{i-1}(1-u)^{N-i}\, du,$$

thus proving (8). By simple differentiation we arrive at the density (9). □

Theorem 4 *Let* $U_N^{(1)} < \ldots < U_N^{(N)}$ *be the order statistics of a set of independent observations* $U_1, \ldots, U_N$*, each distributed uniformly over* $(0, 1)$*.*

Then the density of $U_N^{(i)}$ *exists and equals*

$$f_N^{(i)}(x) = N \binom{N-1}{i-1} x^{i-1}(1-x)^{N-i}, \; 0 < x < 1, \; 1 \le i \le N. \tag{10}$$

Further,

$$\mathsf{E}U_N^{(i)} = \frac{i}{N+1}, \quad 1 \le i \le N, \tag{11}$$

and

$$\mathsf{var}\, U_N^{(i)} = \frac{i(N-i+1)}{(N+1)^2(N+2)}, \quad 1 \le i \le N, \tag{12}$$

hold.

Proof. (10) is a particular case of (9); (11) and (12) follow immediately from the relation

$$\int_0^1 x^a(1-x)^b \, \mathrm{d}x = \frac{a!b!}{(a+b+1)!}, \tag{13}$$

holding for any non-negative integers a and b. □

3.1.3 Hypothesis H_1 (symmetry). Let H_1 denote the family of N-dimensional densities p such that $p = \prod f(x_i)$, $1 \le i \le N$, where $f(x)$ may be an arbitrary symmetric one-dimensional density:

$$f(x) = f(-x), \quad -\infty < x < \infty. \tag{1}$$

Obviously $H_1 \subset H_0$. Consider the function

$$\operatorname{sign} x = \begin{cases} 1, & \text{if } x > 0 \\ 0, & \text{if } x = 0 \\ -1, & \text{if } x < 0 \end{cases} \tag{2}$$

and introduce the following statistics:

The sign statistics $\operatorname{sign} X_1, \dots, \operatorname{sign} X_N$
The absolute values of observations $|X_1|, \dots, |X_N|$
The vector $|X| = (|X_1|, \dots, |X_N|)$
The order statistics $|X|^{(i)} = o_i(|X|)$ for $|X|$
The ranks $R_i^+ = r_i(|X|)$ for $|X|$.

Obviously,

$$X_i = |X|^{(R_i^+)} \operatorname{sign} X_i, \quad 1 \le i \le N, \tag{3}$$

with probability one under each $p \in H_1$.

Furthermore, let us put $|X|^{(\cdot)} = (|X|^{(1)}, \dots, |X|^{(N)})$, $\operatorname{sign} X = (\operatorname{sign} X_1, \dots, \operatorname{sign} X_N)$ and $R^+ = (R_1^+, \dots, R_N^+)$, and denote by $\mathsf{V} = \{v = (v_1, \dots, v_N)\}$ the space of all vectors v such that $v_i = 1$ or -1. Obviously V contains 2^N points.

Theorem 1 *For any* $p \in H_1$, $p = \prod f(x_i)$, *the random vectors* $\operatorname{sign} X$, R^+ *and* $|X|^{(\cdot)}$ *defined above are mutually independent.*

$$P(\operatorname{sign} X = v) = \left(\frac{1}{2}\right)^N, \quad v \in \mathsf{V}, \tag{4}$$

$$P(R^+ = r) = \frac{1}{N!}, \quad r \in \mathsf{R}, \tag{5}$$

hold, and $|X|^{(\cdot)}$ *has the density*

$$2^N N! \prod_1^N f(x_i), \quad 0 < x_1 < \ldots < x_N. \tag{6}$$

Proof. From (1) it follows that $\operatorname{sign} X$ and $|X|$ are independent, as well as that $\operatorname{sign} X$ satisfies (4) and $|X|$ has the density $2^N \prod f(x_i)$ for $x_i > 0$. Now it only remains to apply Theorem 3.1.2.1 to $|X|$, since the density of $|X|$ belongs to H_0. □

3.1.4 Hypothesis H_2 (independence). We shall denote by H_2 the family of $2N$-dimensional densities $p = p(x_1, y_1, \ldots, x_N, y_N)$ such that

$$p = \prod_1^N f(x_i) g(y_i), \tag{1}$$

where f and g may be arbitrary one-dimensional densities.

Under H_2 the partial sets of observations $X = (X_1, \ldots, X_N)$ and $Y = (Y_1, \ldots, Y_N)$ are independent and each is governed by a density from H_0. Introducing the statistics $X^{(i)} = o_i(X)$, $Y^{(i)} = o_i(Y)$, $R_i = r_i(X)$, $Q_i = r_i(Y)$, and putting $X^{(\cdot)} = (X^{(1)}, \ldots, X^{(N)})$, $Y^{(\cdot)} = (Y^{(1)}, \ldots, Y^{(N)})$, $R = (R_1, \ldots, R_N)$ and $Q = (Q_1, \ldots, Q_N)$, we get from Theorem 3.1.2.1 immediately the following

Theorem 1 *Under any density* p *from* H_2, $p = \prod_1^N f(x_i) g(y_i)$, *the random vectors* $X^{(\cdot)}$, $Y^{(\cdot)}$, R *and* Q *are independent.*

$$P(R = r) = P(Q = r) = \frac{1}{N!}, \quad r \in \mathsf{R}, \tag{2}$$

holds and the densities of $X^{(\cdot)}$ *and* $Y^{(\cdot)}$ *are*

$$N! \prod_1^N f(x_i) \quad \textit{and} \quad N! \prod_1^N g(x_i), \quad x_1 < \ldots < x_N, \tag{3}$$

respectively.

3.1.5 Hypothesis H_3 (random blocks). We shall denote by H_3 the family of nk-dimensional densities $p = p(x_{11},\ldots,x_{1k},\ldots, x_{n1},\ldots,x_{nk})$ such that

$$p = \prod_{j=1}^{n}\prod_{i=1}^{k} f_j(x_{ji}), \tag{1}$$

where $f_1,\ldots,f_n$ may be arbitrary one-dimensional densities.

Under H_3 the partial sets of observations $X_1 = (X_{11},\ldots,X_{1k})$, $\ldots$, $X_n = (X_{n1}, \ldots, X_{nk})$ are mutually independent and each is governed by a density from H_0 with $N = k$. Introducing $X_j^{(i)} = o_i(X_j)$ and $R_{ji} = r_i(X_j)$, $1 \le j \le n$, $1 \le i \le k$, and putting $X_j^{(\cdot)} = (X_j^{(1)},\ldots,X_j^{(k)})$ and $R_j = (R_{j1},\ldots,R_{jk})$ we deduce from Theorem 3.1.2.1 the following

Theorem 1 *Under any density p from H_3, $p = \prod_1^n\prod_1^k f_j(x_{ji})$, the random vectors $X_j^{(\cdot)}$ and R_j are mutually independent.*

$$P(R_j = r) = \frac{1}{k!}, \quad r \in \mathcal{R},\ 1 \le j \le n, \tag{2}$$

holds, with $\mathcal{R}$ being the space of all permutations of $(1,\ldots,k)$, and $X_j^{(\cdot)}$, $1 \le j \le n$, has density

$$k!\prod_{i=1}^{k} f_j(x_{ji}), \quad x_{j1} < \ldots < x_{jk}. \tag{3}$$

3.2 FROM PERMUTATION TESTS AND INVARIANT TESTS TO RANK TESTS

Rank tests may be obtained by the process of specialization from either of the two following families of tests:

– permutation tests,

– tests invariant under changes of location and scale.

In what follows, basic information about the two families will be given, and some advantages and shortcomings of rank tests as compared with them will be pointed out.

3.2.1 Permutation tests. Let us consider a function $\psi(x^{(\cdot)}, r)$ on $\mathcal{X}^{(\cdot)} \times \mathcal{R}$ and put

$$\Psi(x_1,\ldots,x_N) = \psi(x^{(1)},\ldots,x^{(N)},r_1,\ldots,r_N), \tag{1}$$
$$\text{if } x_i \ne x_j \text{ for all } i \ne j,$$

and define Ψ arbitrarily on the remaining subset, since it has probability 0 under H_0 (or H_*). We assume that x and $x^{(\cdot)}$ are related by $x^{(i)} = o_i(x)$, $1 \leq i \leq N$, and that $r_i = r_i(x)$. Thus $x^{(1)} < \ldots < x^{(N)}$ are the values $x_1, \ldots, x_N$ rearranged in ascending order, and r_i is the rank of x_i. Obviously, for each $\Psi(x)$ there exists a $\psi(x^{(\cdot)}, r)$ such that (1) holds. Recall that $\mathcal{A}^{(\cdot)}$ denotes the σ-field consisting of Borel subsets of $\mathcal{X}^{(\cdot)}$.

Lemma 1 *The function Ψ is $\mathcal{A}$-measurable if and only if the corresponding function ψ is $\mathcal{A}^{(\cdot)}$-measurable for every fixed $r \in \mathcal{R}$.*

Proof. For every linear Borel set B the inverse image $\Psi^{-1}(B)$ is $\mathcal{A}$-measurable if and only if all intersections $\{R = r\} \cap \Psi^{-1}(B)$ are $\mathcal{A}$-measurable. Denote the mapping $x_i = x^{(r_i)}$, $1 \leq i \leq N$, from $\mathcal{X}^{(\cdot)}$ to $\mathcal{X}$ by $x = \pi_r(x^{(\cdot)})$, and introduce the functions $\psi_r(x^{(\cdot)}) = \psi(x^{(\cdot)}, r)$. Then, under $\{R = r\}$, we have $\Psi\pi_r = \psi_r$ and $\Psi = \psi_r\pi_r^{-1}$, so that the $\mathcal{A}$-measurability of Ψ is equivalent to the $\mathcal{A}^{(\cdot)}$-measurability of ψ_r. □

Definition. Assume that $\psi(x^{(\cdot)}, r)$ is $\mathcal{A}^{(\cdot)}$-measurable for every fixed $r \in \mathcal{R}$, $0 \leq \psi \leq 1$, and that for some $\alpha \in (0, 1)$

$$\frac{1}{N!} \sum_{r \in \mathcal{R}} \psi(x^{(\cdot)}, r) = \alpha, \quad x^{(\cdot)} \in \mathcal{X}^{(\cdot)}. \tag{2}$$

Then the test with critical function Ψ, related to ψ by (1), will be called a *permutation test.*

If the permutation test is non-randomized, i.e. $\psi = 1$ or 0, then α must equal $k/N!$ for some integer k. Note that, on account of Theorem 3.1.2.1, relation (2) may also be written as follows:

$$\mathsf{E}_0\psi(y, R) = \alpha, \quad y \in \mathcal{X}^{(\cdot)} \tag{3}$$

where E_0 denotes the expectation under H_0.

A test with critical function Ψ is called *similar* if

$$\int \Psi \, \mathrm{d}P = \alpha, \quad \text{for all } P \in H. \tag{4}$$

Thus for similar tests the probability of the error of the first kind is constant throughout the whole hypothesis.

Theorem 1 *Under H_0 the family of similar tests coincides with the family of permutation tests.*

Proof. The proof follows from the fact that $X^{(\cdot)}$ is a complete sufficient statistic (see Problems 7 and 8), and may be found in Lehmann (1959, p. 184). □

Theorem 2 *For every q the most powerful test of H_* against q may be found in the family of permutation tests.*

Proof. The proof is left for the reader as an exercise (see Lehmann and Stein 1949).

Theorem 3 *The most powerful level α permutation test of H_0 (or H_*) against a simple alternative $q(x) = q^*(x^{(\cdot)}, r)$ has the following structure:*

$$\psi(x^{(\cdot)}, r) = \begin{cases} 1, & \text{if } q^*(x^{(\cdot)}, r) > k(x^{(\cdot)}), \\ 0, & \text{if } q^*(x^{(\cdot)}, r) < k(x^{(\cdot)}), \end{cases} \tag{5}$$

where $k(x^{(\cdot)})$, $x^{(\cdot)} \in \mathcal{X}^{(\cdot)}$, and $\psi(x^{(\cdot)}, r)$ for r such that $q^*(x^{(\cdot)}, r) = k(x^{(\cdot)})$ should and can be found so as to satisfy (2).

Proof. Since (2) must hold for each $x^{(\cdot)} \in \mathcal{X}^{(\cdot)}$, it suffices to maximize the conditional power

$$\mathsf{E}_q\{\psi(X^{(\cdot)}, R) \mid X^{(\cdot)} = x^{(\cdot)}\} \tag{6}$$

for each $x^{(\cdot)}$ separately. From Theorem 3.1.1.1 it follows that (6) equals

$$(\bar{q})^{-1} \sum_{r \in \mathcal{R}} \psi(x^{(1)}, \ldots, x^{(N)}, r_1, \ldots, r_N) q(x^{(r_1)}, \ldots, x^{(r_N)}). \tag{7}$$

In order to maximize (7) under the side condition (2) it suffices to apply the Neyman-Pearson lemma to the problem of testing that the distribution on $\mathcal{R}$ is uniform against the alternative with $Q(r) = (\bar{q})^{-1} q(x^{(r_1)}, \ldots, x^{(r_N)})$. This, carried out for each $x^{(\cdot)}$ separately, yields (5). □

Example. Let $N = m + n$, and

$$q = (2\pi)^{-\frac{1}{2}N} \exp\Big[-\frac{1}{2}\sum_{i=1}^{m}(x_i - \Delta)^2 - \frac{1}{2}\sum_{i=m+1}^{m+n} x_i^2\Big], \tag{8}$$

where Δ is a positive constant. Then

$$q^*(x^{(\cdot)}, r) = (2\pi)^{-\frac{1}{2}N} \exp\Big[-\frac{1}{2}\sum_{i=1}^{m+n}(x^{(i)})^2 - \frac{1}{2}m\Delta^2 + \Delta\sum_{i=1}^{m} x^{(r_i)}\Big], \tag{9}$$

and the critical region of the best permutation test is determined by

$$\sum_{i=1}^{m} x^{(r_i)} \geq k(x^{(\cdot)}) \tag{10}$$

with possible randomization in the case of equality, where $k(x^{(\cdot)})$ is chosen so that (10) is satisfied by $100\alpha\%$ permutations $r \in \mathsf{R}$. This test is due to Pitman (1937/38).

Let us summarize:

(a) Under H_* and for an arbitrary alternative q, for each non-similar test there exists a better similar test.

(b) Under H_0 the similar tests coincide with permutation tests; a fortiori this is true under H_*.

(c) Most powerful permutation tests may be found quite easily, at least theoretically, since, by the conditional argument, the problem may be carried over from the λ-field of similar events to the σ-field of all subsets of R.

There remains open the question of how useful the non-similar tests may be under H_0, i.e. whether the difference between $\beta(\alpha, H_0, q)$ and the power of the most powerful permutation test is essential or not. What we know is that in the usual models the difference is asymptotically negligible. This is true, for example, for the following alternatives to H_0:

$$q = \prod_{i=1}^{N} f(x_i - d_i) \tag{11}$$

where $I(f)\sum(d_i - \bar{d})^2$ is not too large, say ≤ 25, while $\max(d_i - \bar{d})^2[\sum(d_i - \bar{d})^2]^{-1}$ is sufficiently small, say ≤ 0.1. What is more, it will be shown that the class of permutation tests is unnecessarily large for this purpose, because the same holds even for the family of rank tests, in which $\psi(x^{(\cdot)}, r)$ is independent of $x^{(\cdot)}$.

On the other hand, if the alternative is composite, then the family of rank tests does not contain the asymptotically maximin most powerful test in many important cases where the family of permutation tests does. This may happen even if K contains only two densities, $K = \{q_1, q_2\}$.

In conclusion, let us mention that the permutation tests, dealt with above, are related to H_0 (or H_*), and that their definition may be extended to H_1, H_2 and H_3 as follows: Under H_1, $\psi(|x|^{(\cdot)}, \operatorname{sign} x, r^+)$ is called a permutation test if for each $|x|^{(\cdot)}$

$$\mathsf{E}\psi(|x|^{(\cdot)}, \operatorname{sign} X, R^+) = \alpha. \tag{12}$$

Under H_2, $\psi(x^{(\cdot)}, y^{(\cdot)}, r, r')$ is called a permutation test if for each $(x^{(\cdot)}, y^{(\cdot)})$

$$\mathsf{E}\psi(x^{(\cdot)}, y^{(\cdot)} R, Q) = \alpha. \tag{13}$$

Finally, under H_3, $\psi(x_1^{(\cdot)}, \ldots, x_n^{(\cdot)}, r_1, \ldots, r_N)$ is called a permutation test if for each $(x_1^{(\cdot)}, \ldots, x_n^{(\cdot)})$

$$\mathsf{E}\psi(x_1^{(\cdot)}, \ldots, x_n^{(\cdot)}, R_1, \ldots, R_n) = \alpha. \tag{14}$$

Permutation tests are sometimes also called randomization tests.

3.2.2 Tests invariant under changes of location and scale. An event $B \in \mathcal{A}$ is called invariant under changes of location and scale if for all real u and $\lambda > 0$

$$[(x_1, \ldots, x_N) \in B] \Longleftrightarrow [(\lambda x_1 - u, \ldots, \lambda x_N - u) \in B]. \tag{1}$$

The class $\mathcal{B}_1$ of all events possessing this property is a sub-σ-field of $\mathcal{A}$. For any N-dimensional density $p(x)$ and any real μ and $\sigma > 0$ the two probability distributions P and $P_{\mu,\sigma}$ determined by

$$p = p(x_1, \ldots, x_N) \tag{2}$$

and

$$p_{\mu,\sigma} = \sigma^{-N} p\Big(\frac{x_1 - \mu}{\sigma}, \ldots, \frac{x_N - \mu}{\sigma}\Big) \tag{3}$$

coincide on $\mathcal{B}_1$, i.e.

$$P(B) = P_{\mu,\sigma}(B), \quad \text{for each } B \in \mathcal{B}_1. \tag{4}$$

A test Ψ will be called invariant under changes of location and scale, if Ψ is $\mathcal{B}_1$-measurable. In order to be able to find the most powerful invariant test we have to express the restriction of P to $\mathcal{B}_1$ by a $\mathcal{B}_1$-measurable density p_1 relative to some σ-finite measure $\nu(\cdot)$:

$$P(B) = \int_B p_1(x)\,\mathrm{d}\nu, \quad B \in \mathcal{B}_1. \tag{5}$$

Note that $p_1(x)$ is $\mathcal{B}_1$-measurable if and only if

$$p_1(\lambda x_1 - u, \ldots, \lambda x_N - u) = p_1(x_1, \ldots, x_N), \; -\infty < u < \infty, \; \lambda > 0,$$

and that the integral (5) is not well defined unless $\nu(\cdot)$ is σ-finite on $\mathcal{B}_1$. Since the Lebesgue measure of each $B \in \mathcal{B}_1$ is either 0 or ∞, $\nu(\cdot)$ cannot be Lebesgue measure. We can take, however, for ν any probability distribution P_0, the density of which satisfies $p_0(x_1, \ldots, x_N) > 0$ everywhere. For example, we may put

$$p_0 = (2\pi)^{-\frac{1}{2}N} \exp\Big(-\frac{1}{2}\sum_{i=1}^{N} x_i^2\Big). \tag{6}$$

On $\mathcal{A}$, the density of P relative to P_0 equals $p(x)/p_0(x)$. Actually, this follows from

$$P(A) = \int_A p\,\mathrm{d}x = \int_A (p/p_0)\,\mathrm{d}P_0, \tag{7}$$

where we have utilized the fact that $p_0 > 0$ everywhere. Further, if $\mathcal{B}_2$ is the sub-σ-field generated by $(x_3, \ldots, x_N)$ and

$$\tilde{p}(x_1, x_2, \ldots, x_n) = \int_{-\infty}^{\infty} \int_{-\infty}^{\infty} p(t, s, x_3, \ldots, x_N) \,\mathrm{d}t \,\mathrm{d}s \tag{8}$$

$$\tilde{p}_0(x_1, x_2, \ldots, x_n) = \int_{-\infty}^{\infty} \int_{-\infty}^{\infty} p_0(t, s, x_3, \ldots, x_N) \,\mathrm{d}t \,\mathrm{d}s \tag{9}$$

then $\tilde{p}/\tilde{p}_0$ represents the density of P relative to P_0 on $\mathcal{B}_2$. Actually, $\tilde{p}$ and $\tilde{p}_0$ are both $\mathcal{B}_2$-measurable (i.e. $\tilde{p}(x_1, x_2, x_3, \ldots, x_N) = \tilde{p}(y_1, y_2, x_3, \ldots, x_N)$, $\tilde{p}_0(x_1, x_2, x_3, \ldots, x_N) = \tilde{p}_0(y_1, y_2, x_3, \ldots, x_N)$ for any real $(y_1, y_2, x_1, x_2, \ldots, x_N)$), thus $\tilde{p}/\tilde{p}_0$ also is $\mathcal{B}_2$-measurable. Moreover, for each $B \in \mathcal{B}_2$

$$P(B) = \int_B (\tilde{p}/\tilde{p}_0) \,\mathrm{d}P_0, \quad B \in \mathcal{B}_2, \tag{10}$$

holds, as may be easily seen from (8) and (9).

Making use of this technique in combination with a change of variables, we get the following

Theorem 1 *Assume that* $p_0 > 0$ *holds everywhere. Put*

$$\bar{p}(x_1, \ldots, x_N) = \int_0^{\infty} \int_{-\infty}^{\infty} p(\lambda x_1 - u, \ldots, \lambda x_N - u)\lambda^{N-2} \,\mathrm{d}u \,\mathrm{d}\lambda \tag{11}$$

and define $\bar{p}_0$ *similarly.*

Then $\bar{p}/\bar{p}_0$ *represents the density of* P *relative to* P_0 *on* $\mathcal{B}_1$, *i.e.*

$$\frac{\bar{p}(x_1, \ldots, x_N)}{\bar{p}_0(x_1, \ldots, x_N)} = \frac{\bar{p}(\lambda x_1 - u, \ldots, \lambda x_N - u)}{\bar{p}_0(\lambda x_1 - u, \ldots, \lambda x_N - u)}, \quad 0 < \lambda, \ -\infty < u < \infty, \tag{12}$$

and for each $B \in \mathcal{B}_1$

$$P(B) = \int_B (\bar{p}/\bar{p}_0) \,\mathrm{d}P_0, \quad B \in \mathcal{B}_1, \tag{13}$$

holds.

Proof. Let C be an orthogonal $N \times N$ matrix with the vector $(1, 1, \ldots, 1)$ in the first row, and put $y = Cx$. Then

$$y_1 = x_1 + \ldots + x_N, \tag{14}$$

and

$$(x_1, \ldots, x_N) \leftrightarrow (y_1, y_2, \ldots, y_N) \tag{15}$$

implies

$$(x_1 - u, \ldots, x_N - u) \leftrightarrow (y_1 - Nu, y_2, \ldots, y_N), \quad -\infty < u < \infty. \tag{16}$$

This means that location changes in x are reflected as y_1-changes in y. Furthermore, put

$$\begin{aligned} y_1 &= z_1 \\ y_2 &= z_2 \cos z_3 \\ y_3 &= z_2 \sin z_3 \cos z_4 \\ &\vdots \\ y_{N-1} &= z_2 \sin z_3 \ldots \sin z_{N-1} \cos z_N \\ y_N &= z_2 \sin z_3 \ldots \sin z_{N-1} \sin z_N \end{aligned} \tag{17}$$

which amounts to introducing polar coordinates in $(y_2, \ldots, y_N)$-space. Assume $-\infty < z_1 < \infty$, $0 < z_2 < \infty$, $0 < z_3, \ldots, z_{N-1} < \pi$ and $0 \leq z_N < 2\pi$. The transformation (17), say $y = \tau(z)$, establishes a one-to-one correspondence between y-space and z-space, up to a subset of Lebesgue measure 0. Now the compound map $z = \tau^{-1}Cx$ establishes a one-to-one correspondence between the x-space and the z-space such that

$$(x_1, \ldots, x_N) \leftrightarrow (z_1, z_2, z_3, \ldots, z_N) \tag{18}$$

implies

$$(\lambda x_1 - u, \ldots, \lambda x_N - u) \leftrightarrow (\lambda z_1 - Nu, \lambda z_2, z_3, \ldots, z_N), \tag{19}$$

when to each real u and $\lambda > 0$ there corresponds just one point (t, s), $-\infty < t < \infty$, $0 < s < \infty$, such that $t = \lambda z_1 - Nu$ and $s = \lambda z_2$. Consequently, invariance with respect to location and scale changes in x-space corresponds to invariance with respect to (z_1, z_2)-changes in z-space. Thus it suffices to compute the respective density in z-space, to apply the formulas (8) and (9), and finally to substitute $z = \tau^{-1}Cx$.

The Jacobian of the transformation C equals $N^{\frac{1}{2}}$ and the Jacobian of the transformation (17) is known to be

$$z_2^{(N-2)}(\sin z_3)^{(N-3)}(\sin z_4)^{(N-4)} \ldots \sin z_{N-1}, \tag{20}$$

so that the density in z-space equals

$$p[(C^{-1}\tau(z)]z_2^{N-2}k(z_3, \ldots, z_N), \tag{21}$$

where $k(z_3, \ldots, z_N)$ does not depend on p. Now, on account of (18) and (19), $x = C^{-1}\tau(z)$ implies

$$C^{-1}\tau(t, s, z_3, \ldots, z_N) = (\lambda x_1 - u, \ldots, \lambda x_N - u) \tag{22}$$

where

$$\lambda = s/z_2 \quad \text{and} \quad u = sz_1/(z_2N) - t/N. \tag{23}$$

Consequently

$$\begin{aligned}\tilde{p} &= \int_0^\infty \int_{-\infty}^\infty p[C^{-1}\tau(t,s,z_3,\ldots,z_N)] \\ &\qquad s^{N-2}k(z_3,\ldots,z_N)\,\mathrm{d}t\,\mathrm{d}s \\ &= k(z_3,\ldots,z_N)Nz_2^{N-1} \\ &\qquad \cdot\int_0^\infty \int_{-\infty}^\infty p(\lambda x_1 - u,\ldots,\lambda x_N - u)\lambda^{N-2}\,\mathrm{d}u\,\mathrm{d}\lambda.\end{aligned} \tag{24}$$

Similarly,

$$\begin{aligned}\tilde{p}_0 &= k(z_3,\ldots,z_N)z_2^{N-1} \\ &\qquad \cdot\int_0^\infty \int_{-\infty}^\infty p_0(\lambda x_1 - u,\ldots,\lambda x_N - u)\lambda^{N-2}\,\mathrm{d}u\,\mathrm{d}\lambda.\end{aligned}$$

Consequently, $\tilde{p}/\tilde{p}_0 = \bar{p}/\bar{p}_0$ with $\bar{p}$ being given by (11) and $\bar{p}_0$ defined similarly. □

As an immediate consequence of applying the Neyman-Pearson lemma to the densities $\bar{p}/\bar{p}_0$ and $\bar{p}/\bar{q}_0$ we get the following

Theorem 2 *The most powerful test of p against q, invariant under location and scale changes, is given by Ψ which equals 1, if*

$$\begin{aligned}&\int_0^\infty \int_{-\infty}^\infty q(\lambda x_1 - u,\ldots,\lambda x_N - u)\lambda^{N-2}\,\mathrm{d}u\,\mathrm{d}\lambda \\ &\qquad > k\int_0^\infty \int_{-\infty}^\infty p(\lambda x_1 - u,\ldots,\lambda x_N - u)\lambda^{N-2}\,\mathrm{d}u\,\mathrm{d}\lambda,\end{aligned} \tag{25}$$

and equals 0 if the opposite strict inequality holds. The constant k, and Ψ in the remaining points, should and can be chosen so that $\int \Psi\,\mathrm{d}P = \alpha$, where α is the prescribed significance level.

Example. Take

$$p = (2\pi)^{-\frac{1}{2}N}\exp\Big(-\frac{1}{2}\sum_{i=1}^N x_i^2\Big) \tag{26}$$

and the alternative q given by (3.2.1.8). We have

$$\begin{aligned}\bar{p} &= (2\pi)^{-\frac{1}{2}N}\int_0^\infty\int_{-\infty}^\infty \exp\Big[-\frac{1}{2}\sum_{i=1}^N(\lambda x_i-u)^2\Big]\lambda^{N-2}\,\mathrm{d}u\,\mathrm{d}\lambda\\ &= N^{-\frac{1}{2}}(2\pi)^{-\frac{1}{2}N+\frac{1}{2}}\int_0^\infty \exp\Big[-\frac{1}{2}\lambda^2\sum_{i=1}^N(x_i-\overline{\overline{x}})^2\Big]\lambda^{N-2}\,\mathrm{d}\lambda\\ &= \frac{1}{2}N^{-\frac{1}{2}}\pi^{-\frac{1}{2}N+\frac{1}{2}}\Gamma\Big(\frac{1}{2}N-\frac{1}{2}\Big)\Big[\sum_{i=1}^N(x_i-\overline{\overline{x}})^2\Big]^{-\frac{1}{2}N+\frac{1}{2}},\end{aligned}$$

with

$$\overline{\overline{x}} = (1/N)\sum_{i=1}^N x_i,$$

and

$$\begin{aligned}\bar{q} &= (2\pi)^{-\frac{1}{2}N}\int_0^\infty\int_{-\infty}^\infty \exp\Big[-\frac{1}{2}\sum_{i=1}^m(\lambda x_i-u-\Delta)^2\\ &\qquad -\frac{1}{2}\sum_{i=m+1}^N(\lambda x_i-u)^2\Big]\lambda^{N-2}\,\mathrm{d}u\,\mathrm{d}\lambda\\ &= N^{-\frac{1}{2}}(2\pi)^{-\frac{1}{2}N+\frac{1}{2}}\int_0^\infty \exp\Big[-\frac{1}{2}\sum_{i=1}^m\Big(\lambda x_i-\lambda\overline{\overline{x}}-\frac{n}{N}\Delta\Big)^2\\ &\qquad -\frac{1}{2}\sum_{i=m+1}^N\Big(\lambda x_i-\lambda\overline{\overline{x}}+\frac{m}{N}\Delta\Big)^2\Big]\lambda^{N-2}\,\mathrm{d}\lambda\\ &= N^{-\frac{1}{2}}(2\pi)^{-\frac{1}{2}N+\frac{1}{2}}\Big[\sum_{i=1}^N(x_i-\overline{\overline{x}})^2\Big]^{-\frac{1}{2}N+\frac{1}{2}}\exp\Big(-\frac{1}{2}\frac{mn}{N}\Delta^2\Big)\\ &\qquad\cdot\int_0^\infty \lambda^{N-2}\exp(-\tfrac{1}{2}\lambda^2+\lambda\Delta t^*)\,\mathrm{d}\lambda,\end{aligned}$$

where

$$t^* = \frac{mn}{N}\frac{\bar{x}-\bar{y}}{(\sum_{i=1}^N(x_i-\overline{\overline{x}})^2)^{\frac{1}{2}}} \tag{27}$$

with

$$\bar{y} = \frac{1}{n}\sum_{i=m+1}^N x_i \quad\text{and}\quad \bar{x} = \frac{1}{m}\sum_{i=1}^m x_i.$$

Consequently

$$\frac{\bar{q}(x)}{\bar{p}(x)} = \mathrm{const}(\Delta)\int_0^\infty \lambda^{N-2}\exp(-\tfrac{1}{2}\lambda^2+\lambda\Delta t^*)\,\mathrm{d}\lambda,$$

which is, for $\Delta > 0$, an increasing function of t^*, and t^*, in turn, is an increasing function of

$$t = \left[\frac{mn(N-2)}{N}\right]^{\frac{1}{2}} \frac{\bar{x} - \bar{y}}{[\sum_{i=1}^{m}(x_i - \bar{x})^2 + \sum_{j=m+1}^{N}(x_j - \bar{y})^2]^{\frac{1}{2}}}. \tag{28}$$

Thus the solution is the classical Student's t-test.

Since the distribution of invariant test statistics is invariant under location and scale changes, we could assume the composite hypothesis consisting of all densities

$$p_{\mu,\sigma} = (2\pi\sigma^2)^{-\frac{1}{2}N} \exp\Big[-\frac{1}{2}\sigma^{-2}\sum_{i=1}^{N}(x_i - \mu)^2\Big], \tag{29}$$
$$-\infty < \mu < \infty,\ \sigma > 0,$$

and the composite alternative consisting of all densities

$$q_{\mu,\sigma} = (2\pi)^{-\frac{1}{2}N}\sigma^{-N} \exp\Big\{-\frac{1}{2}\sigma^{-2}\Big[\sum_{i=1}^{m}(x_i - \mu - \Delta\sigma)^2 \tag{30}$$
$$- \sum_{i=m+1}^{N}(x_i - \mu)^2\Big]\Big\}, \quad -\infty < \mu < \infty,\ \sigma > 0,$$

without affecting the size and power of the test.

Rank tests are invariant not only under increasing linear transformations but under all strictly increasing continuous transformations, and thus constitute a subfamily of tests invariant under location and scale changes. For many important alternatives the most powerful invariant tests may be well approximated by rank tests. In the above example, the t-test may be approximated by the van der Waerden test or the Fisher-Yates-Terry-Hoeffding test (cf. Subsection 4.1.1).

On the other hand,in some situations the rank tests are useless,whereas the invariant tests are very expedient. For example, consider the problem of distinguishing two types of densities

$$\{\sigma^{-1}f((x-\mu)/\sigma)\} \quad \text{and} \quad \{\sigma^{-1}g(x-\mu)/\sigma)\}, \tag{31}$$
$$-\infty < \mu < \infty,\ \sigma > 0,$$

on the basis of N independent observations. This may be formulated as testing $H = \{p_{\mu,\sigma}\}$, against $K = \{q_{\mu,\sigma}\}$ where

$$p_{\mu,\sigma} = \prod_{i=1}^{N}\frac{1}{\sigma}f\Big(\frac{x_i - \mu}{\sigma}\Big), \quad \infty < \mu < \infty,\ \sigma > 0, \tag{32}$$

and

$$q_{\mu,\sigma} = \prod_{i=1}^{N} \frac{1}{\sigma} g\Big(\frac{x_i - \mu}{\sigma}\Big), \quad \infty < \mu < \infty,\ \sigma > 0, \tag{33}$$

and the discrimination of H from K may be accomplished by the most powerful invariant test with critical region

$$\int_0^\infty \int_{-\infty}^\infty [\lambda^{N-2} \prod_{i=1}^{N} g(\lambda x_i - u)]\, \mathrm{d}u\, \mathrm{d}\lambda \tag{34}$$
$$\geq k \int_0^\infty \int_{-\infty}^\infty [\lambda^{N-2} \prod_{i=1}^{N} f(\lambda x_i - u)]\, \mathrm{d}u\, \mathrm{d}\lambda.$$

It is clear that H and K are completely indistinguishable by rank tests because R is uniformly distributed under H as well as under K.

The invariant tests just considered are most frequently used for testing a sub-hypothesis of H_0 against some alternative. For sub-hypotheses of H_1 and H_2 corresponding invariant tests may be introduced as follows:

Since H_1 is not invariant under location shifts, we have to restrict ourselves to scale changes only. Denote by $\mathcal{B}_3$ the sub-σ-field of events invariant under the transformation $(y_1, \dots, y_N) = (\lambda x_1, \dots, \lambda x_N)$, $\lambda > 0$. Obviously $\mathcal{B}_3 \subset \mathcal{B}_1$. Introducing

$$\hat{p}(x_1, \dots, x_N) = \int_0^\infty p(\lambda x_1, \dots, \lambda x_N) \lambda^{N-1}\, \mathrm{d}\lambda \tag{35}$$

and defining $\hat{p}_0$ similarly, we could show that $\hat{p}/\hat{p}_0$ represents the density of P relative to P_0 on $\mathcal{B}_3$. Most powerful invariant tests are then based on $\hat{p}$ analogously as $\mathcal{B}_1$-measurable tests were based on $\bar{p}$.

In connection with H_2 we may consider the location and scale changes for each of the points $(x_1, \dots, x_N)$ and $(y_1, \dots, y_N)$ separately. Denote the corresponding sub-σ-field by $\mathcal{B}_4$. If the density on $\mathcal{A}$ is given by $p = p(x_1, y_1, \dots, x_N, y_N)$, and

$$\bar{\bar{p}}(x_1, y_1, \dots, x_N, y_N) \tag{36}$$
$$= \int_0^\infty \int_0^\infty \int_{-\infty}^\infty \int_{-\infty}^\infty p(\lambda x_1 - u, \kappa y_1 - v, \dots$$
$$\dots, \lambda x_N - u, \kappa y_N - v) \lambda^{N-2} \kappa^{N-2}\, \mathrm{d}u\, \mathrm{d}v\, \mathrm{d}\lambda\, \mathrm{d}\kappa,$$

then $\bar{\bar{p}}/\bar{\bar{p}}_0$ is the density of P relative to P_0 on $\mathcal{B}_4$. Again $\bar{\bar{p}}$ is the basis for most powerful invariant tests.

We omit the corresponding considerations for H_3.

3.2.3 Rank tests. The definition of rank tests is very simple but must also be given separately for each null hypothesis H_0, H_1, H_2 and H_3.

Under H_0 a randomized test is called a rank test if its critical function Ψ is a function of the vector of ranks R, $\Psi = \Psi(R)$. The critical region of non-randomized rank tests is a union of some of the following events:

$$\begin{aligned} B_r &= \{R = r\} \\ &= \{(x_1, \ldots, x_N) : \operatorname{sign}(x_i - x_j) = \operatorname{sign}(r_i - r_j), \\ &\qquad 1 \le i \ne j \le N\}, \quad r \in \mathsf{R}. \end{aligned} \tag{1}$$

The sub-σ-field generated by R, say $\mathcal{B}_0$, consists of all such unions, while the rest of the space X where some coordinates coincide and R is not defined may be neglected, since its probability is zero under any distribution determined by a density.

Under H_0 we have

$$P(R = r) = P(B_r) = \frac{1}{N!}, \quad r \in \mathsf{R}. \tag{2}$$

Under some alternative Q, $\mathrm{d}Q = q\,\mathrm{d}x$, we have probabilities $Q(R = r) = Q(B_r)$. Introducing on the sub-σ-field $\mathcal{B}_0$ the so-called counting measure $\nu(\cdot)$, which is defined by $\nu(B_r) = 1$, then $P(B_r)$ and $Q(B_r)$ may be regarded as densities of $P(\cdot)$ and $Q(\cdot)$ relative to the sub-σ-field $\mathcal{B}_0$ and with respect to the counting measure ν. Actually, if

$$q_0(x) = Q(B_r), \quad \text{if } x \in B_r, \tag{3}$$

then, obviously,

$$Q(B) = \int_B q_0(x)\,\mathrm{d}\nu, \quad \text{for } B \in \mathcal{B}_0. \tag{4}$$

The following theorem is an immediate consequence of this fact and of the Neyman-Pearson lemma.

Theorem 1 *The most powerful rank test of H_0 (or H_*) against some simple alternative q is given by*

$$\Psi(r) = \begin{cases} 1, & \text{if } Q(R = r) > k \\ 0, & \text{if } Q(R = r) < k, \end{cases} \tag{5}$$

where $\mathrm{d}Q = q\,\mathrm{d}x$, and k and $\Psi(r)$ for r such that $Q(R = r) = k$ should and can be determined so that $\mathsf{E}\Psi(R) = \alpha$ under H_0.

In practice, however, the exact evaluation of $\Psi(r)$ is rarely possible because $Q(R = r)$ is difficult to compute. For this reason we shall develop locally most powerful rank tests in the last part of this chapter. One of the few exceptions is given in the following

Example a. Let $m < N$ and

$$q(x_1, \ldots, x_N) = 2^m \prod_{i=1}^{m} x_i, \quad 0 < x_1, \ldots, x_N < 1, \tag{6}$$

which means that the observations $X_1, \ldots, X_m$ have the density $2x$, $0 < x < 1$, and that $X_{m+1}, \ldots, X_N$ are uniformly distributed. Put $p(x_1, \ldots, x_N) = 1$, $0 < x_1, \ldots, x_N < 1$, which means that all observations are uniformly distributed. According to Theorem 3.1.2.2 we have

$$Q(R = r) = \mathsf{E}\Big\{2^m \prod_{i=1}^{m} X^{(r_i)}\Big\} \frac{1}{N!} \tag{7}$$

where the expectation is taken with respect to the uniform density p. Now since the distribution of $X^{(\cdot)}$ is also uniform over the region $0 < x^{(1)} < \ldots < x^{(N)} < 1$ under p, in view of Theorem 3.1.2.1 (7) equals

$$2^m \int \cdots \int_{0<x^{(1)}<\ldots<x^{(N)}<1} \Big[\prod_{i=1}^{m} x^{(r_i^*)}\Big] \,\mathrm{d}x^{(1)} \ldots \mathrm{d}x^{(N)}, \tag{8}$$

where $r_1^* < \ldots < r_m^*$ are the ranks $r_1, \ldots, r_m$ rearranged in ascending order. After easy computation (8) yields

$$Q(R = r) = [(N + m)!]^{-1} 2^m r_1^*(r_2^* + 1) \ldots (r_m^* + m - 1). \tag{9}$$

Even this explicit result does not yield a test statistic the distribution of which could be easily established under H_0.

Example b. Let $m < N$ and

$$q_\Delta(x_1, \ldots, x_N) = \begin{cases} 1, & \text{if } \Delta < x_1, \ldots, x_m < 1 + \Delta, \\ & \quad 0 < x_{m+1}, \ldots, x_N < 1, \\ 0, & \text{otherwise, } 0 < \Delta < 1. \end{cases} \tag{10}$$

This is a special case of the alternative (3.4.4.1) in the sequel, with f being the uniform distribution over $(0, 1)$. Introducing

$$\begin{aligned} a &= N - \max\{r_{m+1}, \ldots, r_N\}, \\ b &= \min\{r_1, \ldots, r_m\} - 1, \end{aligned} \tag{11}$$

we shall prove that

$$Q_\Delta\{R = r\} = \sum_{j=0}^{a} \sum_{k=0}^{b} \Delta^{j+k}(1 - \Delta)^{N-j-k} \frac{1}{j!k!(N - j - k)!}. \tag{12}$$

Proof. (10) and (11) entail that

$$Q_\Delta\{R=r\} = \int\cdots\int_{\substack{0<z_1<\ldots<z_N<1+\Delta \\ \Delta<z_{b+1},z_{N-a}<1}} \mathrm{d}z_1\ldots\mathrm{d}z_N \tag{13}$$

$$= \frac{1}{a!b!}\int\cdots\int_{\Delta<z_{b+1}<\ldots<z_{N-a}<1} z_{b+1}^b(1+\Delta-z_{N-a})^a \, \mathrm{d}z_{b+1}\ldots\mathrm{d}z_{N-a}$$

$$= \frac{1}{a!b!}\sum_{j=0}^{a}\binom{a}{j}\Delta^j\int\cdots\int_{\Delta<z_{b+1}<\ldots<z_{N-a}<1} z_{b+1}^b(1-z_{N-a})^{a-j} \, \mathrm{d}z_{b+1}\ldots\mathrm{d}z_{N-a}$$

$$= \sum_{j=0}^{a}\frac{\Delta^j}{j!b!(N-b-j-1)!}\int_\Delta^1 z_{b+1}^b(1-z_{b+1})^{N-b-j-1}\,\mathrm{d}z_{b+1}$$

$$= \sum_{j=0}^{a}\frac{\Delta^j}{j!(n-j)!}\sum_{k=0}^{b}\binom{N-j}{k}\Delta^k(1-\Delta)^{N-j-k}$$

$$= \sum_{j=0}^{a}\sum_{k=0}^{b}\Delta^{j+k}(1-\Delta)^{N-j-k}\frac{1}{j!k!(N-j-k)!}. \qquad \square$$

From (12) it follows that $Q_\Delta\{R=r\}$ depends on r through (a,b) only and that it is an increasing function of both a and b. Thus, if (a,b) corresponds to r and (a',b') to r', then, for $0<\Delta<1$,

$$[(a'\leq a,\, b'<b) \text{ or } (a'<a,\, b'\leq b)] \implies [Q_\Delta(R=r)>Q_\Delta(R=r')]. \tag{14}$$

The relative magnitude of $Q_\Delta(R=r)$ and $Q_\Delta(R=r')$ for pairs (a,b) and (a',b') such that $(a'<a,\, b'>b)$ depends on the particular value of Δ. For Δ close to 0 or to 1 see Problem 13.

In conclusion we define rank tests for H_1, H_2 and H_3: Under H_1 a randomized test is called a rank test if its critical function Ψ is a function of sign X and R^+, $\Psi=\Psi(\operatorname{sign}X, R^+)$. Under H_2 we require $\Psi=\Psi(R,Q)$, and under H_3 it should be $\Psi=\Psi(R_1,\ldots,R_n)$. We omit the respective analogues of the previous theorem.

3.2.4 Lehmann's alternatives. Let $\mathcal{B}$ be a sub-σ-field and let Ψ be the most powerful $\mathcal{B}$-measurable test of H against Q_1. Then Ψ is also most powerful for H against any Q coinciding with Q_1 on $\mathcal{B}$:

$$Q(B)=Q_1(B), \quad B\in\mathcal{B}. \tag{1}$$

Consequently Ψ is uniformly most powerful for H against any $K=\{Q\}$ consisting merely of Q that satisfy (1).

Let t be an arbitrary map of X to X such that $tB \subset B$ for all $B \in \mathcal{B}$, i.e. such that

$$[t(x) \in B] \Longleftrightarrow [x \in B], \quad B \in \mathcal{B}. \tag{2}$$

Then (1) is satisfied by Q given as follows:

$$Q(X \in A) = Q_1(t(X) \in A), \quad A \in \mathcal{A}. \tag{3}$$

Note that the domain of definition of t may be restricted to some set C such that $Q_1(C) = 1$. We shall here consider the maps

$$t(x) = \big(F^{-1}(x_1), \ldots, F^{-1}(x_N)\big), \tag{4}$$

where F^{-1} is the inverse of a one-dimensional distribution function. This inverse will be defined as follows:

$$F^{-1}(x) = \inf\{y : F(y) \geq x\}. \tag{5}$$

Theorem 1 *If F is continuous, then F^{-1} given by (5) is strictly increasing and t given by (4) satisfies (2) for $B \in \mathcal{B}_0$, $\mathcal{B}_0$ being the sub-σ-field generated by the vector of ranks.*

Proof. If $F(x)$ is continuous, then for every $0 < x_1 < x_2 < x_3 < 1$ there are points $y_1 < y_2 < y_3$ such that $F(y_i) = x_i$, $i = 1, 2, 3$. Then $F^{-1}(x_1) \leq y_1 < y_2 \leq F^{-1}(x_3)$, which shows that $x_1 < x_3$ entails $F^{-1}(x_1) < F^{-1}(x_3)$, i.e. that F^{-1} is strictly increasing. Thus $x_i < x_j$ entails $F^{-1}(x_i) < F^{-1}(x_j)$, $1 \leq i, j \leq N$, and in turn $R(x) = r$ entails $R(t(x)) = r$. □

In view of the above theorem, having a probability distribution Q_1 concentrated on the unit cube, i.e. such that $Q_1(0 < X_i < 1, 1 \leq i \leq N) = 1$, any distribution given by (3) and (4) assigns to the event $\{R = r\}$ the same probabilities as Q_1:

$$Q(R = r) = Q_1(R = r), \quad r \in \mathsf{R}. \tag{6}$$

Let us denote t in (4) by t_F, Q in (3) by Q_F, and the family of one-dimensional continuous distribution functions by F.

Definition. The composite alternatives $K = \{Q_F, F \in \mathsf{F}\}$ are called Lehmann's alternatives. They were introduced by Lehmann (1953).

If Q_1 corresponds to the distribution function

$$Q_1(X_i \leq x_i, \ 1 \leq i \leq N) = \prod_{i=1}^{N} D_i(x_i), \tag{7}$$

then Q_F corresponds to the distribution function

$$Q_F(X_i \leq x_i,\ 1 \leq i \leq N) = \prod_{i=1}^{N} D_i\big(F(x_i)\big). \tag{8}$$

This proposition is a consequence of the following

Theorem 2 *Let F be an arbitrary distribution function, and let F^{-1} be given by (5). Then for any random variable U with distribution function $D(u)$ such that $D(0) = 0$ and $D(1) = 1$,*

$$P\big(F^{-1}(U) \leq x\big) = D\big(F(x)\big) \tag{9}$$

holds.

Proof. It is sufficient to show that $[F^{-1}(u) \leq x]$ is equivalent to $[u \leq F(x)]$. Now $[F^{-1}(u) \leq x]$ entails $[F(x+\varepsilon) \geq u]$ for every $\varepsilon > 0$, which, in view of the continuity of $F(x)$ from the right, yields $[F(x) \geq u]$. The opposite implication $[u \leq F(x)] \Longrightarrow [F^{-1}(u) \leq x]$ follows directly from (5).

Example a. Let Q_1 be the distribution given by the density q in (3.2.3.6). Comparing (3.2.3.6) with (7), we obtain $D_i(x) = x^2$, $1 \leq i \leq m$ and $D_i(x) = x$, $m < i \leq N$, $0 < x < 1$. Thus the corresponding Lehmann's alternative consists of distribution functions

$$Q_F(X_i \leq x_i,\ 1 \leq i \leq N) = \prod_{i=1}^{m} F^2(x_i) \prod_{i=m+1}^{N} F(x_i), \quad F \in \mathsf{F}. \tag{10}$$

If $F(x)$ is absolutely continuous, then Q_F corresponds to the density

$$q_F = 2^m \prod_{i=1}^{m} f(x_i)F(x_i) \prod_{i=m+1}^{N} f(x_i). \tag{11}$$

In seeking the optimum rank test, a Lehmann's alternative may be replaced by any single one of its members. Since the translation alternatives can be regarded as simplest, it is of interest to investigate whether or not a given Lehmann's alternative includes a simple alternative of this kind.

Example b. *Proportional hazards model.* Consider Lehmann's alternative $K = \{Q_F\}$ with

$$Q_F\{X_i \leq x_i,\ 1 \leq i \leq N\} = \prod_{i=1}^{N} [F(x_i)]^{1+c_i}, \quad F \in \mathsf{F}, \tag{12}$$

which includes the previous one as a special case for $c_1 = \ldots = c_m = 1$, $c_{m+1} = \ldots = c_N = 0$. Choosing $F_0(x) = \exp(-e^{-x})$, we obtain

$$q_{F_0} = \prod_{i=1}^{N} f_0[x_i - \log(1 + c_i)], \tag{13}$$

with f_0 and q_{F_0} denoting the densities corresponding to F_0 and Q_{F_0}, respectively. The alternative (13) is obviously of translation type. More generally, let us have N distribution functions F_i, $1 \leq i \leq N$, with densities f_i and survival functions $\bar{F}_i = 1 - F_i$. If for the hazard functions $\mu_i(x) = -(\mathrm{d}/\mathrm{d}x)\log \bar{F}_i(x)$ we have

$$\mu_i(x) = \mu(x) \cdot \exp\{1 + c_i\}, \quad 1 \leq i \leq N,$$

where $\mu(x)$ is some non-negative function, then this is known as the *proportional hazards model* and it is due to Cox (1972). A still more general model of this kind is investigated in Subsection 4.9.5.

3.3 EXPECTATIONS AND VARIANCES OF LINEAR RANK STATISTICS

3.3.1 Linear rank statistics for H_0. A statistic T which is a function of R, $T = t(R)$, will be called a rank statistic. An important subfamily of rank statistics are statistics expressible by

$$S = \sum_{i=1}^{N} a(i, R_i), \tag{1}$$

where $\{a(i,j)\}$ is an arbitrary $N \times N$-matrix. They will be called *linear rank statistics.* Put

$$a(.\,,j) = \frac{1}{N}\sum_{i=1}^{N} a(i,j), \tag{2}$$

$$a(i,.\,) = \frac{1}{N}\sum_{j=1}^{N} a(i,j), \tag{3}$$

$$\bar{a} = \frac{1}{N^2}\sum_{i=1}^{N}\sum_{j=1}^{N} a(i,j). \tag{4}$$

Theorem 1 *Under H_0 (or H_*)*

$$\mathsf{E}S = N\bar{a} \tag{5}$$

and

$$\operatorname{var} S = \frac{1}{N-1} \sum_{i=1}^{N} \sum_{j=1}^{N} [a(i,j) - a(.\,,j) - a(i,.\,) + \bar{a}]^2 \tag{6}$$

hold.

Proof. The relation (5) follows immediately from $P(R_i = j) = 1/N$, $1 \leq i, j \leq N$. Now

$$\begin{aligned} \operatorname{var} S &= \mathsf{E}(S - \mathsf{E}S)^2 \\ &= \mathsf{E}\Big\{ \sum_{i=1}^{N} [a(i, R_i) - a(.\,, R_i) - a(i,.\,) + \bar{a}] \Big\}^2 \\ &= \sum_{i=1}^{N} \mathsf{E}\{a(i, R_i) - a(.\,, R_i) - a(i,.\,) + \bar{a}^2\} \\ &\quad + \sum_{i \neq j}\sum \mathsf{E}\{[a(i, R_i) - a(.\,, R_i) - a(i,.\,) + \bar{a}] \\ &\quad \cdot [a(j, R_j) - a(.\,, R_j) - a(j,.\,) + \bar{a}]\}. \end{aligned} \tag{7}$$

Since

$$P(R_i = k,\ R_j = h) = \begin{cases} \dfrac{1}{N(N-1)}, & i \neq j,\ h \neq k \\ 0, & i \neq j,\ h = k \end{cases} \tag{8}$$

we have from (7)

$$\begin{aligned} \operatorname{var} S &= \frac{1}{N} \sum_{i=1}^{N} \sum_{j=1}^{N} [a(i,j) - a(.\,,j) - a(i,.\,) + \bar{a}]^2 \\ &\quad + \frac{1}{N(N-1)} \sum_{i \neq j}\sum \sum_{h \neq k}\sum [a(i,h) - a(.\,,h) - a(i,.\,) + \bar{a}] \\ &\quad \cdot [a(j,k) - a(.\,,k) - a(j,.\,) + \bar{a}] \end{aligned}$$

which gives (6) after making use of

$$\begin{aligned} &\sum_{h \neq k}\sum [a(i,h) - a(.\,,h) - a(i,.\,) + \bar{a}] \\ &\quad \cdot [a(j,k) - a(.\,,k) - a(j,.\,) + \bar{a}] \\ &= - \sum_{h=1}^{N} [a(i,h) - a(.\,,h) - a(i,.\,) + \bar{a}] \\ &\quad \cdot [a(j,h) - a(.\,,h) - a(j,.\,) + \bar{a}] \end{aligned}$$

and of

$$\sum\sum_{i \neq j} [a(i,h) - a(.\,,h) - a(i,.\,) + \bar{a}]$$
$$\cdot [a(j,h) - a(.\,,h) - a(j,.\,) + \bar{a}]$$
$$= -\sum_{i=1}^{N} [a(i,h) - a(.\,,h) - a(i,.\,) + \bar{a}]^2. \qquad \square$$

If a rank statistic T is not linear in the above sense, we may try to find its *projection* $\hat{T}$ into the family of linear statistics, i.e. a linear rank statistic $\hat{T}$ for which $\mathsf{E}(T - \hat{T})^2$ is minimized under H_0. Very often $\hat{T}$ approximates T quite well and the difference is asymptotically negligible as $N \to \infty$.

Theorem 2 *Consider a rank statistic $T = t(R_1, \ldots, R_N)$ and put*

$$\hat{a}(i,j) = \mathsf{E}\{T \mid R_i = j\}, \quad 1 \leq i,\ j \leq N. \tag{9}$$

Then the statistic

$$\hat{T} = \frac{N-1}{N} \sum_{i=1}^{N} \hat{a}(i, R_i) - (N-2)\mathsf{E}T \tag{10}$$

is the projection of T into the family of linear rank statistics.

Proof. Without loss of generality, we may assume $\mathsf{E}T = 0$. We shall show that under H_0 the last term in

$$\mathsf{E}(T - S)^2 = \mathsf{E}(T - \hat{T})^2 + \mathsf{E}(\hat{T} - S)^2 + 2\mathsf{E}(T - \hat{T})(\hat{T} - S)$$

equals 0 for any linear rank statistic (1). Since

$$\mathsf{E}[(T - \hat{T})(\hat{T} - S)] \sum_{i=1}^{N} \sum_{j=1}^{N} \frac{1}{N} \Big[\frac{N-1}{N} \hat{a}(i,j) - a(i,j)\Big] \mathsf{E}\{T - \hat{T} \mid R_i = j\},$$

it suffices to show $\mathsf{E}\{T - \hat{T} \mid R_i = j\} = 0$, $1 \leq i,\ j \leq N$, i.e. to show

$$\mathsf{E}\{\hat{T} \mid R_i = j\} = \hat{a}(i,j), \quad 1 \leq i,\ j \leq N. \tag{11}$$

We have

$$\begin{aligned} \mathsf{E}\{\hat{T} \mid R_i = j\} &= \frac{N-1}{N} \sum_{h=1}^{N} \mathsf{E}\{\hat{a}(h, R_h) \mid R_i = j\} \qquad (12) \\ &= \frac{N-1}{N} \hat{a}(i,j) + \frac{N-1}{N} \sum_{h \neq i} \frac{1}{N-1} \sum_{k \neq j} \hat{a}(h,k). \end{aligned}$$

Note that events $\{R_h = k\}$, $1 \le k \le N$, represent a partition of $\mathcal{X}$ for every h, as also do the events $\{R_h = j\}$, $1 \le h \le N$, for every j. Thus $\mathsf{E}T = 0$ entails

$$\sum_{k=1}^{N} \hat{a}(h,k) = \sum_{k=1}^{N} \mathsf{E}\{T \mid R_h = k\} = N\mathsf{E}T = 0, \quad 1 \le h \le N, \tag{13}$$

and

$$\sum_{h=1}^{N} \hat{a}(h,j) = \sum_{h=1}^{N} \mathsf{E}\{T \mid R_h = j\} = N\mathsf{E}T = 0, \quad 1 \le j \le N. \tag{14}$$

On utilizing first (13) and then (14), we get

$$\frac{N-1}{N} \sum_{h \ne i} \frac{1}{N-1} \sum_{k \ne j} \hat{a}(h,k) = -\frac{1}{N} \sum_{h \ne i} \hat{a}(h,j) = \frac{1}{N} \hat{a}(i,j),$$

which substituted into (12) gives (11). □

Example. (Kendall's and Spearman's statistics.) M. G. Kendall introduced the statistic

$$\tau = \frac{1}{N(N-1)} \sum\sum_{i \ne h} \operatorname{sign}(i-h) \operatorname{sign}(R_i - R_h) \tag{15}$$

which may serve as a measure of correlation between the two rankings $(1, \ldots, N)$ and $R_1, \ldots, R_N)$. This statistic is not a linear rank statistic; let us then try to compute its projection into the family of linear rank statistics according to the theorem. Obviously, $\mathsf{E}\tau = 0$ and

$$\mathsf{E}\{\operatorname{sign}(R_i - R_h) \mid R_k = j\} = \begin{cases} 0, & k \ne i,\ k \ne h, \\ \dfrac{2j-N-1}{N-1}, & k = i,\ 1 \le j \le N, \\ \dfrac{N+1-2j}{N-1}, & k = h,\ 1 \le j \le N, \end{cases}$$

so that

$$\begin{aligned} \mathsf{E}\{\tau \mid R_k = j\} &= \frac{1}{N(N-1)} \sum_h \operatorname{sign}(k-h) \frac{2j-N-1}{N-1} \\ &\quad + \frac{1}{N(N-1)} \sum_i \operatorname{sign}(i-k) \frac{N+1-2j}{N-1} \\ &= \frac{8}{N(N-1)^2} \left(k - \frac{N+1}{2}\right)\left(j - \frac{N+1}{2}\right) = \hat{a}(k,j). \end{aligned}$$

Thus an application of (10) gives

$$\hat{\tau} = \frac{8}{N^2(N-1)} \sum_{i=1}^{N} \Big(i - \frac{N+1}{2}\Big)\Big(R_i - \frac{N+1}{2}\Big). \tag{16}$$

On the other hand, the Spearman rank correlation coefficient ρ is defined by

$$\rho = \frac{12}{N(N^2-1)} \sum_{i=1}^{N} \Big(i - \frac{N+1}{2}\Big)\Big(R_i - \frac{N+1}{2}\Big). \tag{17}$$

Consequently, up to a multiplicative constant, the Spearman coefficient is the projection of the Kendall coefficient into the family of linear rank statistics. This fact may be utilized for an easy computation of the correlation coefficient of τ and ρ. Actually this coefficient equals the positive square root of $\operatorname{var}\hat{\tau}/\operatorname{var}\tau$. Making use of Problems 19 through 23, we thus obtain

$$\frac{\operatorname{var}\hat{\tau}}{\operatorname{var}\tau} = \frac{\dfrac{4}{9}\Big(\dfrac{N+1}{N}\Big)^2 \dfrac{1}{N-1}}{\dfrac{2(2N+5)}{9N(N-1)}} = \frac{2(N+1)^2}{N(2N+5)}. \tag{18}$$

Since the last expression converges to 1 as $N \to \infty$, the two statistics are asymptotically equivalent. A different, rather complicated, proof may be found in M. G. Kendall (1948).

A linear rank statistic S will be called *simple* if

$$S = \sum_{i=1}^{N} c_i a(R_i), \tag{19}$$

where $(c_1, \ldots, c_N)$ and $(a(1), \ldots, a(N))$ are some vectors. Put

$$\bar{a} = \frac{1}{N}\sum_{i=1}^{N} a(i), \quad \bar{c} = \frac{1}{N}\sum_{i=1}^{N} c_i \tag{20}$$

and

$$\sigma_a^2 = \frac{1}{N-1}\sum_{i=1}^{N} \big(a(i) - \bar{a}\big)^2. \tag{21}$$

Theorem 1 may then be specialized as follows.

Theorem 3 *Under H_0 we have for a statistic S given by* (19)

$$\mathsf{E}S = \bar{a}\sum_{i=1}^{N} c_i \tag{22}$$

and

$$\operatorname{var} S = \sigma_a^2 \sum_{i=1}^{N} (c_i - \bar{c})^2. \tag{23}$$

Finally, if there are two vectors $(a(1), \ldots, a(N))$ and $(b(1), \ldots, b(N))$, put

$$\sigma_{ab} = \frac{1}{N-1} \sum_{i=1}^{N} (a(i) - \bar{a})(b(i) - \bar{b}). \tag{24}$$

Theorem 4 *If S is given by (19) and $S^* = \sum_{i=1}^{N} d_i b(R_i)$, then*

$$\operatorname{cov}(S, S^*) = \sigma_{ab} \sum_{i=1}^{N} (c_i - \bar{c})(d_i - \bar{d}). \tag{25}$$

Terminological note. Subsequently the numbers $a(i)$ will be called the *scores* and the numbers c_i will be called the *regression constants.*

3.3.2 Linear rank statistics for H_1, H_2 and H_3. Antiranks. With H_1 we shall consider linear statistics of the following type:

$$S = \sum_{i=1}^{N} a(R_i^+) \operatorname{sign} X_i. \tag{1}$$

Obviously, under H_1, $\mathsf{E}S = 0$ and

$$\operatorname{var} S = \sum_{i=1}^{N} a^2(i) \tag{2}$$

hold.

With H_2 we shall concentrate on the statistics

$$S = \sum_{i=1}^{N} a(R_i) b(Q_i). \tag{3}$$

From Theorem 3.3.1.1 it follows that we have, under H_2,

$$\mathsf{E}S = N\bar{a}\bar{b} \tag{4}$$

and

$$\operatorname{var} S = (N-1)\sigma_a^2 \sigma_b^2, \tag{5}$$

with σ_a^2 defined by (3.3.1.21) and σ_b^2 defined similarly.

Finally, in connection with H_3 we shall consider the statistics

$$S_i = \sum_{j=1}^{n} a(R_{ji}), \quad 1 \le i \le k, \tag{6}$$

where $a(1), \ldots, a(k)$ are some numbers. It may easily be shown that under H_3

$$\mathsf{E}S_i = n\bar{a}, \tag{7}$$

$$\operatorname{var} S_i = \frac{n}{k}\sum_{j=1}^{k}(a(j) - \bar{a})^2, \tag{8}$$

and

$$\operatorname{cov}(S_i, S_j) = -\frac{n}{k(k-1)}\sum_{h=1}^{k}(a(h) - \bar{a})^2, \quad i \ne j. \tag{9}$$

In conclusion, let $(d_1, \ldots, d_N)$ denote the inverse permutation with respect to $(r_1, \ldots, r_N)$, i.e.

$$r_{d_i} = d_{r_i} = i, \quad 1 \le i \le N. \tag{10}$$

Let $D = (D_1, \ldots, D_N)$ be the inverse in this sense of $R = (R_1, \ldots, R_N)$. Then since the correspondence $(d_1, \ldots, d_N) \leftrightarrow (r_1, \ldots, r_N)$ is one-to-one, the distribution of D is uniform under H_0 (or H_*) as well:

$$P(D = r) = \frac{1}{N!}, \quad r \in \mathsf{R}. \tag{11}$$

Moreover the vector $R^0 = (R_1^0, \ldots, R_N^0)$ where

$$R_i^0 = Q_{D_i} \tag{12}$$

will also be uniformly distributed over R under H_2:

$$P(R^0 = r) = \frac{1}{N!}, \quad r \in \mathsf{R}. \tag{13}$$

Thus (3) may be represented as

$$S = \sum_{i=1}^{N} a(i)b(R_i^0), \tag{14}$$

and then Theorem 3.3.1.3 is directly applicable with $c_i = a(i)$ and $R_i = R_i^0$. The statistics $D_1, \ldots, D_N$ will be called *antiranks*.

3.4 LOCALLY MOST POWERFUL RANK TESTS

3.4.1 Definition of locally most powerful tests. Consider an indexed set of densities $\{q_\Delta\}$, $\Delta \geq 0$, and assume that $q_0 \in H$. A test will be called *locally most powerful* for H against $\Delta > 0$ at some level α if it is uniformly most powerful at level α for H against $K_\varepsilon = \{q_\Delta, 0 < \Delta < \varepsilon\}$ for some $\varepsilon > 0$.

A correspondingly modified terminology will be used if we restrict ourselves to $\mathcal{B}$-measurable tests, where $\mathcal{B}$ is a sub-σ-field. Thus we shall speak about *locally most powerful rank tests*, for example.

If Δ may also be negative, we shall call a test locally most powerful for $\Delta = 0$ against $\Delta \neq 0$ if it is uniformly most powerful for H against $K_\varepsilon = \{q_\Delta,\ 0 < |\Delta| < \varepsilon\}$ for some $\varepsilon > 0$.

In the remainder of the book, we shall sometimes use the abbreviation LMP test for a locally most powerful test, and LMPR test for a locally most powerful rank test.

3.4.2 A convergence theorem. We shall repeatedly use the following, somewhat abstractly formulated, theorem:

Theorem 1 *Let $(\Omega, \mathcal{A}, \nu)$ be a measure space with σ-finite measure and $\Omega = \{\omega\}$. Consider a sequence of $\mathcal{A}$-measurable functions $h(\omega), h_1(\omega)$, $h_2(\omega), \ldots$ such that*

$$\lim_{j\to\infty} h_j(\omega) = h(\omega), \tag{1}$$

almost everywhere, and

$$\limsup_{j\to\infty} \int |h_j| \,\mathrm{d}\nu \leq \int |h| \,\mathrm{d}\nu < \infty. \tag{2}$$

Then for each $A \in \mathcal{A}$

$$\lim_{j\to\infty} \int_A h_j \,\mathrm{d}\nu = \int_A h \,\mathrm{d}\nu. \tag{3}$$

Proof. Decomposing $h, h_1, \ldots$ into positive parts $h^+, h_1^+, \ldots$ and negative parts $h^-, h_1^-, \ldots$ we easily conclude from (1) that

$$\lim_{j\to\infty} h_j^+(\omega) = h^+(\omega), \tag{4}$$

and

$$\lim_{j\to\infty} h_j^-(\omega) = h^-(\omega), \tag{5}$$

almost everywhere.

Moreover, according to Fatou's lemma

$$\liminf_{j\to\infty} \int h_j^+ \,\mathrm{d}\nu \geq \int h^+ \,\mathrm{d}\nu \tag{6}$$

and

$$\liminf_{j\to\infty} \int h_j^- \,\mathrm{d}\nu \geq \int h^- \,\mathrm{d}\nu. \tag{7}$$

Now (6) and (7) are compatible with (2) only if

$$\lim_{j\to\infty} \int h_j^+ \,\mathrm{d}\nu = \int h^+ \,\mathrm{d}\nu \tag{8}$$

and

$$\lim_{j\to\infty} \int h_j^- \,\mathrm{d}\nu = \int h^- \,\mathrm{d}\nu. \tag{9}$$

Furthermore, the same lemma implies

$$\liminf_{j\to\infty} \int_A h_j^+ \,\mathrm{d}\nu = \int_A h^+ \,\mathrm{d}\nu, \quad a \in \mathcal{A}, \tag{10}$$

which is compatible with (8) only if

$$\lim_{j\to\infty} \int_A h_j^+ \,\mathrm{d}\nu = \int_A h^+ \,\mathrm{d}\nu, \quad A \in \mathcal{A}. \tag{11}$$

Similar relations must hold for h_j^-. Together these imply (3). □

The usefulness of this kind of convergence theorem for statistics was first recognized by Scheffé (1947).

3.4.3 Scores. Let us return to the φ-functions defined in Subsection 2.2.4. Further, let $U_N^{(1)} < \ldots < U_N^{(N)}$ be an ordered sample from the uniform distribution on $[0, 1]$. We shall call the numbers

$$a_N(i, f) = \mathsf{E}\varphi(U_N^{(i)}, f), \quad 1 \leq i \leq N, \tag{1}$$

the *scores* corresponding to the density f. In view of Theorem 3.1.2.4, (1) may be rewritten as follows:

$$a_N(i, f) = N \binom{N-1}{i-1} \int_0^1 \varphi(u, f) u^{i-1} (1-u)^{N-i} \,\mathrm{d}u. \tag{2}$$

If $X_N^{(1)} < \ldots < X_N^{(N)}$ is an ordered sample from the distribution with density f, we can also write

$$\begin{aligned} a_N(i, f) &= \mathsf{E}\Big\{ -\frac{f'(X_N^{(i)})}{f(X_N^{(i)})} \Big\} \\ &= N \binom{N-1}{i-1} \int_{-\infty}^{\infty} f'(x)[F(x)]^{i-1}[1-F(x)]^{N-i} \,\mathrm{d}x. \end{aligned} \tag{3}$$

In problems concerning scale we shall need scores generated by $\varphi_1(u, f)$:

$$a_{1N}(i, f) = \mathsf{E}\varphi_1(U_N^{(i)}, f) = \mathsf{E}\Big\{-1 - X_N^{(i)} \frac{f'(X_N^{(i)})}{f(X_N^{(i)})}\Big\}. \tag{4}$$

In testing H_1 against location alternatives we shall need

$$\varphi^+(u, f) = \varphi(\tfrac{1}{2} + \tfrac{1}{2}u, f) \tag{5}$$

and the scores

$$a_N^+(i, f) = \mathsf{E}\varphi^+(U_N^{(i)}, f) = \mathsf{E}\Big\{-\frac{f'(|X|_N^{(i)})}{f(|X|_N^{(i)})}\Big\}. \tag{6}$$

Similarly, in testing H_1 against scale alternatives we need

$$\varphi_1^+(u, f) = \varphi_1(\tfrac{1}{2} + \tfrac{1}{2}u, f) \tag{7}$$

and the scores

$$a_{1N}^+(i, f) = \mathsf{E}\varphi_1^+(U_N^{(i)}, f) = \mathsf{E}\Big\{-1 - |X|_N^{(i)} \frac{f'(|X|_N^{(i)})}{f(|X|_N^{(i)})}\Big\}. \tag{8}$$

Theorem 1 *If $\int_{-\infty}^{\infty} |f'(x)|\, \mathrm{d}x < \infty$, then*

$$\sum_{i=1}^{N} a_N(i, f) = 0 \tag{9}$$

and

$$\frac{1}{N}\sum_{i=1}^{N} a_N^2(i, f) \le I(f). \tag{10}$$

Similarly, $\int_{-\infty}^{\infty} |xf'(x)|\, \mathrm{d}x < \infty$ entails

$$\sum_{i=1}^{N} a_{1N}(i, f) = 0 \tag{11}$$

and

$$\frac{1}{N}\sum_{i=1}^{N} a_{1N}^2(i, f) \le I_1(f). \tag{12}$$

Proof. On account of Theorem 3.1.2.2 and the lemmas of Subsection 2.2.4, we have

$$\begin{aligned}\sum_{i=1}^{N} a_N(i,f) &= \sum_{i=1}^{N} \mathsf{E}\varphi(U_N^{(i)},f) = \sum_{i=1}^{N} \mathsf{E}\{\varphi(U_1,f)|R_1 = i\} \\ &= N\mathsf{E}\varphi(U_1,f) = N\int_0^1 \varphi(u,f)\,\mathrm{d}u = 0,\end{aligned}$$

and similarly for $a_{1N}(i,f)$. Further

$$\begin{aligned}\sum_{i=1}^{N} a_N^2(i,f) &= \sum_{i=1}^{N} [\mathsf{E}\varphi(u_N^{(i)},f)]^2 = \sum_{i=1}^{N} [\mathsf{E}\{\varphi(U_1,f)|R_1 = i\}]^2 \\ &\leq \sum_{i=1}^{N} \mathsf{E}\{\varphi^2(U_1,f)|R_1 = i\} = N\mathsf{E}\varphi^2(U_1,f) \\ &= N\int_0^1 \varphi^2(u,f)\,\mathrm{d}u = NI(f)\end{aligned}$$

and similarly for $a_{1N}^2(i,f)$. □

We shall show in Chapter 6 that for $N \to \infty$ the limit of the left-hand side of (10) equals the right-hand side and similarly for (12).

If N is sufficiently large and $\varphi(u,f)$ sufficiently smooth in the neighbourhood of $i/(N+1)$, we have approximately

$$a_N(i,f) \cong \varphi(\mathsf{E}U_N^{(i)},f)$$

i.e., on account of Theorem 3.1.2.4,

$$a_N(i,f) \cong \varphi\Big(\frac{i}{N+1},f\Big). \tag{13}$$

Another approximation could be

$$a_N(i,f) \cong N\int_{(i-1)/N}^{i/N} \varphi(u,f)\,\mathrm{d}u. \tag{14}$$

We shall call (13) *approximate scores* corresponding to f. Similarly, $\varphi_1\big(i/(N+1),f\big)$ are called approximate scores for scale problems concerning f.

3.4.4 H_0 against two samples differing in location. Let us test H_0 against

$$q_\Delta = \prod_{i=1}^{m} f(x_i - \Delta) \prod_{i=m+1}^{m+n} f(x_i), \quad \Delta > 0, \tag{1}$$

where f is known. In words, (1) means that the first sample is shifted to the right with respect to the second sample $X_{m+1}, \ldots, X_{m+n}$, denoted usually as $Y_1, \ldots, Y_n$.

We shall assume that f is absolutely continuous and that

$$\int_{-\infty}^{\infty} |f'(x)| \,\mathrm{d}x < \infty, \tag{2}$$

in order that the scores $a_N(i, f)$ given by (3.4.3.1) be well defined.

Note that

$$\int_{-\infty}^{\infty} |f'(x)| \,\mathrm{d}x = \int_{-\infty}^{\infty} \left|\frac{f'(x)}{f(x)}\right| f(x) \,\mathrm{d}x \leq \sqrt{I(f)},$$

so that $I(f) < \infty$ implies (2), but not necessarily the opposite way round.

Theorem 1 *Under assumption* (2) *the test with critical region*

$$\sum_{i=1}^{m} a_N(R_i, f) \geq k \tag{3}$$

is the locally most powerful rank test for H_0 *against* $\{q_\Delta,\ \Delta > 0\}$ *at the respective level.*

Proof. The theorem is a special case of Theorem 3.4.6.1. □

Terminological note. The respective level mentioned in the theorem equals

$$\alpha = P\Big(\sum_{i=1}^{m} a_N(R_i, f) \geq k\Big), \tag{4}$$

where P refers to H_0.

Remark 1. The test considered in the theorem is non-randomized, so that we are able to find the locally most powerful rank tests only for α expressible by (4). Levels with this property are sometimes called *natural levels* for the test statistic $\sum_{i=1}^{m} a_N(R_i, f)$.

Locally most powerful tests for a richer set of levels could be found by considering, in addition, derivatives of higher orders, if they exist, in the development we shall make in Subsection 3.4.8, and by randomization.

Remark 2. From (3) it may be easily seen that the resulting test depends on the type of f only. Actually, putting $f_{\mu,\sigma}(x) = \sigma^{-1} f[(x-\mu)/\sigma]$, (2.2.4.14) and (3.4.3.1) entail

$$a_N(i, f_{\mu.\sigma}) = \frac{1}{\sigma} a_N(i, f). \tag{5}$$

Consequently, the replacement of f by $f_{\mu,\sigma}$ does not change the test statistic except for multiplication by a positive constant. This does not change, however, the test itself, if k in (3) is adjusted correspondingly.

This remark pertains to all locally most powerful tests to be given in the sequel.

3.4.5 H_0 against two samples differing in scale. Let us test H_0 against

$$q_\Delta = e^{-m\Delta} \prod_{i=1}^{m} f[(x_i - \mu)e^{-\Delta}] \prod_{i=m+1}^{m+n} f(x_i - \mu), \quad \Delta > 0, \tag{1}$$

where f is known and μ is an arbitrary real number. In words, (1) means that the first sample $X_1, \ldots, X_m$ is more dispersed about μ than the second sample $X_{m+1}, \ldots, X_{m+n}$. We shall assume that f is absolutely continuous and

$$\int_{-\infty}^{\infty} |xf'(x)| \, \mathrm{d}x < \infty. \tag{2}$$

Let the scores $a_{1N}(i, f)$ be given by (3.4.3.4).

Theorem 1 *Under the assumption* (2) *the test with critical region*

$$\sum_{i=1}^{m} a_{1N}(R_i, f) \geq k \tag{3}$$

is the locally most powerful rank test for H_0 against $\{q_\Delta,\ \Delta > 0\}$ at the respective level.

Proof. The theorem is a special case of Theorem 3.4.7.1. □

3.4.6 H_0 against regression in location. Put

$$q_\Delta = \prod_{i=1}^{N} f(x_i - \Delta c_i), \quad \Delta > 0, \tag{1}$$

where f is a known density and $c_1, \ldots, c_N$ are known regression constants. This situation includes the two-sample case from Subsection 3.4.4 as a special case for

$$c_i = \begin{cases} 1, & i = 1, \ldots, m \\ 0, & i = m+1, \ldots, m+n. \end{cases} \tag{2}$$

Theorem 1 *If f is absolutely continuous and* (3.4.4.2) *holds, then the test with critical region*

$$\sum_{i=1}^{N} c_i a_N(R_i, f) \geq k \tag{3}$$

is the locally most powerful rank test for H_0 against $\{q_\Delta,\ \Delta > 0\}$ at the respective level.

Proof. The proof follows from Theorem 3.4.8.1 and Lemma 3.4.8.1. □

Note the following counterexample: If f is the density of a rectangular distribution on some finite interval (a, b), then (3.4.4.2) holds but f is not absolutely continuous. Therefore in Theorem 1 absolute continuity of f must be assumed.

3.4.7 H_0 against regression in scale.

Put

$$q_\Delta = \exp\Big(-\Delta \sum_{i=1}^{N} c_i\Big) \prod_{i=1}^{N} f[(x_i - \mu)e^{-\Delta c_i}], \quad \Delta > 0, \tag{1}$$

where f is a known density, $c_1, \ldots, c_N$ are known regression constants, and μ is an arbitrary real number. This situation includes the two-sample case considered in Subsection 3.4.5 as a special case for $c_i = 1$ or 0 if $i \leq m$ or $i > m$, respectively.

Theorem 1 *If f is absolutely continuous and* (3.4.5.2) *holds, then the test with critical region*

$$\sum_{i=1}^{N} c_i a_{1N}(R_i, f) \geq k \tag{2}$$

is the locally most powerful rank test for H_0 against $\{q_\Delta,\ \Delta > 0\}$ at the respective level.

Proof. The proof follows from Theorem 3.4.8.1 and Lemma 3.4.8.2. □

3.4.8 H_0 against a general alternative.

We shall now prove a general theorem, embracing Theorems 3.4.4.1–3.4.7.1 as special cases.

Definition. Let J be an open interval containing 0. A family of densities $d(x, \Theta)$, $\Theta \in J$, will be said to satisfy *condition A_1* if

(i) $d(x, \Theta)$ is absolutely continuous in Θ for almost every x;

(ii) the limit

$$\dot{d}(x, 0) = \lim_{\Theta \to 0} \frac{1}{\Theta}[d(x, \Theta) - d(x, 0)] \tag{1}$$

exists for almost every x;

(iii)

$$\lim_{\Theta\to 0}\int_{-\infty}^{\infty}|\dot{d}(x,\Theta)|\,\mathrm{d}x=\int_{-\infty}^{\infty}|\dot{d}(x,0)\,\mathrm{d}x<\infty \tag{2}$$

holds, where $\dot{d}(x,\Theta)$ is the partial derivative with respect to Θ.

Note that the existence of $\dot{d}(x,\Theta)$ for almost every Θ is ensured at every point x such that $d(x,\Theta)$ is absolutely continuous in Θ. This, however, does not make the condition (ii) superfluous.

Now consider the alternative

$$q_\Delta=\prod_{i=1}^{N}d(x_i,\Delta c_i) \tag{3}$$

and introduce the scores

$$a_N(i,d)=\mathsf{E}\Big\{\frac{\dot{d}(X_N^{(i)},0)}{d(X_N^{(i)},0)}\Big\}, \tag{4}$$

with $X_N^{(i)}$ denoting the i-th order statistic from a sample of size N from the distribution with density $d(x,0)$. We also have

$$a_N(i,d)=\mathsf{E}\{\varphi(U_N^{(i)},d,0\}, \tag{5}$$

with φ defined by (2.2.4.4) and $U_N^{(i)}$ denoting the i-th order statistic from the uniform distribution.

Theorem 1 *Let the family of densities $d(x,\Theta)$, $\Theta\in J$, satisfy condition A_1 formulated above. Then the test with critical region*

$$\sum_{i=1}^{N}c_i a_N(R_i,d)\geq k \tag{6}$$

is the locally most powerful rank test for H_0 against $\{q_\Delta,\Delta>0\}$ at the respective level.

Proof. We have

$$Q_\Delta(R=r)=\int\cdots\int_{R=r}q_\Delta\,\mathrm{d}x_1\ldots\mathrm{d}x_N \tag{7}$$

$$=\int\cdots\int_{R=r}[\prod_1^N d(x_i,0)]\,\mathrm{d}x_1\ldots\mathrm{d}x_N$$

$$+\Delta\int\cdots\int_{R=r}\frac{1}{\Delta}[\prod_1^N d(x_i,\Delta c_i)-\prod_1^N d(x_i,0)]\,\mathrm{d}x_1\ldots\mathrm{d}x_N$$

$$= 1/N! + \Delta \sum_{k=1}^{N} \int \cdots \int_{R=r} \Big[\frac{d(x_k, \Delta c_k) - d(x_k, 0)}{\Delta} \cdot \prod_{i=k+1}^{N} d(x_i, 0) \prod_{j=1}^{k-1} d(x_j, \Delta c_j) \Big] \, \mathrm{d}x_1 \ldots \mathrm{d}x_N .$$

Now, in view of (1),

$$\lim_{\Delta \to 0} \frac{d(x_k, \Delta c_k) - d(x_k, 0)}{\Delta} \prod_{i=k+1}^{N} d(x_i, 0) \prod_{j=1}^{k-1} d(x_j, \Delta c_j) = c_k \dot{d}(x_k, 0) \prod_{i \neq k}^{N} d(x_i, 0), \quad 1 \leq k \leq N, \tag{8}$$

holds almost everywhere in x. Moreover, if, say, $\Delta c_k > 0$,

$$\begin{aligned} \int \cdots \int \Big[\frac{|d(x_k, \Delta c_k) - d(x_k, 0)|}{\Delta} & \prod_{i=k+1}^{N} d(x_i, 0) \\ & \cdot \prod_{j=1}^{k-1} d(x_j, \Delta c_j) \Big] \, \mathrm{d}x_1 \ldots \mathrm{d}x_n \\ = \int_{-\infty}^{\infty} \frac{|d(x_k, \Delta c_k) - d(x_k, 0)|}{\Delta} \, \mathrm{d}x_k & \\ = \int_{-\infty}^{\infty} \frac{1}{\Delta} \Big| \int_0^{\Delta c_k} \dot{d}(x_k, \Theta) \, \mathrm{d}\Theta \Big| \, \mathrm{d}x_k & \\ \leq \int_{-\infty}^{\infty} \frac{1}{\Delta} \int_0^{\Delta c_k} |\dot{d}(x_k, \Theta)| \, \mathrm{d}\Theta \, \mathrm{d}x_k & \\ = \frac{1}{\Delta} \int_0^{\Delta c_k} \int_{-\infty}^{\infty} |\dot{d}(x_k, \Theta)| \, \mathrm{d}x_k \, \mathrm{d}\Theta & \end{aligned}$$

and a similar result may be obtained for $\Delta c_k \leq 0$.

Consequently, in view of (2),

$$\limsup_{\Delta \to 0} \int \cdots \int \Big[\frac{|d(x_k, \Delta c_k) - d(x_k, 0)|}{\Delta} \prod_{i=k+1}^{N} d(x_i, 0) \cdot \prod_{j=1}^{k-1} d(x_j, \Delta c_j) \Big] \, \mathrm{d}x_1 \ldots \mathrm{d}x_N \leq |c_k| \int_{-\infty}^{\infty} |\dot{d}(x_k, 0)| \, \mathrm{d}x_k . \tag{9}$$

This means, however, that the least upper bound of the integral is for $\Delta \to 0$ dominated by the integral of the absolute value of the right side of (8). Hence Theorem 3.4.2.1 is applicable, and

$$\lim_{\Delta \to 0} \sum_{k=1}^{N} \int \cdots \int_{R=r} \Big[\frac{d(x_k, \Delta c_k) - d(x_k, 0)}{\Delta} \prod_{i=k+1}^{N} d(x_i, 0) \cdot \tag{10}$$

$$\cdot \prod_{j=1}^{k-1} d(x_j, \Delta c_j) \Big] \, \mathrm{d}x_1 \ldots \mathrm{d}x_N$$

$$= \sum_{k=1}^{N} \int \cdots \int_{R=r} [c_k \dot{d}(x_k, 0) \prod_{i \neq k}^{N} d(x_i, 0)] \, \mathrm{d}x_1 \ldots \mathrm{d}x_N.$$

Now, since $d(x, \Theta) \geq 0$ and J is open, $d(x, 0) = 0$ together with (1) entails $\dot{d}(x, 0) = 0$. Consequently, $d(x, 0) = 0$ and $\dot{d}(x, 0) \neq 0$ are compatible on a set of Lebesgue measure zero only. Thus, making use of Theorem 3.1.2.2 we have

$$\sum_{k=1}^{N} \int \cdots \int_{R=r} [c_k \dot{d}(x_k, 0) \prod_{i \neq k}^{N} d(x_i, 0)] \, \mathrm{d}x_1 \ldots \mathrm{d}x_N \tag{11}$$

$$= \sum_{k=1}^{N} c_k \int \cdots \int_{R=r} \Big[\frac{\dot{d}(x_k, 0)}{d(x_k, 0)} \prod_{i=1}^{N} d(x_i, 0) \Big] \, \mathrm{d}x_1 \ldots \mathrm{d}x_N$$

$$= (1/N!) \sum_{k=1}^{N} c_k \mathsf{E} \Big\{ \frac{\dot{d}(X_k, 0)}{d(X_k, 0)} | R = r \Big\}$$

$$= (1/N!) \sum_{k=1}^{N} c_k \mathsf{E} \Big\{ \frac{\dot{d}(X^{(r_k)}, 0)}{d(X^{(r_k)}, 0)} \Big\}$$

$$= (1/N!) \sum_{k=1}^{N} c_k a_n(r_k, d).$$

On combining (7), (10) and (11), we can easily see that there exists an $\varepsilon > 0$ such that for $0 < \Delta < \varepsilon$

$$\sum_{i=1}^{N} c_k a_N(r_k, d) > \sum_{i=1}^{N} c_k a_N(r_k', d)$$

implies

$$Q_\Delta(R = r) > Q_\Delta(R = r').$$

This, in view of Theorem 3.2.3.1, concludes the proof. □

Lemma 1 *If a density $f(x)$ is absolutely continuous and satisfies* (3.4.4.2), *then the family $d(x, \Theta) = f(x - \Theta)$ satisfies condition A_1.*

Proof. (i) and (ii) are obvious. Further

$$\int_{-\infty}^{\infty} |\dot{d}(x,\Theta)|\,\mathrm{d}x = \int_{-\infty}^{\infty} |f'(x-\Theta)|\,\mathrm{d}x = \int_{-\infty}^{\infty} |f'(x)|\,\mathrm{d}x < \infty,$$

so that (iii) obviously holds. □

Lemma 2 *If a density $f(x)$ is absolutely continuous and satisfies* (3.4.5.2), *then the family $d(x,\Theta) = e^{-\Theta} f[(x-\mu)e^{-\Theta}]$ satisfies condition A_1.*

Proof. Here we have

$$\begin{aligned}
&\int_{-\infty}^{\infty} |\dot{d}(x,\Theta)|\,\mathrm{d}x \\
&= \int_{-\infty}^{\infty} |e^{-\Theta} f[(x-\mu)e^{-\Theta}] + e^{-2\Theta}(x-\mu) f'[(x-\mu)e^{-\Theta}]|\,\mathrm{d}x \\
&= \int_{-\infty}^{\infty} |f(x) + x f'(x)|\,\mathrm{d}x < \infty.
\end{aligned}$$

so that (iii) is satisfied trivially again. □

We must also note that Janssen and Mason (1990) developed a new, more general approach to locally most powerful rank tests, when the standard regular assumptions do not hold. Namely, the assumptions on derivatives of densities can be substituted by simpler conditions, e.g. for regression in location the densities must be only square integrable. The theory is mainly convenient for rank tests for lifetime models, such as Weibull location models. It is interesting that they arrive at different scores compared to those given here, but their relevant proof is only a clear extension of Theorem 3.4.8.1 here, and under condition A_1 in 3.4.8, both scores coincide. Janssen–Mason scores were also used by Puri and Sen (1971).

3.4.9 H_1 against the location shift.

Put

$$q_\Delta = \prod_{i=1}^{N} f(x_i - \Delta), \quad \Delta \geq 0, \tag{1}$$

where f is a known symmetric density.

Let the scores $a_N^+(i,f)$ be given by (3.4.3.6).

Theorem 1 *Under assumption* (3.4.4.2) *the test with critical region*

$$\sum_{i=1}^{N} a_N^+(R_i^+, f)\,\mathrm{sign}\,X_i \geq k \tag{2}$$

is the locally most powerful rank test for H_1 against $\{q_\Delta\}$ at the respective level.

Remark. The statistics in (2) are also often called *signed rank statistics.*

Proof. We have

$$\lim_{\Delta\to 0}\frac{1}{\Delta}[2^N N! Q_\Delta(\operatorname{sign} X = v, R^+ = r) - 1]$$

$$= \lim_{\Delta\to 0} 2^N N! \sum_{k=1}^{N} \int \cdots \int_{\substack{\operatorname{sign} X = v \\ R^+ = r}} \Big[\frac{f(x_k-\Delta)-f(x_k)}{\Delta}$$

$$\cdot \prod_{i=k+1}^{N} f(x_i) \prod_{j=1}^{k-1} f(x_j - \Delta)\Big] \cdot \mathrm{d}x_1 \ldots \mathrm{d}x_N$$

$$= 2^N N! \sum_{k=1}^{N} \int \cdots \int_{\substack{\operatorname{sign} X = v \\ R^+ = r}} [-f'(x_k)] \prod_{i\neq k}^{N} f(x_i)\,\mathrm{d}x_i \ldots \mathrm{d}x_N$$

$$= 2^N N! \sum_{k=1}^{N} \int \cdots \int_{\substack{\operatorname{sign} X = v \\ R^+ = r}} \Big[-\operatorname{sign} x_k \frac{f'(|x_k|)}{f(|x_k|)}$$

$$\cdot \prod_{i=1}^{N} f(x_i)\Big]\,\mathrm{d}x_i \ldots \mathrm{d}x_N$$

$$= \sum_{k=1}^{N} v_k a_N^+(r_k, f),$$

where the limit under the integral sign could be justified in the same way as in the proof of Theorem 3.4.8.1. □

Here we can add a similar remark on Janssen and Mason (1990) developments as in Subsection 3.4.8.

3.4.10 H_1 against two samples differing in scale.

Put

$$q_\Delta = \prod_{i=1}^{m} e^{-\Delta} f(x_i e^{-\Delta}) \prod_{i=m+1}^{m+n} f(x_i), \quad \Delta \geq 0, \tag{1}$$

where f is a known *symmetric density.* Since $H_1 \subset H_0$ and the above alternative is a subhypothesis of the alternative considered in Subsection 3.4.5, we could use the test of Theorem 3.4.5.1. However, that test may not be optimum in our specialized situation, where H_0 is narrowed to H_1.

Let the scores $a_{1N}^+(i, f)$ be given by (3.4.3.8).

Theorem 1 *Under assumption* (3.4.5.2) *the test with critical region*

$$\sum_{i=1}^{m} a_{1N}^{+}(R_i^{+}, f) \geq k \tag{2}$$

is the locally most powerful rank test of H_1 against $\{q_\Delta\}$ at the respective level.

Proof. We have

$$\lim_{\Delta\to 0}\frac{1}{\Delta}[2^N N! Q_\Delta(\operatorname{sign} X = v, R^+ = r) - 1] \tag{3}$$

$$= \lim_{\Delta\to 0} 2^N N! \sum_{k=1}^{m} \int \cdots \int_{\substack{\operatorname{sign} X = v\\ R^+ = r}} \Big[\frac{e^{-\Delta} f(x_k e^{-\Delta}) - f(x_k)}{\Delta}$$

$$\cdot \prod_{i=k+1}^{N} f(x_i) \prod_{j=1}^{k-1} e^{-\Delta} f(x_j e^{-\Delta})\Big] \cdot \mathrm{d}x_1 \ldots \mathrm{d}x_N$$

$$= 2^N N! \sum_{k=1}^{m} \int \cdots \int_{\substack{\operatorname{sign} X = v\\ R^+ = r}} \Big[-1 - x_k \frac{f'(x_k)}{f(x_k)}\Big]$$

$$\cdot \prod_{i=1}^{N} f(x_i)\,\mathrm{d}x_1 \ldots \mathrm{d}x_N$$

$$= 2^N N! \sum_{k=1}^{m} \int \cdots \int_{\substack{\operatorname{sign} X = v\\ R^+ = r}} \Big[-1 - |x_k| \frac{f'(|x_k|)}{f(|x_k|)}\Big]$$

$$\cdot \prod_{i=1}^{N} f(x_i)\,\mathrm{d}x_1 \ldots \mathrm{d}x_N$$

$$= \sum_{k=1}^{m} a_{1N}^{+}(r_k, f),$$

where the limit under the integral sign could be justified as in the proof of Theorem 3.4.8.1. □

3.4.11 ***H_2* against dependence.** Put

$$q_\Delta = \prod_{i=1}^{N} h_\Delta(x_i, y_i), \quad -\infty < \Delta < \infty, \tag{1}$$

where

$$h_\Delta(x, y) = \int_{-\infty}^{\infty} f(x - \Delta z) g(y - \Delta z)\,\mathrm{d}M(z) \tag{2}$$

with $M(z)$ being a distribution function such that

$$0 < \int_{-\infty}^{\infty} (z-\zeta)^2 \,\mathrm{d}M(z) < \infty, \qquad \text{with } \zeta = \int_{-\infty}^{\infty} z \,\mathrm{d}M(z). \tag{3}$$

Assume that f and g are known densities or densities of known types.

In words, we assume under this alternative that

$$X_i = X_i^* + \Delta Z_i \tag{4}$$

$$Y_i = Y_i^* + \Delta Z_i, \tag{5}$$

where X_i^*, Y_i^* and Z_i are mutually independent, the types of the distributions of X_i^* and Y_i^* are known, and the distribution of Z_i is arbitrary.

Theorem 1 *Assume that* (3.4.4.2) *holds for both the densities f and g and that the derivatives $f'(x)$ and $g'(x)$ are continuous almost everywhere.*

Then the test with critical region

$$\sum_{i=1}^{N} a_N(R_i, f) a_N(Q_i, g) \geq k \tag{6}$$

is the locally most powerful rank test for H_2 against $\{q_\Delta\}$ at the respective level.

Proof. Put

$$f_\Delta(x) = \int f(x - \Delta z) \,\mathrm{d}M(z)$$

and

$$g_\Delta(x) = \int g(x - \Delta z) \,\mathrm{d}M(z).$$

We first show that

$$\lim_{\Delta\to 0} \frac{1}{\Delta^2}[h_\Delta(x,y) - f_\Delta(x) g_\Delta(y)] = f'(x) g'(y) \int_{-\infty}^{\infty} (z-\zeta)^2 \,\mathrm{d}M(z) \tag{7}$$

at each point (x,y) such that x is a continuity point of $f'(.)$ and y is a continuity point of $g'(.)$.

We can write

$$\begin{aligned} &\frac{1}{\Delta^2}[h_\Delta(x,y) - f_\Delta(x) q_\Delta(y)] \\ &= \frac{1}{\Delta^2}\frac{1}{2} \int\int [(f(x) - \Delta z) - f(x - \Delta z')] \\ &\qquad \cdot [g(y-\Delta z) - g(y - \Delta z')] \,\mathrm{d}M(z) \,\mathrm{d}M(z') \\ &= \frac{1}{2}\frac{1}{\Delta^2} \int\int_{\substack{|\Delta z| \leq \delta \\ |\Delta z'| \leq \delta}} [f(x-\Delta z) - f(x - \Delta z')] \cdot \\ &\qquad \cdot [g(y-\Delta z) - g(y - \Delta z')] \,\mathrm{d}M(z) \,\mathrm{d}M(z') + R_{\Delta\delta}. \end{aligned} \tag{8}$$

Now

$$\begin{aligned} R_{\Delta\delta} &\le 4 \max_t f(t) \max_t g(t) \frac{1}{\Delta^2} P(|\Delta Z| > \delta) \\ &\le 4 \max_t f(t) \max_t g(t) \delta^{-2} \int_{|z|>\delta/\Delta} z^2 \, \mathrm{d}M(z) \end{aligned}$$

from which it is easily seen that for each $\delta > 0$

$$\lim_{\Delta \to 0} R_{\Delta\delta} = 0. \tag{9}$$

On the other hand, on account of the continuity of $f'(.)$ at the point x, for every $\varepsilon > 0$ there exists a $\delta > 0$ such that $|\Delta z| \le \delta$, $|\Delta z'| \le \delta$ implies

$$\begin{aligned} &\Big|\frac{1}{\Delta(z-z')}[f(x-\Delta z) - f(x - \Delta z')] - f'(x)\Big| \\ &\quad = \Big|\frac{1}{\Delta(z-z')} \int_{x-\Delta z'}^{x-\Delta z} f'(t) \, \mathrm{d}t - f'(x)\Big| < \varepsilon, \end{aligned} \tag{10}$$

and similarly for g. However (8), (9) and (10) entail (7). The relation (8) also implies

$$\begin{aligned} &\frac{1}{\Delta^2} \int_{-\infty}^{\infty} \int_{-\infty}^{\infty} |h_\Delta(x,y) - f_\Delta(x) g_\Delta(y)| \, \mathrm{d}x \, \mathrm{d}y \\ &= \frac{1}{2} \frac{1}{\Delta^2} \int\int \Big| \int\int \Big[\int_{\Delta z'}^{\Delta z} f'(x-t) \, \mathrm{d}t\Big] \\ &\qquad \cdot \Big[\int_{\Delta z'}^{\Delta z} g'(y-s) \, \mathrm{d}s\Big] \, \mathrm{d}M(z) \, \mathrm{d}M(z') \Big| \, \mathrm{d}x \, \mathrm{d}y \\ &\le \frac{1}{2} \frac{1}{\Delta^2} \int\int\int\int \int_{\Delta z'}^{\Delta z} \int_{\Delta z'}^{\Delta z} |f'(x-t) g'(y-s)| \\ &\qquad \mathrm{d}t \, \mathrm{d}s \, \mathrm{d}M(z) \, \mathrm{d}M(z') \, \mathrm{d}x \, \mathrm{d}y \\ &= \frac{1}{2} \frac{1}{\Delta^2} \int\int \int_{\Delta z'}^{\Delta z} \int_{\Delta z'}^{\Delta z} \int\int |f'(x-t) g'(y-s)| \\ &\qquad \mathrm{d}x \, \mathrm{d}y \, \mathrm{d}t \, \mathrm{d}s \, \mathrm{d}M(z) \, \mathrm{d}M(z') \\ &= \frac{1}{2} \frac{1}{\Delta^2} \int\int (\Delta z - \Delta z')^2 \, \mathrm{d}M(z) \, \mathrm{d}M(z') \\ &\qquad \cdot \int_{-\infty}^{\infty} \int_{-\infty}^{\infty} |f'(x) g'(y)| \, \mathrm{d}x \, \mathrm{d}y \\ &= \int_{-\infty}^{\infty} (z-\zeta)^2 \, \mathrm{d}M(z) \int_{-\infty}^{\infty} \int_{-\infty}^{\infty} |f'(x) g'(y)| \, \mathrm{d}x \, \mathrm{d}y. \end{aligned} \tag{11}$$

Now we can proceed as in the preceding proofs:

$$\lim_{\Delta\to 0}\frac{1}{\Delta^2}[(N!)^2 Q_\Delta(R=r, Q=r')-1]$$

$$=\lim_{\Delta\to 0}(N!)^2\sum_{k=1}^{N}\int\cdots\int_{\substack{R=r\\Q=r}}\Big[\frac{h_\Delta(x_k,y_k)-f_\Delta(x_k)g_\Delta(y_k)}{\Delta^2}$$

$$\cdot\prod_{i=k+1}^{N} f_\Delta(x_i)g_\Delta(x_i)\prod_{j=1}^{k-1}h_\Delta(x_j,y_j)\Big]\,\mathrm{d}x_1\ldots\mathrm{d}x_n\,\mathrm{d}y_1\ldots\mathrm{d}y_N$$

$$=\sigma^2(N!)^2\sum_{k=1}^{N}\int\cdots\int_{\substack{R=r\\Q=r}}\Big[\frac{f'(x_k)g'(y_k)}{f(x_k)g(y_k)}\prod_{i=1}^{N}f(x_i)g(y_i)\Big]$$

$$\mathrm{d}x_1\ldots\mathrm{d}x_N\,\mathrm{d}y_1\ldots\mathrm{d}y_N$$

$$=\sigma^2\sum_{k=1}^{N}\Big[N!\int\cdots\int_{R=r}\frac{f'(x_k)}{f(x_k)}\prod_{i=1}^{N}f(x_i)\,\mathrm{d}x_1\ldots\mathrm{d}x_N\Big]$$

$$\cdot\Big[N!\int\cdots\int_{Q=r'}\frac{g'(y_k)}{g(y_k)}\prod_{i=1}^{N}g(y_i)\,\mathrm{d}y_1\ldots\mathrm{d}y_N\Big]$$

$$=\sigma^2\sum_{i=1}^{N}a_N(r_i,f)a_N(r_i',f)$$

where $\sigma^2=f(z-\zeta)^2\,\mathrm{d}M(z)$; the limit under the integral sign is justified by (11). □

Again, Janssen and Mason (1990) developed a more general theory than given here. Specifically, they start with the two-dimensional densities

$$p_t(x,y)=\prod f(x,tz)g(y,tz)\,\mathrm{d}M(z)$$

with some weight function M, arrive at different scores (similarly as remarked in 3.4.8), but their proof much resembles our proof in Theorem 3.4.11.1.

The two-dimensional problem and its solution were extended to the general case of $k\geq 2$ simultaneous random variables by Ciesielska and Ledwina (1983) and Shirahata (1974).

3.5 STATISTICAL FUNCTIONALS

In the preceding section, LMPR test statistics are characterized as suitable (signed) linear rank statistics. In some specific cases, they may also be expressed in terms of U-statistics (Hoeffding, 1948a) or V-statistics (von Mises, 1947), or their generalizations. Moreover, the alternative hypotheses considered in the last section are of specific parametric forms; in nonparametric hypotheses testing problems, often such alternative hypotheses may not be of specific parametric form. For example, in the two-sample model, treated in Subsections 3.4.4 and 3.4.5, the null hypothesis of homogeneity relates to $H_0 : F = G$, and we may be interested in the following:

$$H_1 : F(x) \geq G(x), \text{ for all } -\infty < x < \infty,$$

with strict inequality holding at least on a set of x with a positive measure; this is referred to as the *stochastic ordering* of X and Y. Another related alternative hypothesis is

$$H_2 : P\{X \leq Y\} = \int_{-\infty}^{\infty} F(x) dG(x) > 1/2,$$

which is also referred to as stochastic ordering of X and Y. Although linear rank statistics based on appropriate scores can be validly used for such hypotheses testing problems, they may no longer be LMPR. In fact, as these alternatives may not be of the parametric type, the notion of LMPR test may not be very appealing here. For this reason, often the null and alternative hypotheses are expressed in terms of some regular functionals of the underlying distributions, and their sample counterparts are then incorporated in the construction of appropriate test statistics; if the null hypothesis corresponds to a hypothesis of invariance, EDF tests can be obtained in this manner, and often they are rank tests as well. Such statistics, expressible as statistical functionals, occupy a prominent place in nonparametrics, and hence we provide here an outline of the relevant methodology with due emphasis on U-, V-statistics and differentiable statistical functions.

3.5.1 ***U*-statistics and *V*-statistics.** Let $X_1, \ldots, X_N$ be a sample where each X_i has the distribution function F, defined on p-dimensional Euclidean space. Let $\theta(F)$ be a functional of F, and let $\mathcal{F}$ be the domain of F. Then (Hoeffding, 1948a), $\theta(F)$ is said to be an *estimable parameter* or a *regular functional* of F, if there exists a *kernel* $\psi(x_1, \ldots, x_m)$ of *degree* m (≥ 1), such that

$$\begin{aligned} \theta(F) &= \mathsf{E}_F \psi(X_1, \ldots, X_m) \\ &= \int \cdots \int \psi(x_1, \ldots, x_m) \, \mathrm{d}F(x_1) \ldots \, \mathrm{d}F(x_m), \end{aligned} \tag{1}$$

for all $F \in \mathcal{F}$. Without loss of generality we may take $\varphi(x_1, \ldots, x_m)$ to be a symmetric kernel (in the sense that it is a symmetric function of its m arguments). As before, we denote the sample (empirical) distribution funftion, based on $X_1, \ldots, X_N$, by F_N. Then a natural (i.e., plug-in) estimator of $\theta(F)$ is the following (von Mises' functional):

$$V_N = \int \cdots \int \psi(x_1, \ldots, x_m) \, \mathrm{d}F_N(x_1) \ldots \, \mathrm{d}F_N(x_m) \tag{2}$$

$$= N^{-m} \sum_{i_1=1}^{N} \cdots \sum_{i_m=1}^{N} \psi(X_{i_1}, \ldots, X_{i_m}).$$

For $m = 1$, V_N is a linear functional and is unbiased for $\theta(F)$. However, for $m \geq 2$, in general, V_N is not unbiased for $\theta(F)$. An unbiased estimator of $\theta(F)$, due to Hoeffding (1948a), is the U-statistic

$$U_N = \binom{N}{m}^{-1} \sum_{\{1 \leq i_1 < \cdots < i_m \leq N\}} \psi(X_{i_1}, \ldots, X_{i_m}), \tag{3}$$

where it is tacitly assumed that $N \geq m$, a condition that is not necessary for V_N. For both U_N and V_N it is assumed that the kernel $\psi(.)$ is of finite degree m (≥ 1), which may not always be true. As a simple illustration, consider the population α-quantile

$$\theta(F) = F^{-1}(\alpha) = \inf\{x : F(x) \geq \alpha\}, \ F \in \mathcal{F}, \tag{4}$$

where $\alpha \in (0, 1)$. For an arbitrary (even continuous) F, there may not be a kernel of finite degree that unbiasedly estimates $F^{-1}(\alpha)$. In such a case, we may not be able to define U_N or V_N as in (2) or (3), but may still consider the plug-in estimator (the sample quantile)

$$T_N = \theta(F_N) = F_N^{-1}(\alpha) = \inf\{x : F_N(x) \geq \alpha\}. \tag{5}$$

T_N may not share the regular properties of U_N or V_N, and we shall make more comments on it later on.

Suppose now that we have a regular case where the kernel is of finite degree m $(\leq n)$. If $\psi(x_1, \ldots, x_m)$ remains invariant under any strictly monotone transformation on $x_1, \ldots, x_m$, then clearly, $\psi(.)$ depends on the x_i only through their ranks, so that U_N and V_N are effectively rank statistics — though they need not be linear rank statistics. A similar characterization holds when the kernel remains invariant with respect to some other group of transformations that maps the sample space onto itself. We illustrate this point with the following examples.

Wilcoxon signed-rank statistic: Let R_{Ni}^{+} be the rank of $|X_i|$ among $|X_1|, \ldots, |X_N|$, for $i = 1, \ldots, N$, and let

$$W_N = \frac{1}{N(N+1)} \sum_{i=1}^{N} (\text{sign}\, X_i) R_{Ni}^{+}. \tag{6}$$

Some routine computations yield that

$$W_N = \frac{N-1}{N+1} U_N^{(1)} + \frac{2}{N+1} U_N^{(2)}, \tag{7}$$

where

$$U_N^{(1)} = \binom{N}{2}^{-1} \sum_{\{1 \le i < j \le N\}} \operatorname{sign}(X_i + X_j), \tag{8}$$

$$U_N^{(2)} = N^{-1} \sum_{i=1}^{N} \operatorname{sign}(X_i), \tag{9}$$

both of which remain invariant under the group of transformation: $X \to X^* = (\operatorname{sign} X) g(|X|)$, where $g(.)$ is an arbitrary, strictly monotone function on $0 \le X < \infty$. On the contrary, if we examine (5), we may observe that T_N is not a rank statistic, but an order statistic that has the *equivariance* property that for an arbitrary strictly monotone $g(.)$ on the real line, we have $T_N(g(X_1), \ldots, g(X_N)) = g(T_N(X_1, \ldots, X_N))$.

If we consider, as in Subsection 3.4.9, a general signed-rank statistic

$$S_N = \sum_{i=1}^{N} (\operatorname{sign} X_i) a_N^+(R_{Ni}^+), \tag{10}$$

where the scores $a_N^+(i)$ need not be linear in $i/(N+1)$, then it may not be expressible in terms of one or more U- (or V-) statistics, though it is still a statistical functional. For this reason, in the next subsection we present an outline of some other statistical functionals. We consider here two other illustrations of U-statistics that are relevant to rank tests.

Kendall's tau statistic. Let (X_i, Y_i), $i = 1, \ldots, N$, be a sample where each (X_i, Y_i) has a bivariate distribution function $F(x, y)$. For testing the null hypothesis of independence of X and Y, that is, $H_o : F(x, y) = F(x, \infty) F(\infty, y)$, an appropriate test statistic is

$$K_N = \binom{N}{2}^{-1} \sum_{1 \le i < j \le N} \operatorname{sign}(X_i - X_j) \operatorname{sign}(Y_i - Y_j). \tag{11}$$

This is clearly a U-statistic, and the corresponding kernel $\psi(a, b) = \operatorname{sign}(a_1 - a_2) \operatorname{sign}(b_1 - b_2)$, of degree 2, is invariant under any strictly monotone transformations: $x \to x^* = g_1(x), y \to y^* = g_2(y)$, where g_1, g_2 are monotone. Therefore, K_N qualifies as a rank statistic but not a linear rank statistic. We refer to Subsections 3.3.1 (and 4.6.2) where K_N was introduced as a rank statistic and some related projection results are presented.

Spearman's rank correlation. Side by side, we consider the following statistic

$$S_N = \frac{12}{N(N^2 - 1)} \sum_{i=1}^{N} \left(R_i - \frac{n+1}{2} \right) \left(Q_i - \frac{N+1}{2} \right), \tag{12}$$

where R_i is the rank of X_i among $X_1, \ldots, X_N$, and Q_i is the rank of Y_i

among $Y_1, \ldots, Y_N$. We again refer to Subsections 3.3.1 and 4.6.1 where S_N has been identified as a linear rank statistic; it can also be expressed as a linear combination of two U-statistics.

In fact, in some cases, a linear rank statistic can be expressed as a function of two or more U- (or V-) statistics, and in some other cases it can well be approximated by a linear combination of U-statistics. This was the line of attack prior to the development of the general theory of linear rank statistics (see for example Dwass (1956)), and we shall discuss this aspect in later chapters as well. Nevertheless, these illustrations bring the relevance of U- and V-statistics to the theory of rank tests. No wonder scores of EDF/ADF tests based on U-statistics have appeared in the statistical research literature; we refer to Puri and Sen (1971) for details.

3.5.2 Differentiable statistical functionals. We have noticed in the preceding subsection that rank statistics can be expressed as statistical functionals that need not be U- or V-statistics, or a linear rank statistic; (3.5.1.4) and (3.5.1.11) are appropriate illustrative examples. Basically, in nonparametrics, a parameter of interest is expressed as a functional $\theta(F)$ of the underlying d.f. F, so that a natural estimator of $\theta(F)$ is the corresponding sample functional $T_n = \theta(F_n)$ based on the sample distribution function F_n. Let us consider a group of transformations $\mathcal{G}$, and for $g \in \mathcal{G}$, we write $X^g = g \cdot X$, and denote the distribution function of X^g by F^g. Typically, such a g relates to strictly monotone transformations or sign-invariant strictly monotone ones, treated before (3.5.1.10). Then $\theta(F)$ is g-invariant if $\theta(F^g) = \theta(F)$, for all $g \in \mathcal{G}$. A general signed-rank statistic S_N, as described in (3.5.1.10), is a g-invariant statistical functional; though they are not necessarily of linear type, under appropriate smoothness conditions on the scores, S_N may be approximable by a linear one. In general, however, this need not be the case for statistical functionals. As an example, consider the one-sample Kolmogorov-Smirnov type test statistics (for testing symmetry of F about 0), considered by Butler (1969) and Chatterjee and Sen (1973b). Let

$$\begin{aligned} Z_n^+ &= \sup_{x \geq 0}\{F_N(x) + F_n(-x-) - 1\}, \qquad (1) \\ Z_n &= \sup_{x \geq 0}\{|F_n(x) + F_n(-x-) - 1|\}. \end{aligned}$$

Either one is a functional of the sample distribution function F_n, and their population counterparts vanish when F is symmetric about 0. These are signed rank statistics, but not linear ones.

In general, for statistical functionals, suitable modes of *differentiability* have been investigated to provide convenient means for the study of related asymptotics. This will be elaborated in later chapters. Basically, we write

$$T_N = \theta(F_N) = \theta(F + (F_N - F)), \qquad (2)$$

where we note that $F_N - F$ has generally many nice statistical properties, as will be seen in later chapters. Thus, if $\theta(\cdot)$ is smooth enough, we might be tempted to use Taylor's expansion, though in a functional space, and express

$$T_N = \theta(F) + \theta_1(F; F_N - F) + R_N, \tag{3}$$

where $\theta_1(F; F_N - F)$ is a linear functional, and R_N is the remainder term. Note that if T_N is itself a rank statistic, and the first-order expansion in (3) is a valid one (in the sense that R_N is small compared to $\theta_1(F; F_N - F)$, whenever $\|F_N - F\|$ is small in a well defined norm), then $\theta_1(F; F_N - F)$ corresponds to a linear rank statistic; this brings the relevance of statistical functionals to rank test statistics. Because of the asymptotic flavour, we postpone the discussion of some of these aspects to later chapters.

3.5.3 Generalized statistical functionals. In the context of two-sample problems, we encounter a functional of two distribution functions, say, F and G. For example, for the Wilcoxon two-sample statistic, we have $\theta(F, G) = \int F(x)\,\mathrm{d}G(x) = P\{X \le Y\}$, so that if F_m and G_n are the two sample distribution functions, respectively, then the sample counterpart of $\theta(F, G)$ is

$$\begin{aligned} \theta(F_m, G_n) &= \int F_m(x)\,\mathrm{d}G_n(x) \\ &= (mn)^{-1} \sum_{i=1}^{m} \sum_{j=1}^{n} I(X_i \le Y_j). \end{aligned} \tag{1}$$

Motivated by this classical example, we may consider, in general, a kernel $\psi(X_1, \ldots, X_{m_1}; Y_1, \ldots, Y_{n_1})$ of degree (m_1, n_1) $(\ge (1,1))$, such that $\psi(.;.)$ is symmetric in its first m_1 arguments (and also in its last n_1 ones), though the roles of these two sets may not be the same. If then

$$\theta(F, G) = \mathsf{E}_{F,G}\psi(X_1, \ldots, X_{m_1}; Y_1, \ldots, Y_{n_1}), \tag{2}$$

we term $\theta(F, G)$ a generalized estimable parameter or a regular functional. The corresponding generalized U-statistic is given by

$$U_{m,n} = \binom{m}{m_1}^{-1} \binom{n}{n_1}^{-1} \sum_{1}^{*} \sum_{2}^{*} \psi(X_{i_1}, \ldots, X_{i_{m_1}}; Y_{j_1}, \ldots, Y_{j_{n_1}}), \tag{3}$$

where $m \ge m_1$, $n \ge n_1$, and the summation $\sum_{1}^{*}$ $(\sum_{2}^{*})$ extends over all $1 \le i_1 < \ldots < i_{m_1} \le m$ $(1 \le j_1 < \ldots < j_{n_1} \le n)$. Likewise, the generalized von Mises functional is

$$V_{m,n} = m^{-m_1} n^{-n_1} \sum_{1}^{o} \sum_{2}^{o} \psi(X_{i_1}, \ldots, X_{i_{m_1}}; Y_{j_1}, \ldots, Y_{j_{n_1}}), \tag{4}$$

where the summation $\overset{o}{\sum_1}(\overset{o}{\sum_2})$ extends over all $i_k = 1, \ldots, m$, $1 \leq k \leq m_1$ ($j_k = 1, \ldots, n$, $1 \leq k \leq n_1$), not necessarily all distinct. (Such generalized U- and V-statistics have also been worked out for more general multi-sample models). Here also, if

$$\psi(X_{i_1}, \ldots, X_{i_{m_1}}; Y_{j_1}, \ldots, Y_{j_{n_1}}) \\ = \psi(g.\, X_{i_1}, \ldots, g.\, X_{i_{m_1}}; g.\, Y_{j_1}, \ldots, g.\, Y_{j_{n_1}}), \tag{5}$$

for all $g \in \mathcal{G}$, of strictly monotone transformations, then (3) and (4) are both g-invariant, and are therefore rank statistics. Thus whenever a hypothesis of invariance under $\mathcal{G}$ can be framed in terms of $\theta(F, G)$, the corresponding U- and V-statistics (that need not be linear rank statistics) can be incorporated to construct suitable rank tests. It is also possible to have EDF/ADF tests based on appropriate generalized U- or V-statistics that may not be of the rank-type. A very simple example of this type is the Lehmann (1951) two-sample scale test based on a generalized U-statistic for which the kernel is of degree $(2, 2)$ and is given by

$$\Psi = \psi(x_1, x_2; y_1, y_2) \;=\; 1 \text{ or } 0, \text{ according as } \\ |x_1 - x_2| \leq |y_1 - y_2| \text{ or not.} \tag{6}$$

The test based on (6) is neither a rank test nor is EDF. During the 1950s and 1960s, a large number of nonparametric tests based on generalized U- and V-statistics were proposed. These tests are consistent against some appropriate classes of alternatives that are not necessarily of the location/scale type, although they pertain well to such parametric type alternatives. Dwass (1956) even tried to approximate a two-sample linear rank statistic in terms of generalized U-statistics, so that distribution theory of such generalized statistics, already developed on a strong footing, could be used for linear rank statistics as well. Similar characterizations and approximations by linear functionals hold for general (differentiable) functionals $\theta(F, G)$ of two distribution functions. Viewed from this perspective, multisample statistical functionals remain pertinent to the development of the theory of rank tests. We refer to Puri and Sen (1971) and Jurečková and Sen (1996) for detailed accounts of some related developments.

PROBLEMS AND COMPLEMENTS TO CHAPTER 3

Section 3.1

1. Put $f^{(\cdot)}(y) = N! \prod_1^N f(y_i)$ for $y \in \mathcal{X}^{(\cdot)}$ and $f^{(\cdot)}(y) = 0$ otherwise. Then show that (3.1.2.9) is equivalent to

$$f_N^{(i)}(x^{(i)}) \;=\; \underbrace{\int_{-\infty}^{\infty} \cdots \int_{-\infty}^{\infty}}_{(N-1)\text{-times}} f^{(\cdot)}(y_1, \ldots, y_{i-1}, x^{(i)}, y_{i+1}, \ldots, y_N)$$

$$\cdot\, dy_1 \dots dy_{i-1}\, dy_{i+1} \cdot dy_N$$

and, consequently, (3.1.2.9) is in accord with the usual manner of computing marginal densities.

2. Recall the properties of the incomplete Beta function ratio $I_z(a, b)$, and use them to elaborate in detail the proof of Theorem 3.1.2.3, in particular how the sum (3.1.2.7) in this Theorem implies the integral (3.1.2.8).

3. Show that the density of $X^{(\cdot)}$ equals

$$N! \prod_1^N f(x^{(i)}), \quad x^{(\cdot)} \in \mathcal{X}^{(\cdot)}. \tag{1}$$

[Hint: Use (3.1.2.2) and (3.1.2.4).]

4. Put $Y_i = X_N^{(s_i)}$, $1 \le s_1 < \dots < s_n \le N$, with $X_N^{(i)}$ being the order statistics of a sample from a distribution with density $f(x)$. Then the density of $(Y_1, \dots, Y_n)$ exists and equals

$$\frac{N! f(y_1) \dots f(y_n)}{(s_1 - 1)!(s_2 - s_1 - 1)! \dots (N - s_n)!} [F(y_1]^{s_1 - 1} \tag{2}$$
$$\cdot [F(y_2) - F(y_1)]^{s_2 - s_1 - 1} \dots [1 - F(y_n)]^{N - s_n}$$

for $y_1 < \dots < y_n$.

5. Let $X_N^{(1)} < \dots < X_N^{(N)}$ be an ordered sample from the distribution with exponential density e^{-x}, $x \ge 0$. Then $X_N^{(1)}$ has density Ne^{-Nx}, $x \ge 0$, and the conditional distribution of $Y_1 = X_N^{(2)} - X_N^{(1)}, \dots, Y_{N-1} = X_N^{(N)} - X_N^{(1)}$ given $X_N^{(1)} = x$ is the same as of the ordered sample $X_{N-1}^{(1)} < \dots < X_{N-1}^{(N-1)}$ from the distribution with the same density e^{-x}, $x \ge 0$. Further, the statistics $X_N^{(1)}, X_N^{(2)} - X_N^{(1)}, \dots, X_N^{(N)} - X_N^{(N-1)}$ are independent, and the density of $X_N^{(i+1)} - X_N^{(i)}$ is $(N - i)e^{-x(N-i)}$, $x \ge 0$. Consequently,

$$\mathsf{E} X_N^{(i)} = \sum_{j=0}^{i-1} \frac{1}{N - j}.$$

6. Let $U_N^{(1)} < \dots < U_N^{(N)}$ be an ordered sample from the uniform distribution, and put $Y_i = U_N^{(N-i)} / U_N^{(N-i+1)}$, $1 \le i \le N$, $U_N^{(N+1)} = 1$. Then the random variables Y_i are independent. [Hint: $U_N^{(N-i+1)} = \exp(-X_N^{(i)})$, where $X_N^{(i)}$ are order statistics from the distribution with density e^{-x}, $x \ge 0$.]

Definition. A class of probability distributions $\boldsymbol{P} = \{P\}$ will be said to be complete on a sub-σ-field $\mathcal{B}$ if $[t(x)$ is $\mathcal{B}$-measurable and $\int t(x)\,\mathrm{d}P = 0$ for all $P \in \boldsymbol{P}]$ entails $[P(t(X) = 0) = 1$ for $P \in \boldsymbol{P}]$.

Definition. A statistic $T = t(X)$ (or a sub-σ-field $\mathcal{B}$) is called sufficient for a class of probability distributions $\boldsymbol{P} = \{P\}$, if there is a version of conditional probabilities $P_0(A|T = t)$ such that $P(A|T = t) = P_0(A|T = t)$ for all $P \in \boldsymbol{P}$.

7. $X^{(\cdot)}$ is sufficient for H_0 (or H_*), and for any random variable $Y = y(X)$, with finite expectation, we have

$$\mathsf{E}(Y|X^{(\cdot)} = x^{(\cdot)}) = \frac{1}{N!} \sum_{r \in R} y(x^{(r_1)}, \ldots, x^{(r_N)}).$$

The formula holds even for the set of all distributions (including the discrete ones) that are invariant with respect to the permutation of arguments.

8. The class of probability distributions of $X^{(\cdot)}$ induced by H_0 is complete on $\mathcal{A}^{(\cdot)}$. (The individual distributions are given by densities (1) with f running through the set of all one-dimensional densities.)

9. $|X|^{(\cdot)}$ is sufficient for H_1, and the class of its distributions under H_1 is complete.

10. $(X^{(\cdot)}, Y^{(\cdot)})$ is sufficient for H_2, and the class of its distributions under H_2 is complete.

11. $(X_1^{(\cdot)}, \ldots, X_n^{(\cdot)})$ is sufficient for H_3 and the class of its distributions under H_3 is complete.

Section 3.2

12. Show that t^* and t given by (3.2.2.27) and (3.2.2.28), respectively, are related by $t = (N-2)^{\frac{1}{2}} t^* \left(mn/N) - t^{*2}\right)^{-\frac{1}{2}}$.

13. Putting $c = \min(a, b)$, and $d = \max(a, b)$, (3.2.3.12) may be rewritten as follows:

$$\begin{aligned} Q_\Delta(R = r) &= \sum_{i=0}^{c} \Delta^i \frac{1}{i!(N-i)!} - \frac{\Delta^{d+1}}{(d+1)!(N-d-1)!} \\ &\quad + \sum_{i=d+2}^{N} \Delta^i K_i(a, b), \end{aligned}$$

i.e., the coefficient of Δ^{c+1} either equals 0 or is negative. Utilizing this fact, prove the existence of $\varepsilon > 0$ such that $[\min(a, b) > \min(a', b')]$ or

$[\min(a, b) = \min(a', b'), a + b > a' + b']$ entails $Q_\Delta(R = r) > Q_\Delta(R = r')$ for $0 < \Delta < \varepsilon$.

Further establish the existence of $\varepsilon > 0$ such that for $1 - \varepsilon < \Delta < 1$, $[a + b > a' + b']$ or $[a + b = a' + b', \min(a, b) > \min(a', b')]$ entails $Q_\Delta(R = r) > Q_\Delta(R = r')$.

14. Let $\mathcal{G} = \{g\}$ be a group of one-to-one maps of $\mathcal{X}$ to itself. Then the class of events B such that $gB = B$ is a σ-field.

15. Consider the group of maps $g = (u, v)$, $-\infty < u$, $v < \infty$, such that $g(x_1, \ldots, x_N) = (e^v x_1 + u, \ldots, e^v x_N + u)$. Then the composition of $g_1 = (u_1, v_1)$ and $g_2 = (u_2, v_2)$ equals $g = g_2 g_1 = (u_2 + u_1 e^{v_2}, v_2 + v_1)$ and defines multiplication in the group.

Let B be a Borel set in the (u, v)-plane and $Bg = \{g'' : g'' = g'g, g' \in B\}$. Then $\nu(Bg) = \nu(B)$ for all B and g, ν denoting Lebesgue measure in the plane. [Hint: The Jacobian $\partial(x, y)/\partial(u', v')$ of the transformation $x = u' + ue^{v'}$, $y = v' + v$ equals 1.]

16. (Continuation.) If X has density $p = p(x_1, \ldots, x_N)$, then gX with $g = (u, v)$ has density $p_{u,v} = e^{-Nv} p[(x_1 - u)e^{-v}, \ldots, (x_N - u)e^{-v}]$. Show that $\bar{p}$ in (3.2.2.11) equals $\bar{p} = \int_{-\infty}^{\infty} \int_{-\infty}^{\infty} p_{u,v}(x)\, du\, dv$. Thus $\bar{p}$ equals the integral of $p_{u,v}(x)$ with respect to the invariant measure on the (u, v)-group.

17. (Continuation.) Let ν_n denote the uniform distribution on the rectangle $|u| < ne^n$ and $|v| < n$. Show that for any Borel set B in the (u, v)-plane and for any $g_1 = (u_1, v_1)$ the following relation holds:

$$\lim_{n \to \infty} |\nu_n(Bg_1) - \nu_n(B)| = 0.$$

18. The test with critical region (3.2.2.34) is maximin most powerful for $H = \{p_{\mu,\sigma}\}$ against $K = \{g_{\mu,\sigma}\}$ at the respective level. [Hint: Make use of the Hunt-Stein theorem and of Theorem 4 of Chapter VI in Lehmann (1959), and of the previous exercises.]

Section 3.3

19. τ given by (3.3.1.15) may also be written as $\tau = 2/\big(N(N-1)\big) \cdot \sum_{i>h} \operatorname{sign}(R_i - R_h)$.

20. Let $R_1, \ldots, R_N$ and τ_N correspond to $(X_1, \ldots, X_N)$, and let τ_{N-1} correspond to $(X_1, \ldots, X_{N-1})$. Then

$$\tau_N = \frac{N-2}{N} \tau_{N-1} + \frac{4}{N(N-1)} \Big(R_N - -\frac{1}{2}(N+1) \Big), \tag{3}$$

and the two terms on the right-hand side are independent.

21. Utilizing (3), prove

$$[N(N-1)]^2 \operatorname{var} \tau_N = [(N-1)(N-2)]^2 \operatorname{var} \tau_{N-1} + \frac{4}{3}(N^2-1) \tag{4}$$

and show that

$$\operatorname{var} \tau_N = \frac{2(2N+5)}{9N(N-1)}. \tag{5}$$

22. Let $c_i = 1$ for $1 \le i \le m$ and $c_i = 0$ for $m < i \le N$. Show that

$$\sum_{i=1}^{m} (c_i - \bar{c})^2 = \frac{mn}{N} \tag{6}$$

and

$$\sum_{i=1}^{N} \left(i - \frac{1}{2}(N+1)\right)^2 = \frac{1}{12} N(N-1)(N+1). \tag{7}$$

Further, prove that

$$\operatorname{var}\left(\sum_{i=1}^{m} R_i\right) = \frac{1}{12} mn(N+1). \tag{8}$$

[Hint: $\sum_1^N i(i-1)\ldots(i-K+1) = [1/(K+1)](N+1)N(N-1)\ldots(N-K+1)$.]

23. Let ϱ and $\hat{\tau}$ be given by (3.3.1.17) and (3.3.1.16), respectively. Prove that

$$\operatorname{var} \varrho = \frac{1}{N-1}, \quad \operatorname{var} \hat{\tau} = \frac{4}{9}\left(\frac{N+1}{N}\right)^2 \frac{1}{N-1}. \tag{9}$$

Definition. (k-statistics). Put $k_{1c} = \bar{c}$, $k_{2c} = \sigma_c^2$ and

$$k_{3c} = \frac{N}{(N-1)(N-2)} \sum_{i=1}^{N} (c_i - \bar{c})^3, \tag{10}$$

$$k_{4c} = \frac{1}{(N-1)(N-2)(N-3)} \Big\{ N(N+1) \cdot \sum_{i=1}^{N} (c_i - \bar{c})^4 - 3(N-1) \Big[\sum_{i=1}^{N} (c_i - \bar{c})^2 \Big]^2 \Big\}. \tag{11}$$

24. Let $c_i = 1$ for $1 \le i \le m$ and $c_i = 0$ for $m < i \le N$. Then

$$k_{3c} = \frac{mn(n-m)}{N(N-1)(N-2)}, \tag{12}$$

$$k_{4c} = \frac{m^2n^2}{N(N-1)(N-2)(N-3)}\Big[\frac{N^2+N}{mn} - 6\Big]. \tag{13}$$

25. Put $S = \sum_{i=1}^N c_i a(R_i)$. Then

$$\mathsf{E}(S - \mathsf{E}S)^3 = k_{3a}\sum_{i=1}^N (c_i - \bar{c})^3, \tag{14}$$

$$\begin{aligned}\mathsf{E}(S - \mathsf{E}S)^4 &= 3\frac{(N-1)^3}{N+1}\sigma_a^4\sigma_c^4 \\ &+\frac{(N-1)(N-2)(N-3)}{N(N+1)}k_{4a}k_{4c}.\end{aligned} \tag{15}$$

26. Show that

$$\begin{aligned}&\mathsf{E}\Big(\sum_{i=1}^m R_i - m\cdot\frac{1}{2}(N+1)\Big)^4 = \frac{mn(m+n+1)}{240} \\ &\cdot[5(m^2n+mn^2) - 2(m^2+n^2) + 3mn - 2(m+n)],\end{aligned} \tag{16}$$

and, consequently, the excess of $S = \sum_{i=1}^m R_i$ equals

$$\gamma_2 = -\frac{6}{5}\frac{m^2+n^2+mn+m+n}{mn(m+n+1)}. \tag{17}$$

(Recall that the excess equals $\mathsf{E}(S - \mathsf{E}S)^4/[\mathsf{var}\, S]^2 - 3$.) The sixth moment may be found in Fix and Hodges (1955).

27. The third and fourth moments of general linear statistics. Put $S = \sum_{i=1}^N a(i, R_i)$ and $d_{ij} = a(i,j) - a(.,j) - a(i,.) + \bar{a}$, with $a(.,j)$, $a(i,.)$ and $\bar{a}$ defined by (3.3.1.2), (3.3.1.3) and (3.3.1.4), respectively. Show that

$$\mathsf{E}(S - \mathsf{E}S)^3 = \frac{N}{(N-1)(N-2)}\sum_{i=1}^N\sum_{j=1}^N d_{ij}^3, \tag{18}$$

$$\begin{aligned}\mathsf{E}(S - \mathsf{E}S)^4 &= 3\frac{N^2-3N+1}{N(N-1)(N-2)(N-3)}\Big(\sum_{i=1}^N\sum_{j=1}^N d_{ij}^2\Big)^2 \\ &-\frac{3}{(N-2)(N-3)}\sum_{i=1}^N\sum_{j=1}^N\sum_{k=1}^N (d_{ij}^2 d_{ik}^2 + d_{ij}^2 d_{kj}^2) + \\ &+\frac{N(N+1)}{(N-1)(N-2)(N-3)}\sum_{i=1}^N\sum_{j=1}^N d_{ij}^4 \\ &+\frac{6}{N(N-1)(N-2)(N-3)}\sum_{i=1}^N\sum_{j=1}^N\sum_{k=1}^N\sum_{h=1}^N d_{ik}d_{ih}d_{jk}d_{jh}.\end{aligned} \tag{19}$$

[M. Fořtová, unpublished.] Check the agreement with (14) and (15).

28. Show that

$$\sum_{i=1}^{N}\Big(i-\frac{1}{2}(N+1)\Big)^4=\frac{1}{240}N(N-1)(N+1)(3N^2-7). \tag{20}$$

29. Show that the excess γ_2 of $S=\sum_{i=1}^{m}a(R_i)$ is always negative when $m=\frac{1}{2}N$, independently of the choice of the a's.

Section 3.4

30. Verify condition A_1 for the following families:

$$d(u,\Theta) = 1-\Theta+2\Theta u, \tag{21}$$
$$D(u,\Theta) = [1-(1-u)^{1/(1+\Theta)}]^{1+\Theta}, \tag{22}$$
$$D(u,\Theta) = u^{1+\Theta}, \tag{23}$$
$$D(u,\Theta) = 1-(1-u)^{1/(1+\Theta)}, \quad u,\Theta\in[0,1]. \tag{24}$$

Show that the φ-functions at $\Theta=0$ equal $2u-1$, $\log u/(1-u)$, $1+\log u$, $-1-\log(1-u)$, respectively; $D(u,\Theta)=\int_0^u d(x,\Theta)\,\mathrm{d}x$.

31. The critical region $\sum_{i=1}^{m}R_i\geq k$ belongs to the locally most powerful rank test of H_0 against $Q_\Delta(X_i\leq x_i, 1\leq i\leq N)=\prod_{i=1}^{m}[(1-\Delta)F(x_i)+\Delta F^2(x_i)]\prod_{i=m+1}^{N}F(x_i)$, with F being an arbitrary continuous distribution function. [Lehmann (1953).]

32. Let $U_N^{(1)}<\ldots<U_N^{(N)}$ be an ordered sample from the uniform distribution. Show that

$$\mathsf{E}\{-\log(1-U_N^{(i)})\}=\sum_{j=0}^{i-1}\frac{1}{N-j}, \tag{25}$$

$$\mathsf{E}\{-\log U_N^{(i)}\}=\sum_{j=0}^{N-i}\frac{1}{N-j}. \tag{26}$$

33. The critical region $\sum_{i=1}^{m}a_N(R_i)\geq k$ with

$$a_N(i)=\sum_{j=0}^{i-1}\frac{1}{N-j}-\sum_{j=0}^{N-i}\frac{1}{N-j} \tag{27}$$

belongs to the locally most powerful test of H_0 against $Q_\Delta(X_i \le x_i, 1 \le i \le N) = \prod_{i=1}^{m} [F(x_i)]^{1+\Delta} \prod_{i=m+1}^{N} \left[1-\left(1-F(x_i)\right)^{1+\Delta}\right]$, with F being some arbitrary continuous distribution function. (The so-called Psi test suggested by Gibbons (1964a), (1964b).)

34. Putting $\psi(x) = [\mathrm{d}/\,\mathrm{d}x] \ln \Gamma(x)$, we have, for integers $k \ge 2$,

$$\psi(k) = -C + \frac{1}{1} + \frac{1}{2} + \ldots + \frac{1}{k-1}, \qquad C = 0 \cdot 577\ldots$$

Thus $a_N(i)$ of (25) may be expressed as follows:

$$a_N(i) = \psi(i) - \psi(N-i+1).$$

35. Verify A_1 for $d(x,\Theta) = f(x-\Theta)$, if $f(x) = \frac{1}{2}(\beta+1)(1-|x|)^\beta$ for $|x| \le 1$ and $f(x) = 0$ otherwise, $\beta > 0$.

36. (Translation of the uniform distribution.) Put $A = N - \max(R_{m+1}, \ldots, R_N)$ and $B = \min(R_1, \ldots, R_m) - 1$. Then the critical region

$$\min(A,B) + \frac{1}{N}(A+B) \ge k$$

belongs to the locally most powerful test for H_0 against $\{q_\Delta, \Delta > 0\}$ given by (3.2.3.10). Moreover, for some $\varepsilon > 0$, the critical region

$$A + B + \frac{1}{N}\min(A,B) \ge k$$

belongs to the uniformly most powerful test for H_0 against the same q_Δ for Δ such that $1-\varepsilon < \Delta < 1$. (The locally most powerful test in the vicinity of $\Delta = 1$.) [Hint: See Problem 13.]

37. (Regression in the uniform distribution.) Put $q_\Delta = 1$ if $\Delta c_i < x_i < 1 + \Delta c_i$, $1 \le i \le N$, and $q_\Delta = 0$ otherwise. Let $d = (d_1, \ldots, d_N)$ be the vector of antiranks corresponding to the vector of ranks $r = (r_1, \ldots, r_N)$, i.e. $r_{d_i} = i$, $1 \le i \le N$. Similarly let d' correspond to r'. Prove that there exists an $\varepsilon > 0$ such that $c_{d_N} - c_{d_1} > c_{d'_N} - c_{d'_1}$ entails $Q_\Delta(R = r) > Q_\Delta(R = r')$ for $0 < \Delta < \varepsilon$.

38. (Generalization of Theorem 3.4.11.1.) Assume that $d_1(x,\Theta)$ and $d_2(x,\Theta)$ both satisfy condition A_1 formulated in Subsection 3.4.8. Moreover, assume that both $\dot{d}_1(x,\Theta)$ and $\dot{d}_2(x,\Theta)$ are continuous at $\Theta = 0$ for almost every x. Put $h_\Delta(x,y) = \int d_1(x,\Delta z)\, d_2(y,\Delta z)\ \mathrm{d}M(z)$ and prove that Theorem 3.4.11.1 still holds with $a_N(i,f)$ and $a_N(i,g)$ replaced by $a_N(i,d_1)$ and $a_N(i,d_2)$, respectively.

39. (Negative correlation.) Show that the critical region

$$\sum_{i=1}^{N} a(R_i, f)a(Q_i, g) \leq k$$

belongs to the locally most powerful test for H_2 against q_Δ given by (3.4.11.1), where

$$h_\Delta(x, y) = \int g(x - \Delta z)g(y + \Delta z)\,\mathrm{d}M(z).$$

Chapter 4

Selected rank tests

From the multitude of rank tests suggested by various authors, we are going first (in Sections 1–7) to present here those which fit well into our theoretical framework and are of interest for applications. The tests are classified primarily according to the null hypotheses (Sections 1–4 correspond to H_0, Section 5 to H_1, Section 6 to H_2 and Section 7 to H_3) and secondarily according to the alternatives.

The description of each test will include its definition, the expectation and variance of the test statistic under the hypothesis if of interest, the limiting distribution under the hypothesis, and the indication of the alternative for which it is optimal. As we know from Subsection 3.4.1 the property of being a locally most powerful rank test is attached to a certain fixed N, assumes that a measure of distance between the hypothesis and the alternative is *close to zero*, and refers to the class of rank tests only. On the other hand, the notion of an asymptotically optimum test, which will be treated in Section 8.1 (the more exact term is an 'asymptotically maximin most powerful test' — see Definition in 8.1.3), refers to the class of all possible tests, is attached to $N \to \infty$, and assumes that the measure of distance is *bounded*. In the two-sample problem of location, for example, the distance may be measured by $I(f_0)mn(m+n)^{-1}\Delta^2$. If $I(f_0) < \infty$, then all locally most powerful rank tests are also asymptotically optimum tests. If a test is asymptotically optimum for a density f and the true density is g, then its asymptotic efficiency (see Section 8.2) equals $\rho^2 = [\int_0^1 \gamma(u,f)\gamma(u,g)\,du]^2$, γ being defined by (2.2.5.2).

The letters c, c_u, c_ℓ will be used as a generic notation for critical values, i.e. for such suitably chosen numbers that yield a test at the preassigned level of significance.

4.1 TWO-SAMPLE TESTS OF LOCATION

Let $X_1, \dots, X_m$ and $Y_1, \dots, Y_n$ be random samples with densities f_1 and f_2 respectively. We wish to test the hypothesis H_0 that f_1 and f_2 are identical but otherwise arbitrary. Setting $X_{m+j} = Y_j$, $j = 1, \dots, n$, and $N = m + n$, and thus forming a pooled sample $X_1, \dots, X_N$, we see that, in fact, this hypothesis H_0 coincides with H_0 defined in Subsection 3.1.2. The present section deals with the difference in location of the two samples as the alternative to H_0. This alternative is expressed by $f_1(x) = f(x - \Delta)$, $f_2(x) = f(x)$, $\Delta \neq 0$, where the type of f is more or less known. In other words, under the alternative both densities are identical but for location. If $\Delta > 0$, then f_1 is shifted to the right as compared to f_2, for example.

This model, with f of known type, is well suited for theoretical investigations; in particular, it is possible to determine an optimal rank test for such an alternative. In practice, however, f is usually not known.

Often it happens that f_1 and f_2 differ also in their variances in addition to the difference in location (Behrens-Fisher problem), and there were some attempts in the literature to adapt the usual rank tests for this practical situation. However, when using a straightforward method of estimating the variances and plugging these estimates into the tests, the resulting tests are usually no longer distribution-free (i.e. their distributions depend on the underlying distributions).

Generally, if even the type of f is not known, it may at best be only approximately guessed, or estimated, or chosen in some reasonable way from some fixed appropriate family of tests. (Cf. Section 8.5 on adaptive rank tests.) A similar remark applies to all testing situations in all subsequent sections.

4.1.1 Linear rank tests. As usual, let R_i $(i = 1, \dots, N)$ denote the rank of the observation X_i in the ordered sequence $X^{(1)} < X^{(2)} < \dots < X^{(N)}$. In particular, observe that $R_1, \dots, R_m$ are the ranks corresponding to the first sample in the pooled sample $X_1, \dots, X_N$.

The Fisher-Yates-Terry-Hoeffding (normal scores) test. Trying to use Theorem 3.4.4.1 for f normal, we are led by formula (3.4.3.1) and Table 1 of 2.2.2 to the scores $a_{m+n}(i) = \mathsf{E}\Phi^{-1}(U_{m+n}^{(i)})$ where Φ^{-1} is the inverse function of $\Phi(x) = (2\pi)^{-\frac{1}{2}} \int_{-\infty}^{x} \exp[-\frac{1}{2}t^2]\, dt$ and $U_{m+n}^{(1)} < U_{m+n}^{(2)} < \dots < U_{m+n}^{(m+n)}$ is an ordered sample from the uniform distribution on $[0, 1]$. Putting $\Phi^{-1}(U_{m+n}^{(i)}) = V_{m+n}^{(i)}$ we have

$$a_{m+n}(i) = \mathsf{E}V_{m+n}^{(i)}, \tag{1}$$

where $V_{m+n}^{(1)} < V_{m+n}^{(2)} < \ldots < V_{m+n}^{(m+n)}$ is obviously an ordered sample of size $m+n$ from the standardized normal distribution. The two-sample test statistic based on these scores is

$$S = \sum_{i=1}^{m} a_{m+n}(R_i), \tag{2}$$

and the test with critical region $\{S \geq c_u\}$ is the locally most powerful rank test for H_0 against $\Delta > 0$ if f is of normal type. The Fisher-Yates-Terry-Hoeffding test is also asymptotically optimum for such an f.

The statistic S has a symmetric distribution about $\mathsf{E}S = 0$, its variance being

$$\mathsf{var}\, S = \frac{mn}{(m+n)(m+n-1)} \sum_{i=1}^{m+n} [\mathsf{E}V_{m+n}^{(i)}]^2. \tag{3}$$

This test was proposed originally by Fisher and Yates (1938), and later derived and studied in detail by Terry (1952) on the basis of a paper by Hoeffding (1950).

The van der Waerden test employs a similar statistic, based on the approximate scores (3.4.3.13), namely

$$S = \sum_{i=1}^{m} \Phi^{-1}\Big(\frac{R_i}{m+n+1}\Big). \tag{4}$$

The statistic (4) is asymptotically equivalent to (2), so that the van der Waerden test is also asymptotically optimum for the alternatives with f normal. The distribution of S is again symmetric about $\mathsf{E}S = 0$, and

$$\mathsf{var}\, S = \frac{mn}{(m+n)(m+n-1)} \sum_{i=1}^{m+n} \Big[\Phi^{-1}\Big(\frac{i}{m+n+1}\Big)\Big]^2. \tag{5}$$

This test was introduced and studied by van der Waerden (1952/53), (1953).

The Wilcoxon test. Probably the best known and most frequently used among the two-sample rank tests is the Wilcoxon test based on the statistic

$$S = \sum_{i=1}^{m} R_i. \tag{6}$$

Proceeding as in the case of the Fisher-Yates-Terry-Hoeffding test, it can be seen that the test with critical region $\{S \geq c_u\}$ is the locally most powerful rank test for H_0 against $\Delta > 0$ in the case of a density f of logistic type.

The Wilcoxon test is also asymptotically optimum for this density. The distribution of the Wilcoxon statistic S is symmetric about

$$\mathsf{E}S = \tfrac{1}{2}m(m+n+1), \tag{7}$$

its variance being

$$\operatorname{var} S = \tfrac{1}{12}mn(m+n+1). \tag{8}$$

The test statistic (6) was suggested by Wilcoxon (1945). Another form of this test (sometimes called the Mann-Whitney test) is due to Mann and Whitney (1947): the statistic U is defined as the number of pairs X_i, Y_j ($i = 1, \ldots, m$; $j = 1, \ldots, n$) with $X_i < Y_j$. These two tests are equivalent (see Problem 1).

Some older tables for them are quoted in Hájek and Šidák (1967); refer also to Sen and Krishnaiah (1984), pp. 939–945.

The Wilcoxon test is also sometimes used for testing against the following more general alternatives: with the critical region $\{S \geq c_u\}$ against $F_1 \leq F_2$, $F_1 \not\equiv F_2$ (then the random variable X is said to be stochastically larger than Y), and with the critical region $\{S \geq c_u \text{ or } S \leq c_\ell\}$ against $F_1 \leq F_2$, $F_1 \not\equiv F_2$, or $F_1 \geq F_2$, $F_1 \not\equiv F_2$. Cf. Problem 4.

The median test. Similarly, starting from the double-exponential density, but applying the approximate scores (3.4.3.13), we should arrive at a test whose statistic would be

$$S' = \sum_{i=1}^{m} \operatorname{sign}\left(R_i - \tfrac{1}{2}(m+n+1)\right). \tag{9}$$

However, simpler and more appealing for applications is an equivalent test statistic S equal to the number of observations from the first sample exceeding the median of the pooled sample, and increased by $\frac{1}{2}$ if and only if $m+n$ is odd and this median belongs to the first sample. More formally,

$$S = \sum_{i=1}^{m} \tfrac{1}{2}\left[\operatorname{sign}\left(R_i - \tfrac{1}{2}(m+n+1)\right) + 1\right] = \tfrac{1}{2}(S' + m). \tag{10}$$

The median test is asymptotically optimum in the case of a density f of double exponential type. The distribution of S is symmetric about its expectation $\mathsf{E}S = \frac{1}{2}m$, and

$$\operatorname{var} S = \begin{cases} \dfrac{mn}{4(m+n-1)} & \text{for } m+n \text{ even,} \\[2ex] \dfrac{mn}{4(m+n)} & \text{for } m+n \text{ odd.} \end{cases} \tag{11}$$

Some authors define the median test statistic simply as the number of X_i's exceeding the median of the pooled sample (without adding $\frac{1}{2}$). The formulas for its distribution are then simpler, but this distribution for $m+n$ odd is not symmetric. Some other forms of the test statistic are also employed, e.g. those originating in the expression of the problem as a 2×2 contingency table.

It is difficult to trace the authorship of the median test, since it has been used for a long time.

Common properties. For each test statistic S of this subsection, the standardized variable $(S - \mathsf{E}S)(\mathsf{var}\, S)^{-\frac{1}{2}}$ is, under H_0, asymptotically normally distributed with expectation 0 and variance 1 whenever $m \to \infty$, $n \to \infty$ in an arbitrary manner (see 6.1.5 and 6.1.6).

Finally, let us point out clearly the distinction between one-sided and two-sided tests. With each of the statistics S, when we are testing H_0 against the one-sided alternative $\Delta > 0$, meaning that the first density is shifted to the right, we employ a one-sided critical region $\{S \geq c_u\}$. In testing against the two-sided alternative $\Delta \neq 0$ we employ a two-sided critical region $\{S \geq c_u \text{ or } S \leq c_\ell\}$.

4.1.2 Tests based on exceeding observations. Let A, and B', denote the number of observations among $X_1, \dots, X_m$ larger than $\max\limits_{1\leq j\leq n} Y_j$, or smaller than $\min\limits_{1\leq j\leq n} Y_j$, respectively, and let A', and B, denote the number of observations among $Y_1, \dots, Y_n$ larger than $\max\limits_{1\leq i\leq m} X_i$, or smaller than $\min\limits_{1\leq i\leq m} X_i$, respectively. Naturally, only one of the numbers A, A' (or B, B') is positive, while the other must be zero.

The Haga test is based on the statistic

$$T = A + B - A' - B', \tag{1}$$

and the *E-test*, newly proposed here, on the statistic

$$E = \min(A, B) - \min(A', B'). \tag{2}$$

These statistics may be used against both two-sided and one-sided alternatives. Against the one-sided alternatives $\Delta > 0$ we may use the simpler statistics $A + B$, or $\min(A, B)$.

The statistic $A + B$ generates the locally most powerful rank test for H_0 against a shift Δ of the uniform distribution over (α, β) for Δ close to $\beta - \alpha$, that is for $\beta - \alpha - \varepsilon < \Delta < \beta - \alpha$. An analogous assertion holds for the statistic $\min(A, B)$ with the shift Δ close to 0, that is for $0 < \Delta < \varepsilon$. (See Problem 3.13.) These properties are preserved also for the statistics $A + B + N^{-1}\min(A, B)$, and $\min(A, B) + N^{-1}(A + B)$, respectively, giving a richer family of possible critical regions. (See Problem 3.36.) None of the above statistics, however, generates an asymptotically optimum test within the class of all possible tests; as a matter of fact, it may be shown that no rank test can be asymptotically optimum for uniform densities.

If the underlying distribution is not uniform over (α, β), or if the shift Δ is not close to $\beta - \alpha$, then the statistic $T = A + B - A' - B'$ should be preferred to $A + B$.

All the above statistics are non-linear rank statistics and are not asymptotically normally distributed.

Critical regions for one-sided and two-sided alternatives for both test statistics T and E are defined similarly as those for S in the preceding section.

The Haga test was introduced by Haga (1959/60).

4.1.3 Tests of Kolmogorov-Smirnov types. *The Kolmogorov-Smirnov test.* In this subsection we shall need the antiranks $D_1, \ldots, D_{m+n}$ defined in (3.3.2.10) for the pooled sample $X_1, \ldots, X_{m+n}$. Recall that $D_k = j$ if and only if the k-th order statistic $X^{(k)}$ is the observation X_j. Now put

$$c_i = \begin{cases} 1 & \text{for } i = 1, \ldots, m, \\ 0 & \text{for } i = m+1, \ldots, m+n, \end{cases} \tag{1}$$

and introduce the statistic

$$K^+ = \left(\frac{m+n}{mn}\right)^{\frac{1}{2}} \max_{1 \le k \le m+n} \left(k\frac{m}{m+n} - c_{D_1} - \ldots - c_{D_k}\right). \tag{2}$$

More briefly, $c_{D_k} = 1$ whenever the k-th order statistic $X^{(k)}$ belongs to the first sample, and $c_{D_k} = 0$ whenever $X^{(k)}$ belongs to the second sample. A statistic for testing two-sided differences is

$$K^\pm = \left(\frac{m+n}{mn}\right)^{\frac{1}{2}} \max_{1 \le k \le m+n} \left|k\frac{m}{m+n} - c_{D_1} - \ldots - c_{D_k}\right|. \tag{3}$$

Both K^+ and $K^\pm$ will be called the Kolmogorov-Smirnov statistics.

Let us point out the following relation to linear rank tests. We might define a general quantile test statistic

$$B_k = c_{D_1} + \ldots + c_{D_k}, \tag{4}$$

that is, define B_k as the number of observations from the first sample among $X^{(1)}, \ldots, X^{(k)}$. Then K^+ is equal (up to a multiplicative constant) to the maximum of all the $m+n$ linear rank statistics $\mathsf{E}B_k - B_k$. Observe also that, in particular, the median test statistic for $m+n$ even is equivalent to $m - B_{\frac{1}{2}(m+n)}$.

There are also more traditional forms of the statistics K^+ and $K^\pm$

expressed as follows. Denote by $F_{1,m}$ and $F_{2,n}$ the empirical distribution functions of the samples $X_1, \ldots, X_m$ and $Y_1, \ldots, Y_n$, respectively. That is, $F_{1,m}(x)$ equals the number of observations $X_1, \ldots, X_m$ that are less than or equal to x, and similarly for $F_{2,n}$. Then

$$K^+ = \Big(\frac{mn}{m+n}\Big)^{\frac{1}{2}} \max_{-\infty<x<\infty} \big(F_{2,n}(x) - F_{1,m}(x)\big), \tag{5}$$

and

$$K^\pm = \Big(\frac{mn}{m+n}\Big)^{\frac{1}{2}} \max_{-\infty<x<\infty} |F_{2,n}(x) - F_{1,m}(x)|. \tag{6}$$

Clearly, in (5) and (6) it suffices to seek the maximum only among the $m+n$ values obtained by inserting $x = X^{(1)}, \ldots, X^{(m+n)}$ into the difference $F_{2,n}(x) - F_{1,m}(x)$.

In testing the difference in location of two samples, as described at the beginning of this section, one employs the critical region $\{K^+ \geq c\}$ against the one-sided alternative $\Delta > 0$, and $\{K^\pm \geq c\}$ against the two-sided alternative $\Delta \neq 0$.

However, these two statistics are capable of, and are also used for, revealing much more general alternatives. In fact, since $F_{1,m}(x)$ converges for each x with probability 1 to the distribution function $F_1(x)$ of X_i $(i = 1, \ldots, m)$, and similarly $F_{2,n}(x)$ converges to the distribution function $F_2(x)$ of Y_j $(j = 1, \ldots, n)$, the test with the critical region $\{K^+ \geq c\}$ is consistent against the alternative that $F_2(x) > F_1(x)$ for at least one x. Observe that this general alternative includes the special case of the difference in location $F_1(x) = F_2(x - \Delta)$, $\Delta > 0$. An analogous consistency conclusion is valid for $\{K^\pm \geq c\}$ and the alternative $F_2(x) \neq F_1(x)$ for at least one x.

The Kolmogorov-Smirnov test introduced here is usually treated in the literature as a test against these general alternatives. We have broken this tradition, since we shall later introduce in Sections 4.2.3 and 4.3.3 also its analogues for the difference in scale and for regression, and the present test is more sensitive to difference in location when compared to those latter analogues.

Naturally, when dealing with the case of finite sample sizes it suffices to define the test statistics only by terms expressed by 'max' in formulas (2), (3) and (5), (6). Multiplicative constants before these expressions have been introduced here only in order that the statistics might have asymptotic distributions. These distributions will be shown in Subsection 6.3.7.

Investigation of statistics similar to (6) was begun by Kolmogorov (1933), who found the limiting distribution of the statistic $N^{\frac{1}{2}} \sup |F_N(x) - F(x)|$, suitable for testing the goodness-of-fit problem of whether $X_1, \ldots, X_N$ have a known continuous distribution function F. Later Smirnov (1939) treated the two-sample statistics (5) and (6). Some references to tables for (5) and (6) may be found in

Hájek and Šidák (1967), the tables themselves e.g. in Sen and Krishnaiah (1984), pp. 951.

The Rényi test is based on similar statistics, namely on

$$R_a^+ = \left(\frac{m+n}{mn}\right)^{\frac{1}{2}} \cdot \max_{a(m+n)\leq k\leq m+n} \left\{\frac{m+n}{k}\left(k\frac{m}{m+n} - c_{D_1} - \ldots - c_{D_k}\right)\right\} \tag{7}$$

and

$$R_a^\pm = \left(\frac{m+n}{mn}\right)^{\frac{1}{2}} \cdot \max_{a(m+n)\leq k\leq m+n} \left\{\frac{m+n}{k}\left|k\frac{m}{m+n} - c_{D_1} - \ldots - c_{D_k}\right|\right\} \tag{8}$$

with a being some number between 0 and 1. Another form of these statistics is

$$R_a^+ = \left(\frac{mn}{m+n}\right)^{\frac{1}{2}} \max \frac{N(F_{2,n}(x) - F_{1,m}(x))}{mF_{1,m}(x) + nF_{2,n}(x)} \tag{9}$$

and

$$R_a^\pm = \left(\frac{mn}{m+n}\right)^{\frac{1}{2}} \max \frac{N|F_{2,n}(x) - F_{1,m}(x)|}{mF_{1,m}(x) + nF_{2,n}(x)} \tag{10}$$

where the maximum is always taken over all x such that

$$N^{-1}[mF_{1,m}(x) + nF_{2,n}(x)] \geq a.$$

The same considerations regarding critical regions, and modifications for finite sample sizes alternatives as for the Kolmogorov-Smirnov statistics, also apply to these statistics of Rényi. Asymptotic distributions of R_a^+ and $R_a^\pm$ will be given in 6.3.8.

Perhaps one practical remark should be added. With the Rényi statistics the difference $F_{2,n}(x) - F_{1,m}(x)$ is weighted by the reciprocal of the 'pooled' empirical distribution function $N^{-1}[mF_{1,m}(x) + nF_{2,n}(x)]$ and thus the difference in low observations receives more weight. On the other hand, if we wished to stress primarily the difference in large observations we should perform the whole test procedure with the values $-X_i$ and $-Y_j$ in place of the original X_i and Y_j.

The Rényi statistics for testing goodness of fit, analogous to (9) and (10), were introduced by Rényi (1953a); this paper also contains tables of the limiting distribution of (10).

The Cramér-von Mises test is the third test of a similar type which we shall mention here. Its test statistic is

$$M = \frac{1}{mn}\sum_{k=1}^{m+n}\left(k\frac{m}{m+n} - c_{D_1} - \ldots - c_{D_k}\right)^2, \tag{11}$$

or expressed by another formula,

$$M = \frac{mn}{m+n} \int_{-\infty}^{\infty} [F_{2,n}(x) \\ -F_{1,m}(x)]^2 \, \mathrm{d}\Big[\frac{mF_{1,m}(x) + nF_{2,n}(x)}{m+n}\Big]. \tag{12}$$

As is immediately seen from this last expression, the Cramér-von Mises test with critical region $\{M \geq c\}$ is designed only for two-sided alternatives: in the case of location difference against $\Delta \neq 0$, in the general case against $F_2(x) \neq F_1(x)$ for at least one x.

The asymptotic distribution of M will be shown in 6.3.9. Of course, just the sum in (11) can be considered for the finite sample case, and the factor $(mn)^{-1}$ may be left out in the definition.

The Cramér-von Mises test can be given in a still different simple form. Let $R_1^* < R_2^* < \ldots < R_m^*$ denote the increasingly ordered ranks $R_1, R_2, \ldots, R_m$ corresponding to the first sample, and similarly let $R_{m+1}^* < R_{m+2}^* < \ldots < R_{m+n}^*$ denote the increasingly ordered ranks $R_{m+1}, R_{m+2}, \ldots, R_{m+n}$ corresponding to the second sample. Setting

$$T = m\sum_{i=1}^{m}(R_i^* - i)^2 + n\sum_{j=1}^{n}(R_{m+j}^* - j)^2, \tag{13}$$

it can be shown that

$$M = T\frac{1}{mn(m+n)} - \frac{4mn-1}{6(m+n)}. \tag{14}$$

Thus, the test statistic M is equivalent to T, and the Cramér-von Mises test can be performed also making use of the critical region $\{T \geq c\}$.

The test is named after Cramér (1928) and von Mises (1931), who suggested a goodness-of-fit statistic $\int [F_N(x) \neq F(x)]^2 \, \mathrm{d}H(x)$, were F_N is an empirical distribution function of N observations from F, and H is some non-decreasing function.

Let us mention two more tests against general alternatives $F_1(x)+F_2(x)$. The first of these was suggested by Lehmann (1951) and later studied by Sundrum (1954), and Wegner (1956). It is based on the number T of quadruples X_i, X_j, Y_k, Y_ℓ, $i < j$, $k < \ell$, such that either X_i, $X_j < Y_k$, Y_ℓ or X_i, $X_j > Y_k$, Y_ℓ. (See Problem 6.)

Rényi (1953b) proposed a similar test based on triplets only, but it is equivalent to the above test of Lehmann. (See Problem 7.)

The second test is the well-known run test studied by Wald and Wolfowitz (1940). It is based on the number of runs, i.e. the number of maximal ν interrupted sequences of X's, or of Y's, in $X^{(1)}, X^{(2)}, \ldots, X^{(m+n)}$. It is consistent against general alternatives $F_1(x) \neq F_2(x)$ under very general conditions, but its asymptotic efficiency in common situations with normal observations is 0.

4.2 TWO-SAMPLE TESTS OF SCALE

The experimental situation under the null hypothesis H_0 is the same as described at the beginning of Section 1 for two-sample tests of location, and we shall continue to use the notation introduced there. However, the alternative hypothesis is now the difference in scale of the two samples. It may be expressed by

$$f_1(x) = \frac{1}{\tau} f\Big(\frac{x-\mu}{\tau}\Big), \quad f_2(x) = f(x-\mu),$$

where f is some density, μ is a nuisance parameter, and τ is the parameter to be tested. In other words, we require both densities to be of the same type and to have the same location parameter μ, the null hypothesis being $\tau = 1$, the two-sided alternative being $\tau \neq 1$ (both densities differ only in their variance), the one-sided alternatives being $\tau > 1$ (the first density has a larger variance than the second one) and $\tau < 1$ (the first density has a smaller variance).

The assumption of equal location of both densities is essential; without it no rank test for testing difference of scale can behave in a satisfactory manner, as has been pointed out by Moses (1963). (See Problem 16.)

4.2.1 Linear rank tests.

The Capon (normal scores) test is a certain analogue of the Fisher-Yates-Terry-Hoeffding test. It uses the statistic

$$S = \sum_{i=1}^{m} a_{1,m+n}(R_i), \tag{1}$$

where the scores are defined by

$$a_{1,m+n}(i) = \mathsf{E}[V_{m+n}^{(i)}]^2 \tag{2}$$

with $V_{m+n}^{(1)} < \ldots < V_{m+n}^{(m+n)}$ being an ordered sample of size $m+n$ from the standardized normal distribution again. It can be seen from Theorem 3.4.5.1, formula (3.4.3.4), and Table 1 of 2.2.2 that the test with critical region $\{S \geq c_u\}$ is the locally most powerful rank test for H_0 against $\tau > 1$ whenever f is a density of normal type. The Capon test is also asymptotically optimum for such a density.

The expectation and variance of S are $\mathsf{E}S = m$ and

$$\operatorname{var} S = \frac{mn}{(m+n)(m+n-1)} \sum_{i=1}^{m+n} [\mathsf{E}(V_{m+n}^{(i)})^2]^2 - \frac{mn}{m+n-1}. \tag{3}$$

This test was obtained by Capon (1961) from optimality considerations for the scale of normal densities.

The Klotz test is a similar analogue of the van der Waerden test. Based on approximate scores, it employs the test statistic

$$S = \sum_{i=1}^{m} \Big[\Phi^{-1}\Big(\frac{R_i}{m+n+1}\Big)\Big]^2. \tag{4}$$

Again, both statistics (1) and (4) are asymptotically equivalent, and (4) yields an asymptotically optimum test for a density f of normal type. The expectation and variance of S are

$$\mathsf{E}S = \frac{m}{m+n} \sum_{i=1}^{m+n} \Big[\Phi^{-1}\Big(\frac{i}{m+n+1}\Big)\Big]^2, \tag{5}$$

$$\operatorname{var} S = \frac{mn}{(m+n)(m+n-1)} \sum_{i=1}^{m+n} \Big[\Phi^{-1}\Big(\frac{i}{m+n+1}\Big)\Big]^4 - \frac{n}{m(m+n-1)}(\mathsf{E}S)^2. \tag{6}$$

This test is due to Klotz (1962).

The Ansari-Bradley test. From the intuitive point of view, the following test appears to be an analogue of the Wilcoxon test. Take the pooled ordered sample $X^{(1)} < X^{(2)} < \ldots < X^{(m+n)}$, and assign the 'ranks' (in a sense used only here) $1, 2, \ldots, [\frac{1}{2}(m+n+1)]$ to these observations starting from the smallest $X^{(1)}$ towards the median, and assigning the same 'ranks' in the opposite direction from the largest $X^{(m+n)}$ towards the median. Now, the test statistic S is the sum of those 'ranks' corresponding to the observations from the first sample. More formally, using the usual ranks R_i,

$$S = \sum_{i=1}^{m} [\tfrac{1}{2}(m+n+1) - |R_i - \tfrac{1}{2}(m+n+1)|]. \tag{7}$$

Looking at Table 1 in 2.2.2, we find that S employs the approximate scores for the scale problem associated with the density

$$f(x) = \tfrac{1}{2}(1+|x|)^{-2}. \tag{8}$$

The present test is asymptotically optimum for this problem. It is found that

$$\mathsf{E}S = \begin{cases} \frac{1}{4}m(m+n+2) & \text{for } m+n \text{ even}, \\ \dfrac{m(m+n+1)^2}{4(m+n)} & \text{for } m+n \text{ odd}, \end{cases} \tag{9}$$

$$\operatorname{var} S = \begin{cases} \dfrac{mn(m+n-2)(m+n+2)}{48(m+n-1)} & \text{for } m+n \text{ even}, \\ \dfrac{mn(m+n+1)[(m+n)^2+3]}{48(m+n)^2} & \text{for } m+n \text{ odd}. \end{cases} \tag{10}$$

The present test was studied by Ansari and Bradley (1960), and further by Moses (1963).

The quartile test is an intuitive analogue of the median test. It is based on the statistic

$$S = \frac{1}{2}\sum_{i=1}^{m}\left[\operatorname{sign}\left(\left|R_i - \frac{m+n+1}{2}\right| - \frac{m+n+1}{4}\right) + 1\right]. \tag{11}$$

In fact, the analogy and the name stems from the fact that S is approximately equal to the number of observations from the first sample which lie outside the first and the third quartiles of the pooled sample. Precisely, S is obtained by counting the observations X_i, $i = 1, \ldots, m$, for which $R_i < \frac{1}{4}(m+n+1)$ or $R_i > \frac{3}{4}(m+n+1)$, and, if $m+n+1$ is divisible by 4, by adding $\frac{1}{2}$ whenever $R_i = \frac{1}{4}(m+n+1)$ or $R_i = \frac{3}{4}(m+n+1)$ for some $i = 1, \ldots, m$, or by adding 1 whenever both the last mentioned equalities occur for two different indices i.

Observe that S is derived from the approximate scores for the scale problem with the underlying density

$$f(x) = \begin{cases} 1 & \text{for } |x| \leq \frac{1}{4}, \\ 1/(16x^2) & \text{for } |x| > \frac{1}{4}, \end{cases} \tag{12}$$

and the quartile test is an asymptotically optimum test for this situation.

The first two moments of the quartile test statistic are

$$\mathsf{E}S = \begin{cases} \dfrac{m(m+n-1)}{2(m+n)} & \text{for } m+n = 4k-1, \\ \dfrac{2km}{m+n} & \text{for } m+n = 4k \text{ or } 4k+1 \text{ or } 4k+2, \end{cases} \tag{13}$$

$$\operatorname{var} S = \begin{cases} \dfrac{mn[(m+n)^2 - 2(m+n) - 1]}{2(m+n)^2(m+n-1)} & \text{for } m+n = 4k-1, \\ \dfrac{2kmn(m+n-2k)}{(m+n)^2(m+n-1)} & \text{for } m+n = 4k \\ & \text{or } 4k+1 \text{ or } 2k+2. \end{cases} \tag{14}$$

An analogue of our quartile test, based simply on the number of X_i's lying outside the first and the third quartiles of the pooled sample, is mentioned e.g. by Westenberg (1948).

The Savage test. So far all tests listed in this section are proper for densities which are positive on the whole real line $-\infty < x < \infty$. On the other hand, the Savage test having the statistic

$$S = \sum_{i=1}^{m} \sum_{j=m+n-R_i+1}^{m+n} 1/j \tag{15}$$

is proper for densities positive on the half-line $0 < x < \infty$ only. In particular, it can be proved that the test with critical region $\{S \geq c_u\}$ is the locally most powerful rank test for H_0 against $\tau > 1$ if the underlying density is of exponential type, that is if

$$f(x) = e^{-x} \quad \text{for } 0 < x < \infty. \tag{16}$$

The Savage test is also asymptotically optimum for this situation.

Using the transformation $x = e^z$ for the exponential density (16) we arrive at the density

$$g(z) = \exp[z - e^z] \quad \text{for } -\infty < z < \infty. \tag{17}$$

Now, scale changes in x are reflected as location changes in z. Thus, the Savage test also possesses analogous optimality properties for testing H_0 against a difference in location of two samples, if the underlying density is (17).

The expectation of the Savage test statistic is $\mathsf{E}S = m$, and its variance is

$$\mathsf{var}\, S = \frac{mn}{m+n-1}\Big(1 - \frac{1}{m+n}\sum_{j=1}^{m+n}\frac{1}{j}\Big). \tag{18}$$

This test is essentially due to I.R. Savage (1956). More precisely, denoting by $R_{m+1}, \ldots, R_{m+n}$ the ranks of the observations in the second sample, Savage obtained the test with the critical region $\{\sum_{i=1}^{n}\sum_{j=R_{m+i}}^{m+n} 1/j \leq c_\ell\}$ as a locally optimal one against Lehmann's alternatives $F_1(x) = [F(x)]^{\Delta_1}$, $F_2(x) = [F(x)]^{\Delta_2}$, $\Delta_2 > \Delta_1 > 0$.

Common properties. If S is any test statistic of the present subsection, the limiting distribution of the standardized variable $(S - \mathsf{E}S)(\mathsf{var}\, S)^{-\frac{1}{2}}$ for $\min(m, n) \to \infty$ is under H_0 the standardized normal distribution.

When we are testing H_0 against the one-sided alternative $\tau > 1$, i.e. the alternative that the first density is more dispersed than the second one, we use a one-sided critical region $\{S \geq c_u\}$ for the statistics (1), (4) (11) and (15), but the region $\{S \leq c_\ell\}$ for the Ansari-Bradley test statistic (7). We use the critical regions $\{S \geq c_u \text{ or } S \leq c_\ell\}$ when testing H_0 against the two-sided alternative $\tau \neq 1$.

Remarks. For practical purposes the following method of converting tests of location into tests of scale may be convenient. It has been employed e.g. by Siegel and Tukey (1960) in the case of the Wilcoxon test. Let us rearrange the pooled ordered sample $X^{(1)}, X^{(2)}, \ldots, X^{(N)}$ into a new sequence

$$\begin{aligned} &X^{(1)}, X^{(N)}, X^{(N-1)}, X^{(2)}, \\ &\qquad X^{(3)}, X^{(N-2)}, X^{(N-3)}, X^{(4)}, X^{(5)}, \ldots, \end{aligned} \tag{19}$$

and denote this sequence by $\tilde{X}^{(1)}, \tilde{X}^{(2)}, \tilde{X}^{(3)}, \ldots$. More formally, we put

$$\tilde{X}^{(1)} = X^{(1)};\ \tilde{X}^{(4j-2)} = X^{(N+2-2j)};\ \tilde{X}^{(4j-1)} = X^{(N+1-2j)}; \qquad (20)$$
$$\tilde{X}^{(4j)} = X^{(2j)}; \tilde{X}^{(4j+1)} = X^{(2j+1)} \text{ for } j = 1, 2, \ldots.$$

Thus, clearly, if for example the first sample is more dispersed than the second one, it must be shifted to the left in the sequence (19) relative to the second sample. Now, it is possible to employ formally any of the location tests for this newly ordered sequence (19). In more detail, let us assign successively the ranks $1, 2, \ldots, N$ to the observations ordered as in (19). Denoting this 'modified rank' of X_i by $\tilde{R}_i$, we have obviously

$$\tilde{R}_i = 1 \Longleftrightarrow R_i = 1;\ \tilde{R}_i = 4j - 2 \Longleftrightarrow R_i = N + 2 - 2j; \qquad (21)$$
$$\tilde{R}_i = 4j - 1 \Longleftrightarrow R_i = N + 1 - 2j;\ \tilde{R}_i = 4j \Longleftrightarrow R_i = 2j;$$
$$\tilde{R}_i = 4j + 1 \Longleftrightarrow R_i = 2j + 1 \text{ for } j = 1, 2, \ldots.$$

Tests for scale differences of two samples can now be derived by replacing R_i by $\tilde{R}_i$ in formulas for the location test statistics. This procedure has the practical advantage that such a scale test statistic is governed, under H_0, by the same distribution as the corresponding location test statistic, so that the same tables of critical values may be used. This assertion follows immediately from (21), which shows there is a one-to-one correspondence between the $\tilde{R}_i$'s and R_i's.

Let us mention a few other tests for scale. Mood (1954) proposed the test statistic $\sum_{i=1}^{m}[R_i - \frac{1}{2}(m + n + 1)]^2$, which was later studied by Sukhatme (1957). Sukhatme (1957), (1958b) suggested two other tests, for the use of which, however, the medians of both densities must be supposed to be zero. His first test is based on the statistic $T = (mn)^{-1} \sum_{i=1}^{m} \sum_{j=1}^{n} \psi(X_i, Y_j)$ where $\psi(X, Y) = 1$ for $0 < X < Y$ or $Y < X < 0$, $\psi(X, Y) = 0$ otherwise; his second test employs a similar but more complicated statistic. Note that T is not a rank statistic for H_0 in our sense, but it is a rank statistic for H_1. (See Problem 22.) Sukhatme (1958a) also studied modifications of his statistics where the medians are unknown and are only estimated from the samples.

4.2.2 Tests based on exceeding observations. We shall now again use the quantities A, B, A', B' defined in Subsection 4.1.2.

The Kamat test is an intuitive scale analogue of the Haga test. It is based on the statistic

$$T = A + B' - A' - B. \qquad (1)$$

The E-test. Similarly, as an analogue of the E-test statistic (4.1.2.2), we can take here

$$E = \min(A, B') - \min(A', B). \tag{2}$$

For both these tests, critical regions against the one-sided alternative $\tau > 1$ are $\{T \geq c_u\}$, or $\{E \geq c_u\}$.

Kamat (1956) originally introduced the test statistic $A + B' - A' - B + n$, which is always non-negative.

4.2.3 Tests of Kolmogorov-Smirnov types. *The test of the Kolmogorov-Smirnov type.* Starting from an idea similar to that used in Subsection 4.1.3 we can construct the statistics

$$K^+ = \left(\frac{m+n}{mn}\right)^{\frac{1}{2}} \max_{1\leq k\leq[\frac{1}{2}(m+n)]} \left(2k\frac{m}{m+n} - c_{D_1} - c_{D_{m+n}} - \ldots - c_{D_k} - c_{D_{m+n-k+1}}\right) \tag{1}$$

and

$$K^{\pm} = \left(\frac{m+n}{mn}\right)^{\frac{1}{2}} \max_{1\leq k\leq[\frac{1}{2}(m+n)]} \left|2k\frac{m}{m+n} - c_{D_1} - c_{D_{m+n}} - \ldots - c_{D_k} - c_{D_{m+n-k+1}}\right| \tag{2}$$

where the meaning of the symbols c_{D_i} is the same as in Subsection 4.1.3. In contrast to the Kolmogorov-Smirnov location statistics, in the present formulas we successively subtract simultaneously pairs of regression constants corresponding to the k-th smallest and the k-th largest observations. Clearly, these statistics will be sensitive to the difference in scale of two densities with the same location.

In the testing problem described at the beginning of this section we make use of the critical region $\{K^+ \geq c\}$ against the alternative $\tau < 1$ that the first density has a smaller variance, and of $\{K^{\pm} \geq c\}$ against the two-sided alternative $\tau \neq 1$.

Of course, it would be possible to introduce also similar modifications of the Rényi statistics and the Cramér-von Mises statistic, but we hope the whole matter is already sufficiently clear.

All statistics given here have the same asymptotic distributions as their counterparts in Subsection 4.1.3.

4.3 REGRESSION

In this section we shall deal with the problem of regression in location, which appears to be a generalization of the problem of the location difference of two samples. Thus our exposition may be brief, being to a high degree parallel to that of Section 4.1 on two-sample tests of location.

Take a random sample of independent observations $X_1, \ldots, X_N$, where X_i has density f_i $(i = 1, \ldots, N)$. The null hypothesis H_0 is $f_1 = f_2 = \ldots = f_N$, where the common density is otherwise arbitrary. The alternatives of interest will be $f_i(x) = f(x - \Delta c_i)$ for $i = 1, \ldots, N$ with $\Delta \neq 0$ (the two-sided alternative) or $\Delta > 0$ (the one-sided alternative). Here the c_i's are some known numbers, called regression constants. The case of two samples is obviously obtained on setting $c_i = 1$ for $i = 1, \ldots, m$, and $c_i = 0$ for $i = m+1, \ldots, N$. As usual, $R_1, \ldots, R_N$ will denote the ranks of $X_1, \ldots, X_N$, respectively.

The problem of regression in scale will not be dealt with here for the sake of saving space. Note only that corresponding locally most powerful rank tests for this problem can be generated by using Theorem 3.4.7.1.

Let us also mention the problem of trend alternatives, which are defined as regression alternatives either simply with $c_i = i$, $i = 1, \ldots, N$, or, more generally, with $c_1 < c_2 < \ldots < c_N$ but otherwise unknown c_i's. All tests of the present section may be used for testing against trend on putting e.g. $c_i = i$. The Spearman and Kendall correlation coefficients are also frequently applied. There are still several other tests and numerous papers concerning the problem of trend, but we shall go into no further details.

4.3.1 Linear rank tests. *The test of the Fisher-Yates-Terry-Hoeffding (normal scores) type* is based on the statistic

$$S = \sum_{i=1}^{N} c_i a_N(R_i) \tag{1}$$

where the scores $a_N(i)$ are equal to $a_{m+n}(i)$ of formula (4.1.1.1). By Theorem 3.4.6.1 we conclude that the test with critical region $\{S \geq c_u\}$ is the locally most powerful rank test for H_0 against the alternative $\Delta > 0$ provided that f is a density of normal type. This test is also asymptotically optimum for such a density if certain conditions concerning the limiting behaviour of regression constants are satisfied.

This test was studied by Terry (1952) on the basis of Hoeffding's (1950) paper.

The test of the van der Waerden type uses similarly the statistic

$$S = \sum_{i=1}^{N} c_i \Phi^{-1}\Big(\frac{R_i}{N+1}\Big). \tag{2}$$

As far as the property of being asymptotically optimum is concerned, the remark on the preceding test holds also for the present test.

The test of the Wilcoxon type employs a simple statistic

$$S = \sum_{i=1}^{N} c_i R_i. \tag{3}$$

The test having the critical region $\{S \geq c_u\}$ is the locally most powerful rank test, and also the asymptotically optimum test for H_0 against $\Delta > 0$, if the underlying density f is of the logistic type.

The test of the Wilcoxon type for the problem of trend with $c_i = i$, $i = 1, \ldots, N$, coincides essentially with the Spearman statistic (4.6.1.10) for testing independence.

The test of the median type. This regression test is based on the statistic

$$S = \sum_{i=1}^{N} c_i \operatorname{sign}\left(R_i - \tfrac{1}{2}(N+1)\right) \tag{4}$$

which is a generalization of the median test statistic for testing location. The test is asymptotically optimum for densities of the double-exponential type.

Common properties. Expectations and variances of the statistics mentioned here can be simply found on applying Theorem 3.3.1.3, and so will not be given here explicitly. It will be proved in 6.1.5 that under certain assumptions on regression constants, and if H_0 is true, all the statistics S of the present subsection are asymptotically normally distributed.

When testing against $\Delta > 0$ a proper critical region is one-sided, namely $\{S \geq c_u\}$, whereas against $\Delta \neq 0$ a proper critical region is $\{S \geq c_u$ or $S \leq c_\ell\}$.

4.3.2 The E-test. Let us say that the vector of antiranks $d = (d_1, \ldots, d_N)$ is more significant than $d^* = (d_1^*, \ldots, d_N^*)$ if, for some k,

$$\begin{aligned}
&\min(c_{d_N}, \ldots, c_{d_{N-j+1}}) - \max(c_{d_1}, \ldots, c_{d_j}) \\
&\quad = \min(c_{d_N^*}, \ldots, c_{d_{N-j+1}^*}) - \max(c_{d_1^*}, \ldots, c_{d_j^*}), \quad 1 \leq j \leq k-1, \\
&\min(c_{d_N}, \ldots, c_{d_{N-k+1}}) - \max(c_{d_1}, \ldots, c_{d_k}) \\
&\quad > \min(c_{d_N^*}, \ldots, c_{d_{N-k+1}^*}) - \max(c_{d_1^*}, \ldots, c_{d_k^*}).
\end{aligned}$$

Then the E-test may be defined as follows. Each of its critical regions (for varying α) consists of some vector of antiranks d^* and of all vectors of antiranks d that are more significant than d^*. (We use here a different type of definition of a test, defining it by means of a nested family of its critical regions instead of defining it by a test statistic generating such a nested family.)

The mentioned critical regions are one-sided, and the test may be used against the alternatives $\Delta > 0$.

4.3.3 Tests of Kolmogorov-Smirnov types. *The test of the Kolmogorov-Smirnov type.* It is possible to introduce also the regression analogues of the two-sample Kolmogorov-Smirnov statistics of Subsection 4.1.3. The one-sided statistic of this kind is

$$K^+ = \frac{\max\limits_{1 \le k \le N} (k\bar{c} - c_{D_1} - \ldots - c_{D_k})}{\left[\sum\limits_{i=1}^{N} (c_i - \bar{c})^2\right]^{\frac{1}{2}}}; \tag{1}$$

the two-sided statistic is

$$K^\pm = \frac{\max\limits_{1 \le k \le N} |k\bar{c} - c_{D_1} - \ldots - c_{D_k}|}{\left[\sum\limits_{i=1}^{N} (c_i - \bar{c})^2\right]^{\frac{1}{2}}}. \tag{2}$$

Here, as usual, D_i is the antirank of the order statistic $X^{(i)}$, $i = 1, \ldots, N$ and $\bar{c} = N^{-1} \sum\limits_{i=1}^{N} c_i$.

When testing against one-sided alternatives $\Delta > 0$ one makes use of K^+, against $\Delta \neq 0$ of $K^\pm$. In each case large values of the statistic are significant.

Both these statistics for the regression problem have under H_0 the same asymptotic distributions as their two-sample counterparts; these distributions and relevant convergence theorems will be given in Section 6.3.7.

The test of the Rényi type for the regression problem is defined analogously by means of the statistics

$$R_a^+ = \frac{\max\limits_{aN \le k \le N} \left\{\frac{N}{k}(k\bar{c} - c_{D_1} - \ldots - c_{D_k})\right\}}{\left[\sum\limits_{i=1}^{N} (c_i - \bar{c})^2\right]^{\frac{1}{2}}} \tag{3}$$

and

$$R_a^\pm = \frac{\max\limits_{aN \le k \le N} \left\{\frac{N}{k}|k\bar{c} - c_{D_1} - \ldots - c_{D_k}|\right\}}{\left[\sum\limits_{i=1}^{N} (c_i - \bar{c})^2\right]^{\frac{1}{2}}} \tag{4}$$

where a is some arbitrary fixed number satisfying $0 < a < 1$.

Previous remarks concerning the test of the Kolmogorov-Smirnov type are valid here also. Asymptotic distributions for the statistics of the Rényi type are shown in Section 6.3.8.

The test of the Cramér-von Mises type. One more analogy: the generalization of the Cramér-von Mises two-sample statistic to the regression problem is

$$M = \frac{\sum_{k=1}^{N} (k\bar{c} - c_{D_1} - \ldots - c_{D_k})^2}{N \sum_{i=1}^{N} (c_i - \bar{c})^2}. \tag{5}$$

It is designed only against two-sided alternatives $\Delta \neq 0$, and the required convergence theorem will be presented in Section 6.3.9.

4.4 THREE OR MORE SAMPLES

Here we shall present some rank tests for testing H_0 against differences in location of k samples, $k > 2$. Let the first sample consist of the observations $X_1, \ldots, X_{n_1}$, all having the density f_1, the second sample of the observations $X_{n_1+1}, \ldots, X_{n_1+n_2}$, all having the density f_2, etc., generally let the j-th sample $(j = 1, 2, \ldots, k)$ consist of the observations $X_{n_1+\ldots+n_{j-1}+1}, \ldots, X_{n_1+\ldots+n_{j-1}+n_j}$, all having the density f_j. Thus the sizes of the samples are $n_1, n_2, \ldots, n_j, \ldots, n_k$ and we write $N = \sum_{j=1}^{k} n_j$. For simplicity, we shall use the symbol s_j $(j = 1, \ldots, k)$ for the set of indices corresponding to the j-th sample, that is $s_j = \{n_1 + \ldots + n_{j-1} + 1, n_1 + \ldots + n_{j-1} + 2, \ldots, n_1 + \ldots + n_{j-1} + n_j\}$. After forming the total pooled sample $X_1, \ldots, X_N$, we may in the usual manner find the ranks $R_1, \ldots, R_N$ of these observations. Then, e.g., if the ranks corresponding to the members of the j-th sample are R_i with $i \in s_j$, their sum will be denoted accordingly by $\sum_{i \in s_j} R_i$, etc. Under H_0 all densities are equal to same common but arbitrary density. The alternatives of differences in location, against which the subsequent tests are designed, may be expressed by $f_j(x) = f(x - \Delta_j)$, for $j = 1, \ldots, k$, where the parameters Δ_j are not all equal and unknown. All tests described in this section possess only one-sided critical regions; if the test statistic computed from experimental data is larger than the respective critical value, then H_0 is rejected in favour of the above-mentioned alternative.

4.4.1 Rank tests of χ^2-types. *The test of the Fisher-Yates-Terry-Hoeffding (normal scores) type* for the k-sample problem can be defined using the statistic

$$Q = \frac{N-1}{\sum_{i=1}^{N} a_N^2(i)} \sum_{j=1}^{k} \frac{1}{n_j} \Big[\sum_{i \in s_j} a_N(R_i) \Big]^2, \tag{1}$$

where the scores $a_N(i)$ coincide with $a_{m+n}(i)$ of formula (4.1.1.1). Similarly, *the test of the van der Waerden type* can be based on

$$Q = \frac{N-1}{\sum_{i=1}^{N} \left[\Phi^{-1}\left(\frac{i}{N+1}\right)\right]^2} \sum_{j=1}^{k} \frac{1}{n_j} \Big[\sum_{i \in s_j} \Phi^{-1}\Big(\frac{R_i}{N+1}\Big)\Big]^2. \tag{2}$$

Both these tests are asymptotically optimum for the k-sample problem with the underlying density f of normal type. However, to the best of our knowledge, these tests have not yet been treated in the literature. On the contrary,

the Kruskal-Wallis test is a very well known test for this problem. It employs the statistic

$$\begin{aligned} Q &= \frac{12}{N(N+1)} \sum_{j=1}^{k} \frac{1}{n_j} \Big[\sum_{i \in s_j} R_i - n_j \frac{N+1}{2} \Big]^2 \qquad (3) \\ &= \frac{12}{N(N+1)} \sum_{j=1}^{k} \frac{1}{n_j} \Big[\sum_{i \in s_j} R_i \Big]^2 - 3(N+1), \end{aligned}$$

and it is asymptotically optimum for the density f of logistic type. Clearly, in the two-sample case this test is equivalent to the two-sided Wilcoxon test.

The Kruskal-Wallis test was introduced by Kruskal (1952) and Kruskal and Wallis (1952). Tables of critical values for it may be found in Sen and Krishnaiah (1984), pp. 945–948.

The median test. Denote now

$$A_j = \sum_{i \in s_j} \tfrac{1}{2}[\text{sign}\left(R_i - \tfrac{1}{2}(N+1)\right) + 1] \quad \text{for } j = 1, \ldots, k. \tag{4}$$

That is, we obtain A_j as the number of observations from the j-th sample which are larger than the median of the pooled sample, adding $\frac{1}{2}$ to this number if and only if N is odd and if this median belongs to the j-th sample. Then we can introduce the median test statistic

$$Q = 4 \sum_{j=1}^{k} \frac{1}{n_j} [A_j - \tfrac{1}{2} n_j]^2 = 4 \sum_{j=1}^{k} \frac{1}{n_j} A_j^2 - N. \tag{5}$$

The median test is asymptotically optimum for the underlying density f of double-exponential type.

A similar median test where $\frac{1}{2}$ is not added has been described by Mood (1950), Brown and Mood (1950).

Common properties. It can be shown on the basis of Theorem 6.2.2.1 that all statistics Q of the present subsection have, under H_0, asymptotically a χ^2 distribution with $k-1$ degrees of freedom whenever $\min(n_1, \ldots, n_k) \to \infty$.

The form of all these statistics has been inspired by the classical χ^2 statistic. The general form of such a rank statistic is

$$Q = \frac{N-1}{\sum_{i=1}^{N}[a_N(i) - \bar{a}_N]^2} \sum_{j=1}^{k} \frac{1}{n_j}[S_j - \mathsf{E}S_j]^2 \tag{6}$$

$$= \frac{N-1}{\sum_{i=1}^{N}[a_N(i) - \bar{a}_N]^2} \Big\{ \sum_{j=1}^{k} \frac{1}{n_j} S_j^2 - N\bar{a}_N^2 \Big\},$$

where

$$S_j = \sum_{i \in s_j} a_N(R_i), \tag{7}$$

and $a_N(i)$ are some given scores.

4.4.2 Other rank tests and some general ideas. In this subsection we shall mention three ideas for constructing k-sample test statistics from simpler statistics. In particular, these ideas will be illustrated by statistics of the Kolmogorov-Smirnov type.

For this purpose, denote in the present subsection the j-th sample $\{X_i,\ i \in s_j\}$ by the symbol $X_{(j)}$, its empirical distribution function by F_{j,n_j}, and the vector of corresponding ranks (i.e. ranks of observations in $X_{(j)}$ considered within the total sample $X_1, \ldots, X_N$) by $R_{(j)}$.

First, as we have seen, it may be useful to take the linear rank statistics S_j in (4.4.1.7) and form a χ^2-like statistic Q as in (4.4.1.6). More generally, we may start from any function $S = S(R_{i_1}, \ldots, R_{i_n})$ appropriately defined for an arbitrary vector of ranks $(R_{i_1}, \ldots, R_{i_n})$, $\{i_1, \ldots, i_n\} \subset \{1, \ldots, N\}$, and take for the test statistic some convenient weighted sum of squares (or perhaps a quadratic form) of k quantities $S(R_{(j)}) - \mathsf{E}S(R_{(j)})$, $j = 1, \ldots, k$, as in (4.4.1.6), provided such a statistic is sensitive to the alternatives in question. A closely related idea applied to the Kolmogorov-Smirnov test gives rise to the test statistic

$$\max_{-\infty < x < \infty} \sum_{j=1}^{k} n_j \Big[F_{j,n_j}(x) - \frac{1}{N} \sum_{j=1}^{k} n_j F_{j,n_j}(x) \Big]^2. \tag{1}$$

Still more closely, (1) can also be written as

$$\max_{t=1,\ldots,N} \sum_{j=1}^{k} \frac{1}{n_j} \Big\{ \sum_{i \in s_j} \tfrac{1}{2}[\text{sign}^*(t - R_i) + 1] - t\frac{n_j}{N} \Big\}^2, \tag{2}$$

where

$$\text{sign}^*(x) = \begin{cases} 1 & \text{for } x \geq 0, \\ -1 & \text{for } x < 0. \end{cases} \tag{3}$$

This shows that (1), or (2), equals (up to multiplicative constants) the maximum of certain statistics (4.4.1.6), and also shows its relation to the median test. (See also the remark to formula (4.1.3.4).)

In principle a similar statistic as (1), but in which the basic idea is inspired by the Cramér-von Mises statistic, was also investigated.

All these statistics have been studied by Gihman (1957), Fisz (1960), Kiefer (1959). A concise survey with tables was given by Sen and Krishnaiah (1984), pp. 954–957.

The second general idea starts from employing some two-sample statistic $T = T(Y, Z)$ which is a function of samples $Y = (Y_1, \ldots, Y_m)$ and $Z = (Z_1, \ldots, Z_n)$. Then such a function generates a k-sample statistic

$$\max_{i,j;i \neq j} T(X_{(i)}, X_{(j)}). \tag{4}$$

In particular, such a statistic of the Kolmogorov-Smirnov type is expressed by

$$\max_{i,j;i \neq j} \max_{-\infty < x < \infty} |F_{i,n_i}(x) - F_{j,n_j}(x)|. \tag{5}$$

Further information concerning this statistic may be found in David (1958) for $k = 3$, Fisz (1960), and Birnbaum and Hall (1960).

For a simple statement of the third general idea let the symbol $X_{(1)} \cup \ldots \cup X_{(j)}$ denote the pooled sample consisting of the samples $X_{(1)}, \ldots, X_{(j)}$ that is $X_{(1)} \cup \ldots \cup X_{(j)} = (X_1, \ldots, X_{n_1+\ldots+n_j})$. Using a two-sample statistic T as before we can perform the k-sample test by means of $k - 1$ statistics

$$T(X_{(2)}, X_{(1)});\ T(X_{(3)}, X_{(1)} \cup X_{(2)}); \ldots; \\ T(X_{(k)}, X_{(1)} \cup \ldots \cup X_{(k-1)}), \tag{6}$$

or simply by means of their maximum

$$\max_{j=2,\ldots,k} T(X_{(j)}, X_{(1)} \cup \ldots \cup X_{(j-1)}). \tag{7}$$

For the Kolmogorov-Smirnov test we obtain

$$\max_{j=2,\ldots,k} \max_{-\infty<x<\infty} \left| F_{j,n_j}(x) - \frac{1}{\sum_{i=1}^{j-1} n_i} \sum_{i=1}^{j-1} n_i F_{i,n_i}(x) \right|. \tag{8}$$

If the original two-sample statistic $T(Y, Z)$ is a function only of the *set* of ranks of $Y_1, \ldots, Y_m$ within the pooled sample $Y_1, \ldots, Y_m, Z_1, \ldots, Z_n$ (i.e. it is a rank statistic which does not change under an arbitrary permutation of indices in $Y_1, \ldots, Y_m$) then all $k-1$ random variables in (6) are under H_0 independent (see Subsection 5.1.3). This yields a very simple method of deriving k-sample tests from the two-sample ones. Observe that the linear rank statistics $\sum_{i=1}^{m} a_{m+n}(R_i)$ of Theorem 3.4.4.1 and of Subsection 4.1.1, as well as the usual Kolmogorov-Smirnov statistics of Subsection 4.1.3, are of this type, so that, in particular, (8) is the maximum of $k-1$ independent random variables.

Statistics like (8) have been treated by Chang and Fisz (1957), Fisz (1957), (1960), and Dwass (1960).

Thus, we can conclude that an arbitrary two-sample test statistic may be used to construct a k-sample test by utilizing either of the two latter general ideas. Note only, in addition, that formulas like (4), (5), (7) and (8) are sometimes introduced with certain weights for the individual two-sample statistics.

For survey papers concerning the ideas described see Fisz (1960) and Dwass (1960).

We could also mention a specific area of statistics, called selecting and ordering populations. A typical problem in this area is to select the 'best' from among k populations, where the word 'best' may mean that population with the largest expectation, with the smallest variance, and similarly. There are many ramifications of the problem, for details cf. the books by Gibbons et al. (1977), Gupta and Panchapakesan (1979), Gupta and Huang (1981), and Büringer et al. (1980). Of course, there are also some relations to rank tests. A pioneering paper trying to select the population with the largest expectation by means of rank procedures is Lehmann (1963). Unfortunately, the paper is wrong because he determined erroneously the so-called 'least favourable configuration' (the same error is reproduced in Bartlett and Govindarajulu (1968)), cf. Rizvi and Woodworth (1970).

A special problem of this sort is the so-called 'slippage problem', where we know in advance that exactly one density has a larger expectation, and naturally we wish to find it, cf. Mosteller (1948) (cf. also Problem 5.7), Doornbos and Prins (1958). For an analogous problem concerning variances see Sadowski (1960/61).

Rank tests for the comparison of several populations with one control were introduced by Steel (1959a), (1959b) and Lukaszewicz and Sadowski (1960/61).

A problem related to this area is to compare all $\frac{1}{2}k(k-1)$ pairs of densities by the Wilcoxon test, cf. Steel (1960), (1961), and Dunn (1964).

4.5 TESTS OF SYMMETRY

Consider independent random variables $X_1, \dots, X_N$, where X_i has density f_i. The null hypothesis H_1 is that all the variables have the same density f, symmetric about zero (i.e. $f(x) = f(-x)$ for all x), which otherwise may be arbitrary. We shall restrict ourselves to the alternatives called 'shift in location'. They are expressed by the formula $f_i(x) = f(x - \Delta)$ for all i, f symmetric about zero, $\Delta \neq 0$, or in words, by the statement that the 'original' density, symmetric about 0, is shifted by Δ (e.g. for $\Delta > 0$ the shift is to the right).

4.5.1 Linear rank tests. Since we are dealing with the hypothesis H_1, we must make use of the ranks R_i^+. Recall their definition. Taking the absolute values of the observations $|X_1|, \dots, |X_N|$, and rearranging them in increasing order $|X|^{(1)} < |X|^{(2)} < \dots < |X|^{(N)}$, the rank R_i^+ ($i = 1, \dots, N$) is equal to the usual rank of $|X_i|$ in this increasing sequence $|X|^{(1)}, \dots, |X|^{(N)}$.

In order to find locally optimal rank tests for the present problem, let us recall Theorem 3.4.9.1. Consider the alternatives of shift in location with the underlying density f of some specific type, and try to determine the relevant scores $a_N^+(i, f)$.

Note that

$$\sum_{X_i > 0} a_N^+(R_i^+, f) = \frac{1}{2}\Big[\sum_{i=1}^{N} a_N^+(R_i^+, f) \operatorname{sign} X_i + \sum_{i=1}^{N} a_N^+(i, f)\Big] \qquad (1)$$

with probability 1, so that we may use instead of the statistics of Theorem 3.4.9.1 the simpler statistics (1) which are equivalent. We might recall that the statistics in the middle of (1) are often called *signed rank statistics.*

The Fraser (normal scores) test. If f is the standardized normal density, we write simply $a_N^+(i, f) = a_N^+(i)$, and we obtain from (3.4.3.6) and Table 1 of 2.2.2 that $a_N^+(i) = \mathsf{E}\Phi^{-1}(\frac{1}{2} + \frac{1}{2}U_N^{(i)})$, with Φ being the standardized normal distribution function and $U_N^{(1)} < \dots < U_N^{(N)}$ being the ordered sample from the uniform distribution on $[0, 1]$. However, if $V_1, \dots, V_N$ is a random sample from the standardized normal distribution, i.e. each V_i has distribution function Φ, then each absolute value $|V_i|$ has distribution function $2\Phi - 1$. Consequently, upon ordering the absolute values $|V_1|, \dots, |V_N|$

into an increasing sequence $|V|^{(1)} < \ldots < |V|^{(N)}$, we see that the variables $2\Phi(|V|^{(i)}) - 1$, $i = 1, \ldots, N$, may be taken as $U_N^{(i)}$. Thus

$$a_N^+(i) = \mathsf{E}|V|^{(i)}, \tag{2}$$

and the test which we are looking for can be performed by means of the statistic

$$S^+ = \sum_{X_i > 0} a_N^+(R_i^+). \tag{3}$$

Theorem 3.4.9.1 asserts that the test with critical region $\{S^+ \geq c_u\}$ is the locally most powerful rank test for H_1 against $\Delta > 0$, whenever the underlying density f is of normal type. The Fraser test is also asymptotically optimum for such a density. The expectation and variance of S^+ under the hypothesis H_1 are

$$\mathsf{E}S^+ = N(2\pi)^{-\frac{1}{2}}, \tag{4}$$

$$\operatorname{var} S^+ = \frac{1}{4}\sum_{i=1}^{N}[\mathsf{E}|V|^{(i)}]^2. \tag{5}$$

The test based on the statistic (3) was derived by Fraser (1957b). Later, it was studied by Klotz (1963).

The test of the van der Waerden type is another asymptotically optimum test for densities of normal type. Its test statistic is

$$S^+ = \sum_{X_i > 0} \Phi^{-1}\Big(\frac{1}{2} + \frac{1}{2}\frac{R_i^+}{N+1}\Big), \tag{6}$$

which has under H_1 the expectation and variance

$$\mathsf{E}S^+ = \frac{1}{2}\sum_{i=1}^{N} \Phi^{-1}\Big(\frac{1}{2} + \frac{1}{2}\frac{i}{N+1}\Big), \tag{7}$$

$$\operatorname{var} S^+ = \frac{1}{4}\sum_{i=1}^{N} \Big[\Phi^{-1}\Big(\frac{1}{2} + \frac{1}{2}\frac{i}{N+1}\Big)\Big]^2. \tag{8}$$

This test was mentioned by van Eeden (1963).

The Wilcoxon one-sample test employs the simple statistic

$$S^+ = \sum_{X_i > 0} R_i^+. \tag{9}$$

By Theorem 3.4.9.1, the test with the critical region $\{S^+ \geq c_u\}$ is the locally most powerful rank test for H_1 against $\Delta > 0$ if the density f is of logistic type. This Wilcoxon test is also asymptotically optimum for such a density. It can be easily shown that

$$\mathsf{E}S^+ = \tfrac{1}{4}N(N+1), \tag{10}$$
$$\operatorname{var} S^+ = \tfrac{1}{24}N(N+1)(2N+1). \tag{11}$$

This test was introduced by Wilcoxon (1945). Some tables for it are quoted in Hájek and Šidák (1967); refer also to Sen and Krishnaiah (1984), pp. 938–939.

The simplest test is the old and famous

sign test based on the statistic

$$S^+ = \sum_{X_i > 0} 1, \tag{12}$$

or, in words, on S^+ equal to the number of positive values among $X_1, \ldots,$ X_N. The one-sided sign test is the locally most powerful rank test for H_1 against $\Delta > 0$ if the underlying density f is of double-exponential type. The sign test is also asymptotically optimum for these densities. Obviously, $\mathsf{E}S^+ = \frac{1}{2}N$, and $\operatorname{var} S^+ = \frac{1}{4}N$.

The sign test has been used for a very long time. More detailed information on it can be found e.g. in Dixon and Mood (1946), and in van der Waerden and Nievergelt (1956).

The sign test is often treated and used as a test of the wider hypothesis that the median of f is 0, without any additional assumption on the form of f (in such a case it is called a test of location rather than of symmetry).

Common properties. It will be shown in 6.1.7 that all statistics $(S^+ - \mathsf{E}S^+)\cdot(\operatorname{var} S^+)^{-\frac{1}{2}}$ have under the hypothesis H_1, for $N \to \infty$ asymptotically the standardized normal distribution.

If we wish to test H_1 against $\Delta > 0$, that is, against a shift to the right, we make use of one-sided critical regions $\{S^+ \geq c_u\}$. In testing against $\Delta \neq 0$ we make use of two-sided regions $\{S^+ \geq c_u \text{ or } S^+ \leq c_\ell\}$.

4.5.2 Kolmogorov-type tests. Recall that the hypothesis of symmetry of F around the origin asserts that $F(x) + F(-x-) - 1 = 0$, for all x, and whenever F is continuous, in the above characterization, $F(-x-)$ may also be replaced by $F(-x)$. Therefore, in the same spirit as in Subsection 4.1.3 and following the lead by Butler (1969) and Chatterjee and Sen (1973b), we consider some Kolmogorov-type tests that are based on the empirical distribution function.

We denote the empirical distribution function for $X_1, \ldots, X_N$ by $F_N(x)$, $-\infty < x < \infty$. Let then

$$K_N^+ = \sup_{-\infty<x<\infty} \{F_N(x) + F_N(-x-) - 1\}, \tag{1}$$

$$K_N^- = \sup_{-\infty<x<\infty} \{1 - F_N(x) - F_N(-x-)\}, \tag{2}$$

$$K_N = \sup_{-\infty<x<\infty} \{|F_N(x) + F_N(-x-) - 1|\}. \tag{3}$$

As in Subsection 4.1.3, these statistics are more in the spirit of goodness of fit, and are appropriate even if the alternative hypotheses are not of the shift-type (as considered in the preceding subsection). Let us denote, as before, the ordered values of $|X_1|, \ldots, |X_N|$ by $|X|^{(1)} < \ldots < |X|^{(N)}$; as F is assumed to be continuous, ties are neglected with probability one. Then note that by definition

$$X_i = |X|^{(R_i^+)} \operatorname{sign} X_i, \; i = 1, \ldots, N. \tag{4}$$

Let us denote

$$U_{Ni} = F_N(|X|^{(i)}) + F_N(-|X|^{(i)}-) - 1, \; i = 1, \ldots, N. \tag{5}$$

Then we have on letting $U_{N0} = 2F_N(0) - 1$,

$$K_N^+ = \max_{0\le i\le N} \{U_{Ni}\}; \qquad K_N^- = \max_{0\le i\le N} \{-U_{Ni}\}; \tag{6}$$

$$K_N = \max\{K_N^+, K_N^-\} = \max_{0\le i\le N} \{|U_{Ni}|\}. \tag{7}$$

Let us also denote $\bar{F}_N(x) = 1 - F_N(x)$ and $\bar{F}(x) = 1 - F(x)$. Then note that from the above, we have

$$\bar{F}_N(|X|^{(i)}) = \frac{1}{N}\sum_{j=1}^{N} I(R_j^+ > i)(1 + \operatorname{sign} X_j)/2, \tag{8}$$

$$F_N(-|X|^{(i)}-) = \frac{1}{N}\sum_{j=1}^{N} I(R_j^+ > i)(1 - \operatorname{sign} X_j)/2, \tag{9}$$

for every $i = 0, 1, \ldots, N$. Therefore, rewriting $U_{Ni} = F_N(-|X|^{(i)}-) -\bar{F}_N(|X|^{(i)})$, we obtain that

$$U_{Ni} = -\frac{1}{N}\sum_{j=1}^{N} \operatorname{sign} X_j I(R_j^+ > i), \; i = 0, 1, \ldots, N. \tag{10}$$

Let us recall that under the hypothesis of symmetry of F, the vectors $\operatorname{sign} X = (\operatorname{sign} X_1, \ldots, X_N)$ of signs and $R^+ = (R_1^+, \ldots, R_N^+)$ of absolute ranks are stochastically independent. Hence, under H_1, these Kolmogorov-type tests are all EDF. A simple martingale characterization that provides an easy access to the distribution theory of these statistics under the null hypothesis (even when the X_i are distributed symmetrically but not necessarily identically) was formulated by Chatterjee and Sen (1973b). For the particular case of a sample with identically distributed random observations, Butler's (1969) approach exploits a simple random walk model related to (10).

Remarks. At the end of this section let us remark in addition that one can also easily find suitable rank tests for testing H_1 against other alternatives. For example, for testing two samples with asymmetric basic density against difference in scale, the locally most powerful rank tests can be generated by means of Theorem 3.4.10.1.

Tests of symmetry are frequently applied to differences of paired random variables, for testing the effects of two treatments in random blocks — see the conclusion of Subsection 4.7.1.

4.6 TESTS OF INDEPENDENCE

Very old and very well known among rank tests are the so-called rank correlation coefficients for testing independence of two random variables. They will be described in this section. Let us observe N independent pairs of random variables $(X_1, Y_1), \ldots, (X_N, Y_N)$, and suppose generally that all pairs (X_i, Y_i), $i = 1, \ldots, N$, have the same two-dimensional density $h(x, y)$. We wish to test the null hypothesis H_2 that the two variables X_i and Y_i are independent for each i, that is $h(x, y) = f(x)g(y)$ with arbitrary one-dimensional densities f and g, against various alternatives, under which they are dependent. If h is a two-dimensional normal density with its correlation coefficient denoted by ρ, the problem may be described very simply. In fact, this density belongs to the hypothesis H_2 for $\rho = 0$, and to the alternatives for $\rho \neq 0$. The case of a general density h is more complicated, and there may be different general models of dependence. Later, in Subsection 1, we shall specify the kind of dependence against which the tests described possess certain optimality properties.

Much information on rank tests of independence can be found in Kendall (1948). For a survey paper see also Kruskal (1958).

4.6.1 Linear rank tests.

Consider separately the first coordinates $X_1, \ldots, X_N$ of the two-dimensional sample, and denote by R_i the rank of X_i among these observations reordered in an increasing sequence. Similarly, denote by Q_i the rank of Y_i among the separately taken observations $Y_1, \ldots, Y_N$. Rank statistics for testing H_2 are then functions of the R_i's and Q_i's.

Rank statistics which are invariant under permutations applied simultaneously to the two vectors of ranks $(R_1, \ldots, R_N)$ and $(Q_1, \ldots, Q_N)$, may be defined equivalently as follows. First, let us rearrange all N pairs of observations according to the magnitude of their first coordinate into the sequence $(X_{d_1}, Y_{d_1}), (X_{d_2}, Y_{d_2}), \ldots, (X_{d_N}, Y_{d_N})$ in such a way that $X_{d_1} < X_{d_2} < \ldots < X_{d_N}$. Then put R_i^0 equal to the rank of Y_{d_i} among the observations $Y_{d_1}, \ldots, Y_{d_N}$, where the X_i's are reordered increasingly. Now, rank statistics with the above-mentioned property are simply functions of R_i^0's. This amounts to considering some suitable permutation of indices in the sample of all N observations. Using the antiranks D_i, $i = 1, \ldots, N$, defined in 3.3.2 by $R_{D_i} = i$, we have obviously $R_i^0 = Q_{D_i}$.

We now specify the general model of dependence which we shall have in mind throughout the present subsection. Let

$$X_i = X_i^* + \Delta Z_i, \quad Y_i = Y_i^* + \Delta Z_i, \quad i = 1, \ldots, N, \tag{1}$$

where all the variables X_i^*, Y_i^*, Z_i are mutually independent and their distributions do not depend on i, and Δ is a real non-negative parameter. In this general model with arbitrary distributions of X_i^*, Y_i^*, Z_i, $\Delta = 0$ obviously expresses the hypothesis H_2 of independence of X_i and Y_i, whereas for $\Delta > 0$ the variables X_i and Y_i are dependent. Their correlation coefficient is

$$\frac{\Delta^2 \operatorname{var} Z_i}{[(\operatorname{var} X_i^* + \Delta^2 \operatorname{var} Z_i)(\operatorname{var} Y_i^* + \Delta^2 \operatorname{var} Z_i)]^{\frac{1}{2}}}, \tag{2}$$

provided all variances involved are finite. In order to find optimal tests, e.g. locally most powerful rank tests by means of Theorem 3.4.11.1, we shall specialize further the model for alternative hypotheses, requiring that X_i^*, and Y_i^* have densities of known types f, and g, respectively, and that Z_i have an arbitrary unknown distribution with $0 < \operatorname{var} Z_i < \infty$. If we assume negative correlation between X_i and Y_i under the alternatives, we put $X_i = X_i^* + \Delta Z_i$, $Y_i = Y_i^* - \Delta Z_i$.

Other general models of dependence were treated e.g. by Konijn (1956), Farlie (1961), and Janssen and Mason (1990). As for the last book, different scores were introduced in it, which, however, under some assumptions coincide with ours, cf. Subsection 3.4.11.

The Fisher-Yates (normal scores) test. For the case where the two-dimensional density h is normal, or, more generally, where both f and g are of normal type, we are led by Theorem 3.4.11.1 to the test statistic

$$S = \sum_{i=1}^{N} a_N(R_i) a_N(Q_i) = \sum_{i=1}^{N} a_N(i) a_N(R_i^0) \tag{3}$$

with

$$a_N(i) = \mathsf{E}V_N^{(i)}, \tag{4}$$

$V_N^{(1)} < V_N^{(2)} < \ldots < V_N^{(N)}$ being an ordered sample of N observations from the standardized normal distribution. The statistic

$$S / \sum_{i=1}^{N} [\mathsf{E}V_N^{(i)}]^2 \tag{5}$$

is called the Fisher-Yates correlation coefficient, since it shares some properties of the ordinary empirical correlation coefficient: its value lies between -1 and 1, and its expectation is 0 for X_i and Y_i independent.

In the case of f and g of normal type, the test with the critical region $\{S \geq c_u\}$ is the locally most powerful rank test for H_2 against the alternatives $\Delta > 0$.

Under H_2, we have $\mathsf{E}S = 0$, and

$$\operatorname{var} S = \frac{1}{N-1}\Big\{\sum_{i=1}^{N} [\mathsf{E}V_N^{(i)}]^2\Big\}^2. \tag{6}$$

The Fisher-Yates correlation coefficient was discussed by Fieller and Pearson (1961).

The test of the van der Waerden type. Upon employing the approximate scores in the foregoing problem with f and g normal, we obtain the statistic

$$\begin{aligned} S &= \sum_{i=1}^{N} \Phi^{-1}\Big(\frac{R_i}{N+1}\Big)\Phi^{-1}\Big(\frac{Q_i}{N+1}\Big) \\ &= \sum_{i=1}^{N} \Phi^{-1}\Big(\frac{i}{N+1}\Big)\Phi^{-1}\Big(\frac{R_i^0}{N+1}\Big), \end{aligned} \tag{7}$$

where Φ is the standardized normal distribution function. Since its scores are similar to those of the van der Waerden two-sample statistic (4.1.1.4), the quantity

$$S / \sum_{i=1}^{N} \Big[\Phi^{-1}\Big(\frac{i}{N+1}\Big)\Big]^2 \tag{8}$$

may be called the van der Waerden correlation coefficient. The present coefficient, and the Fisher-Yates correlation coefficient are, of course, asymptotically equivalent.

Under H_2, we have similarly $\mathsf{E}S = 0$, and

$$\operatorname{var} S = \frac{1}{N-1}\Big\{\sum_{i=1}^{N} \Big[\Phi^{-1}\Big(\frac{i}{N+1}\Big)\Big]^2\Big\}^2. \tag{9}$$

The Spearman correlation coefficient. Using for scores simply the consecutive integers $1, 2, \ldots, N$ we obtain the statistic

$$S = \sum_{i=1}^{N} R_i Q_i = \sum_{i=1}^{N} i R_i^0. \tag{10}$$

Obviously, for testing purposes this statistic is equivalent to the very well-known Spearman rank correlation coefficient, which is usually denoted by ρ and is given by

$$\begin{aligned} \rho &= \frac{12}{N^3 - N} \sum_{i=1}^{N} \Big(R_i - \frac{N+1}{2}\Big)\Big(Q_i - \frac{N+1}{2}\Big) \\ &= \frac{12}{N^3 - N} \sum_{i=1}^{N} \Big(i - \frac{N+1}{2}\Big)\Big(R_i^0 - \frac{N+1}{2}\Big). \end{aligned} \tag{11}$$

Another frequently used formula for ρ is

$$\rho = 1 - \frac{6}{N^3 - N} \sum_{i=1}^{N} (R_i - Q_i)^2 = 1 - \frac{6}{N^3 - N} \sum_{i=1}^{N} (i - R_i^0)^2. \tag{12}$$

From this last formula it is finally clear that for testing we may apply also a simple equivalent statistic

$$S' = \sum_{i=1}^{N} (R_i - Q_i)^2 = \sum_{i=1}^{N} (i - R_i^0)^2. \tag{13}$$

Theorem 3.4.11.1 shows that the test having the critical region $\{S \geq c_u\}$ (or $\{\rho \geq c_u\}$, or $\{S' \leq c_\ell\}$) is the locally most powerful rank test for testing H_2 against $\Delta > 0$ in the case when both f and g are of logistic type.

The expectation and variance of S under H_2 are

$$\mathsf{E}S = \tfrac{1}{4} N (N+1)^2, \tag{14}$$

$$\operatorname{var} S = \frac{N^2 (N+1)^2 (N-1)}{144}, \tag{15}$$

or equivalently for ρ, $\mathsf{E}\rho = 0$, $\operatorname{var}\rho = 1/(N-1)$.

Further information may be found in Kendall (1948). For quotations of some tables see Hájek and Šidák (1967); (moreover, tables are contained e.g. in Sen and Krishnaiah (1984). pp. 952–954.

The quadrant test. On utilizing approximate scores in the case of f and

g of double-exponential type for constructing a similar kind of a statistic we arrive at

$$\begin{aligned} S^* &= \sum_{i=1}^{N} \operatorname{sign}\left(R_i - \tfrac{1}{2}(N+1)\right) \operatorname{sign}\left(Q_i - \tfrac{1}{2}(N+1)\right) \qquad (16) \\ &= \sum_{i=1}^{N} \operatorname{sign}\left(i - \tfrac{1}{2}(N+1)\right) \operatorname{sign}\left(R_i^0 - \tfrac{1}{2}(N+1)\right). \end{aligned}$$

For the simplification of the following formula, let the median of $X_1, \ldots, X_N$ be denoted by $\operatorname{med} X$, the median of $Y_1, \ldots, Y_N$ by $\operatorname{med} Y$. Now it is easy to see that a suitable simpler statistic for testing, equivalent to (16), is

$$S = S_1 + S_2 + S_3 + S_4, \qquad (17)$$

where S_1 is the number of pairs (X_j, Y_j) for which $X_j > \operatorname{med} X$ and simultaneously $Y_j > \operatorname{med} Y$; $S_2 = \frac{1}{2}$ if $X_j = \operatorname{med} X$ and $Y_j > \operatorname{med} Y$ for some j, $S_2 = 0$ otherwise; $S_3 = \frac{1}{2}$ if $X_j > \operatorname{med} X$ and $Y_j = \operatorname{med} Y$ for some j, and $S_3 = 0$ otherwise; $S_4 = \frac{1}{4}$ if $X_j = \operatorname{med} X$ and $Y_j = \operatorname{med} Y$ for some j, and $S_4 = 0$ otherwise. Thus, for instance, for N even $S = S_1$. Since the test is essentially based on the number of observations in the quadrants into which the whole plane is divided by the lines $x = \operatorname{med} X$, $y = \operatorname{med} Y$, it may be called the quadrant test.

Under H_2, the expectation is $\mathsf{E}S = \frac{1}{4}N$, and the variance is

$$\operatorname{var} S = \begin{cases} \dfrac{1}{16}\dfrac{N^2}{N-1} & \text{for } N \text{ even,} \\ \dfrac{1}{16}(N-1) & \text{for } N \text{ odd.} \end{cases} \qquad (18)$$

Elandt (1957), (1958) has discussed a simpler form of the quadrant test based on $U = U_1 + U_2$, where U_1, U_2 are defined as follows. U_1 is the number of pairs (X_j, Y_j) such that $(X_j - \operatorname{med} X)(Y_j - \operatorname{med} Y) > 0$; denoting by a, b two indices such that $Y_a = \operatorname{med} Y$, $X_b = \operatorname{med} X$, we put $U_2 = 1$ whenever $(X_a - \operatorname{med} X)(Y_b - \operatorname{med} Y) > 0$, $U_2 = 0$ whenever $(X_a - \operatorname{med} X)(Y_b - \operatorname{med} Y) \leq 0$. This definition permits us to reduce the case $N = 2n+1$ to the case $N = 2n$ (see Problem 5.3). For an equivalent statistic see also Blomqvist (1950).

Common properties. Each of the statistics $(S - \mathsf{E}S)(\operatorname{var} S)^{-\frac{1}{2}}$ has under H_2 asymptotically for $N \to \infty$ the standardized normal distribution, as we shall see in 6.1.8.

For testing H_2 against the alternative that there is a positive correlation between the two variables X_i and Y_i, we use the one-sided critical regions $\{S \geq c_u\}$ (or $\{\rho \geq c_u\}$ or $\{S' \leq c_\ell\}$ for S' given by (13)). On the other hand, we can test against the alternative that there is a non-zero correlation between X_i and Y_i (though this more general correlation is not embraced by our model (1)) on using two-sided critical regions $\{S \geq c_u \text{ or } S \leq c_\ell\}$.

4.6.2 Other rank tests. *The Kendall rank correlation coefficient* is another very well known rank statistic for testing independence, which is, however, non-linear. It is usually denoted by τ and is expressed by the formula

$$\begin{aligned}\tau &= \frac{1}{N(N-1)} \sum\sum_{i \neq j} \operatorname{sign}(R_i - R_j) \operatorname{sign}(Q_i - Q_j) \qquad (1)\\ &= \frac{1}{N(N-1)} \sum\sum_{i \neq j} \operatorname{sign}(i - j) \operatorname{sign}(R_i^0 - R_j^0).\end{aligned}$$

If T is defined as the number of pairs of indices (i, j), $i < j$, for which either $X_i < X_j$, $Y_i < Y_j$, or $X_i > X_j$, $Y_i > Y_j$, (that is, for which $R_i^0 < R_j^0$) the following relation can be easily found:

$$\tau = \frac{4T}{N(N-1)} - 1. \qquad (2)$$

Thus, for testing, the statistics τ and T are equivalent, and, in fact, the simpler quantity T is often used rather than τ.

Note also that in Subsection 3.3.1 it was proved that by projecting τ into the space of linear rank statistics we obtain a multiple of the Spearman correlation coefficient ρ.

It can be shown that, under H_2, $\mathsf{E}\tau = 0$,

$$\operatorname{var} \tau = \frac{2(2N+5)}{9N(N-1)}, \qquad (3)$$

and the distribution of $\tau / (\operatorname{var} \tau)^{\frac{1}{2}}$ tends to the standardized normal distribution.

Critical regions for τ are similar to those for S, or ρ, in the preceding subsection.

The Kendall correlation coefficient was introduced by Kendall (1938); for much more information cf. Kendall (1948) including the partial rank correlation. References to pertaining tables may be found in Hájek and Šidák (1967), the tables themselves in Sen and Krishnaiah (1984), pp. 952–953.

4.6.3 Kolmogorov-type tests. We may note that the stochastic independence of X and Y can be restated as

$$D(x, y) = F(x, y) - F(x, \infty)F(\infty, y) = 0, \quad \text{for all } (x, y). \qquad (1)$$

Motivated by this feature, Hoeffding (1948b) considered a non-negative functional

$$\Delta = \Delta(F) = \int\int_{\substack{-\infty<x<\infty \\ -\infty<y<\infty}} D^2(x, y) dF(x, y) \qquad (2)$$

and used his U-statistic approach to formulate a test statistic for testing the null hypothesis of independence against general alternative hypotheses that $\Delta(F) > 0$. For this define

$$\psi(a, b, c) = I(a \geq b) - I(b \geq c), \text{for real } a, b, c, \tag{3}$$

and consider the kernel of degree 5 :

$$\begin{aligned} \gamma((x_1, y_1), \ldots, (x_5, y_5)) &= \frac{1}{4}\psi(x_1, x_2, x_3)\psi(x_1, x_4, x_5) \\ &\quad \cdot\psi(y_1, y_2, y_3)\psi(y_1, y_4, y_5); \end{aligned} \tag{4}$$

note that the kernel is not a symmetric kernel, and hence, we use the following definition of U-statistic that is apparently different from the one prescribed in Subsection 3.5.1. We let

$$D_N = N^{-[5]} \sum{}' \gamma((X_{i_1}, Y_{i_1}), \ldots, (X_{i_5}, Y_{i_5})), \tag{5}$$

where the summation $\sum'$ extends over all possible ($N^{[5]}$) subsets $1 \leq i_1 \neq \ldots \neq i_5 \leq N$. Under the hypothesis of independence, $\Delta(F) = 0$, and hence its unbiased estimator D_N also has mean 0. Hoeffding (1948b) showed that under the hypothesis of independence,

$$\text{var}\, D_n = \frac{2(N^2 + 5N - 32)}{9N(N-1)(N-2)(N-3)(N-4)}, \tag{6}$$

and incorporating the same permutation principle as in the preceding subsection, obtained the exact null hypothesis distribution of D_N for very small values of N (≥ 5).

Blum et al. (1961) considered some variants of the test statistic D_N based on the empirical joint and marginal distribution functions. For this purpose, we consider first the joint empirical distribution function.

$$F_N(x, y) = N^{-1} \sum_{i=1}^{N} I(X_i \leq x, Y_i \leq y). \tag{7}$$

The two marginal empirical distribution functions are then defined as

$$F_{N1}(x) = N^{-1} \sum_{i=1}^{N} I(X_i \leq x); \; F_{N2}(x) = N^{-1} \sum_{i=1}^{N} I(Y_i \leq x). \tag{8}$$

Then in the spirit of the Cramér-von Mises tests, they considered the test statistic

$$B_N = \int\int_{\substack{-\infty<x<\infty \\ -\infty<y<\infty}} \{F_N(x, y) - F_{N1}(x)F_{N2}(y)\}^2 \, \mathrm{d}F_N(x, y). \tag{9}$$

It is quite clear that B_N and D_N are quite close to each other, and in fact they have the same asymptotic properties. They also considered a Kolmogorov-type test statistic

$$A_N = \sup_{\substack{-\infty<x<\infty \\ -\infty<y<\infty}} \{|F_N(x,y) - F_{N1}(x)F_{N2}(y)|\}. \tag{10}$$

The null hypothesis distribution of A_N can be enumerated, for small values of N, by reference to the same permutation principle considered in the preceding subsection. However, none of these test statistics is of the linear rank statistic type.

4.7 RANDOM BLOCKS

Suppose that we wish to test the effects of k treatments, and for this purpose we observe random variables in n blocks, each treatment having one observation in each block. In mathematical notation, we observe in the j-th block ($j = 1, \ldots, n$) the random variables $X_{j1}, \ldots, X_{jk}$, where X_{ji} is taken under the i-th treatment ($i = 1, \ldots, k$). Let the density of X_{ji} be generally f_{ji}. The null hypothesis H_3 is that all treatments have the same effect, or, mathematically, that $f_{j1} = f_{j2} = \ldots = f_{jk}$ holds for each j. Otherwise these densities may be arbitrary, and, in particular, they may depend on the block index j. We shall restrict ourselves to the alternatives of location shifts which may be expressed by $f_{ji}(x) = f_j(x-\Delta_i)$ for $j = 1, \ldots, n$ and $i = 1, \ldots, k$, where f_j are some densities depending on the block index j, and the parameters Δ_i are not all equal. Intuitively speaking, under the alternative the treatments have unequal additive effects $\Delta_1, \ldots, \Delta_k$.

The test statistics described below are used only for one-sided tests, their critical regions consisting of values larger than the respective critical value.

The present problem of random blocks, viewed from a different point of view, is also called 'the problem of n rankings', and is treated in detail e.g. by Kendall (1948).

4.7.1 Rank tests of χ^2-types. For the definition of ranks, take each block separately, and denote by R_{ji} the rank of X_{ji} among the observations $X_{j1}, \ldots, X_{jk}$ in the j-th block.

The Friedman test is probably the best known among the rank tests for random blocks. Its statistic is

$$Q = \frac{12}{nk(k+1)} \sum_{i=1}^{k} \Big[\sum_{j=1}^{n} R_{ji} - \tfrac{1}{2}n(k+1) \Big]^2 \tag{1}$$

$$= \frac{12}{nk(k+1)} \sum_{i=1}^{k} \Big(\sum_{j=1}^{n} R_{ji} \Big)^2 - 3n(k+1).$$

The statistic Q has, under H_3, for k fixed and $n \to \infty$ asymptotically a χ^2 distribution with $k-1$ degrees of freedom, as will be proved in 6.2.3.

This test was discussed by Friedman (1937) and (1940), where some tables for it are included; for other tables cf. Sen and Krishnaiah (1984), pp. 948–949, or quotations in Hájek and Šidák (1967).

The median test is another simple test for random blocks using the statistic

$$Q_0 = \sum_{i=1}^{k} \Big[\sum_{j=1}^{n} A_{ji} - \tfrac{1}{2} n \Big]^2 = \sum_{i=1}^{k} \Big(\sum_{j=1}^{n} A_{ji} \Big)^2 - \tfrac{1}{4} n^2 k, \tag{2}$$

where

$$A_{ji} = \begin{cases} 1 & \text{if } X_{ji} \text{ is larger than the median of } X_{j1}, \ldots, X_{jk}, \\ \frac{1}{2} & \text{if } X_{ji} \text{ is equal to this median,} \\ 0 & \text{if } X_{ji} \text{ is smaller than this median.} \end{cases}$$

The statistic

$$Q = \begin{cases} \dfrac{4(k-1)}{nk} Q_0 & \text{for } k \text{ even,} \\[2ex] \dfrac{4}{n} Q_0 & \text{for } k \text{ odd,} \end{cases} \tag{3}$$

has, under H_3, for k fixed and $n \to \infty$, asymptotically a χ^2 distribution with $k-1$ degrees of freedom.

A similar median test, in whose statistic the A_{ji}'s take only the values 1 or 0, is discussed by Mood (1950), and Brown and Mood (1950).

Remarks. Naturally, it would also be possible to define in a similar way the tests of the Fisher-Yates-Terry-Hoeffding and of the van der Waerden types, utilizing the scores $\mathsf{E}V_k^{(i)}$ and $\Phi^{-1}(i/(k+1))$, respectively.

The general form of all the rank statistics for random blocks mentioned in this section is

$$\begin{aligned} Q &= \frac{k-1}{k \operatorname{var} S_i} \sum_{i=1}^{k} (S_i - \mathsf{E}S_i)^2 \\ &= \frac{k-1}{n \sum_{j=1}^{k} [a_k(j) - \bar{a}_k]^2} \Big\{ \sum_{i=1}^{k} S_i^2 - kn^2 \bar{a}_k^2 \Big\}, \end{aligned} \tag{4}$$

where

$$S_i = \sum_{j=1}^{n} a_k(R_{ji}) \tag{5}$$

with some scores $a_k(j)$.

In addition, let us briefly discuss the case of two treatments, $k = 2$. Since in this case $A_{ji} = R_{ji} - 1$, the Friedman test and the median test are equivalent. Moreover, $A_{j1} = 1$ if $X_{j1} - X_{j2} > 0$, $A_{j1} = 0$ if $X_{j1} - X_{j2} < 0$ and $\sum_{j=1}^{n} A_{j2} = n - \sum_{j=1}^{n} A_{j1}$, so that both of these test are also equivalent to the test whose statistic is the number of positive values among the differences $X_{j1} - X_{j2}$ $(j = 1, \ldots, n)$, that is, to the sign test for these differences. On the other hand, the difference $X_{j1} - X_{j2}$ has, under H_3, a distribution symmetric about zero, while under the alternative this distribution is shifted, and thus the use of the sign test is really justified also from this point of view. If, in addition, all densities f_{ji} $(j = 1, \ldots, n;\ i = 1, \ldots, k)$ are equal under H_3, all differences $X_{j1} - X_{j2}$ $(j = 1, \ldots, n)$ have the same distribution symmetric about zero, so that any test of Section 4.5 may be applied for these differences. The problem of locally optimal tests based on *intra-block* ranking was discussed by Sen (1968a).

The problem of testing H_3 against the alternative that exactly one treatment has a larger effect than the remaining ones (the slippage problem) was treated by Doornbos and Prins (1958).

We may point out in this context that the intra-block ranking procedures described above have the main advantage that they can be used for ranking data as well, and they do not need the assumption of additivity of the block effects. On the other hand, in many experimental setups, the block effects convey some statistical information that can be recovered by *inter-block* comparisons, and thereby may enhance the precision of statistical conclusions to be made from such experimental setups. One simple way to utilize such interblock comparisons is to appeal to *ranking after alignment* made with respect to estimated block effects. The original idea is due to Hodges and Lehmann (1962) and developed systematically, using a multivariate ranking procedure, by Sen (1968b). Unlike the intra-block rank tests considered above, such aligned rank tests are generally not EDF but CDF. We will provide an outline of aligned rank tests in Section 10.1.

4.8 TREATMENT OF TIES

The whole theory of rank tests is based on the essential assumption that all the observed random variables are governed by continuous distributions. Random variables observed in practice, however, are always discrete, either by their nature (e.g. integer-valued variables) or as a consequence of rounding-off.

The present section is designed to give a brief survey of ideas dealing with certain consequences of this discrepancy between theory and practice. It has only an informative character and omits discussion of details. Note that this is the only section of the book where discrete distributions of basic variables are admitted into considerations.

4.8.1 Randomization, average statistics and scores, mid-ranks.

Let us begin with the *hypothesis* H_0. Now, however, in accordance with the above introduction, the hypothesis must be generalized to H_0' under which all variables $X_1, \ldots, X_N$ are independent and have the same arbitrary distribution function F, possibly discontinuous. In this model there is a positive probability of some of the observations being equal. For definiteness, let the observations $X_1, \ldots, X_N$, rearranged in increasing order, be

$$\begin{aligned} X^{(1)} = \ldots = X^{(\tau_1)} \quad &< \quad X^{(\tau_1+1)} = \ldots = X^{(\tau_1+\tau_2)} < \ldots \\ &< \quad X^{(\tau_1+\ldots+\tau_{g-1}+1)} = \ldots = X^{(\tau_1+\ldots+\tau_g)}, \end{aligned} \tag{1}$$

so that the sample decomposes into g groups of equal observations, the j-th group ($j = 1, \ldots, g$) containing τ_j observations, $\sum_{j=1}^{g} \tau_j = N$. Naturally, g and $\tau_1, \ldots, \tau_g$ are random variables, and we shall write simply $\tau = (\tau_1, \ldots, \tau_g)$. The groups with $\tau_j \geq 2$ are called ties, and the observations in such groups are called tied observations, whereas the untied observations are those in groups with $\tau_j = 1$.

The ranks of untied observations are well defined but those of tied observations are not. Thus, having a rank statistic T, its value is not determined for a sample containing tied observations, and it is necessary to adopt some supplementary modifications of its definition. Essentially, we must work under the condition of some definite structure of ties, so the resulting tests are now only CDF rank tests.

Randomization. In principle, for ranks of observations in the j-th group any permutation of the numbers

$$\tau_1 + \ldots + \tau_{j-1} + 1, \ldots, \tau_1 + \ldots + \tau_j \tag{2}$$

might be taken. Thus, for fixed τ and fixed ranks of the untied observations, there are altogether $\tau_1!\tau_2!\ldots\tau_g!$ possible assignments of ranks to all observations in the sample. These assignments produce the same number of possible values t_k (not necessarily distinct) of the statistic T. Now, the method of randomization consists in a random selection of one of these assignments, or, equivalently, of one of the values t_k, each having the probability $(\tau_1!\tau_2!\ldots\tau_g!)^{-1}$ of being selected.

This selection must be done by means of a supplementary experiment, the structure of which depends, of course, on the value of τ. Denoting the result of this experiment by Ξ, the final ranks of the observations are functions of X and Ξ, that is $R_i = R_i(X, \Xi)$.

It may be seen that, under H_0' and subsequent randomization given τ, the ranks $R_1(X,\Xi), \ldots, R_N(X,\Xi)$ take on an arbitrary permutation of the numbers $1, \ldots, N$ with equal probabilities $(N!)^{-1}$, which shows that the distributions of the rank statistics $T = T(X,\Xi)$ are the same as those under H_0. In particular, the usual tables of critical values may be applied, and the levels of significance are not changed. On the other hand, since we are bringing into the problem some supplementary irrelevant random effect, the behaviour of the test under the alternatives may be somewhat less favourable than with other approaches.

Average statistics. Another method is to base the test on the 'average statistic' $\bar{T}$ defined as follows. On observing $X_1, \ldots, X_N$, and thus having τ and the ranks of the untied observations fixed, we define $\bar{T}$ as the average of the quantities t_k mentioned above. (In this subsection a letter with a bar will always denote such an average statistic.) In other words, if E_Ξ is the operation of expectation with respect to the above-mentioned randomization experiment, we have, for fixed τ and fixed ranks of the untied observations,

$$\bar{T} = \mathsf{E}_\Xi T(X,\Xi) = (\tau_1!\tau_2!\ldots\tau_g!)^{-1} \sum_k t_k, \tag{3}$$

where the summation extends over $k = 1, 2, \ldots, \prod_{j=1}^{g} \tau_j!$.

For the distribution of $\bar{T}$, we can show only the following more modest general result. For brevity, let $\mathsf{E}'\{.\mid\tau\}$ and $\mathsf{var}'\{.\mid\tau\}$ denote the conditional expectation and the conditional variance, respectively, under H_0' for fixed τ, while E and var will be used, as usual, for the unconditional expectation and variance under H_0. We noted before that under H_0' randomization entails that the ranks $R_1, \ldots, R_N$ take on permutations of $1, \ldots, N$ with probabilities $(N!)^{-1}$, so that two successive operations of expectations $\mathsf{E}'\{\mathsf{E}_\Xi(.)\mid\tau\}$ are equivalent to the operation E. Consequently, for every τ,

$$\mathsf{E}'\{\bar{T}\mid\tau\} = \mathsf{E}'\{\mathsf{E}_\Xi T\mid\tau\} = \mathsf{E}T, \tag{4}$$

and

$$\begin{aligned}\mathsf{var}'\{\bar{T}\mid\tau\} &= \mathsf{E}'\{(\mathsf{E}_\Xi T - \mathsf{E}T)^2\mid\tau\} \\ &\le \mathsf{E}'\{\mathsf{E}_\Xi(T - \mathsf{E}T)^2\mid\tau\} = \mathsf{E}\{(T-\mathsf{E}T)^2\} = \mathsf{var}\,T.\end{aligned} \tag{5}$$

As a practical consequence of (5), we may expect that the usual critical values of T, used also for the average statistic $\bar{T}$, give a test of a smaller significance level than the nominal one, at least if $\bar{T}$ is approximately normally distributed.

Average scores. Alternatively, if the test statistic in question is defined by means of some scores $a(k)$, $k = 1, \ldots, N$, the averaging by means of the expectation E_{Ξ} may be applied directly to the scores $a(R_i(X, \Xi))$. For rank statistics of χ^2-type the present method of average scores differs from the preceding method of average statistics, while for simple linear rank statistics of the form $S = \sum_{i=1}^{N} c_i a(R_i)$ these two methods coincide since

$$\bar{S} = \mathsf{E}_{\Xi} S(X, \Xi) = \sum_{i=1}^{N} c_i \mathsf{E}_{\Xi} a\big(R_i(X, \Xi)\big). \tag{6}$$

This formula for $\bar{S}$ may be written as

$$\bar{S} = \sum_{i=1}^{N} c_i a(R_i, \tau), \tag{7}$$

with the 'average scores' $a(k, \tau)$ depending on τ, and for fixed τ being given by

$$a(k, \tau) = \tau_j^{-1}[a(\tau_1 + \ldots + \tau_{j-1} + 1) + \ldots + a(\tau_1 + \ldots + \tau_j)] \tag{8}$$
$$\text{for } \tau_1 + \ldots + \tau_{j-1} < k \leq \tau_1 + \ldots + \tau_j .$$

Thus, in $\bar{S}$ we are simply making use of the scores $a(k, \tau)$ in place of the original scores $a(k)$. The fact that the ranks R_i in (7) are not defined for the tied observations X_i, does not matter here: if X_i belongs to the j-th group of equal observations in (1), any of the numbers (2) may be taken for R_i, since the corresponding score (8) is the same for each of these numbers. Naturally, for untied observations we have $a(k, \tau) = a(k)$.

Obviously, the conditional expectation of $\bar{S}$ in (7) for given τ equals the usual $\mathsf{E}S$, and the conditional variance of $\bar{S}$ for given τ may be computed using the formula of Theorem 3.3.1.3.

Mid-ranks. For the two-sample Wilcoxon test and the regression test of the Wilcoxon type, the method of average scores leads to the so-called midranks, which are the best known and most frequently applied method of dealing with ties. More generally, mid-ranks might be used in all statistics whose scores $a(k)$ have meaning also when k is a halved integer (e.g. for statistics of the van der Waerden type, but not for those of the Fisher-Yates-Terry-Hoeffding type). The essence of this method is to apply the expectation E_{Ξ}, for τ fixed, directly to the ranks $R_i(X, \Xi)$, obtaining in this manner for each X_i from the j-th group in (1) the mid-rank $\tau_1 + \ldots + \tau_{j-1} + \frac{1}{2}(\tau_j + 1)$, and more generally, the 'mid-rank score' $a(\tau_1 + \ldots + \tau_{j-1} + \frac{1}{2}(\tau_j + 1))$ instead of $a(R_i)$.

Similarly, as in the case of the scores (8), if the mid-rank scores are used in S, the conditional expectation and conditional variance of S, given τ, may be found by applying the formulas of Theorem 3.3.1.3.

Finally, let us note that in the two-sample and k-sample problems, if the j-th group contains tied observations, but only from the same sample, we obviously need not bother about it and may take for the corresponding ranks an arbitrary permutation of the numbers (2).

For the *hypothesis* H_2 and H_3, we might repeat the same basic principles with only obvious modifications. These principles are applied separately to each coordinate in the case of H_2, and separately to each block in the case of H_3, the results being afterwards combined for both coordinates, or for all the n blocks.

The case of the *hypothesis* H_1 of symmetry is somewhat different. We shall concentrate on the statistics S^+ of Subsection 4.5.1, and it is necessary to distinguish two kinds of discontinuities of the underlying distribution function $F(x)$ of the observations X_i.

The first kind are discontinuities at points $x \neq 0$, which may produce a group of several observations $X_i \neq 0$ with equal absolute values $|X_i|$, so that the corresponding ranks R_i^+ are not defined. If such a group contains only observations of the same signs, the choice of ranks R_i^+ within this group is clearly irrelevant. On the other hand, if such a group contains both positive and negative values, the definitions of ranks R_i^+ or of the statistic S^+ may be modified analogously as described for the case of H_0.

The second source of difficulties is related to a possible discontinuity of $F(x)$ at $x = 0$. To clarify the problem, let us recall that the statistic of the locally most powerful rank test has, by Theorem 3.4.9.1, the form $\sum_{i=1}^{N} a_N^+(R_i^+)$ sign X_i. In Subsection 4.5.1, however, we have given the statistics in the more convenient form

$$S^+ = \sum_{X_i > 0} a_N^+(R_i^+). \tag{9}$$

The two statistics are equivalent whenever $F(x)$ is continuous at $x = 0$, since then

$$S^+ = \frac{1}{2}\Big[\sum_{i=1}^{N} a_N^+(R_i^+) \operatorname{sign} X_i + \sum_{i=1}^{N} a_N^+(i)\Big] \tag{10}$$

with probability 1. However, if there are some observations X_i equal to zero, (10) is no longer true, so that the definition of S^+ in (9) should be modified. The first modification, suggested by (10), is to consider the statistic

$$\sum_{X_i>0} a_N^+(R_i^+) + \frac{1}{2} \sum_{X_i=0} a_N^+(R_i^+). \tag{11}$$

As for the undefined ranks R_i^+ in the case of $\tau_0 \geq 2$ observations X_i equal to 0, we may simply take for these ranks any permutation of the numbers $1, 2, \ldots, \tau_0$, or apply any of the principles of dealing with ties mentioned for H_0.

Another procedure for dealing with zero observations is randomization: with probability $\frac{1}{2}$ act as if a zero observation X_i were positive and include the corresponding score $a_N^+(R_i^+)$ into the sum in (9), and with the same probability act in the opposite way, the choices for zero observations with distinct indices being independent. Obviously, (11) is the expectation of (9) with respect to this randomization.

4.8.2 Other methods of dealing with ties. Here we shall add several brief remarks of lesser generality and importance.

One of the possible methods of dealing with ties is to take for testing the 'least favourable' value of the statistic T. For example, when we are making use of a one-sided critical region $\{T \geq c\}$ in testing H_0, we may take (in the notation of Subsection 4.8.1) the smallest value among all t_k, $k = 1, \ldots, \prod_{j=1}^{g} \tau_j!$. Similarly, for two-sided regions $\{|T| \geq c\}$ we may take the smallest of the absolute values $|t_k|$. Of course, the significance level of this procedure never exceeds the nominal level, but such a procedure may be too conservative and too 'pessimistic'.

Formulas (4.1.3.5) and (4.1.3.6) for the Kolmogorov-Smirnov statistics may serve with no change also in the presence of ties, and, in connection with the usual critical values, they also give a test having a significance level not exceeding the nominal one. Actually, we can imagine that the observations X_i, $i = 1, \ldots, m+n$, have arisen by rounding off upwards some observations X_i^*, having a continuous distribution function. Then the values of $F_{2,n}(x) - F_{1,m}(x)$, $-\infty < x < \infty$, form a subset of the values of $F_{2,n}^*(x) - F_{1,m}^*(x)$, where $F_{2,n}^*$ and $F_{1,m}^*$ are the corresponding empirical distribution functions of the X_i^*'s. Hence,

$$\max_{-\infty<x<\infty} \left(F_{2,n}(x) - F_{1,m}(x)\right) \leq \max_{-\infty<x<\infty} \left(F_{2,n}^*(x) - F_{1,m}^*(x)\right), \tag{1}$$

and similarly for the maxima of the absolute values. (For details see Noether (1963) and Walsh (1963).)

If there is only a small number of tied observations, we may also omit them entirely from the sample. Of course, this can cause some loss of information. In the case of the sign test, however, it is advisable to omit zero observations, since the resulting test is more powerful than the test based on the statistic $\sum_{x_i>0} 1 + \frac{1}{2} \sum_{x_i=0} 1$. This follows from the fact, noted by Hemelrijk (1952), that the critical region for the latter test forms a subset of the critical region for the former test (see Problem 29).

In some papers also the conditional approach is used, namely, the ties are supposed fixed and all developments are made under this condition.

Finally, we wish to emphasize that quite a lot of papers were published on tied observations (in particular since Hájek and Šidák (1967) appeared) but here in the present section, because of space limitations, we had to present only the basic information and omit all other details. After a more thorough examination of the literature, it seemed that the most attention has been paid to a) the method of randomization, b) the method of average scores, c) the method of conditional fixing of ties.

4.9 RANK TESTS FOR CENSORED DATA

Rank tests adapt well in various *censoring schemes* that commonly arise in practice, particularly in *life testing* and *survival analysis* models. The following censoring schemes are most commonly encountered in practice:

(i) right and/or left truncation (Type I censoring);
(ii) right and/or left censoring (Type II censoring);
(iii) progressive censoring schemes;
(iv) interval censoring or grouped data;
(v) random censoring.

EDF rank tests exist in cases (ii) and (iii), while in the remaining three cases, CDF rank tests prevail. The Cox (1972) *proportional hazards* models and *partial likelihood* functions have important roles to play in case (v), and the so-called *log-rank statistics* have a special significance in this context. The present section relates to a finite sample treatise of rank tests under censoring schemes; the related asymptotics will be outlined in later chapters.

4.9.1 Type I censoring. For N independent real random variables $X_1, \ldots, X_N$ with continuous distribution functions $F_1, \ldots, F_N$ respectively, consider the simple regression model:

$$F_j(x) = F(x - \theta - \beta c_j), \; j = 1, \ldots, N, \tag{1}$$

where the c_j are known regression constants, not all equal, (θ, β) refers to the intercept and regression parameter, while F is an unknown, continuous distrubiton function. In a Type I (right) censoring scheme, for a given $T(< \infty)$, the observable random variables are

$$X_i^* = \min(X_i, T), \quad \delta_i = I(X_i \le T), \quad i = 1, \ldots, N. \tag{2}$$

Similarly, for left truncation at a point L ($> -\infty$), the observable random variables are $\max(X_i, L)$, $\delta_i = I(X_i \geq L)$, $i = 1, \ldots, N$. It is also possible to have a double truncation (at (L, T)) model. We consider here only the case of right truncation, as the other two cases follow on parallel lines.

Suppose that based on (2) we intend to test the null hypothesis $H_0 : \beta = 0$ (i.e., no regression) against $H_1 : \beta > 0$ (or $\beta < 0$ or $\beta \neq 0$). Note that under H_0, the (X_i^*, δ_i) are independent identically distributed random vectors. Since F is unspecified (while T is given), we intend that the hypothesis testing problem remains invariant under strictly monotone transformations of the observations. Hence rank tests have a natural appeal in this context. Let

$$N^* = \sum_{i=1}^{N} \delta_i = \text{number of uncensored observations}, \tag{3}$$

so that the remaining $N - N^*$ observations are censored (at T); because of the assumed continuity of F, ties among the N^* uncensored observations can be neglected with probability 1. This corresponds to a special case of the model for tied observations treated in the preceding section; in the current case, the associated order statistics are $X_N^{(1)} < \ldots < X_N^{(N^*)} < T = X_N^{(N^*+1)} = \ldots = X_N^{(N)}$. Thus, if we use the tie-adjusted linear rank statistic of the preceding section based on the *averaged scores* approach, we would have

$$\begin{aligned} S_I &= \sum_{i=1}^{N^*} c_{D_i} a_N(i) + a_N^*(N^*) \sum_{j=N*+1}^{N} c_{D_j} \\ &= N \bar{a}_N \bar{c}_N + \sum_{i=1}^{N^*} (c_{D_i} - \bar{c}_N)[a_N(i) - a_N^*(N^*)], \end{aligned} \tag{4}$$

where $D = (D_1, \ldots, D_N)$ stands for the vector of *antiranks* (that is, $R_{D_i} = D_{R_i} = i$, $i = 1, \ldots, N$), $\bar{a}_N = N^{-1} \sum_{i=1}^{N} a_N(i)$, $\bar{c}_N = N^{-1} \sum_{i=1}^{N} c_i$, and

$$a_N^*(N^*) = \frac{1}{N - N^*} \sum_{j=N^*+1}^{N} a_N(j), \text{ for } N^* < N, \tag{5}$$

and we let it be 0 for $N = N^*$. Since the R_i are translation-invariant, without any loss of generality we may put $\bar{a}_N = 0 = \bar{c}_N$. Moreover, we note that $D_1, \ldots, D_{N^*}$ are observable, while the complementary set $(D_{N^*+1}, \ldots, D_N)$ has indistinguishable elements; but that does not affect S_I. Thus, we write

$$S_I = \sum_{i=1}^{N^*} c_{D_i} [a_N(i) - a_N^*(N^*)], \tag{6}$$

and conventionally we let $S_I = 0$ when $N^* = 0$, i.e., all the N observations lie to the right of T. Note that for $N^* = 0$, $S_I(= 0)$ is non-informative, and hence is not a suitable test statistic. Further, N^* is a non-negative integer valued random variable whose distribution depends on the underlying $F_i(T), i = 1, \ldots, N$. Under the null hypothesis, letting $p = P(T) = F(T - \theta)$, we have

$$P_0\{N^* = k\} = \binom{N}{k}[p(T)]^k[1 - P(T)]^{N-k}, \; k = 0, 1, \ldots, N, \tag{7}$$

which depends on the unknown distribution function F through $F(T - \theta)$. Therefore, a test based on S_I in (6) may not be in general EDF. On the other hand, under H_0, given $N^* = n^*(1 \leq n* \leq N)$, $D_1, \ldots, D_{n^*}$ takes on each realization $(i_1, \ldots, i_{n^*})$, $1 \leq i_1 \neq \ldots \neq i_{n^*} \leq N$, with the common (conditional) probability $N^{-[n^*]}$. Therefore, based on the randomization principle considered in the previous section, and given $N^* = n^*$, S_I is CDF under H_0. Instead of the average scores, we might have gone for the *mid-ranks* (where we may note that for the $N - N^*$ censored observations, the mid-rank is $(N-N^*)(N+N^*+1)/2$, so that instead of the $a_N^*(N^*)$ defined by (5) we might have used $a_N((N-N^*)(N+N^*+1)/2)$). But none of these choices may be strictly justified on the LMPR grounds discussed in Section 3.4. Nevertheless, (5) has some nice interpretations and satisfies a basic martingale property [Chatterjee and Sen (1973a)] that will be elaborated later on.

Let us look at the likelihood function for Type I (right) censoring,

$$L_{N,T}(\beta) = \prod_{i=1}^{N}\{f(X_i^* - \beta c_i)\}^{\delta_i}\{1 - F(T - \beta c_i)\}^{1-\delta_i}, \tag{8}$$

where in view of translation-invariance, we let $\theta = 0$ without loss of generality. The probability law for N^* is

$$p_{\beta,T}(n^*) = \sum_{N}^{*}\prod_{i=1}^{N}\{[F(T - \beta c_i)]^{\delta_i}[1 - F(T - \beta c_i)]^{1-\delta_i}\}, \tag{9}$$

where the summation $\sum_N^*$ extends over all possible $\delta_1, \ldots, \delta_N$, such that $\sum_{j=1}^N \delta_j = n^*$. Therefore, the conditional likelihood, given $N^* = n^*$, is

$$L_N^*(\beta|n^*) = L_{N,T}(\beta)/p_{\beta,T}(n^*). \tag{10}$$

The presence of the denominator in (10) introduces additional complications in the computation of the (conditional) LMPR test statistic. Nevertheless, in the next subsection, we shall simplify this for Type II censoring, and later on justify the asymptotic LMPR structure in a simple way.

4.9.2 Type II censoring. In a Type II right censoring scheme, typically arising in a lifetesting model, the experiment is curtailed when a prescribed number of reponses is obtained. With respect to the same model in (4.9.1.1) in the Type I censoring scheme, denote by $n^*(\leq N)$ a prefixed number, so that censoring takes place after the n^*th order statistic $X_N^{(n^*)}$ is observed; here we denote the ordered observations by $X_N^{(1)} < \ldots < X_N^{(N)}$, and note that by the formulation observations having values larger than $X_N^{(n^*)}$ are censored. In contrast to the Type I censoring case (where T is fixed but N^* is stochastic), here n^* is prefixed, but the censoring point $X_N^{(n^*)}$ is random. Thus the observable random elements are

$$(X_N^{(i)} = X_{D_i}, c_{D_i}), \quad i = 1, \ldots, n^* \tag{1}$$

while the remaining $N - n^*$ observations $(X_{D_j}, j > n^*)$ are all censored. As in the preceding subsection, we may consider here the linear rank statistic

$$S_{\text{II}} = \sum_{i=1}^{n^*} c_{D_i}[a_N(i) - a_N^*(n^*)] \tag{2}$$

and note that here additionally n^* is non-stochastic. Recalling that we suppose $\bar{c}_N = 0$ (see after formula (4.9.1.5)), and putting $b_{n^*,N}(i) = a_N(i)$ for $i = 1, \ldots, n^*$, and $b_{n^*,N}(i) = a_N^*(n^*)$ for $i = n^* + 1, \ldots, N$, we may rewrite

$$\begin{aligned}
\sum_{j=1}^{N} c_j b_{n^*,N}(R_j) &= \sum_{i=1}^{N} c_{D_i} b_{n^*,N}(i) \\
&= \sum_{i=1}^{n^*} c_{D_i} a_N(i) + \sum_{i=n^*+1}^{N} c_{D_i} a_N^*(n^*) \\
&= \sum_{i=1}^{n^*} c_{D_i} a_N(i) + a_N^*(n^*) \sum_{i=n^*+1}^{N} c_{D_i} \\
&= \sum_{i=1}^{n^*} c_{D_i}[a_N(i) - a_N^*(n^*)] = S_{\text{II}},
\end{aligned}$$

where under $H_0 : \beta = 0$, S_{II} is EDF, and its mean and variance can be computed as in earlier sections. Left censoring or double censoring under the Type II scheme can be worked out analogously, and in either case, the EDF characterization remains valid.

Let us now consider the likelihood function for the Type II censoring scheme, where without loss of generality we set $\theta = 0$. Then

$$L_{N,n^*}(\beta) = \prod_{i=1}^{n^*} f(X_{D_i} - \beta c_{D_i}) \prod_{j=n^*+1}^{N} [1 - F(X_N^{(n^*)} - \beta c_{D_j})]. \tag{3}$$

Under $H_0 : \beta = 0$, we have

$$L_{N,n^*}(0) = \prod_{i=1}^{n^*} f(X_{D_i}) \cdot [1 - F(X_N^{(n^*)})]^{N-n^*}. \tag{4}$$

Therefore, writing $\mathcal{X}^{(\cdot)*} = \{(x_1, \ldots, x_{n^*}) : -\infty < x_1 < \ldots < x_{n^*} < \infty\}$, and proceeding as in Section 3.4, we obtain that

$$Q_\beta(D = d) = \int \cdots \int_D = dL_{N,n^*}(\beta)\ dx_1 \ldots dx_{n^*}, \tag{5}$$

where $D = (D_1, \ldots, D_N)$, $d = (d_1, \ldots, d_N)$ and $X_{D_i} \to x_i$, $i = 1, \ldots, n^*$.

We proceed as in Section 3.4 confined to the contemplated simple regression model, and introduce the following notation:

$$\begin{aligned} a_N^0(i) &= \mathsf{E}_0\{-f'(X_N^{(i)})/f(X_N^{(i)})\} \\ &= -i\binom{N}{i}\int_{-\infty}^{\infty} [F(x)]^{i-1}[1 - F(x)]^{N-i}\, \mathrm{d}f(x), \end{aligned} \tag{6}$$

for $i = 1, \ldots, N$;

$$\begin{aligned} a_N^{*0}(n^*) &= \mathsf{E}_0\{f(X_N^{(n^*)})/[1 - F(X_N^{(n^*)}]\} \\ &= n^*\binom{N}{n^*}\int_{-\infty}^{\infty} f^2(x)[F(x)]^{n^*-1}[1 - F(x)]^{N-n^*}\, \mathrm{d}x \\ &= \frac{N}{N - n^*}\int_{-\infty}^{\infty} f(x) \\ &\qquad \cdot \mathrm{d}[\sum_{j=n^*}^{N-1}\binom{N-1}{j}[F(x)]^j[1 - F(x)]^{N-j-1}] \\ &= \sum_{j=n^*}^{N-1}\{\frac{N}{N - n^*}\int_{-\infty}^{\infty}\binom{N-1}{j}[F(x)^j[1 - F(x)]^{N-j-1} \\ &\qquad \cdot \frac{\{-f'(x)\}}{f(x)} f(x)\, \mathrm{d}x\} \\ &= \frac{1}{N - n^*}\sum_{j=n^*}^{N-1}(j+1)\binom{N}{j+1}\int_{-\infty}^{\infty}\frac{-f'(x)}{f(x)}[F(x)]^j\ \cdot \\ &\qquad \cdot [1 - F(x)]^{N-j-1} f(x)\, \mathrm{d}x \\ &= \frac{1}{N - n^*}\sum_{j=n^*+1}^{N} j\binom{n}{j}\int_{-\infty}^{\infty}\frac{-f'(x)}{f(x)}[F(x)]^{j-1} \\ &\qquad \cdot [1 - F(x)]^{N-j} f(x)\, \mathrm{d}x \\ &= \frac{1}{N - n^*}\sum_{j=n^*+1}^{N} a_N^0(j). \end{aligned} \tag{7}$$

Therefore, proceeding as in the proof of Theorem 3.4.8.1, we obtain that a LMPR test statistic under Type II (right) censoring is given by $\sum_{i=1}^{n^*} c_{D_i} a_N^0(i)$ $+ a_N^{*0}(n^*) \sum_{j=n^*+1}^{N} c_{D_j} = \sum_{i=1}^{n^*} c_{D_i} [a_N^0(i) - a_N^{*0}(n^*)]$, and this corresponds to S_{II} whenever $a_N(i) = a_N^0(i)$, $i = 1, 2, \ldots, N$. This also provides a characterization of the scores used on the grounds of the LMPR test property.

4.9.3 Progressive censoring. Various types of *interim analsysis* schemes are commonly encountered in clinical trials and life-testing studies. In an interim analysis, provision is made for having *statistically meaningful looks* into the accumulating dataset, either periodically on the basis of *calendar time* or on *information time*, usually with the intention of an *early termination* of the study, depending on the accumulating statistical (and clinical) evidence. In this setup, the failure time (primary response variable) (X) is non-negative. Thus, if we adopt the same regression model as in Subsection 4.9.1, at a calendar time t, we denote by n_t the number of failures that have occurred prior to t (> 0). Thus, n_t is a non-negative integer-valued random variable, and is non-decreasing in t (> 0); it is bounded from above by N, the total number of experimental units that are being followed through over time. Using the same notation as before, we denote the sample ordered failures by $X_N^{(1)} \leq \ldots \leq X_N^{(N)}$. Then we have

$$n_t = k \quad \text{for} \quad X_N^{(k)} \leq t < X_N^{(k+1)}, \quad 0 \leq k \leq N, \tag{1}$$

where, conventionally, we let $X_N^{(0)} = 0$, $X_N^{(N+1)} = +\infty$. Therefore, at timepoint t, the accumulated dataset relates to the observable random elements

$$\mathcal{D}_N^{(t)} = \{X_i^*(t) = (\min X_i, t),\ \delta_i^*(t) = I(X_i \leq t);\ 1 \leq i \leq N\}. \tag{2}$$

Since the data-cloud changes only at the successive ordered failure points, we may equivalently consider the following discretized data-sequence

$$\mathcal{D}_{N,k} = \{X_{i,k}^* = \min(X_i, X_N^{(k)}),\ \delta_{i,k}^* = I(X_i \leq X_N^{(k)}); i \leq k\}, \tag{3}$$

for $k = 1, \ldots, N$.

Now, if we have a prefixed number (say, $r(\geq 1)$) of time points $t_1 < \ldots < t_r$, then we obtain the accumulating dataset (in a *time-sequential* manner) as

$$\mathcal{D}_{N,n_1^*} \subseteq \ldots \subseteq \mathcal{D}_{N,n_r^*}, \tag{4}$$

where $n_j^* = n_{t_j}$, $j = 1, \ldots, r$, are possibly stochastic. This scenario corresponds to a *multiple* Type I censoring scheme — a natural extension of the single-point Type I censoring considered in Subsection 4.9.1. On the other hand, on an information time basis, we have a prefixed subset $\{n_1 < \ldots < n_r\}$ of positive integers ($n_r \leq N$), and we decide to look into the accumulating dataset only at the (stochastic) timepoints $X_N^{(n_j)}$, $j = 1, \ldots, r$. Thus, we consider here the partial sequence

$$\mathcal{D}_{N,n_1} \subseteq \ldots \subseteq \mathcal{D}_{N,n_r}, \tag{5}$$

where we note that the n_j are not stochastic. This corresponds to a multiple Type II censoring scheme. These two multiple censoring schemes differ in the same way as the original single-point censoring schemes described in the preceding two subsections.

In the multiple Type I censoring scheme, at each (stochastic) n_j^*, we may compute the statistic S_I, with N^* replaced by n_j^*, and we denote this by $S_I^{(j)}$, for $j = 1, \ldots, r$. As in Subsection 4.9.1, it is easy to show that under $H_0 : \beta = 0$ (i.e., $F_1 = \ldots = F_N = F$), given $\{n_1^*, \ldots, n_r^*\}$, $(S_I^{(1)}, \ldots, S_I^{(r)})$ is CDF (not EDF), and hence, suitable CDF rank tests can be based on this partial set $\{S_I^{(1)}, \ldots, S_I^{(r)}\}$. In the construction of such a (possibly multi-stage) rank test, the accumulating nature of the $\mathcal{D}_{N,n_j^*}$, as well as the possibility of an early stopping (at one of the prefixed timepoints $t_1, \ldots, t_r$) are to be properly taken into account.

The situation is comparatively simpler for the multiple Type II censoring scheme. Here, based on the dataset $\mathcal{D}_{N,n_j}$, we may compute the statistic S_{II} that we denote by $S_{II}^{(j)}$, for $j = 1, \ldots, r$. Since the n_j are all nonstochastic, it follows from the general results in Subsection 4.9.2, that under the null hypothesis H_0, the $S_{II}^{(j)}, j = 1, \ldots, r$ are all EDF; the (discrete uniform) permutation distribution of the antirank vector $D = (D_1, \ldots, D_N)$ over the permutations of $\{1, \ldots, N\}$ generates the joint distribution of the $S_{II}^{(j)}$. From this perspective, multiple Type II censoring has a tactical advantage over the multiple Type I censoring scheme.

Given the time-sequential nature of the $\mathcal{D}_{N,k}$, $k \leq N$, and provision for an early stopping, it is quite natural to draw an analogy with multi-stage or sequential tests that can be adapted to this partial sequence, and can be based either on the $S_I^{(j)}$ or the $S_{II}^{(j)}$. Moreover, in many clinical studies, particularly in exploratory ones, it may be difficult to choose r properly, as well as $t_1, \ldots, t_r$ (or $n_1, \ldots, n_r$) in advance. An optimal choice of r, and the n_j or t_j, may depend quite intricately on the underlying distribution function F (which is generally not precisely known), as well as on the class of alternative hypothesis which may not be that precisely formulated either. Further, in many medical studies involving human beings as subjects, *medical ethics* may often prompt a more or less continuous monitoring of the trial. This amounts to looking at the Type II censored statistics $S_{\text{II}}^{(k)}$ at the successive failure points $X_N^{(k)}, 1 \leq k \leq N$. As has been noted earlier, there is no change in the accumulating data-cloud in between successive failure points, and hence continuous monitoring can be conveniently discretized in this manner without any essential loss of statistical information. This motivates the basic formulation of progressive censoring schemes (Chatterjee and Sen, 1973a). As we shall see, in these schemes both Type I and Type II censoring can be effectively merged into a more general scheme that is essentially time-sequential in nature; though a progressive censoring scheme has a discrete time-parameter, it adopts very well to a continuous time-parameter setup. Keeping these factors in mind, we adopt S_{II} when $n^* = k$, and denote it by $S_{N,k}$, for $k = 0, 1, \ldots, N$. Note that $S_{N,0} = 0$, while $S_{N,N-1} = S_{N,N} = \sum_{i=1}^{N} c_{D_i} a_N(i)$. Defining the antirank vector D as before, we let

$$D^{(k)} = (D_1, \ldots, D_k), \quad 1 \leq k \leq N; D^{(0)} = \{0\}. \tag{6}$$

Note that under $H_0 : \beta = 0$ (or $F_1 = \ldots = F_N = F$), we have

$$P_0\{D^{(k)} = (i_1, \ldots, i_k)\} = N^{-[k]} = \{N \ldots (N-k+1)\}^{-1}, \tag{7}$$

for every $1 \leq i_1 \neq \ldots \neq i_k \leq N$, and every $1 \leq k \leq N$. Let us also denote by $D^{(k)c} = D \setminus D^{(k)} = (D_{k+1}, \ldots, D_N)$, the complementary subset of antiranks (for $k = 0, 1, \ldots, N$). Then given $D^{(k)}$, its complementary $D^{(k)c}$ is given too, though the ordering of its $N-k$ elements is not known. Nevertheless, under H_0, all the $(N-k)!$ possible orderings of its elements are conditionally equally likely, and hence,

$$P_0\{\, D_{k+j} = i_{k+j} \mid D^{(k)}\} = (N-k)^{-1}, \quad 1 \leq j \leq N-k, \tag{8}$$

for every $i_{k+j} \in \{1, \ldots, N\} \setminus \{i_1, \ldots, i_k\}$. We denote by $\mathcal{B}_{N,k}$ the sub-σ-field generated by $D^{(k)}$, for $k = 0, 1, \ldots, N$; $\mathcal{B}_{N,0}$ is the trivial σ-field, and $\mathcal{B}_{N,k}$ is non-decreasing in k ($\leq N$). Then we have on noting that $\sum_{i=1}^{N} c_i = 0$,

$$\begin{aligned} \mathsf{E}_0\{S_{N,N} \mid \mathcal{B}_{N,k}\} &= \sum_{i=1}^{k} c_{D_i} a_N(i) + \sum_{i=k+1}^{N} a_N(i) \mathsf{E}_0\{c_{D_i} \mid \mathcal{B}_{N,k}\} \\ &= \sum_{i=1}^{k} c_{D_i} a_N(i) + \sum_{i=k+1}^{N} a_N(i) \{\frac{1}{N-k} \sum_{j=k+1}^{N} c_{D_j}\} \\ &= \sum_{i=1}^{k} c_{D_i} a_N(i) - \{\frac{1}{N-k} \sum_{i=k+1}^{N} a_N(i)\} \sum_{i=1}^{k} c_{D_i} \\ &= \sum_{i=1}^{k} c_{D_i} [a_N(i) - a_N^*(k)] = S_{N,k}, \quad 0 \leq k \leq N. \end{aligned} \tag{9}$$

This shows that under H_0, for every $N(\geq 1)$, the array $\{S_{N,k}; 0 \leq k \leq N\}$ is a martingale array, closed on the right by $S_{N,N}$. Let us denote

$$A_{N,k}^2 = \frac{1}{N-1}\{\sum_{i=1}^{k} a_N^2(i) + (N-k)(a_N^*(k))^2 - N\bar{a}_N^2\}, \tag{10}$$

for $k = 0, 1, \ldots, N$; note that $A_{N,0}^2 = 0,\ A_{N,N}^2 = A_N^2$, where

$$A_N^2 = \frac{1}{N-1}\{\sum_{i=1}^{N} a_N^2(i) - N\bar{a}_N^2\}. \tag{11}$$

Then it follows from the standard formulas for the moments of linear rank statistics, considered before, that under H_0, $\mathsf{E}_0\{S_{N,k}\} = 0$, and

$$\mathsf{E}_0\{S_{N,k}^2\} = C_N^2 A_{N,k}^2, \text{ for } k = 0, 1, \ldots, N. \tag{12}$$

Therefore, incorporating the above (Chatterjee and Sen, 1973a) martingale characterization, we obtain that for every $N \geq 1$,

$$0 = A_{N,0}^2 \leq A_{N,1}^2 \leq \ldots \leq A_{N,N}^2 = A_N^2. \tag{13}$$

Motivated by the martingale and variance inequality results, in a progressive censoring scheme, we may consider the following test statistics:

(i) One-sided alternatives ($\beta > 0$),

$$K_N^+ = \max_{0 \leq k \leq N} S_{N,k}/\{A_N C_N\}; \tag{14}$$

(ii) Two-sided alternatives ($\beta \neq 0$),

$$K_N = \max_{0 \leq k \leq N} |S_{N,k}|/\{A_N C_N\}. \tag{15}$$

Some other variants of K_N, K_N^+, based on suitable weight-functions or truncation of k values at some Np for a prefixed $p \in (0, 1]$, have also been proposed. We shall discuss them in later chapters.

4.9.4 Interval censoring. In practice, even if the underlying distributions were continuous, mostly, due to rounding off the observations, the process of data collection may result in a large number of small class intervals. For other reasons, we may often encounter a prefixed number of non-overlapping class intervals whose widths may not be small compared to the range of the observations. In the first case, the methodology introduced in Section 4.8 applies, while the situation with the latter case is somewhat different. Taking clues from the theory of LMPR tests, a general class of rank tests for such grouped data has been developed by Sen (1967), and extended further by Ghosh (1973a, b). Such rank tests are also usable for ordinal data with polychotomous responses (that are at least partially ordered). For example, in an opinion survey, the response may be recorded on a five-point scale: strongly in favour, mildly in favour, no preference, mildly opposed, and strongly opposed. Thus, if we conceive of an underlying (unobservable) trait having a continuous distribution, then these ordered categories may be conceived as appropriate class intervals on the trait variable. In what follows, interval censoring relates to this composite situation where the response variable X may not be precisely measurable; rather it corresponds to a set of (partially) ordered class intervals. Though in many cases the boundaries of these non-overlapping class intervals are either given or observable in some way, in many other contexts, particularly arising in psychometric and educational testing problems, they may be only vaguely ordered in some sense. For simplicity of presentation, we consider first the specified boundaries case, and then show how the same methodology applies to the other case as well.

We conceive of a finite set $J = \{1, \ldots, m\}$ of non-overlapping class intervals $I_j = (a_j, a_{j+1})]$, $j = 1, \ldots, m$, where

$$-\infty \leq a_1 < a_2 < \ldots < a_m < a_{m+1} \leq \infty. \tag{1}$$

[Ghosh (1973a, b) showed that a countable set J, considered by Sen (1973), can be essentially replaced by a finite set.] Let us also introduce the unobservable random variables as $X_1, \ldots, X_N$, and as in (4.9.1.1) we assume that they follow a regression model. The observable random variables are then defined as

$$X_i^* = \sum_{j=1}^{m} Z_{ij} I_j, \; i = 1, \ldots, N, \tag{2}$$

where for each $i = 1, \ldots, N$,

$$Z_{ij} = 1 \text{ or } 0, \text{ according as } X_i \in I_j \text{ or not}, \tag{3}$$

for $j = 1, \ldots, m$. Thus, the X_i^* represent the (random) class intervals where the unobservable X_i belong to.

Note that in the above formulation, ties among the X_i^* may no longer be neglected, with probability one; hence, as in Section 4.8, based on suitable randomization principles, we need to make adjustments for ties among the X_i^*. Nevertheless, we may appeal to suitable LMP criteria to motivate and formulate appropriate (CDF) rank tests for such interval censoring schemes. As in Section 3.4, we assume that the distribution function F in (4.9.1.1) admits an absolutely continuous density f with finite Fisher information for location, viz., $I(f) = \int_{-\infty}^{\infty} \{f'(x)/f(x)\}^2 dF(x) < \infty$. Also, let

$$F_j = F(a_j), \; P_j = F(a_{j+1}) - F(a_j). \; j = 1, \ldots, m. \tag{4}$$

Then the likelihood function of the X_i^* under the regression model in (4.9.1.1) is

$$\prod_{i=1}^{N}\Big\{\prod_{j\in J}[F(a_{j+1}-\beta c_i)-F(a_j-\beta c_i)]^{Z_{ij}}\Big\}, \tag{5}$$

so that the log-likelihood ratio statistic for testing $H_0:\beta=0$ against $H_1:\beta>0$, comes out as

$$\sum_{i=1}^{N}\sum_{j\in J} Z_{ij}\log\{[F(a_{j+1}-\beta c_i)-F(a_j-\beta c_i)]/P_j\}. \tag{6}$$

Then confining β to a small interval $(0,\gamma]$, for a small $\gamma(>0)$, by direct expansion, we obtain that (6) reduces to

$$\beta\sum_{i=1}^{N}\sum_{j\in J} Z_{ij}c_i\Delta_j^o \;+\; o(\beta), \quad \text{for all } 0<\beta<\gamma, \tag{7}$$

where

$$\begin{aligned}\Delta_j^o &= \{f(a_j)-f(a_{j+1})\}/P_j \\ &= \Big(\int_{F_j}^{F_{j+1}}\psi_F(u)\,\mathrm{d}u\Big)\Big/\Big(\int_{F_j}^{F_{j+1}}\mathrm{d}u\Big), \quad j\in J,\end{aligned} \tag{8}$$

and

$$\psi_F(u)=-f'(F^{-1}(u))/f(F^{-1}(u)), \quad 0<u<1, \tag{9}$$

is the Fisher score function. Thus, a LMP test for $H_0:\beta=0$ against $H_1:\beta>0$ is based on the test statistic

$$T_N=\sum_{i=1}^{N}c_i\Big(\sum_{j\in J}Z_{ij}\Delta_j^o\Big). \tag{10}$$

Now the distribution function F and the density f (or f') are generally not known, so that $\psi_F(u)$ or the Δ_j^o are not known either. We proceed to replace the Δ_j^o by their sample counterparts as follows.

Let $N_j=\sum_{i=1}^{N}Z_{ij}$ and $P_{Nj}=N_j/N$, $j\in J$. Also let $F_{N,j+1}=\sum_{r\le j}P_{Nr}$, $j\in J$; $F_{N,1}=0$. Finally, consider a score generating function $\varphi=\{\varphi(u),\ 0<u<1\}$ in the same way as in the case of continuous distribution functions, treated in earlier sections, and define

$$\hat{\Delta}_{Nj}=P_{Nj}^{-1}\int_{F_{N,j}}^{F_{N,j+1}}\varphi(u)\,\mathrm{d}u, \quad j\in J; \tag{11}$$

if $P_{Nj} = 0$ for some $j \in J$ (that corresponds to an empty cell), we let conventionally $\hat{\Delta}_{Nj} = \varphi(F_{N,j+1})$. This does not affect the picture as $P_{Nj} = 0 \Rightarrow Z_{ij} = 0$, $i = 1, \ldots N$. Then we define an interval censored linear rank statistic as

$$L_{N,IC} = \sum_{i=1}^{N} c_i \Big(\sum_{j \in J} \hat{\Delta}_{Nj} Z_{ij} \Big). \tag{12}$$

Note that the $\hat{\Delta}_{Nj}$ depend on the adopted score function φ and the cell frequencies N_j, $j \in J$, so that the test statistic in (12) covers also the case of ordered categorical data models. With this motivation, next we proceed to present the moments of $L_{N,IC}$ under a randomization model.

Note that under H_0, the X_i^* are independent identically distributed, and their permutation distribution $\mathcal{P}_N$ is generated by the $(N!/\prod_{j \in J} N_j!)$ conditionally (given the $N_j, j \in J$) equally likely distinct permutations. Further, given the N_j, the $\hat{\Delta}_{Nj}$ are fixed, so that these estimated scores are $\mathcal{P}_N$-invariant. Moreover,

$$\mathsf{E}_{\mathcal{P}_N} Z_{ij} = \mathsf{E}_{\mathcal{P}_N} Z_{ij}^2 = P_{Nj}, \quad \mathsf{E}_{\mathcal{P}_N} Z_{ij} Z_{ij'} = 0, \tag{13}$$

for every $j \neq j' \in J$, $i = 1, \ldots, N$, and

$$E_{\mathcal{P}_N} Z_{ij} Z_{i'j'} = N_j (N_{j'} - \delta_{jj'})/\{N(N-1)\}, \tag{14}$$

where $i \neq i'$, $j, j' \in J$, and $\delta_{jj'}$ is the Kronecker delta. Therefore,

$$\begin{aligned} \mathsf{E}_{\mathcal{P}_N} L_{N,IC} &= \sum_{i=1}^{N} c_i \Big(\sum_{j \in J} P_{Nj} \hat{\Delta}_{NJ} \Big) \\ &= \sum_{i=1}^{N} c_i \Big(\int_0^1 \varphi(u) du \Big) = 0, \end{aligned} \tag{15}$$

as $\bar{c}_N = N^{-1} \sum_{i=1}^{N} c_i = 0$. Similarly,

$$\mathsf{E}_{\mathcal{P}_N} L_{N,IC}^2 = N(N-1)^{-1} C_N^2 A_{N(J)}^2, \tag{16}$$

where, as before $C_N^2 = \sum_{i=1}^{N} (c_i - \bar{c}_N)^2 = \sum_{i=1}^{N} c_i^2$, and

$$A_{N(J)}^2 = \sum_{j \in J} P_{Nj} \hat{\Delta}_{Nj}^2 - \bar{\varphi}^2; \; \bar{\varphi} = \int_0^1 \varphi(u) du. \tag{17}$$

Therefore, in this interval censored case, it may be quite intuitive to consider the following standardized linear rank statistic:

$$L^*_{N,IC} = (\frac{N-1}{N})^{1/2} C_N^{-1} A_{N(J)}^{-1} L_{N,IC}. \tag{18}$$

A two-sided test statistic can be obtained by using $|L^*_{N,IC}|$ instead of $L^*_{N,IC}$.

At this stage, we may note that if the class intervals were chosen on the basis of the order statistics $X_N^{(k)}$, $k = N_1, \ldots, N_m$, for some prefixed $N_j, j \in J$, then interval censoring would have corresponded to the multiple Type II censoring , treated in the previous subsection; hence we would have an EDF test. However, in the given case, the $\hat{\Delta}_{Nj}$ are all stochastic (though $\mathcal{P}_N$-invariant), and hence, we have CDF (not EDF) tests. Other properties of interval censored rank tests will be studied later on.

4.9.5 Random censoring. In clinical trials and medical studies involving a follow-up scheme, subjects or patients may drop out of the scheme due to various reasons. For example, they may move away from the clinic area or die due to some other causes. For this reason, it is necessary to take into account the *compliance* of the subjects in the study plan, and this is usually done by introducing a non-negative random variable (say, T) representing the withdrawal or drop-out time. Thus, corresponding to the (possibly unobservable) failure times $X_1, \ldots, X_N$ of the N subjects in a follow-up scheme, we conceive of a set $\{T_1, \ldots, T_N\}$ of withdrawal times, and the observable random elements are

$$X_i^* = \min\{X_i, T_i\} \text{ and } \delta_i = I(X_i^* = X_i),\ i = 1, \ldots, N. \tag{1}$$

The main difference between this scheme and Type I or II censoring schemes is that here the censoring time T_i is not assumed to be the same for all i; rather, they are allowed to be stochastically different. However, to accommodate this flexibility, there is a basic assumption in the present context that deserves a careful appraisal. It is generally assumed that the T_i are independent identically distributed random variables, and that the T_i are independent of the X_i which may not always be tenable in practice. For example, if we consider the simple regression model in (4.9.1.1) (for the X_i), when $\beta \neq 0$, the X_i carry information on the regression parameter. On the other hand, if $G(x) = P\{T_i \leq x\}$, $0 \leq x < \infty$, is the distribution function of T_i, then by assumption, G is the same for all i, and moreover, T_i being independent of X_i, does not convey any information on β. That is why T_i is termed a *non-informative censoring* variable, and the scheme is termed a random censoring scheme.

Recall that by definition in (1), and by the assumptions made above, we have for $H_i(x) = P\{X_i^* \leq x\}$,

$$\bar{H}_i(x) = 1 - H_i(x) = [1 - F_i(x)][1 - G(x)] = \bar{F}_i(x)\bar{G}(x), \tag{2}$$

for $i = 1, \ldots, N$, where $\bar{F}_i, \bar{G}, \bar{H}_i$ stand for the respective survival functions. Therefore, the null hypothesis that $F_1 = \ldots = F_N = F$ (unknown) implies that $H_1 = \ldots = H_N = H = 1 - \bar{F}\bar{G}$ holds. Hence, one simple solution to this situation (involving the nuisance distribution function G) is to ignore completely the tagging variables δ_i and base the conventional rank tests solely on the X_i^*. This will result in an EDF test for the homogeneity of the F_i. But by (2), for every $i \neq j (= 1, \ldots, N)$,

$$\begin{aligned} |H_i(x) - H_j(x)| &= |\bar{H}_i(x) - \bar{H}_j(x)| = |\bar{G}(x)[\bar{F}_i(x) - \bar{F}_j(x)]| \\ &\leq |\bar{F}_i(x) - \bar{F}_j(x)| = |F_i(x) - F_j(x)|, \text{ for all } x, \end{aligned} \tag{3}$$

as $\bar{G}(x) \leq 1$, for all x. Therefore the H_i carry less discriminatory power than the original F_i. Thus, a rank test based solely on the X_i^* may entail some loss of power-efficiency, particularly if there is *heavy censoring* (which occurs when $\bar{G}(x)$ is small for the normal range of support of the $F_i(x)$). Note that if G is assumed to have a density $g(.)$, then $h_i(x) = (\mathrm{d}/\mathrm{d}x)H_i(x)$ can be expressed as

$$h_i(x) = \bar{G}(x)f_i(x) + \bar{F}_i(x)g(x), \quad i = 1, \ldots, N, \tag{4}$$

so that under the contemplated regression model (4.9.1.1), we have

$$h_i(x) = \bar{G}(x)f(x - \beta c_i) + g(x)\bar{F}(x - \beta c_i), \quad i = 1, \ldots, N. \tag{5}$$

Thus, we have

$$\begin{aligned} (\partial/\partial\beta)h_i(x) &= g(x)c_i f(x - \beta c_i) + c_i\{-f'(x - \beta c_i)\}\bar{G}(x) \\ &= c_i f(x - \beta c_i)\bar{G}(x)\{\frac{g(x)}{\bar{G}(x)} + \frac{-f'(x - \beta c_i)}{f(x - \beta c_i)}\}, \end{aligned} \tag{6}$$

for $i = 1, \ldots, N$. This invalidates the LMPR characterization of the rank test based on the Fisher score function $\psi_F(.)$, defined in (4.9.4.9); the optimal score function depends in a very intricate manner on $g, \bar{G}$, and their proximity to f and $\bar{F}$. For these reasons, the information in the tagging variables δ_i is generally utilized in the formulation of suitable rank tests in such a random censoring scheme.

Note that the joint density for (X_i^*, δ_i) is given by

$$[f_i(x^*)\bar{G}(x^*)]^{\delta_i}[g(x^*)\bar{F}_i(x^*)]^{1-\delta_i}, \quad \delta_i = 0, 1, \ x^* \geq 0, \tag{7}$$

for $i = 1, \ldots, N$. If we let $\mu_i(x) = f_i(x)/\bar{F}_i(x)$, $i = 1, \ldots, N$, and $\nu(x) = g(x)/\bar{G}(x)$, $x \geq 0$, be the *hazard rates* corresponding to F_i and G respectively, then we may rewrite (7) as

$$\bar{H}_i(x^*)\{\mu_i(x^*)\}^{\delta_i}\{\nu(x^*)\}^{1-\delta_i}, \ x^* \geq 0, \tag{8}$$

for $i = 1, \ldots, N$. Even in this form, the derived likelihood function is not amenable for construction of EDF rank tests. Cox (1972) eliminated this impasse by extending the Lehmann (1953) alternative to a regression model, based on the hazard functions $\mu_i(x), i \geq 1$. Since, by definition, the $\mu_i(x)$ are non-negative, Cox (1972) introduced the following *proportional hazards model*:

$$\mu_i(x) = \mu(x) \cdot \exp\{\beta c_i\}, \quad i = 1, \ldots, N, \tag{9}$$

where $\mu(\cdot)$ is an arbitrary non-negative function, termed the *base line hazard function*, and it is treated as a nonparametric function. On the other hand, the dependence of the $\mu_i(\cdot)$ on the regression constants c_i is of a specified parametric form, involving the (unknown) regression parameter β. Therefore, this model is also termed a *semiparametric model*. The ingenuity of Cox (1972) lies in the formulation of the so-called *partial likelihood function* that leads to suitable tests for $H_0 : \beta = 0$ against $H_1 : \beta \neq 0$ (or $> $ or < 0) that are essentially rank type tests.

To formulate such tests, we denote the order statistics related to the X_i^* by $X_N^{*(1)} \leq \ldots \leq X_N^{*(N)}$; ties among them may be neglected with probability one. For any $x \geq 0$, let

$$\mathcal{R}_N(x) = \{i : X_i^* \geq x, \ i = 1, \ldots, N\} \tag{10}$$

be the set of observations who are at risk at timepoint x, and hence we term the $\mathcal{R}_N(x)$ as the *risk sets*; for notational simplicity, we let $\mathcal{R}_N(X_N^{*(k)}) = \mathcal{R}_{Nk}$, $k = 1, \ldots, N$. Note that

$$\mathcal{R}_{N1} \supseteq \mathcal{R}_{N2} \supseteq \ldots \supseteq \mathcal{R}_{NN}. \tag{11}$$

Let $m \ (= m_N) = \sum_{i=1}^{N} \delta_i$ be the number of failure points, so that the remaining $N - m$ timepoints relate to censoring points. Under $H_0 : F_1 = \ldots = F_N = F$ (i.e., $\beta = 0$ in (9)), m has a binomial law with sample size N and parameter $\pi = P\{X < T\} = \int_0^\infty F(x)dG(x)$; we assume that $\pi > 0$. Let us denote the m ordered failure points by $X_{N,1}^* < \ldots < X_{N,m}^*$, and note that by definition

$$X_{N,j}^* = X_N^{*(k_j)}, \ 1 \leq j \leq m; \ 1 \leq k_1 < \ldots < k_m \leq N. \tag{12}$$

Now consider the risk set $\mathcal{R}_{Nj}^* = \mathcal{R}_N(X_{N,j}^*-)$. Then given the set of r_j (= the cardinality of $\mathcal{R}_{Nj}^*$) observations that are in the risk set $\mathcal{R}_{Nj}^*$, the conditional probability that the ith observation culminates in a failure at $X_{N,j}^*$ is given by

$$\mu_i(X_{Nj}^*) / \sum_{r \in \mathcal{R}_{Nj}^*} \mu_r(X_{Nj}^*), \quad \text{for all } i \in \mathcal{R}_{Nj}^*, \tag{13}$$

so that under the proportional hazards model, the above reduces to

$$\exp\{\beta c_i\}/\sum_{r\in\mathcal{R}^*_{Nj}}\exp\{\beta c_r\}, \tag{14}$$

for $j = 1, \ldots, m$. This observation and the nested pattern in (11) led Cox (1972) to formulate the partial likelihood function as

$$L_{NP}(\beta) = \prod_{j=1}^{m} \frac{e^{\beta c_{D^*_j}}}{\sum_{r\in\mathcal{R}^*_{Nj}} e^{\beta c_r}}, \tag{15}$$

where $D^*_j = D_{k_j}$ stands for the anti-ranks for $X_N^{*(k_j)}$, $j = 1, \ldots, m$. We consider then the partial likelihood scores statistic

$$(\partial/\partial\beta)\log L_{NP}(\beta)|_{\beta=0} = \sum_{j=1}^{m}\Big\{c_{D^*_j} - \frac{1}{r_j}\sum_{r\in\mathcal{R}^*_{Nj}} c_r\Big\}, \tag{16}$$

where we note that $r_j = N - k_j + 1$, for $j = 1, \ldots, m$. Recall that $r \in \mathcal{R}^*_{Nj}$ if $r = D_k$, for some $k \geq k_j$, and hence, we rewrite the partial likelihood score statistic as

$$S^*_{N,m} = \sum_{j=1}^{m}\Big\{c_{D_j} - (N - k_j + 1)^{-1}\sum_{k=k_j}^{N} c_{D_k}\Big\}, \tag{17}$$

which for the given $c_1, \ldots, c_N$ and observed $k_1, \ldots, k_m$, is a linear statistic in the antirank scores $\{c_{D_i}, i \geq 1\}$. Hence, it is a linear rank statistic. If there is no censoring, i.e., $m = N$, we have $k_j = j$, $j = 1, \ldots, N$, so that the coefficients of the c_{D_i} in the above expression reduce to $1 - \sum_{r\leq i}(N - r + 1)^{-1} = -a^o_N(i)$, $i = 1, \ldots, N$, where the $a^o_N(i)$ are the so-called *log-rank* or Savage scores for the two-sample problem. This enables us to rewrite $S^*_{N,N}$ in the uncensored case as

$$-\sum_{j=1}^{N} c_{D_j} a^o_N(j) = -\sum_{j=1}^{N} c_j a^o_N(R_j), \tag{18}$$

where the R_j are the ranks of the X_j among themselves, and this is a variant of the classical log-rank statistic. Hence, in the censored case, we term the score statistic the generalized log-rank scores statistic.

Note that this statistic depends on $k_1, \ldots, k_m$ and the anti-rank vector $D = (D_1, \ldots, D_N)$, when the c_i are all prefixed. Therefore, given $(m, k_1, \ldots, k_m)$, this statistic is CDF, and it reduces to a EDF statistic in the uncensored case. In this development the conventional regression model (4.9.1.1) has been replaced by the Cox (1972) model which is a semiparametric model. This explains that conventional rank statistics are also adaptable in some semiparametric models.

4.10 MULTIVARIATE RANK TESTS

In this section, we present the basic features of some multivariate rank statistics that arise in the context of hypothesis testing for *diagonal symmetry* and homogeneity of (several) multivariate distributions; these hypotheses are the generalizations of the hypotheses of symmetry and homogeneity considered in the univariate setups in Sections 4.1 and 4.5. Only the basic *rank-permutation principles* (Chatterjee and Sen (1964), Chatterjee (1966a), and Sen and Puri (1967)) underlying the formulation of CDF tests, and their relevance to the univariate problems, are outlined here. We refer to Puri and Sen (1971) for a more elaborate treatment of this specialized branch of nonparametrics.

4.10.1 Bivariate sign tests. Let $(X_1, Y_1), \ldots, (X_N, Y_N)$ be a random sample from a bivariate distribution function $F(x, y)$. We denote the two marginal distribution functions by

$$F_1(x) = F(x, \infty) \text{ and } F_2(y) = F(\infty, y). \tag{1}$$

Suppose that F_1, F_2 are continuous and strictly monotone at their respective medians which are denoted by θ_1 and θ_2, We frame the null hypothesis as

$$H_0 : \theta_1 = 0, \theta_2 = 0, \text{ or } \theta = \mathbf{0}, \tag{2}$$

against alternatives that $\theta \neq \mathbf{0}$. Note that $(\operatorname{sign} X, \operatorname{sign} Y)$ has four possible realizations $(1,1), (1,-1), (-1,1)$ and $(-1,-1)$, and even under the null hypothesis H_0, the probabilities for these four realizations depend on the unknown parameter $\pi = F(0,0)$, where π lies in $(0, 1/2)$. As a result, a test based on the vector $\mathbf{S}_N = (S_{N1}, S_{N2})' = \sum_{i=1}^{N} (\operatorname{sign} X_i, \operatorname{sign} Y_i)'$ is not generally EDF; ADF tests based on sample estimates of π have been considered by Bennett (1962, 1964), and others.

Chatterjee (1966a) proposed an ingenious conditional setup that generates a simple CDF test. For this, we define (X, Y) to be *concordant* or *discordant* according as $(\operatorname{sign} X)(\operatorname{sign} Y)$ is positive or not. Further, we say that X, Y are concordant of the first and second kind according as $X > 0, Y > 0$ or $X < 0, Y < 0$. Similarly X, Y is discordant of the first and second kind according as $X > 0, Y < 0$ or $X < 0, Y > 0$. Let $P(C_1)$, $P(C_2)$, $P(D_1)$ and $P(D_2)$ be respectively the probabilities of concordance of the first and second kind, and discordance of the first and second kind; in the sample of size N the respective frequencies are denoted by C_1, C_2, D_1 and D_2, so that $C = C_1 + C_2$ and $D = D_1 + D_2$ stand for the number of concordant and discordant observations. Their joint probability law is

$$\frac{N!}{(C_1)!(C_2)!(D_1)!(D_2)!} (P(C_1))^{C_1} (P(C_2))^{C_2} (P(D_1))^{D_1} (P(D_2))^{D_2}. \tag{3}$$

The conditional distribution of C_1, D_1, given C, D, is therefore

$$\binom{C}{C_1}\gamma^{C_1}(1-\gamma)^{C_2} \cdot \binom{D}{D_1}\delta^{D_1}(1-\delta)^{D_2}, \tag{4}$$

where

$$\gamma = P(C_1)/\{P(C_1)+P(C_2)\}, \text{ and } \delta = P(D_1)/\{P(D_1)+P(D_2)\}. \tag{5}$$

The ingenuity of Chatterjee (1966a) lies in showing that

$$H_0 : \theta = \mathbf{0} \ \Leftrightarrow \ \gamma = \delta = 1/2. \tag{6}$$

Therefore, it suffices to test the equivalent null hypothesis $H_0^* : \gamma = \delta = 1/2$, against the set of alternatives that $(\gamma, \delta) \neq (1/2, 1/2)$. Now, under H_0^*, (4) reduces to

$$\binom{C}{C_1}2^{-C} \cdot \binom{D}{D_1}2^{-D}, \tag{7}$$

which is the product of two (conditionally, given C) independent binomial laws that do not involve any nuisance parameter. Based on this characterization, a simple test statistic is

$$T_N = \frac{(C_1 - C_2)^2}{C_1 + C_2} + \frac{(D_1 - D_2)^2}{D_1 + D_2}. \tag{8}$$

The exact (conditional) distribution of T_N, under H_0 (given C) can be obtained from (7) by direct enumeration, and hence T_N is CDF. In this context, Chatterjee (1966a) showed that the test based on T_N is unbiased while other ADF tests may not be so.

4.10.2 Multivariate signed-rank tests. Consider now signed-rank statistics for testing the null hypothesis of diagonal symmetry in a genuine multivariate setup. Recall that a random p-variate vector $\mathbf{X}$ has a distribution function $F(\mathbf{x}; \theta)$ diagonally symmetric about a p-variate constant θ if both $\mathbf{X} - \theta$ and $\theta - \mathbf{X}$ both have the same distribution function. If we write

$$F(\mathbf{x}; \theta) = F(\mathbf{x} - \theta), \tag{1}$$

then whenever F has a density $f(\mathbf{x})$, we have an equivalent characterization of diagonal symmetry:

$$f(\mathbf{x}) = f((-1)\mathbf{x}), \quad \text{for all } p\text{-pvariate } \mathbf{x}. \tag{2}$$

Suppose now that $\mathbf{X}_1, \ldots, \mathbf{X}_N$ is a random sample from the distribution function $F(\mathbf{x}; \theta)$ that is diagonally symmetric about θ. We want to test the null hypothesis

$$H_0 : \theta = \mathbf{0}, \text{ against } H_1 : \theta \neq \mathbf{0}. \tag{3}$$

For the univariate case (i.e., for $p = 1$), signed-rank tests were shown to be EDF for this hypothesis testing problem. However, if the coordinates of $\mathbf{X}$ are not stochastically independent, the vector of coordinatewise signed-ranks may not be EDF even under the null hypothesis. This may easily be verified by considering the particular case of $p = 2, N = 2$. To eliminate this impasse, we consider the following *sign-invariance* principle, due to Sen and Puri (1967), that renders CDF rank tests.

Let us denote the sample point by $\mathbf{E}_N = (\mathbf{X}_1, \ldots, \mathbf{X}_N)$, so that the corresponding sample space is $\mathcal{E}_N = pN$-dimensional Euclidean space. Let then

$$g_N \mathbf{E}_N = ((-1)^{i_1}\mathbf{X}_1, \ldots, (-1)^{i_N}\mathbf{X}_N), \; g_N \in \mathcal{G}_N, \tag{4}$$

where $i_j = 0, 1$, for $j = 1, \ldots, N$, so that $\mathcal{G}_N$ is the group of 2^N possible sign-inversions. Now, for any observed $\mathbf{E}_N$, the group $\mathcal{G}_N$ generates an *orbit* $\mathbf{O}_N$ of 2^N points in $\mathcal{E}_N$, given by the set of realizations in (4). The distribution of $\mathbf{E}_N$ may generally depend on the unknown F through the unknown association pattern (even under H_0). On the other hand, under H_0, each element of $\mathbf{O}_N$ has the same distribution, and hence if we denote the conditional probability law of $\mathbf{E}_N$ on $\mathbf{O}_N$ by $\mathcal{P}_N$, it follows that $\mathcal{P}_N$ has the discrete uniform distribution with the common probability mass 2^{-N} at each of the points of $\mathbf{O}_N$. This conditional probability law (under H_0) does not depend on the unknown F, and hence test statistics based on $\mathcal{P}_N$ will be CDF under H_0.

As in the univariate case, we consider here the jth row of $\mathbf{E}_N$, and denote these elements by $X_{1j}, \ldots, X_{Nj}$, for $j = 1, \ldots, p$. Let then

$$R_{ij}^+ = \sum_{r=1}^{N} I(|X_{rj}| \le |X_{ij}|), \quad i = 1, \ldots, N; \; j = 1, \ldots, p. \tag{5}$$

This way, we generate a *rank-collection* matrix

$$\mathbf{R}_N = ((R_{ij}^+))_{j=1,\ldots,p; i=1,\ldots,N}. \tag{6}$$

Also, we let $S_{ij} = \operatorname{sign} X_{ij}$, for all i, j, and denote the *sign-collection* matrix by

$$\mathbf{S}_N = ((S_{ij}))_{j=1,\ldots,p; i=1,\ldots,N}. \tag{7}$$

Further, for each $j(= 1, \ldots, p)$, we consider a set of scores $a_{Nj}(k)$, $k = 1, \ldots, N$, defined as in the univariate case, and construct a signed-rank statistic

$$T_{Nj} = N^{-1} \sum_{i=1}^{N} S_{ij} a_{Nj}(R_{ij}^+), \quad j = 1, \ldots, p; \tag{8}$$

we let $\mathbf{T}_N = (T_{N1}, \ldots, T_{Np})'$. It follows from the composition of $\mathcal{P}_N$ that

$$\mathsf{E}(\mathbf{T}_N | \mathcal{P}_N) = \mathbf{0} \text{ (a.e. } \mathcal{P}_N); \tag{9}$$

$$N\mathsf{E}(\mathbf{T}_N\mathbf{T}'_N|\mathcal{P}_N) = \mathbf{V}_N = ((v_{Njs}))_{j,s=1,\ldots,p}, \tag{10}$$

where

$$v_{Njs} = N^{-1}\sum_{i=1}^{N} a_{Nj}(R_{ij}^+)a_{Ns}(R_{is}^+)S_{ij}S_{is}, \tag{11}$$

for $j, s = 1, \ldots, p$; note that $v_{Njj} = N^{-1}\sum_{i=1}^{N} a_{Nj}^2(i)$ is non-random, for each $j = 1, \ldots, p$, but the off-diagonal elements are generally random. By analogy with the classical Hotelling's T^2-test (based on the vector of sample means and the sample covariance matrix), we consider here the following quadratic form as a suitable signed-rank test statistic

$$\mathcal{L}_N = N\mathbf{T}'_N\mathbf{V}_N^-\mathbf{T}_N, \tag{12}$$

where $\mathbf{V}_N^-$ is a suitable generalized inverse of $\mathbf{V}_N$. The conditional distribution of $\mathcal{L}_N$, under $\mathcal{P}_N$, over the set $\mathbf{O}_N$ generates a CDF test for H_0 against H_1 in (3). This test statistic is a direct multivariate extension of the univariate linear signed-rank statistic, and for small (to moderate) values of N, the exact conditional distribution of $\mathcal{L}_N$ can be obtained by enumerating all possible 2^N realizations of $\mathbf{T}_N$ (under $\mathcal{P}_N$); note that $\mathbf{V}_N$ is $\mathcal{P}_N$-invariant, and hence it does not vary over $\mathbf{O}_N$.

In passing, we may note that the LMPR test characterization of univariate signed-rank statistics established in Section 3.4 may not be generally tenable in the multivariate case. The main reason for this feature is the fact that θ in (3) is a p-variate vector, so that the power function, even locally, for $\theta \to \mathbf{0}$, depends on p arguments. In that vein, if we consider the scores

$$(\partial/\partial\theta)\log f(\mathbf{X}_i, \theta)|_{\theta=\mathbf{0}}, \quad i = 1, \ldots, N, \tag{13}$$

they may not correspond to a vector of marginal scores (that is, the $(\partial/\partial\theta_j)$ $\log f_j(X_{ij}, \theta_j)|_{\theta_j=0}$) or a linear combination of them; here the f_j stand for the marginal densities.) In fact, unlike the Hotelling T^2-statistic, $\mathcal{L}_N$ is not *affine invariant* (i.e., invariant under arbitrary non-singular linear transformations on the $\mathbf{X}_i$), and hence the best-invariance property may not be tenable for rank statistics. We shall discuss this further later on.

4.10.3 Multivariate linear rank statistics. We now consider some general linear rank statistics that pertain to multivariate several sample problems as well as linear models. Let $\mathbf{X}_1, \ldots, \mathbf{X}_N$ be N independent random p-variate ($p \geq 1$) vectors with continuous distribution functions $F_1, \ldots, F_N$ respectively. The null hypothesis of randomness relates to

$$H_0 : F_1 = \ldots = F_N = F \text{ (unknown)}, \tag{1}$$

against alternatives that they are not all the same. The basic difference between the multivariate case and the univariate one, treated earlier, is that for $p > 1$, the distribution function F may have generally correlated or statistically dependent coordinates, so that the joint distribution of the coordinatewise ranks may depend on such unspecified dependence pattern, even under (1). Thus such rank tests may not be generally EDF. We illustrate this point with the following two specific multivariate problems:

(i) *Multivariate multisample models.* Let $\mathbf{X}_{ij}$, $j = 1, \ldots, N_i$, be independent identically distributed random variables with a continuous p-variate distribution function F_i, for $i = 1, \ldots, c$ (≥ 2), where all these c samples are assumed to be independent, and $N = N_1 + \ldots + N_c$. The null hypothesis of randomness in (1) reduces here to the hypothesis of homogeneity, $H_0 : F_1 = \ldots = F_c = F$ (unknown). As in the univariate case, we may consider here the shift alternatives:

$$H_1 : F_i(\mathbf{x}) = F(\mathbf{x} - \boldsymbol{\Delta}_i), \quad i = 1, \ldots, c, \tag{2}$$

where the $\boldsymbol{\Delta}_i$ are some p-variate vectors, not all equal. It may be possible to consider other forms of alternative hypotheses (Chatterjee and Sen (1964, 1965)). The multivariate multisample location model pertains to (2), and it has generally a greater appeal in practice; it is directly related to the Hotelling-Lawley (generalized) T^2-statistics in the parametric case where F is assumed to be a multinormal distribution function with a positive definite covariance matrix.

(ii) *Multivariate linear models.* We conceive of the following:

$$F_i(\mathbf{x}) = F(\mathbf{x} - \theta - \beta \mathbf{c}_i), \; i = 1, \ldots, N; \; F \text{ unknown}, \tag{3}$$

where $\theta = (\theta_1, \ldots, \theta_p)'$ is an unknown intercept vector, β is a $p \times r$ matrix of unknown regression parameters, and the $\mathbf{c}_i$ are known r-variate vectors of regression constants, not all equal. In this setup, the null hypothesis (1) reduces to

$$H_0 : \beta = \mathbf{0}; \; F \text{ and } \theta \text{ are nuisance}, \tag{4}$$

and the alternatives relate to $\beta \neq \mathbf{0}$ (without loss of generality, we may set $\bar{\mathbf{c}}_N = N^{-1} \sum_{i=1}^{N} \mathbf{c}_i = \mathbf{0}$). Note that the multivariate multisample location model (2) is a special case of (3) when the $\mathbf{c}_i$ can take on only $c - 1$ distinct realizations. Here also, for $p > 1$, there may be a certain impasse in developing LMPR tests.

We denote the sample point by $\mathbf{E}_N = (\mathbf{X}_1, \ldots, \mathbf{X}_N)$. Note that under H_0, the $\mathbf{X}_i$ are independent identically distributed random vectors, so that their joint distribution remains invariant under any permutations of the columns of $\mathbf{E}_N$. Motivated by this basic feature, we consider the following group $\mathcal{G}_N$ of transformations g_N that map the sample space onto itself:

$$g_N \mathbf{E}_N = (\mathbf{X}_{i_1}, \ldots, \mathbf{X}_{i_N}), \tag{5}$$

where $(i_1, \ldots, i_N)$ is a permutation of $(1, \ldots, N)$. Thus there are $N!$ elements of $\mathcal{G}_N$. We denote by $\mathcal{P}_N$ the discrete uniform probability law over these $N!$ elements of $\mathcal{G}_N$; this was incorporated by Chatterjee and Sen (1964) in the formulation of a *rank-permutation* principle that yields CDF rank tests in various multivariate nonparametric problems. We outline this as follows.

For $X_{1j}, \ldots, X_{Nj}$, comprising the jth row of $\mathbf{E}_N$, we denote

$$R_{ij} = \sum_{k=1}^{N} I(X_{kj} \leq X_{ij}), \quad i = 1, \ldots, N;\ j = 1, \ldots, p, \tag{6}$$

and obtain the rank collection matrix

$$\mathbf{R}_N = ((R_{ij}))_{j=1,\ldots,p;i=1,\ldots,N}. \tag{7}$$

For each j $(= 1, \ldots, p)$, consider a set of scores $a_{Nj}(1), \ldots, a_{Nj}(N)$, defined as in the univariate case, and further the linear rank statistics

$$\mathbf{T}_{Nj} = (T_{Nj1}, \ldots, T_{Njr})' = \sum_{i=1}^{N} \mathbf{c}_i a_{Nj}(R_{ij}), \quad j = 1, \ldots, p. \tag{8}$$

Note that marginally, for each j, under H_0, the distribution of $\mathbf{T}_{Nj}$ is generated by the $N!$ equally likely realizations of $(R_{1j}, \ldots, R_{Nj})$, over the permutations of $(1, \ldots, N)$, so that a test based on $\mathbf{T}_{Nj}$ is EDF. But even under the null hypothesis the joint distribution of any pair $\mathbf{T}_{Nj}, \mathbf{T}_{Nr}$, $(j \neq r)$, depends on the unknown dependence pattern of the (X_{ij}, X_{ir}) and hence a test based on the $p \times r$ matrix $\mathbf{T}_N = (\mathbf{T}_{N1}, \ldots, \mathbf{T}_{Np})'$ may not be generally EDF. To eliminate this impasse, Chatterjee and Sen (1964) considered the following CDF test formulation.

We define a reduced rank-collection matrix $\mathbf{R}_N^*$ as the one that is obtained from $\mathbf{R}_N$ by column-permutations in such a way that the top row is in the natural order $(1, \ldots, N)$. Note that each row of $\mathbf{R}_N$ consists of the numbers $1, \ldots, N$, permuted in some way. So the total number of possible realizations of $\mathbf{R}_N$ is $(N!)^p$, and we denote the corresponding set by $\mathcal{R}_N$. This set $\mathcal{R}_N$ can then be partitioned into $(N!)^{p-1}$ subsets, such that each one contains $N!$ members that are reducible to a common $\mathbf{R}_N^*$. By invoking the invariance structure under the group $\mathcal{G}_N$ in (5), we can easily verify that the conditional distribution, under H_0, of $\mathbf{R}_N$ over the $N!$ possible realizations that correspond to a common $\mathbf{R}_N^*$ is discrete uniform, though the unconditional distribution of $\mathbf{R}_N$ over the set $\mathcal{R}_N$ may depend generally on the unknown F. We denote this conditional (permutational) probability law by $\mathcal{P}_N$. Then we have by routine computations,

$$\mathsf{E}_{\mathcal{P}_N} \mathbf{T}_{Nj} = \sum_{i=1}^{N} \mathbf{c}_i \frac{1}{N} \sum_{i=1}^{N} a_{Nj}(i) = \mathbf{0}, \tag{9}$$

$$\mathsf{E}_{\mathcal{P}_N} \mathbf{T}_{Nj} \mathbf{T}'_{Nk} = \mathbf{C}_N \cdot v_{Njk}, \quad j, k = 1, \ldots, p, \tag{10}$$

where

$$\mathbf{C}_N = \sum_{i=1}^{N}(\mathbf{c}_i - \bar{\mathbf{c}}_N)(\mathbf{c}_i - \bar{\mathbf{c}}_N)' = \sum_{i=1}^{N}\mathbf{c}_i\mathbf{c}_i'. \tag{11}$$

Without loss of generality, we may assume that (11) has the full rank r and that $\bar{a}_{Nj} = N^{-1}\sum_{i=1}^{N} a_{Nj}(i) = 0$, $j = 1, \ldots, p$. Then we have

$$v_{Njk} = \frac{1}{N-1}\sum_{i=1}^{N} a_{Nj}(R_{ij})a_{Nk}(R_{ik}), \quad j,k = 1,\ldots,p. \tag{12}$$

We denote $\mathbf{V}_N = ((v_{Njk}))_{j,k=1,\ldots,p}$, and note that the v_{Njj} are non-random but the off-diagonal elements may not be so. Nevertheless the random matrix $\mathbf{V}_N$ is $\mathcal{P}_N$-invariant, in the sense that it remains the same for all $\mathbf{R}_N$ that are permutationally reducible to a common $\mathbf{R}_N^*$. We roll out $\mathbf{T}_N$ into a rp-vector, and denote this by $\mathbf{T}_N^o$. Then, we have from the above results,

$$\mathsf{E}_{\mathcal{P}_N}\mathbf{T}_N^o = \mathbf{0}, \ \mathsf{E}_{\mathcal{P}_N}\mathbf{T}_N^o\mathbf{T}_N^{o\prime} = \mathbf{C}_N \bigotimes \mathbf{V}_N, \tag{13}$$

where $\bigotimes$ stands for the Kronecker product. Let $\mathbf{V}_N^-$ be a generalized inverse of $\mathbf{V}_N$, and $\mathbf{C}_N^{-1}$ be the inverse of $\mathbf{C}_N$. Then, analogous to the Hotelling-Lawley trace statistic in the parametric case, we consider here the following

$$\mathcal{L}_N = \sum_{j=1}^{p}\sum_{k=1}^{p} v_N^{jk}\mathbf{T}_{Nj}'\mathbf{C}_N^{-1}\mathbf{T}_{Nk}, \tag{14}$$

where $\mathbf{V}_N^- = ((v_N^{jk}))$. In particular, for the multisample model, we may rewrite $\mathcal{L}_N$ as

$$\mathcal{L}_N = \frac{N-1}{N}\sum_{i=1}^{c} N_i(\mathbf{T}_N^{(i)})'\mathbf{V}_N^-(\mathbf{T}_N^{(i)}), \tag{15}$$

where

$$\mathbf{T}_N^{(i)} = \frac{1}{N_i}\sum_{k=1}^{N_i}(a_{N1}(R_{k1}^{(i)}), \ldots, a_{Np}(R_{kp}^{(i)}))', \quad i = 1,\ldots,c, \tag{16}$$

and $R_{kj}^{(i)}$ stands for the rank of $X_{kj}^{(i)}$ among all the N observations on the jth variate.

Note that under $\mathcal{P}_N$, $\mathbf{V}_N$ remains invariant, while by reference to the $N!$ conditionally (permutationally) equally likely realizations of $\mathbf{R}_N$, we can enumerate the corresponding $\mathbf{T}_N^o$ (and hence $\mathcal{L}_N$); this generates the exact conditional (permutational) null distribution of $\mathcal{L}_N$, so that the test based on $\mathcal{L}_N$ is CDF. Such a CDF test becomes EDF when $p = 1$. This CDF characterization should not, however, be overemphasized: as N increases, the enumeration of this conditional distribution becomes prohibitively laborious, and hence ADF prospects are to be explored. This will be done in later chapters.

4.10.4 LMPR property for multivariate tests. We conclude this section with some general remarks on LMPR tests in the multivariate case. For simplicity of presentation, we consider the simple regression model in a multivariate case, and a similar picture holds for other models as well. We define the rank-collection matrix $\mathbf{R}_N$ as in (4.10.3.7), and we intend to construct locally optimal tests that are based solely on $\mathbf{R}_N$. Note that here we have

$$F_i(\mathbf{x}) = F(\mathbf{x} - \beta c_i), \quad i = 1, \ldots, N, \tag{1}$$

where $\beta = (\beta_1, \ldots, \beta_p)'$ and the c_i are known constants, not all equal. We confine ourselves to local alternatives where $||\beta|| \leq \Delta$, and $\Delta \to 0$; here $||\mathbf{x}|| = (\mathbf{x}'\mathbf{x})^{1/2}$ stands for the Euclidean norm. Let us denote the probability function for $\mathbf{R}_N$ when β holds by $P_\beta(\mathbf{R}_N)$, and assume that F admits an absolutely continuous density f. Then

$$P_\beta(\mathbf{R}_N) = \int \cdots \int_{\{\mathbf{R}_N\}} \prod_{i=1}^{N} f(\mathbf{x}_i - \beta c_i) d\mathbf{x}_1 \ldots d\mathbf{x}_N, \tag{2}$$

where the integration extends over such a subspace of pN-dimensional Euclidean space for which $\mathbf{E}_N$ corresponds to a given $\mathbf{R}_N$. This subspace is not solely based on the ordering of the coordinatewise elements of $\mathbf{E}_N$. We may define $P_{\mathbf{0}}(\mathbf{R}_N)$ in the same way, so that the likelihood ratio for the rank-collection matrix is given by

$$L_\beta(\mathbf{R}_N) = \frac{\int \cdots \int_{\{\mathbf{R}_N\}} \prod_{i=1}^{N} f(\mathbf{x}_i - \beta c_i) d\mathbf{x}_1 \ldots d\mathbf{x}_N}{\int \cdots \int_{\{\mathbf{R}_N\}} \prod_{i=1}^{N} f(\mathbf{x}_i) d\mathbf{x}_1 \ldots d\mathbf{x}_N}, \tag{3}$$

which depends on $\mathbf{R}_N, \beta$, c_i and the density f. As in the univariate case, we may be tempted in expanding (3) in a Taylor series, neglecting terms that are $o(||\beta||)$. This involves the first order partial derivatives (scores)

$$(\partial/\partial\beta) f(\mathbf{x} - \beta c_i) = -c_i (\partial/\partial\mathbf{x}) f(\mathbf{x} - \beta c_i), \tag{4}$$

and the right hand side may not be generally expressible in terms of the marginal scores $-c_i(\partial/\partial x_j) f_j(x_j - \beta_j c_i)$, $j = 1, \ldots, p$, where the f_j stand for the marginal density. Such a representation holds for a class of multivariate elliptically symmetric distributions that include the multinormal family as a pivotal subclass. Although $\mathbf{R}_N$ is invariant under any strictly monotone (not necessarily linear) transformation on each coordinate variable, it is not affine-invariant (or equivariant). Therefore, a formal expansion of (3), as in Section 3.4, leads to a first order representation:

$$1 + \beta' \Big(\sum_{i=1}^{N} c_i \psi_N(\mathbf{R}_N) \Big) + o(||\beta||), \tag{5}$$

but the vector $\psi_N(\mathbf{R}_N)$ may not conform to a vector of coordinatewise rank scores. Therefore, in a genuine multivariate case, in general, LMPR test statistics may not correspond to a suitable vector of linear rank statistics of the type considered here. In the univariate case, (4) reduces to the linear rank statistic considered in Section 3.4, so that the LMPR characterization holds. Of course, it should be kept in mind that in a general multiparameter hypothesis problem (as in here), where affine-invariance is not imposed as a parameter-dimension reducing criterion, the LMP concept needs to be modified to interpret local optimality in a broader sense. For example, the wellk-nown *A-, D-,* and *E-optimality* criteria can be adapted in a local sense, and in the light of such an extended local optimality property, suitable locally optimal rank tests can be characterized. But again, in a genuine multivariate case, due to the lack of affine-invariance (or equivariance) property of $\mathbf{R}_N$, such an optimality property may not be attainable by the class of coordinatewise linear rank statistics considered here. Nevertheless, we shall see in later chapters that under suitable asymptotic setups, such optimality properties are shared by linear (or signed-) rank statistics, and moreover, they are robust competitors of the classical parametric tests.

PROBLEMS AND COMPLEMENTS TO CHAPTER 4

Section 2.1

1. Let S be the Wilcoxon test statistic (4.1.1.6) and U the Mann-Whitney statistic mentioned in connection with it. Prove that $U = mn + \frac{1}{2}m(m+1) - S$.

2. Let F_1, F_2 be strictly increasing, and put $L(t) = F_2(F_1^{-1}(t))$, $\alpha = P(X < Y) = \int_0^1 t \, \mathrm{d}L(t)$, $\beta^2 = \int_0^1 L^2(t) \, \mathrm{d}t - (1-\alpha)^2$, $\gamma^2 = \int_0^1 t^2 \, \mathrm{d}L(t) - \alpha^2$. If U is the Mann-Whitney statistic, and the X_i's have the distribution function F_1 and the Y_j's the distribution function F_2, show that $\mathsf{E}U = mn\alpha$, $\operatorname{var} U = mn[(m-1)\beta^2 + (n-1)\gamma^2 + \alpha(1-\alpha)]$.

3. Use again the notation of the preceding problem.

(i) Prove that $\int_0^1 [L(t) - t]^2 \, \mathrm{d}t \le \frac{1}{3} - \alpha(1-\alpha)$. [Hint: Use $\int_0^1 (2t-1)L(t) \, \mathrm{d}t = \int_0^1 \int_0^t [L(t) - L(s)] \, \mathrm{d}s \, \mathrm{d}t \ge \int_0^1 \int_0^t [L(t) - L(s)]^2 \, \mathrm{d}s \, \mathrm{d}t$.]

(ii) $\operatorname{var} U \le mn\alpha(1-\alpha)\max(m,n) \le \frac{1}{4} mn \max(m,n)$. [Hint: Use $\beta^2 + \gamma^2 = \int_0^1 [L(t) - t]^2 \, \mathrm{d}t + \frac{2}{3} - \alpha^2 - (1-\alpha)^2$.] (Van Dantzig (1951), Birnbaum and Klose (1957).)

4. Prove that $(mn)^{-1}U$ converges in probability to α, and that the two-sided Mann-Whitney test is consistent against the alternatives $P(X < Y) \neq \frac{1}{2}$. (Van Dantzig (1951).)

5. Prove equality (4.1.3.14).

6. If T is the Lehmann statistic mentioned at the end of 4.1.3, then $\mathsf{E}T = \binom{m}{2}\binom{n}{2}[\frac{1}{3} + 2\int_{-\infty}^{\infty}(F_1 - F_2)^2 \,\mathrm{d}(\frac{1}{2}(F_1 + F_2))]$.

7. Rényi (1953b) proposed the test statistic $W = \frac{1}{m}\binom{n}{2}^{-1} W_1 + \frac{1}{n}\binom{m}{2}^{-1} W_2$, where W_1 is the number of triplets X_i, Y_j, Y_k, $j < k$, such that Y_j, $Y_k < X_i$, and W_2 is the number of triplets X_j, X_k, Y_i, $j < k$, such that X_j, $X_k < Y_i$. If T is the Lehmann statistic, then $\binom{m}{2}^{-1}\binom{n}{2}^{-1} T = 2W - 1$. (Csáki (1959), Zajta (1960).)

If $m = n$, then the Lehmann statistic T is a positive linear function of the Cramér-von Mises statistic M given by (4.1.3.12). (Wegner (1956).)

8. A poor projection. Show that the projection $\hat{T}$ of the Lehmann statistic T into the family of linear rank statistics is a constant, namely $\hat{T} = \mathsf{E}T$.

9. Another poor projection. Let T be the statistic denoting the number of runs. Show that, for $m = n$, $\hat{T} = \mathsf{E}T$ constantly. [Hint: $\hat{a}(1, j) = \hat{a}(i, j)$ for $1 \le i \le m$, and $\hat{a}(N, j) = \hat{a}(i, j)$ for $m < i \le N$, $1 \le j \le N$. Moreover $\hat{a}(1, j) = \hat{a}(N, j)$ for $m = n$. Thus $\hat{a}(i, j) = \hat{a}(j)$, $1 \le i, j \le N$, and $\sum_{i=1}^{N} \hat{a}(i, R_i) = \sum_{i=1}^{N} \hat{a}(i)$.]

10. Denote $A = N - \max(R_{m+1}, \ldots, R_N)$ and $B = \min(R_1, \ldots, R_m) - 1$, and put $E = \min(A, B)$. The projection $\hat{E}$ of E into the family of linear rank statistics equals

$$\hat{E} = \frac{N-1}{n}\sum_{i=1}^{m} \hat{a}(R_i) + \Big(1 - m\frac{N-1}{n}\Big)\mathsf{E}(E) \tag{1}$$

with

$$\hat{a}(j) = \sum_{k=1}^{\min(j-1, N-j, m, n)} \frac{\binom{N-2k-1}{m-k-1}}{\binom{N-1}{m-1}} + \sum_{k=N-j+1}^{\min(m,n)} \frac{\binom{N-2k}{m-k}}{\binom{N-1}{m-1}}. \tag{2}$$

Here $\hat{a}(j) = \hat{a}(i, j)$, $1 \le i \le m$, $1 \le j \le N$. [Hint: Utilize the relation $\mathsf{E}(E \mid R_i = j) = \sum_{k=1}^{\infty} P(E \ge k \mid R_i = j)$.]

11. (Continuation.) Show that $\hat{a}(j)$ given by (2) is a non-decreasing function of j, and that there is a non-decreasing function $t(x)$ such that $E = t(\hat{E})$. Consequently, the family of critical regions generated by E is a subset of the family of critical regions $\{\hat{E} \ge k\}$. The statistic E also may be expressed as a non-decreasing function of linear rank statistics

$$S_\lambda = \sum_{i=1}^{m}(\lambda^{N-R_i} - \lambda^{R_i - 1}), \quad 0 < \lambda \le \frac{1}{2}.$$

12. The logistic density has variance $\sigma^2 = \frac{1}{3}\pi^2$, fourth moment $\mu_4 = \frac{7}{15}\pi^4$, coefficient of excess $\gamma_2 = 1.2$. [Hint: Use integration by parts and $(1+e^{-x})^{-1} = \sum_{k=0}^{\infty}(-e^{-x})^k$ for $x > 0$.]

13. The double-exponential density has $\sigma^2 = 2$, $\mu_4 = 24$, $\gamma_2 = 3$.

14. The uniform density has $\sigma^2 = \frac{1}{12}$, $\mu_4 = \frac{1}{80}$, $\gamma_2 = -1.2$.

15. A generating polynomial. Put $S = \sum_{i=1}^{m} R_i - \frac{1}{2}m(m+1)$ and show that

$$\sum_{k=0}^{mn} t^k P(S = k) = \frac{(1-t^N)(1-t^{N-1})\dots(1-t^{N-m+1})}{(1-t)(1-t^2)\dots(1-t^m)} \binom{N}{m}^{-1}$$

holds. (Hájek (1955).)

Section 4.2

16. Find examples of densities f_1, f_2 with equal variances and transformations t such that $\operatorname{var} t(X) < \operatorname{var} t(Y)$ or $\operatorname{var} t(X) > \operatorname{var} t(Y)$ but the distribution of the ranks remains unchanged under this transformation of observations. (Therefore, in such a case, no rank test can be good for revealing the difference in scale.)

(i) For densities with disjoint supports [such that $F_1(a) = 1$, $F_2(a) = 0$, t increasing].

(ii) For mutually absolutely continuous densities [$f_1(x) = 2(1-x)$, $f_2(x) = 2x$ for $0 < x < 1$, $t(x) = -\log x$ or $t(x) = -\log(1-x)$]. (Moses (1963).)

17. Prove that the Savage test is the locally most powerful rank test for H_0 against difference in scale for the exponential density. [See Problem 3.5.]

18. Prove formula (4.2.1.18) for the variance of the Savage test statistic.

19. Show that the original Savage test against Lehmann's alternatives (mentioned after (4.2.1.18)) leads, for the problem of scale with underlying densities of exponential type, to the test statistic (4.2.1.15). [Hint: Consider the variables $-X_i$, $-Y_j$.]

20. Let S be the Ansari-Bradley test statistic, and let T be the Siegel-Tukey modification of the Wilcoxon statistic, i.e. $T = \sum_{i=1}^{m} \tilde{R}_i$ (see the Remark near the end of 4.2.1).

(i) Show, by an example, that S and T are not equivalent.

(ii) Let T' be the statistic T computed for the variables $-X_1, \dots, -X_m$, $-Y_1, \dots, -Y_n$. Then for N even $S = (T + T' + 1)/4$, for N odd $S =$

$(T + T' + 1)/4$ whenever the median observation does not belong to the first sample, $S = (T + T' + 2)/4$, whenever the median observation belongs to the first sample. Then S and $T + T'$ are equivalent for testing.

(iii) The statistics S and T are asymptotically equivalent in the following sense. If $a_{1N}^{(S)}(i)$ are the scores used for S divided by N, and $a_{1N}^{(T)}(i)$ are the scores used for T divided by $2N$, then

$$\int_0^1 [a_{1N}^{(S)}(1 + [uN]) - a_{1N}^{(T)}(1 + [uN])]^2 \, du \to 0.$$

(For consequences see Chapter 8.)

This clarifies somewhat vague remarks by Klotz (1962) and Moses (1963) that S and T are equivalent.

21. Find the density f such that the corresponding approximate scores $a_{1N}(i) = \varphi_1(i/(N+1), f)$ lead to the Mood test statistic $\sum_{i=1}^{m} [R_i - \frac{1}{2}(N+1)]^2$. [Solution: $f(x) = \{24a^{\frac{1}{2}}[x^{11/18} + a^{-1}x^{7/9}]^{3/2}\}^{-1}$, $F(x) = \frac{1}{2} + \frac{1}{2}[1 - (1 + a^{-1}x^{1/6})^{-1}]^{\frac{1}{2}}$, for $x \geq 0$.]

22. Let T be the Sukhatme test statistic defined at the end of 4.2.1.

(i) Under H_0, $\mathsf{E}T = \frac{1}{2}[\lambda^2 + (1 - \lambda)^2]$ where $\lambda = P(X_i > 0)$. Thus the distribution of T depends on $p \in H_0$.

(ii) T has the same distribution for all $p \in H_0$ with $\lambda = \frac{1}{2}$.

(iii) T has the same distribution, a fortiori, for all $p \in H_1$.

Section 4.6

23. Specify the model (4.6.1.1) for which (X_i, Y_i) have the two-dimensional normal density with a given correlation coefficient ρ, $0 \leq \rho < 1$.

24. Let S be the quadrant test statistic (4.6.1.17), S_1 the statistic used as a summand in this formula, and let U be the Elandt statistic mentioned in the text following (4.6.1.18).

(i) For N even, all statistics S, S_1, U are equivalent.

(ii) Find an example showing that for N odd the statistics S, S_1, U are not equivalent.

25. A generating polynomial. Put $S = \sum\sum_{i<j} \frac{1}{2}[1 + \operatorname{sign}(R_i^0 - R_j^0)]$.

Then

$$\sum_{k=0}^{\frac{1}{2}N(N-1)} t^k P(S = k) = \frac{(1-t)(1-t^2)\dots(1-t^N)}{N!(1-t)^N}.$$

(Hájek (1955).)

Section 4.8

26. Prove the assertion from 4.8.1, paragraph on randomization, that, under H_0' and randomization of ranks of tied observations, $R_1(X, \Xi), \ldots, R_N(X, \Xi)$ take on an arbitrary permutation of $1, \ldots, N$ with probability $(N!)^{-1}$. [Hint: Utilize an auxiliary distribution function in which the jumps of the original distribution function are replaced by continuous lines — see e.g. Fraser (1957a), p. 201.]

27. If $\bar{S}$ is the two-sample Wilcoxon test statistic, where mid-ranks are used for tied observations, then

$$\mathsf{var}'\{\bar{S} \mid \tau\} = \frac{mn(m+n+1)}{12} - \frac{mn}{12(m+n)(m+n-1)} \sum_{j=1}^{g} \tau_j(\tau_j^2 - 1).$$

28. Give a rigorous proof of the inequality (4.8.2.1).

29. Let S^+ be the sign test statistic (4.5.1.12), and S_0 the number of zero observations among $X_1, \ldots, X_N$. Denote by A_1 the equal-tailed two-sided α_1-level critical region for the sign test omitting zero observations, i.e. $A_1 = \{S^+ \leq c(S_0)$ or $S^+ \geq N - S_0 - c(S_0)\}$ with $c(S_0)$ given by $2^{-(N-S_0)} \sum_{\nu=0}^{c(S_0)} \binom{N-S_0}{\nu} \leq \frac{1}{2}\alpha_1$, and denote by A_2 a similar α_2-level region for the sign test counting half of the zero observations as positive, i.e. $A_2 = \{S^+ + \frac{1}{2}S_0 \leq c'$ or $S^+ + \frac{1}{2}S_0 \geq N - c'\}$ with c' given by $2^{-N} \sum_{\nu=0}^{c'} \binom{N}{\nu} \leq \frac{1}{2}\alpha_2$.

(i) If $\alpha_2 \leq \alpha_1 < 1$, then $A_2 \subset A_1$. [Hint: Take $A_{i,k} = A_i \cap \{S_0 = k\}$, $i = 1, 2$, and prove $A_{2,k} \subset A_{1,k}$. In fact, if $A_{2,k} \not\subset A_{1,k}$, by the following lemma (ii) $\alpha_1 < \alpha_2$ would follow.]

(ii) Lemma: If c, k, N are positive integers satisfying $c' = [c + \frac{1}{2}k + \frac{1}{2}] < \frac{1}{2}N - \frac{1}{2}$, then $2^k \sum_{\nu=0}^{c} \binom{N-k}{\nu} < \sum_{\nu=0}^{c'} \binom{N}{\nu}$. [Hint: Proof by induction with respect to k for k even; for k odd take $N+1$ in place of N.] (Hemelrijk (1952).)

Chapter 5

Computation of null exact distributions

5.1 DIRECT USE OF DISTRIBUTION OF RANKS

In every case, the distribution of a rank test statistic may be determined from the distribution of the vector of ranks (or ranks and signs) give in Chapter 3. Under H_0, this direct method consists in computing the values of the rank test statistic $T = t(R_1, \ldots, R_N)$ for all possible vectors $(r_1, \ldots, r_N)$ of ranks. Since, by Theorem 3.1.2.1, $P((R_1, \ldots, R_N) = (r_1, \ldots, r_N)) = 1/N!$ under H_0, we have $P(T = x) = \pi(x)/N!$ where $\pi(x)$ denotes the number of vectors $(r_1, \ldots, r_N)$ such that $t(r_1, \ldots, r_N) = x$. An analogous direct method may be applied to rank statistics associated with the hypotheses H_1, H_2, and H_3.

In the following sections we will mention different methods of computation of the distributions of rank statistics. Some of them may appear somewhat obsolete (possibly taken from Hájek and Šidák (1967)) but we will try to mention briefly more modern ones, related to the contemporary use of computers.

5.1.1 Tests for regression, independence and random blocks. For finding the distribution of a general rank test statistic in all these cases we apply directly the formulas for the probabilities of ranks given in 3.1.

Regression. Consider a regression test statistic

$$S = \sum_{i=1}^{N} c_i a(R_i), \tag{1}$$

or some other statistic of Subsection 4.3.1. Then, in general, the distribution of this statistic is obtained by the procedure described above, i.e. by substituting for $R_1, \ldots, R_N$ all $N!$ permutations $(r_1, \ldots \ldots, r_N)$ of the numbers $1, \ldots, N$. Sometimes, the following theorem may be useful.

Theorem 1 *If*

$$a(k) + a(N - k + 1) = \text{const}, \quad k = 1, \ldots, N, \tag{2}$$

or

$$c_k + c_{N-k+1} = \text{const}, \quad k = 1, \ldots, N, \tag{3}$$

then the distribution of the statistic (1) *is symmetric about* $\mathsf{E}S = \bar{a} \sum_{i=1}^{N} c_i$.

The proof follows immediately by considering along with each vector $(R_1, \ldots, R_N)$ also the vector $(N - R_1 + 1, \ldots, N - R_N + 1)$, which has the same distribution. If (3) holds, we pass to the antiranks. □

The case of testing *independence* is entirely similar. In fact, by formula (3.3.2.13), $P((R_1^0, \ldots, R_N^0) = (r_1, \ldots, r_N)) = 1/N!$ under H_2, which shows the similarity to the case of $(R_1, \ldots, R_N)$ under H_0. Thus, for finding the distribution of a rank statistic the previous result may be used, putting all permutations $(r_1, \ldots, r_N)$ of the numbers $1, \ldots, N$ in place of $(R_1^0, \ldots, R_N^0)$. Probably, for the tests of independence of Fisher-Yates, of the van der Waerden type, and the Spearman correlation coefficient, this is essentially the only method for computing their distributions.

Clearly, the theorem stated above may be used also for tests of independence with a test statistic of the form

$$S = \sum_{i=1}^{N} c(R_i) a(Q_i) = \sum_{i=1}^{N} c(i) a(R_i^0). \tag{4}$$

For the problem of *random blocks* the procedure is changed in an obvious way. The ranks $R_{j1}, \ldots, R_{jk}$ in the j-th block can take as their values all $k!$ permutations of the numbers $1, \ldots, k$. All these possibilities must be combined for all n blocks, giving altogether $(k!)^n$ possibilities, each of which has, of course, the probability $1/(k!)^n$. However, as when using the ranks R_i^0 in tests of independence, we may fix $R_{1i} = i$ for $i = 1, \ldots, k$, and utilize the fact that the respective conditional distribution is the same as the overall distribution. Then we have to substitute all permutations of $1, \ldots, k$ only for the vectors $(R_{j1}, \ldots, R_{jk})$ with $j = 2, \ldots, n$, and there are altogether $(k!)^{n-1}$ possibilities.

5.1.2 Two-sample tests. Here we can lean on the following simple theorem.

Theorem 1 *Consider a random sample* $X_1, \ldots, X_m$ *and a second sample* $Y_1 = X_{m+1}, \ldots, Y_n = X_{m+n}$. *Let* $I(m)$ *be any subset of* m *numbers from* $\{1, 2, \ldots, m+n\}$. *Then, under* H_0, *the probability that the set* $\{R_1, \ldots, R_m\}$ *of ranks corresponding to the first sample coincides with the set* $I(m)$ *is equal to* $1/\binom{m+n}{m}$. *The same result is valid for the* $\tilde{R}_i$*'s introduced in* 4.2.1 *in place of the* R_i*'s.*

The proof is obvious from Theorem 3.1.2.1, since we are, in fact, looking for the probability that the vector $R_1, \ldots, R_m$ will be equal to any of the $m!$ permutations of the numbers in $I(m)$, and that the vector $R_{m+1}, \ldots, R_{m+n}$ will be equal to any of the $n!$ permutations of the n numbers remaining in $\{1, 2, \ldots, m+n\} \setminus I(m)$. □

Occasionally, the following box model is used, very well known and frequently applied. We have $m + n$ boxes labelled $1, 2, \ldots, m+n$ and choose at random a set of m distinct boxes with labels $r_1^* < r_2^* < \ldots < r_m^*$, placing in all these boxes the letter X, and placing the letter Y in the remaining n boxes.

In this model, each such arrangement of X's and Y's has probability $1/\binom{m+n}{m}$, and it corresponds to the event that the ranks $R_1, \ldots, R_m$ of $X_1, \ldots, X_m$, reordered increasingly, coincide with $r_1^*, \ldots, r_m^*$. However, by the theorem above, the latter event has the same probability, which justifies the appropriateness of the box model.

Any reasonable two-sample rank test statistic depends only on the *set* of values of $R_1, \ldots, R_m$, regardless of the order expressed by the indices. Thus the distribution of such a statistic can in general be found by substituting for $R_1, \ldots, R_m$ all $\binom{m+n}{m}$ possible subsets $I(m)$ of m numbers from the set $\{1, 2, \ldots, m+n\}$, assigning to each resulting value the probability $1/\binom{m+n}{m}$, and summing up, if necessary, these probabilities for cases in which the resulting values coincide.

This general method of computation can also be modified for linear rank statistics with scores possessing certain properties of symmetry. (See Hájek and Šidák (1967).)

5.1.3 Tests for three or more samples. For three or more samples a box model similar to that in 5.1.2 may be utilized, and essentially a similar result as Theorem 5.1.2.1 can be proved, of course with necessary changes, e.g. that binomial coefficients are replaced by multinomial coefficients, etc.

Similarly, we can use the third procedure mentioned in Subsection 4.4.2, and prove that the statistics

$$T(X_{(2)}, X_{(1)}); T(X_{(3)}, X_{(1)} \cup X_{(2)}); \ldots ; T(X_{(k)}, X_{(1)} \cup \ldots \cup X_{(k-1)})$$

are independent. (For details see Hájek and Šidák (1967).)

5.1.4 Tests of symmetry

Theorem 1 *Consider a random sample $X_1, \ldots, X_N$, and denote by R_i^+ the rank of $|X_i|$. Let I be any subset (possibly empty) of the set $\{1, 2, \ldots, N\}$. Then under H_1, the probability that the set of ranks R_i^+ of all positive observations X_i will coincide with the set I is equal to $(\frac{1}{2})^N$.*

The proof is easily derived from Theorem 3.1.3.1.

This theorem shows a general method for computing the distribution of a rank statistic for testing symmetry. Looking for the distribution of

$$S^+ = \sum_{X_i>0} a^+(R_i^+), \tag{1}$$

we must select all 2^N possible subsets $I = \{r_1, \ldots, r_j\}$ from $\{1, \ldots \ldots, N\}$, and for each selection associate probability $(\frac{1}{2})^N$ with the value $\sum_{i=1}^{j} a^+(r_i)$; if some of these values coincide then the corresponding probabilities are, of course, summed up.

Theorem 2 *The distribution of the statistic* (1) *is symmetric about* $\mathsf{E}S^+ = \frac{1}{2}\sum_{i=1}^{N} a^+(i)$ *for arbitrary scores* $a^+(i)$.

The proof follows immediately, if to the event that the set of ranks R_i^+ of all positive observations equals I we associate the event that this set of ranks equals the complementary set $\{1, \ldots, N\} \setminus I$.

5.2 EXPLICIT FORMULAS FOR DISTRIBUTIONS

5.2.1 Statistics using the scores 0, $\frac{1}{2}$, 1

Theorem 1 *Let S be the two-sample median test statistic introduced in* (4.1.1.10). *Then*

$$P(S = k) = \binom{[\frac{1}{2}(m+n)]}{[k]}\binom{[\frac{1}{2}(m+n)]}{[m-k]}\binom{m+n}{m}^{-1}, \tag{1}$$

where $[x]$ denotes the largest integer not exceeding x. If $m+n$ is even, formula (1) holds for $k = 0, 1, \ldots, m$ in the case of $m \le n$, and for $k = \frac{1}{2}(m-n), \frac{1}{2}(m-n)+1, \ldots, \frac{1}{2}(m+n)$ in the case of $m \ge n$. If $m+n$ is odd, this formula holds for $k = 0, \frac{1}{2}, \ldots, m-\frac{1}{2}, m$ in the case of $m < n$, and for $k = \frac{1}{2}(m-n), \frac{1}{2}(m-n)+\frac{1}{2}, \ldots, \frac{1}{2}(m+n)-\frac{1}{2}, \frac{1}{2}(m+n)$ in the case of $m > n$.

The proof is easy on making use of the box model from Subsection 5.1.2. (See Hájek and Šidák (1967).) □

Theorem 2 *If S^+ is the sign test statistic given by (4.5.1.12), then*

$$P(S^+ = k) = \binom{N}{k} 2^{-N} \quad \textit{for } k = 0, 1, \ldots, N. \tag{2}$$

Proof. Under H_1, this is simply a special case of the binomial distribution with success probability $\frac{1}{2}$. □

Theorem 3 *Let S be the quadrant test statistic defined by (4.6.1.17). Then for N even*

$$P(S = k) = \binom{N}{\frac{1}{2}N}^{-1} \binom{\frac{1}{2}N}{k}^2 \quad \textit{for } k = 0, 1, \ldots, \frac{1}{2}N, \tag{3}$$

and for N odd

$$\begin{aligned} P(S = k + \tfrac{1}{4}) &= \frac{1}{N}\binom{N-1}{\frac{1}{2}(N-1)}^{-1}\binom{\frac{1}{2}(N-1)}{k}^2 \qquad (4)\\ &\quad \textit{for } k = 0, 1, \ldots, \tfrac{1}{2}(N-1),\\ P(S = k + \tfrac{1}{2}) &= \frac{N-1}{N}\binom{N-1}{\frac{1}{2}(N-1)}^{-1}\binom{\frac{1}{2}(N-3)}{k}\binom{\frac{1}{2}(N-1)}{k}\\ &\quad \textit{for } k = 0, 1, \ldots, \tfrac{1}{2}(N-3),\\ P(S = k) &= \frac{N-1}{N}\binom{N-1}{\frac{1}{2}(N-1)}^{-1}\binom{\frac{1}{2}(N-3)}{k-1}\binom{\frac{1}{2}(N-1)}{k}\\ &\quad \textit{for } k = 1, 2, \ldots, \tfrac{1}{2}(N-1). \end{aligned}$$

Proof. It can be carried out again by means of a special different box model but here we will omit it (for details see Hájek and Šidák (1967)). □

5.2.2 The Haga test and the E-test statistics.

Theorem 1 *Let $T = A + B - A' - B'$ be the Haga statistic introduced in (4.1.2.1). Then*

$$P(T = m + n) = P(T = -m - n) = \binom{m+n}{m}^{-1}, \tag{1}$$

$$P(T = m + n - 1) = P(T = -m - n + 1) = 0,$$

$$P(T = t) = \left\{ \sum_{i=1}^{|t|-1} \binom{m+n-|t|-2}{m-i-1} + \sum_{i=1}^{[(m-|t|)/2]} \binom{m+n-|t|-2i-2}{n-2} + \sum_{i=1}^{[(n-|t|)/2]} \binom{m+n-|t|-2i-2}{m-2} \right\} \binom{m+n}{m}^{-1}$$

$$\text{for } t = -m - n + 2, \ldots, m + n - 2,$$

where the summation $\sum_{i=1}^{j}$ for $j < 1$ stands for 0, and $\binom{j}{k} = 0$ for $j < k$ or $k < 0$.

Proof. Once again we will use the usual two-sample box model, and decompose individual events. E.g., for $t = 2, \ldots, m + n - 2$, we decompose the event $\{T = t\}$ into the events $\{A = i, B = t - i\}$ with $i = 1, \ldots, t - 1$, $\{A = t + i, B' = i\}$ with $i = 1, \ldots, [(n - t)]/2$, and $\{A' = i, B = t + i\}$ with $i = 1, \ldots, ([n - t])/2$, then sum up the probabilities of these events, etc. (For details, see Hájek and Šidák (1967).) □

Theorem 2 *If $E = \min(A, B) - \min(A', B')$ is the E-test statistic given by (4.1.2.2), then*

$$P(E \geq k) = P(E \leq -k) = \binom{m+n-2k}{m-k} \binom{m+n}{m}^{-1} \tag{2}$$

for $k = 1, \ldots, \min(m, n)$.

The proof follows again from the box model on placing into the last k boxes the letter X, into the first k boxes the letter Y, etc. □

5.2.3 Kolmogorov-Smirnov statistics for two samples of equal sizes. As is usual and convenient in connection with the Kolmogorov-Smirnov statistic for $m = n$, we shall employ the following random walk model, different from that of the preceding sections. A particle moves in the plane (x, y) starting from the origin $(0, 0)$, in the k-th step $(k = 1, \ldots, 2n)$ moving from its preceding position $(k-1, y)$ either to $(k, y+1)$ if $X^{(k)}$ comes from the second sample, or to $(k, y-1)$ if $X^{(k)}$ comes from the first sample. In this manner, all possible arrangements of observations from both samples in $X^{(1)}, X^{(2)}, \ldots, X^{(2n)}$ are in one-to-one correspondence to all paths of the particle ending at the point $(2n, 0)$, and there are altogether $\binom{2n}{n}$ such paths. We shall denote by $y(k)$ the y-coordinate of the particle after the k-th step. Clearly, $y(k) = \sum_{i=1}^{k}(1 - 2c_{D_i})$.

Since the factors $(\frac{m+n}{mn})^{\frac{1}{2}}$ in the definition of the Kolmogorov-Smirnov statistics are unimportant for finite samples, we may omit them in the formulas of the following theorem.

Theorem 1 *The distributions of the* $(\frac{nm}{m+n})^{\frac{1}{2}}$*-multiples of the Kolmogorov-Smirnov statistics* (4.1.3.2) *and* (4.1.3.3) *are for* $m = n$ *given by*

$$P\{\max_{1 \le k \le 2n}(\tfrac{1}{2}k - c_{D_1} - \ldots - c_{D_k}) \ge \tfrac{1}{2}h\} \tag{1}$$

$$= \binom{2n}{n}^{-1}\binom{2n}{n-h} \quad \textit{for } h = 1, 2, \ldots, n,$$

$$P\{\max_{1 \le k \le 2n}|\tfrac{1}{2}k - c_{D_1} - \ldots - c_{D_k}| \ge \tfrac{1}{2}h\} \tag{2}$$

$$= 2\binom{2n}{n}^{-1}\sum_{j=1}^{[n/h]}(-1)^{j+1}\binom{2n}{n-jh} \quad \textit{for } h = 2, 3, \ldots, n.$$

Proof. Calling attention to our random walk model, we see that (1) is the probability that a path will touch or cross at some step the line $y = h$. Each path of this kind may be modified so that its initial part up to the moment of its first touching the line $y = h$ is unchanged, but its remaining part after that point is reflected about the line $y = h$. In this way, we establish a one-to-one correspondence between the paths of the former kind and the paths beginning at $(0, 0)$ and ending at $(2n, 2h)$. (Principle of reflection.) Since paths of the latter kind arise when the particle moves in $n + h$ steps upwards and in $n - h$ steps downwards, their total number is $\binom{2n}{n-h}$. This proves (1).

Similarly, we shall compute the number $\pi(\pm h)$ of paths which touch or cross at some step the lines $y = h$ or $y = -h$. Let $\pi(h, j)$ denote the number of paths for which there exist j indices $k_1 < \ldots < k_j$ such that $y(k_1) = h$, but $-h < y(k) < h$ for $1 \le k < k_1$, further $y(k_2) = -h$, but $-h < y(k)$ for $k_1 < k < k_2$, and so on, so that such a path touches or crosses the lines $y = h$, $y = -h$ alternately at least j times. The number $\pi(-h, j)$ is defined analogously, requiring $y(k_1) = -h$, $y(k_2) = h$, etc. Then

$$\pi(\pm h) = \pi(h, 1) + \pi(-h, 1) = 2\pi(h, 1), \tag{3}$$

the last equality being obvious by symmetry. Further, let $\rho(h, j)$ be the number of paths for which there exist j indices $k_1 < \ldots < k_j$ such that $y(k_1) = h$, but $y(k) < h$ for $1 \leq k < k_1$, further $y(k_2) = -h$, but $-h < y(k)$ for $k_1 < k < k_2$, and so on; $\rho(-h, j)$ is defined analogously. Distinguishing the paths by the line they touch or cross first, we obtain

$$\begin{aligned} \rho(h, 1) &= \pi(h, 1) + \pi(-h, 2), \\ \rho(-h, 2) &= \pi(-h, 2) + \pi(h, 3), \\ \ldots & \quad \ldots \\ \rho((-1)^{[n/h]+1}h, [n/h]) &= \pi((-1)^{[n/h]+1}h, [n/h]), \end{aligned} \tag{4}$$

where the last equality follows from the fact that a path can touch (or cross) the two lines $y = h$, $y = -h$ alternately at most $[n/h]$-times.

Now, (3) and (4) imply

$$\pi(\pm h) = 2 \sum_{j=1}^{[n/h]} (-1)^{j+1} \rho((-1)^{j+1}h, j). \tag{5}$$

First, in proving (1) we have already found that $\rho(h, 1) = \binom{2n}{n-h}$. Further, $\rho(-h, 2)$ can be found by a similar method. Namely, each path specified in the definition of $\rho(-h, 2)$ is modified as follows: Its initial part up to the moment of its first touching the line $y = -h$ is unchanged, while its remaining part after that point is reflected about $y = -h$. This modified path clearly touches or crosses the line $y = -3h$, and ends at $(2n, -2h)$. Now, this modified path is modified once more, the part after its first touching the line $y = -3h$ being reflected about this line. Thus we find a one-to-one correspondence between the paths counted in $\rho(-h, 2)$ and the paths beginning at $(0, 0)$ and ending at $(2n, -4h)$. Consequently, their total number is $\binom{2n}{n-2h}$. Continuing in this manner, using j successive reflections for the computation of $\rho((-1)^{j+1}h, j)$, we obtain

$$\rho((-1)^{j+1}h, j) = \binom{2n}{n - jh}, \tag{6}$$

which, in view of (5), implies (2). □

Theorem 1 of the present subsection, including the method of proof of (1), is taken from Gnedenko and Korolyuk (1951) (this paper contains some slight errors and omits the proof of (2)). For other proofs and related results see also Drion (1952), and Carvalho (1959). A good survey paper of many results in this field is Gnedenko (1954).

5.3 RECURRENCE FORMULAS FOR DISTRIBUTIONS

5.3.1 The Wilcoxon and the Kendall test statistics.

Theorem 1 *Let $S_{m,n}$ be the two-sample Wilcoxon test statistic, defined in* (4.1.1.6), *for the samples $X_1, \ldots, X_m$ and $Y_1, \ldots, Y_n$. Then*

$$P(S_{m,n} = k) = \frac{\pi_{m,n}(k)}{\binom{m+n}{m}}, \tag{1}$$
$$\text{for } k = \tfrac{1}{2}m(m+1), \ldots, \tfrac{1}{2}m(m+2n+1),$$

where $\pi_{m,n}(k)$ satisfy the recurrence formula

$$\pi_{m,n}(k) = \pi_{m,n-1}(k) + \pi_{m-1,n}(k-m-n), \tag{2}$$

with the initial and boundary conditions $\pi_{m,0}(\frac{1}{2}m(m+1)) = 1$, $\pi_{m,0}(k) = 0$ otherwise; $\pi_{0,n}(0) = 1$, $\pi_{0,n}(k) = 0$ otherwise; $\pi_{m,n}(k) = 0$ for $k < \frac{1}{2}m(m+1)$.

Proof. Let us denote by $\pi_{m,n}(k)$ the number of arrangements of the observations $\{X_i\}$ and $\{Y_j\}$, regardless of their indices, in $X^{(1)}, \ldots, X^{(m+n)}$ such that $S_{m,n} = k$ (recall the box model of Section 5.1.2). Now, (1) is obvious, and (2) follows easily by discarding the observation $X^{(m+n)}$. Actually, if $X^{(m+n)}$ is some Y_j we get $S_{m,n-1} = k$, and if $X^{(m+n)}$ is some X_i we get $S_{m-1,n} = k - m - n$. □

Theorem 2 *If S_N^+ is the one-sample Wilcoxon test statistic* (4.5.1.9) *for the sample $X_1, \ldots, X_N$, then*

$$P(S_N^+ = k) = \pi_N(k)/2^N, \quad \text{for } k = 0, 1, \ldots, \tfrac{1}{2}N(N+1), \tag{3}$$

with $\pi_N(k)$ satisfying the recurrence formula

$$\pi_N(k) = \pi_{N-1}(k) + \pi_{N-1}(k-N), \tag{4}$$

the initial and boundary conditions being $\pi_0(0) = 1$, $\pi_0(k) = 0$ otherwise; $\pi_N(k) = 0$ for $k < 0$.

Proof. Let $\pi_N(k)$ be the number of subsets $\{r_1, \ldots, r_j\} \subset \{1, \ldots\ldots, N\}$ for which $\sum_{i=1}^{j} r_i = k$. If the set of ranks R_i^+ of all positive observations X_i equals $\{r_1, \ldots, r_j\}$ we obtain $S_N^+ = k$, and (3) follows by Theorem 5.1.4.1. Furthermore, discarding the observation X_i with $R_i^+ = N$, we obtain $S_{N-1}^+ = k$ for $X_i \leq 0$, and $S_{N-1}^+ = k - N$ for $X_i > 0$, which proves (4). □

Theorem 3 *Observe a two-dimensional sample* $(X_1, Y_1), \ldots, (X_N, Y_N)$, *and consider the statistic* $T = T_N$ *related to the Kendall rank correlation coefficient by* (4.6.2.2). *Then*

$$P(T_N = k) = \pi_N(k)/N! \quad \textit{for } k = 0, 1, \ldots, \tfrac{1}{2}N(N-1), \tag{5}$$

with $\pi_N(k)$ *satisfying the recurrence formula*

$$\pi_N(k) = \pi_{N-1}(k) + \pi_{N-1}(k-1) + \ldots + \pi_{N-1}(k-N+1) \tag{6}$$

and with the initial and boundary conditions $\pi_1(0) = 1$, $\pi_1(k) = 0$ *otherwise;* $\pi_N(k) = 0$ *for* $k < 0$.

Proof. Let $\pi_N(k)$ be the number of all vectors $(R_1^0, \ldots, R_N^0)$ yielding the value $T_N = k$. Obviously, (5) holds in view of (3.3.2.13). Now, we leave out the pair (X_i, Y_i) with $Y_i = Y^{(N)}$. If, for example, the corresponding X_i equals $X^{(1)}$ we obtain the value $T_{N-1} = k$. Generally, if X_i equals $X^{(j)}$, $j = 1, \ldots, N$, we obtain $T_{N-1} = k - j + 1$, and (6) follows from this splitting of possibilities. □

5.3.2 Statistics of Kolmogorov-Smirnov types for two samples of unequal sizes. Here we shall again utilize a random walk model similar to that of Subsection 5.2.3. However, the steps of the particle must now have unequal lengths, due to the fact that the two samples under investigation have unequal sizes $m \neq n$. The particle starts from $(0, 0)$, and in the k-th step ($k = 1, \ldots, m+n$) moves from its preceding position $(k-1, y)$ to $(k, y+m)$ whenever $X^{(k)}$ belongs to the second sample $Y_1, \ldots, Y_n$, or to $(k, y-n)$ whenever $X^{(k)}$ belongs to the first sample $X_1, \ldots, X_m$. For convenience, let the y-coordinate of the particle after the k-th step be denoted by $y(k)$. Obviously,

$$y(k) = (m+n)\Big(k\frac{m}{m+n} - c_{D_1} - \ldots - c_{D_k}\Big). \tag{1}$$

Again the factors $(\frac{m+n}{mn})^{\frac{1}{2}}$ in the Kolmogorov-Smirnov statistics will be omitted in subsequent formulas.

Theorem 1 *The distribution of the* $(\frac{mn}{m+n})^{\frac{1}{2}}$*-multiple of the one-sided Kolmogorov-Smirnov statistic* (4.1.3.2) *is given by*

$$P\Big\{\max_{1 \le k \le m+n}\Big(\frac{km}{m+n} - c_{D_1} - \ldots - c_{D_k}\Big) \le \frac{h}{m+n}\Big\} \tag{2}$$
$$= \binom{m+n}{m}^{-1} \sum_{i=h-mn}^{h} \pi_m^+(i),$$

where h is an integer, $0 \leq h \leq mn$, and $\pi_m^+(i)$ may be obtained by the repeated use of

$$\begin{aligned}\pi_j^+(i) &= \pi_{j-1}^+(i+n) + \pi_{j-1}^+(i+n-m) \\ &\quad + \pi_{j-1}^+(i+n-2m) + \ldots + \pi_{j-1}^+(i+n-nm).\end{aligned} \tag{3}$$

The initial and boundary conditions are as follows: $\pi_0^+(h) = 1$, $\pi_0^+(i) = 0$ otherwise; $\pi_j^+(i)$, $j \geq 1$, may be non-zero only for the integers i such that $i - h + nj$ is divisible by m and

$$h - jn \leq i \leq 2h - n. \tag{4}$$

The distribution of the $(\frac{mn}{m+n})^{\frac{1}{2}}$-multiple of the two sided Kolmogorov-Smirnov statistic (4.1.3.3) *is given by*

$$\begin{aligned}P\Big\{ \max_{1 \leq k \leq m+n} \Big| \frac{km}{m+n} - c_{D_1} - \ldots - c_{D_k} \Big| \leq \frac{h}{m+n} \Big\} \\ = \binom{m+n}{m}^{-1} \sum_{i=0}^{h} \pi_m^{\pm}(i),\end{aligned} \tag{5}$$

with h being an integer, $\min(m,n) \leq h \leq mn$, $h \geq \frac{1}{2}\max(m,n)$. *The numbers $\pi_m^{\pm}(i)$ may be found again by the repeated use of a recurrence formula analogous to* (3) *with analogous initial and boundary conditions, except for* (4) *which must be replaced by*

$$0 \leq i \leq 2h - n. \tag{6}$$

Proof. Using the random walk model introduced above, it is immediately seen that the probability in (2), or that in (5), equals $\binom{m+n}{m}^{-1}$ times the number of possible paths with $y(k) \leq h$ for all k, or with $|y(k)| \leq h$ for all k, respectively. Thus everything will become evident after some consideration of possible paths of the particle. For further details see Section IV.3.2 in Hájek and Šidák (1967). This theorem and its proof were adapted from Massey (1951), originally dealing with the case $m = n$.

For the two-sided case, $\pi_j^{\pm}(i)$ are defined analogously as $\pi_j^+(i)$, except that the condition $y(k) \leq h$ is replaced by $|y(k)| \leq h$. The whole reasoning is again analogous except the inequality $0 \leq i$ in (6); but this is again clarified in Hájek and Šidák (1967). □

Theorem 2 *The distribution of the (mn)-multiple of the Cramér-von Mises statistic* (4.1.3.11) *is given by*

$$\begin{aligned}P\Big\{ \sum_{k=1}^{m+n} \Big(\frac{km}{m+n} - c_{D_1} - \ldots - c_{D_k} \Big)^2 = \frac{z}{(m+n)^2} \Big\} \\ = \binom{m+n}{m}^{-1} \pi_{m+n}(0, z),\end{aligned} \tag{7}$$

where $0 < z < \infty$ *and* $\pi_{m+n}(0, z)$ *may be obtained by the repeated use of*

$$\pi_j(i, z) = \pi_{j-1}(i + n, z - i^2) + \pi_{j-1}(i - m, z - i^2), \tag{8}$$

starting from $\pi_0(0,0) = 1$, *with* $\pi_0(i, z) = 0$ *otherwise. Boundary conditions for* (8) *are* $\pi_j(i, z) = 0$ *for* $i \neq jm - g(m + n)$, $g = 0, 1, \ldots, j$.

The proof is based again on the preceding random walk model from Theorem 1, and other details may again be found in Section IV.3.2 in Hájek and Šidák (1967). The original idea for $m = n$ comes from Burr (1963).

A different kind of development for statistics of Kolmogorov-Smirnov types with unequal sizes (though not related to recurrence formulas) originated with the following ideas by Steck (1969). In a two-sample problem $X_1, \ldots, X_m, Y_1, \ldots, Y_n$, let R_i, $i = 1, \ldots, m$, be the rank of X_i in the ordered combined sample, let $D^+(m, n) = \max_x(F_{1,m}(x) - F_{2,n}(x))$, $D^-(m, n) = \max_x(F_{2,n}(x) - F_{1,m}(x))$, $D(m, n) = \max(D^+(m, n), D^-(m, n)$ where $F_{1,m}(x)$ and $F_{2,n}(x)$ are the corresponding empirical distribution functions. The first topic in Steck (1969) is to express the distributions of the mentioned Kolmogorov-Smirnov statistics in terms of certain probabilities for R_i's, e.g. $P(mnD(m, n) \leq r) = P((i(m+n) - r)/m \leq R_i \leq (i(m + n) - n + r)/m,\ 1 \leq i \leq m)$. The second topic is the distribution of these Kolmogorov-Smirnov statistics for Lehmann's alternative $G^k = F$ and its expression as a determinant. For the third topic let us have some fixed numbers b_ν, c_ν, $i \leq \nu \leq j$, and a certain (rather complicated) determinant is derived showing how many times it happens in the combined sample that $b_\nu < R_\nu < c_\nu$, $i \leq \nu \leq j$. These topics with determinants were pursued in further papers in the literature, sometimes under the name 'Steck determinants'.

Certain closed expressions for distributions were obtained also in the case $m = nk$, cf. Korolyuk (1955a), Blackman (1956) (formulas in the last paper are not correct, cf. his Correction (1958)). Steck (1969) also mentions this case.

Vincze (1957), (1959), (1963) studied the joint distributions of the Kolmogorov-Smirnov statistic and of the point at which the maximum is reached for the first time. A survey is presented by Vincze (1960). (See also Problem 20 in Chapter 6.)

The case of m, n slightly different has been studied by Reimann and Vince (1960), see also Vincze (1960).

The methods of Subsections 5.3.1, 5.3.2, and many related ones are elaborated in detail in Mohanty (1979).

5.3.3 The Kruskal-Wallis and the Friedman test statistics. There are probably no simple direct methods for finding the distributions of these statistics. Some more modest results concerning the sums of ranks may be found in Section IV.3.3 in Hájek and Šidák (1967).

5.4 IMPROVEMENTS OF LIMITING DISTRIBUTIONS FOR FINITE SAMPLES

This section contains only some remarks, far from being complete and systematic, since a more thorough treatment would lie beyond the scope of this book. The purpose of the following is only to give the reader some idea about possible improvements for computational work. Note that asymptotic expansions, probably the most important topic in this area, are postponed to Subsection 8.4.1.

5.4.1 Corrections for continuity. In many cases, we should like to approximate finite sample distributions of rank statistics, which are discrete, by their asymptotic distributions, which are continuous. Such replacements of discrete distributions by the continuous ones represent the first source of discrepancies; they may be, however, to some extent reduced by the well-known corrections for continuity.

Assume, as a simple but in this respect most urgent case, that we are dealing with a rank statistic T such that its possible values $t_1, t_2, \dots$ are equidistant, two neighbouring values having the distance λ (usually $\lambda = 1$). Furthermore, let $(T - \mathsf{E}T).(\mathsf{var}\,T)^{-\frac{1}{2}}$ have asymptotically the standardized normal distribution Φ. Now, to bring the distribution of T closer to its continuous limit, we may imagine the probabilities $P(T = t_j)$ spread uniformly over the intervals $(t_j - \frac{1}{2}\lambda, t_j + \frac{1}{2}\lambda)$. Thus we have approximately

$$P(T \leq t_j) = P(T \leq t_j + \tfrac{1}{2}\lambda) \doteq \Phi\Big(\frac{t_j - \mathsf{E}T + \frac{1}{2}\lambda}{(\mathsf{var}\,T)^{\frac{1}{2}}}\Big), \tag{1}$$

and similarly for the upper tail

$$\begin{aligned} P(T \geq t_j) = P(T \geq t_j - \tfrac{1}{2}\lambda) &\doteq 1 - \Phi\Big(\frac{t_j - \mathsf{E}T - \frac{1}{2}\lambda}{(\mathsf{var}\,T)^{\frac{1}{2}}}\Big) \\ &= \Phi\Big(\frac{-t_j + \mathsf{E}T + \frac{1}{2}\lambda}{(\mathsf{var}\,T)^{\frac{1}{2}}}\Big). \end{aligned} \tag{2}$$

Going ahead with this idea, we may even make use of Sheppard's correction for variance, obtaining

$$P(T \leq t_j) \doteq \Phi\Big(\frac{t_j - \mathsf{E}T + \frac{1}{2}\lambda}{(\mathsf{var}\,T - \frac{1}{12}\lambda^2)^{\frac{1}{2}}}\Big) \tag{3}$$

instead of (1), and similarly also for (2).

Formula (1) usually brings a substantial improvement of the simple approximation $\Phi((t_j - \mathsf{E}T)(\mathsf{var}\,T)^{-\frac{1}{2}})$, namely for symmetric and unimodal distributions, and there are some numerical results indicating that (3) may bring a further improvement of (1). As a rough qualitative example, if S is the two-sample Wilcoxon statistic (4.1.1.6) and $P(S \leq s)$ is near the value 0.025 (giving a very frequently used 5% two-sided test), then usually the approximative value $\Phi((s - \mathsf{E}S)(\mathsf{var}\,S)^{-\frac{1}{2}})$ is smaller than the precise value $P(S \leq s)$; using (1) this approximative value is increased so that it is then higher than the precise value, but it can again be somewhat diminished by using (3).

5.4.2 Interpolation for numerical computations. For a computation (and, in particular, for convenient storing in a computer) of critical values of rank tests the following method may be useful. Its origin is in numerical analysis (or theory of approximations).

Let the significance level be fixed so that the critical values are functions only of the sample size N. Then we look for a simple formula for which the maximum absolute difference between its result and the critical value is smaller than 0.5 for all N in the required set of N's. Hence, on rounding off these numbers, we get precise results for integer-valued critical values.

Thus we essentially use the theory of approximations for a discrete problem with the Chebyshev norm (sometimes also called the best approximation or minimax). The pertaining theory may be found e.g. in the book Collatz (1964), §26, together with the necessary theory of Stiefel exchanges, or in Hart (1968).

Let us show the principle by an example. Suppose that the critical values of some test can be well approximated by a normal distribution for large N, and suppose we know the mean value E_N, the standard deviation σ_N for all needed N, and that we know L (≥ 3) exact critical values $c_{i_1}, \ldots, c_{i_L}$. We try to approximate critical values generally by a linear polynomial $A_0 + A_1 E_N + A_2 \sigma_N$. The coefficients A_0, A_1, A_2 are then found by minimizing

$$\max_{N = i_1, \ldots, i_L} |A_0 + A_1 E_N + A_2 \sigma_N - c_N|.$$

Usually, A_2 is the corresponding critical value of the standardized normal distribution.

Altogether, we have arrived at a numerical improvement of the normal approximation.

As can be expected the goodness of the normal approximation is basic for such developments. The description of the method and its use for the sign test, Wilcoxon one-sample and two-sample tests, and for Kendall correlation coefficient τ were given by Bukač (1975), (1995).

5.4.3 Transformations of test statistics and the use of other limiting distributions. Still another method of approximating the distribution of a rank statistic T is to try to find a suitable monotone transformation $h(T)$ whose distribution might be close to some simple known density. Commonly, for such an approximation some suitable type of density is chosen from the Pearson system of density curves, and the values of the free parameters may be determined by equating a necessary number of the lower moments of the approximating and the actual distributions.

This procedure can sometimes be quite successful, even for smaller sample sizes.

For example, if S is the Fisher-Yates-Terry-Hoeffding regression statistic (4.3.1.1), then the distribution of $S_0 = S[(N-1)\operatorname{var} S]^{-\frac{1}{2}}$ may be well approximated by the density $g(x) = [B(\frac{1}{2}, \frac{1}{2}N-1)]^{-1}(1-x^2)^{\frac{1}{2}N-2}$, $-1 \leq x \leq 1$ or equivalently, the distribution of $(N-2)^{\frac{1}{2}}.\, S_0(1-S_0^2)^{-\frac{1}{2}}$ may be approximated by the Student distribution with $N-2$ degrees of freedom. (See Terry (1952) following the suggestion by Pitman (1937/38).)

Similarly concerning the Spearman rank correlation coefficient ρ, the distribution of $(N-2)^{\frac{1}{2}}.\, \rho(1-\rho^2)^{-\frac{1}{2}}$ may be also approximated by the Student distribution with $N-2$ degrees of freedom. (See Kendall, Kendall and Babington Smith (1938), Glasser and Winter (1961).) In the papers by Fieller, Hartley and Pearson (1957) and by Fieller and Pearson (1961), the transformed variable $\frac{1}{2}\log[(1+\rho)(1-\rho)^{-1}]$, which has approximately the normal distribution, is investigated. (A similar transformation for the Fisher-Yates correlation coefficient is studied in the latter paper, that of the Kendall correlation coefficient in both papers mentioned and in Pearson and Snow (1962).) The Pearson curve of the second type for the Spearman correlation coefficient was used by Olds (1938). Further, if Q is the Kruskal and Wallis test statistic, the beta distribution for the variable $Q(\max Q)^{-1}$ was used by Kruskal (1952), Kruskal and Wallis (1952), Rijkoort and Wise (1953), and Wallace (1959) (or, alternatively, for different transformations of Q, also Fisher's z or the F distributions). Similarly, for the Friedman test statistic see Kendall and Babington Smith (1939), Friedman (1940), and Rijkoort and Wise (1953).

A special improvement for the approximate distribution of the van der Waerden statistic, based on the separation of the extreme scores, has been suggested by van der Waerden (1954).

Let us observe generally that the methods mentioned in this subsection belong to classical methods of curve fitting, the detailed description of which can be found in many other books on statistics.

PROBLEMS AND COMPLEMENTS TO CHAPTER 5

Section 5.1

1. Show that the distribution of the Kendall rank correlation coefficient τ is symmetric about 0.

Section 5.2

2. If we have two samples $X_1, \ldots, X_m, Y_1, \ldots, Y_n$, and if S_0 denotes the number of X_i's exceeding the median of the pooled sample (a simplified median test statistic), then

$$P(S_0 = k) = \binom{[\frac{1}{2}(m+n)]}{k} \binom{[\frac{1}{2}(m+n+1)]}{m-k} \binom{m+n}{m}^{-1}$$

for $k = 0, 1, \ldots, m$ in the case of $m \le n$, and for

$$[\tfrac{1}{2}(m-n)], [\tfrac{1}{2}(m-n)] + 1, \ldots, [\tfrac{1}{2}(m+n)]$$

in the case of $m \ge n$.

3. Let U be the Elandt statistic (mentioned in the text following (4.6.1.18)) for testing independence of coordinates in the sample (X_i, Y_i), $i = 1, \ldots, N$. Then both for $N = 2n$ and for $N = 2n+1$ we have

$$P(U = 2k) = \binom{n}{k}^2 \binom{2n}{n}^{-1} \quad \text{for } k = 0, 1, \ldots, n.$$

4. Consider the statistic $A + B$ (a simpler 'one-sided' analogue of the Haga test statistic) with A, B defined in 4.1.2. Then

$$P(A + B = k) = \sum_{a=0}^{m} P(A = a, B = k - a) \quad \text{for } 0 \le k \le m + n,$$

where

$$P(A = a, B = b) = \binom{m+n-a-b-2}{m-a-1} \binom{m+n}{m}^{-1}$$

for

$$\begin{gathered} 0 \le a \le m-1, \quad 0 \le b \le n-1, \\ P(A = m, B < n) = P(A < m, B = n) = 0, \end{gathered}$$

and

$$P(A = m, B = n) = \binom{m+n}{m}^{-1}.$$

5. If T is the Kamat test statistic (4.2.2.1), then

$$\begin{aligned} P(T = t) \quad = \quad & \Big\{ 2 \sum_{i=1}^{\min(m-t,n)-1} \binom{m+n-2i-t-2}{m-i-t-1} \\ & + \gamma(t)(t-1) \binom{m+n-t-2}{n-2} \\ & + 2\delta(m, n+t) \Big\} \binom{m+n}{m}^{-1} \quad \text{for } 0 \le t \le m, \end{aligned}$$

where $\sum_{i=1}^{j}$ for $j < 1$ stands for 0, $\binom{k}{j} = 0$ for $j > k$ or $k < 0$, $\gamma(t) = 0$ for $t = 0$ and $\gamma(t) = 1$ otherwise, $\delta(m, n+t) = 1$ for $m = n+t$ and $\delta(m, n+t) = 0$ otherwise. $P(T = t)$ for $-n \le t \le -1$ is given be an analogous formula with t replaced by $|t|$ and with the roles of m, n interchanged.

6. If E is the E-test statistic (4.2.2.2) for testing scale, then

$$\begin{aligned} P(E \ge k) &= \binom{m+n-2k}{n}\binom{m+n}{m}^{-1} \quad \text{for } 1 \le k \le [\tfrac{1}{2}m], \\ P(E \le -k) &= \binom{m+n-2k}{m}\binom{m+n}{m}^{-1} \quad \text{for } 1 \le k \le [\tfrac{1}{2}n]. \end{aligned}$$

7. Mosteller (1948) suggested the following test statistic T for the slippage problem, i.e. for testing whether one of k densities is translated to the right relative to the remaining ones. Finding first the sample with the largest observation, T equals the number of observations in this sample which exceed all the observations in the $k-1$ remaining samples. Using the notation of Section 4.4, we have, under H_0

$$P(T \ge t) = \sum_{j=1}^{k} \frac{(N-t)!n_j!}{N!(n_j-t)!} \quad \text{for } 1 \le t \le \max(n_1, \dots, n_k),$$

where the summands with $n_j < t$ should be interpreted as 0.

8. Let $F_{1,n}$ and F_{2n} be the empirical distribution functions of the samples $X_1, \dots, X_n$ and $Y_1, \dots, Y_n$, respectively. Then, under H_0,

$$\begin{aligned} &P\{F_{2,n}(x) \le F_{1,n}(x) \quad \text{for} \quad -\infty < x < \infty\} = (n+1)^{-1}, \\ &P\{F_{2,n}(x) \le F_{1,n}(x) \quad \text{for} \quad -\infty < x < \infty, \text{ or } F_{2,n}(x) \ge F_{1,n}(x) \\ &\qquad\qquad\qquad\qquad\quad \text{for} \quad -\infty < x < \infty\} = 2(n+1)^{-1}. \end{aligned}$$

[Hint: Use formula (5.2.3.1) for $h = 1$.]

9. (Continuation.) Prove that

$$\begin{aligned} &P\{F_{2,n}(x) < F_{1,n}(x) \quad \text{for} \quad X^{(1)} \le x < X^{(2n)}\} = \tfrac{1}{2}(2n-1)^{-1}, \\ &P\{F_{2,n}(x) < F_{1,n}(x) \quad \text{for} \quad X^{(1)} \le x < X^{(2n)}, \text{ or } F_{2,n}(x) > F_{1,n}(x) \\ &\qquad\qquad\qquad\qquad\quad \text{for} \quad X^{(1)} \le x < X^{(2n)}\} = (2n-1)^{-1}. \end{aligned}$$

[Hint: Use the random walk model of Section 5.2.3. The number of paths with $y(1) = -1$ is $\binom{2n-1}{n}$. Consider the paths with $y(1) = -1$ which intersect or touch the axis $y = 0$; distinguish those with $y(2n-1) = 1$ and those with $y(2n-1) = -1$. There exists a one-to-one mapping of paths of the latter kind onto paths of the former kind, whose number is $\binom{2n-2}{n}$.] (Drion (1952).)

10. Having two ordered samples $X^{(1)} < \ldots < X^{(n)}$ and $Y^{(1)} < \ldots < Y^{(n)}$, let G be the number of indices k, $1 \leq k \leq n$, such that $Y^{(k)} < X^{(k)}$. (So-called Galton's rank test statistic.) Then, under H_0,

$$P(G = g) = (n+1)^{-1} \quad \text{for } 0 \leq g \leq n.$$

[Hint: In the random walk model of Section 5.2.3, G is the number of k's for which $y(k) > 0$, $y(k) > y(k-1)$. In each path, denote by C_1 its initial part to its first touching the line $y = 1$, by C_2 its subsequent part to its first touching $y = 0$. Interchange C_1 and C_2, attaching C_2 directly to $(0,0)$, etc., obtaining in this manner a one-to-one mapping of paths with $G = g$ onto paths with $G = g - 1$.] (Sarkadi (1961).)

Section 5.3

11. Let $S_{m,n}$ be the Ansari-Bradley test statistic (4.2.1.7). Show that

$$P(S_{m,n} = k) = \frac{\pi_{m,n}(k)}{\binom{m+n}{n}}$$

where

$$\pi_{m,n}(k) = \pi_{m,n-1}(k) + \pi_{m-1,n}\left(k - \left[\tfrac{1}{2}(m+n+1)\right]\right),$$

and find the initial and boundary conditions.

12. Let $S_{m,n}$ be the Savage test statistic (4.2.1.15). Show that

$$P(S_{m,n} = z) = \frac{\pi_{m,n}(z)}{\binom{m+n}{m}}$$

where

$$\pi_{m,n}(z) = \pi_{m,n-1}\left(z - \frac{m}{N}\right) + \pi_{m-1,n}\left(z - \frac{m}{N}\right),$$

and find the initial and boundary conditions.

Chapter 6

Limiting null distributions

6.1 SIMPLE LINEAR RANK STATISTICS

6.1.1 Convergence in indexed sets of statistics. The theory of limiting distributions deals with sequences of probability distributions, which in applications correspond to sequences of some statistics T_N, $N \geq 1$. In our considerations N will denote the sample size, i.e. the number of observations from which T_N is computed. However, the statistics in which we are interested will only exceptionally be indexed by N. For example, in the two-sample problem, they will depend on the sizes m and n of both the partial samples, so that we shall deal with an indexed set $\{T_{m,n}\}$, $m \geq 1$, $n \geq 1$. More generally, in regression problems, the statistics will be indexed by real vectors $c = (c_1, \ldots, c_N)$ of any finite dimensions, so that the subject of our study will be the indexed set $\{T_c\}$.

The convergence statements will concern sequences $\{T_{m_\nu, n_\nu}\}$ or $\{T_{c_\nu}\}$, $c_\nu = (c_{\nu 1}, \ldots, c_{\nu N_\nu})$, selected from the sets $\{T_{m,n}\}$ or $\{T_c\}$, and will hold under some conditions on $\{(m_\nu, n_\nu)\}$ or $\{c_\nu\}$, respectively. For example we shall assert that the statistics T_{m_ν, n_ν} are asymptotically normal for every sequence $\{(m_\nu, n_\nu)\}$ such that

$$\min(m_\nu, n_\nu) \to \infty. \tag{1}$$

However, to simplify the notation, we shall usually drop the index ν and say that the statistics $T_{m,n}$ are asymptotically normal for

$$\min(m, n) \to \infty. \tag{2}$$

A similar meaning is to be attached to the statement that the statistics T_c

are asymptotically normal for

$$\sum_{i=1}^{N}(c_i - \bar{c})^2 / \max_{1 \le i \le N}(c_i - \bar{c})^2 \to \infty. \tag{3}$$

Finally, let us mention that in some cases, including the ones given, the convergence statements make sense even without considering selected sequences. Thus asymptotic normality (μ, σ^2) under (2) may be interpreted as follows: For every $\varepsilon > 0$ there exists an n_0 such that $\min(m, n) > n_0$ implies

$$\sup_x |P(T_{m,n} \le \mu + x\sigma) - \Phi(x)| < \varepsilon. \tag{4}$$

The reader may find it useful to keep this alternative interpretation in mind throughout the present chapter.

6.1.2 A special central limit theorem. Let A be the set of all non-vanishing real vectors $a = (a_1, \ldots, a_N)$ of all finite dimensions $N \ge 1$. We shall say that the statistics T_a, $a \in A$, are asymptotically normal (μ_a, σ_a^2) for

$$\sum_{i=1}^{N} a_i^2 / \max_{1 \le i \le N} a_i^2 \to \infty, \tag{1}$$

if (1) entails

$$F(T_a \le \mu_a + x\sigma_a) \to (2\pi)^{-\frac{1}{2}} \int_{-\infty}^{x} \exp(-\tfrac{1}{2}y^2)\,\mathrm{d}y, \; -\infty < x < \infty. \tag{2}$$

Thus asymptotic normality (μ_a, σ_a^2) is equivalent to convergence in distribution of $(T_a - \mu_a)/\sigma_a$ to a standardized normal random variable. Obviously, asymptotic normality (μ_a, σ_a^2) is equivalent to the same property with (μ_a^*, σ_a^{*2}), if

$$\sigma_a^*/\sigma_a \to 1, \quad (\mu_a^* - \mu_a/\sigma_a \to 0. \tag{3}$$

Theorem 1 *Let Y_1, $Y_2, \ldots$ be independent copies of a random variable with finite expectation μ and finite variance σ^2. Put*

$$T_a = \sum_{i=1}^{N} a_i Y_i, \quad a \in A. \tag{4}$$

Then, for (1), the statistics T_a are asymptotically normal (μ_a, σ_a^2) with

$$\mu_a = \mu \sum_{i=1}^{N} a_i \tag{5}$$

and

$$\sigma_a^2 = \sigma^2 \sum_{i=1}^{N} a_i^2. \tag{6}$$

Proof. The Lindeberg condition (Loève (1955), p. 295) takes on the form

$$\sigma_a^{-2}\sum_{i=1}^{N}\int_{|x|>\varepsilon\sigma_a} x^2\,\mathrm{d}P\big(a_i(Y_i-\mu)\le x\big)\to 0, \tag{7}$$

where σ_a^2 is given by (6). Upon substitution of $a_i y$ for x, we obtain

$$\begin{aligned}\int_{|x|>\varepsilon\sigma_a} & x^2\,\mathrm{d}P\big(a_i(Y_i-\mu)\le x\big) \\ &= a_i^2\int_{|ya_i|>\varepsilon\sigma_a} y^2\,\mathrm{d}P(Y_i-\mu\le y) \\ &\le a_i^2\int_{|y|>\varepsilon\sigma\nu_a} y^2\,\mathrm{d}P(Y_i-\mu\le y)\end{aligned} \tag{8}$$

where

$$\nu_a^2=\sum_{i=1}^{N}a_i^2/\max_{1\le i\le N}a_i^2.$$

Consequently, the Y_i's having the same distribution,

$$\begin{aligned}\sigma_a^{-2}\sum_{i=1}^{N}\int_{|x|>\varepsilon\sigma_a} & x^2\,\mathrm{d}P\big(a_i(Y_i-\mu)\le x\big) \\ &\le \sigma^{-2}\int_{|y|>\varepsilon\sigma\nu_a} y^2\,\mathrm{d}P(Y_1-\mu\le y).\end{aligned} \tag{9}$$

However, the variance of Y_1 is supposed finite and $\nu_a\to\infty$ in view of (1), so that

$$\sigma^{-2}\int_{|y|>\varepsilon\nu\sigma_a} y^2\,\mathrm{d}P(Y_1-\mu\le y)\to 0,\quad \varepsilon>0,$$

which, in accordance with (9), entails (7). □

Remark. The above theorem could be reformulated as follows: For every $\varepsilon>0$ there exists an n_0 such that

$$\sum_{i=1}^{N}a_i^2/\max_{1\le i\le N}a_i^2>n_0 \tag{10}$$

entails

$$\sup_x\Big|P\Big(\sum_{i=1}^{N}a_iY_i\le\mu_a+x\sigma_a\Big)-\Phi(x)\Big|<\varepsilon, \tag{11}$$

where Φ denotes the standardized normal distribution function.

6.1.3 A convergence theorem.

Theorem 1 *Let $(\Omega, \mathcal{A}, \mu)$ be a measure space with a σ-finite measure μ. Consider a sequence $\{h_\nu\}$ of square integrable functions converging almost everywhere to a square integrable function h. Assume that*

$$\limsup_{\nu\to\infty} \int h_\nu^2 \, \mathrm{d}\mu \leq \int h^2 \, \mathrm{d}\mu. \tag{1}$$

Then

$$\lim_{\nu\to\infty} \int (h_\nu - h)^2 \, \mathrm{d}\mu = 0. \tag{2}$$

Proof. Fatou's lemma together with (1) implies

$$\lim_{\nu\to\infty} \int h_\nu^2 \, \mathrm{d}\mu = \int h^2 \, \mathrm{d}\mu. \tag{3}$$

Furthermore, the Schwartz inequality yields

$$\int |h_\nu h| \, \mathrm{d}\mu \leq \left[\int h_\nu^2 \, \mathrm{d}\mu \int h^2 \, \mathrm{d}\mu \right]^{\frac{1}{2}} \tag{4}$$

so that

$$\limsup_{\nu\to\infty} \int |h_\nu h| \, \mathrm{d}\mu \leq \int h^2 \, \mathrm{d}\mu.$$

Consequently, according to Theorem 3.4.2.1

$$\lim_{\nu\to\infty} \int h_\nu h \, \mathrm{d}\mu = \int h^2 \, \mathrm{d}\mu. \tag{5}$$

Now (3) and (5) imply (2). □

6.1.4 Further preliminaries. Consider a probability space $(\Omega, \mathcal{A}, P)$ and a sequence of sub-σ-fields $\mathcal{F}_1 \subset \mathcal{F}_2 \subset \ldots \subset \mathcal{A}$. Denote by $\mathcal{F}_\infty$ the smallest σ-field containing the field $\bigcup_1^\infty \mathcal{F}_N$. For every event $A \in \mathcal{F}_\infty$ and every $\varepsilon > 0$ there exists an N and $A_N \in \mathcal{F}_N$ such that

$$P(A \div A_N) < \varepsilon, \tag{1}$$

where $\div$ denotes the symmetric difference. Actually, the assertion is trivially true for $A \in \bigcup_1^\infty \mathcal{F}_N$, and the events having this property obviously form a σ-field. Denoting by I_A and I_{A_N} the respective indicators, (1) may be rewritten as follows:

$$\mathsf{E}(I_A - I_{A_N})^2 < \varepsilon. \tag{2}$$

If Y is a $\mathcal{F}_\infty$-measurable function such that $\mathsf{E}Y^2 < \infty$, then there exists for every N a $\mathcal{F}_N$-measurable random variable Y_N such that

$$\mathsf{E}(Y_N - Y)^2 \leq \mathsf{E}(Y_N^* - Y)^2 \tag{3}$$

for any other $\mathcal{F}_N$-measurable random variable Y_N^*. It is well known that this property is possessed by the conditional expectation with respect to $\mathcal{F}_N$,

$$Y_N = \mathsf{E}(Y \mid \mathcal{F}_N). \tag{4}$$

If $\mathcal{F}_N$ is generated by a statistic T_N, then $\mathsf{E}(Y \mid \mathcal{F}_N) = \psi(T_N)$, where $\psi(t_N) = \mathsf{E}(Y \mid T_N = t_n)$. Now, if $I_A^N = \mathsf{E}(I_A \mid \mathcal{F}_N)$, then (2), (3) and (4) imply that

$$\mathsf{E}(I_A - I_A^N)^2 < \varepsilon \tag{5}$$

for N sufficiently large, and consequently,

$$\lim_{N\to\infty} \mathsf{E}(I_A - I_A^N)^2 = 0. \tag{6}$$

Before generalizing the relation (6) to all random variables with finite variance, let us recall that

$$\mathsf{E}Y_N^2 \leq \mathsf{E}Y^2 \tag{7}$$

for Y_N given by (4).

Lemma 1 *Let Y be a $\mathcal{F}_\infty$-measurable random variable such that $\mathsf{E}Y^2 < \infty$, and let Y_N be given by* (4). *Then*

$$\lim_{N\to\infty} \mathsf{E}(Y_N - Y)^2 = 0 \tag{8}$$

and

$$\lim_{N\to\infty} \mathsf{E}Y_N^2 = \mathsf{E}Y^2. \tag{9}$$

Proof. Fix an $\varepsilon > 0$ and find a $\mathcal{F}_\infty$-measurable simple function $\sum\limits_{i=1}^{n} c_i I_{A_i}$ such that

$$\mathsf{E}\Big(Y - \sum_{i=1}^{n} c_i I_{A_i}\Big)^2 < \tfrac{1}{6}\varepsilon.$$

Denoting $I_{A_i}^N = \mathsf{E}(I_{A_i} \mid \mathcal{F}_N)$ and noting (7), we have

$$\begin{aligned}\mathsf{E}(Y_N - Y)^2 \;\leq\; & 3\mathsf{E}\Big(Y_N - \sum_{i=1}^{n} c_i I_{A_i}^N\Big)^2 + \\ & + 3\mathsf{E}\Big(Y - \sum_{i=1}^{n} c_i I_{A_i}\Big)^2 + 3\mathsf{E}\Big[\sum_{i=1}^{n} c_i (I_{A_i} - I_{A_i}^N)\Big]^2\end{aligned}$$

$$\begin{aligned} &\leq 6\mathsf{E}\Big(Y - \sum_{i=1}^{n} c_i I_{A_i}\Big)^2 + 3\sum_{i=1}^{n} c_i^2 \sum_{i=1}^{n} \mathsf{E}(I_{A_i} - I_{A_i}^N)^2 \\ &< \varepsilon + 3\sum_{i=1}^{n} c_i^2 \sum_{i=1}^{n} \mathsf{E}(I_{A_i} - I_{A_i}^N)^2. \end{aligned}$$

Since, in view of (6), the last sum converges to 0 as $N \to \infty$, we conclude that

$$\mathsf{E}(Y_N - Y)^2 < \varepsilon$$

for N sufficiently large. This proves (8). Then (9) follows from the well-known relation

$$\mathsf{E}(Y_N - Y)^2 = \mathsf{E}Y^2 - \mathsf{E}Y_N^2, \tag{10}$$

holding for any conditional expectation. □

Now let $U_1, U_2, \ldots$ be independent random variables, each uniformly distributed over $(0, 1)$. Let R_{Ni} denote the rank of U_i, $1 \leq i \leq N$, in the partial sequence $U_1, \ldots, U_N$. Let $\varphi(u)$, $0 < u < 1$, be some square integrable function

$$\int_0^1 \varphi^2(u)\,\mathrm{d}u < \infty, \tag{11}$$

and put

$$a_N^\varphi(i) = \mathsf{E}[\varphi(U_1) \mid R_{N1} = i], \quad 1 \leq i \leq N < \infty. \tag{12}$$

Theorem 1 *Under assumption* (11),

$$\lim_{N\to\infty} \mathsf{E}[a_N^\varphi(R_{N1}) - \varphi(U_1)]^2 = 0, \tag{13}$$

holds, where $a_N^\varphi(i)$ *is defined by* (12).

Proof. Let $\mathcal{F}_N$ be the sub-σ-field generated by $(R_{N1}, \ldots, R_{NN})$. Note that $\mathcal{F}_N \subset \mathcal{F}_{N+1} \subset \ldots$ and recall that $\mathcal{F}_\infty$ denotes the smallest σ-field containing $\bigcup_1^\infty \mathcal{F}_N$. We first show that $\varphi(U_1)$ is equivalent to a $\mathcal{F}_\infty$-measurable random variable. In view of (3.1.2.12), we have

$$\begin{aligned} \mathsf{E}\Big(U_1 - \frac{R_{N1}}{N+1}\Big)^2 &= \frac{1}{N}\sum_{j=1}^{N} \mathsf{E}\Big[\Big(U_1 - \frac{j}{N+1}\Big)^2 \Big| R_{N1} = j\Big] \\ &= \frac{1}{N}\sum_{j=1}^{N} \operatorname{var} U_N^{(j)} = \frac{1}{N}\sum_{j=1}^{N} \frac{j(N-j+1)}{(N+1)^2(N+2)} < \frac{1}{N}, \end{aligned}$$

so that

$$\lim_{\nu\to\infty} \frac{R_{N_\nu 1}}{N_\nu + 1} = U_1$$

with probability 1 for some properly chosen subsequence $\{N_\nu\}$. Consequently U_1, and hence also $\varphi(U_1)$, is equivalent to a $\mathcal{F}_\infty$-measurable random variable. Now it remains to apply the above lemma with $\varphi(U_1) = Y$ and $a_N^\varphi(R_{N1}) = Y_N$. The proof is thus concluded. □

Lemma 2 (D.K. Faddeev.) *Let the functions $f_N(t,u)$, $N \geq 1$, $0 < t$, $u < 1$, be densities in t for each fixed u, such that for every $\varepsilon > 0$*

$$\lim_{N\to\infty} \int_{u-\varepsilon}^{u+\varepsilon} f_N(t,u)\,\mathrm{d}t = 1, \quad 0 < u < 1. \tag{14}$$

Moreover, assume that

$$f_N(t,u) \leq g_N(t,u), \quad N \geq 1,\ 0 < t,\ u < 1, \tag{15}$$

where the functions $g_N(t,u)$ are increasing in $t \in (0,u)$ and decreasing in $t \in (u,1)$ for every fixed $N \geq 1$ and $0 < u < 1$, and

$$\sup_N \int_0^1 g_N(t,u)\,\mathrm{d}t < \infty, \quad 0 < u < 1. \tag{16}$$

Then for every integrable function $\varphi(u)$

$$\lim_{N\to\infty} \int_0^1 \varphi(t) f_N(t,u)\,\mathrm{d}t = \varphi(u) \tag{17}$$

in almost all points $u \in (0,1)$.

Proof. See I.P. Natanson (1957), Theorem 3, §2, Chapter X, and Theorem 5, § 4, Chapter IX.

Theorem 2 *Let $\varphi(u)$, $0 < u < 1$, be square integrable and let $a_N^\varphi(i)$ be given by* (12)*. Then*

$$\lim_{N\to\infty} \int_0^1 [a_N^\varphi(1 + [uN]) - \varphi(u)]^2\,\mathrm{d}u = 0, \tag{18}$$

with $[uN]$ denoting the largest integer not exceeding uN.

Proof. Since, in accordance with (7), where $Y = \varphi(U_1)$,

$$\int_0^1 \left[a_N^\varphi(1 + [uN])\right]^2 \mathrm{d}u \leq \int_0^1 \varphi^2(u)\,\mathrm{d}u, \tag{19}$$

it suffices to prove that

$$\lim_{N\to\infty} a_N^\varphi(1 + [uN]) = \varphi(u) \tag{20}$$

almost everywhere and then apply Theorem 6.1.3.1.

Now (20) follows from Lemma 2, if we put

$$f_N(t,u) = N\binom{N-1}{i-1} t^{i-1}(1-t)^{N-i}, \qquad \frac{i-1}{N} \le u < \frac{i}{N},\ 0<t<1 \tag{21}$$

and

$$\begin{aligned} g_N(t,u) &= N\binom{N-1}{i-1}\left(\frac{i-1}{N-1}\right)^{i-1}\left(\frac{N-i}{N-1}\right)^{N-i}, \qquad \frac{i-1}{N} \le t,\ u < \frac{i}{N}, \\ &= f_N(t,u), \quad \text{otherwise.} \end{aligned}$$

Actually, then (see (3.1.2.10))

$$\begin{aligned} a_N^{\varphi}(1+[uN]) &= \mathsf{E}\varphi(U_N^{(i)}) = \int_0^1 \varphi(t) N\binom{N-1}{i-1} t^{i-1}(1-t)^{N-i}\,\mathrm{d}t \\ &= \int_0^1 \varphi(t) f_N(t,u)\,\mathrm{d}t, \qquad \frac{i-1}{N} \le u < \frac{i}{N}, \end{aligned}$$

while (15) is satisfied since $f_N(t,u)$ is unimodal with mode at $(i-1)/(N-1)$, which lies within the interval $((i-1)/N, i/N)$. Also (16) holds true, since

$$\begin{aligned} &\int_0^1 g_N(t,u)\,\mathrm{d}t \\ &\le \int_0^1 f_N(t,u)\,\mathrm{d}t + \binom{N-i}{i-1}\left(\frac{i-1}{N-1}\right)^{i-1}\left(\frac{N-i}{N-1}\right)^{N-i} \le 2. \end{aligned}$$

Thus (17) holds, which is equivalent to (20). □

6.1.5 Locally optimum rank-test statistics for H_0. Now we are prepared to prove easily all the theorems needed. Consider real vectors $c = (c_1, \ldots, c_N)$ such that

$$\sum_{i=1}^{N}(c_i - \bar{c})^2 > 0 \tag{1}$$

and

$$\bar{c} = \frac{1}{N}\sum_{i=1}^{N} c_i. \tag{2}$$

Let C be the set of real vectors of all finite dimensions $N \geq 1$, satisfying (1). We shall consider limiting distributions of statistics indexed by $c \in C$ for

$$\frac{\sum_{i=1}^{N} (c_i - \bar{c})^2}{\max_{1 \leq i \leq N} (c_i - \bar{c})^2} \to \infty. \tag{3}$$

Take a square integrable function $\varphi(u)$, $0 < u < 1$, and denote by $a_N^{\varphi}(i)$ the scores associated with φ by (6.1.4.12). Put

$$S_c = \sum_{i=1}^{N} c_i a_N^{\varphi}(R_{Ni}), \quad c \in C \tag{4}$$

where R_{Ni} is the rank of X_i in a set of N independent observations $X_1, \ldots, X_N$, each with density f. If $U_i = F(X_i)$, $F(x) = \int_{-\infty}^{x} f(y)\,\mathrm{d}y$, then the random variables U_i will be uniformly distributed and R_{Ni} may be interpreted as the rank of U_i in the set $U_1, \ldots, U_N$ as well. As we know from Section 3.4, the test statistics generating locally most powerful rank tests are just of the type (4).

Theorem 1 *Let the scores $a_N^{\varphi}(i)$ be associated with a square integrable function $\varphi(u)$ by (6.1.4.12). Put $\bar{\varphi} = \int_0^1 \varphi(u)\,\mathrm{d}u$ and assume $\int_0^1 [\varphi(u) - \bar{\varphi}]^2\,\mathrm{d}u > 0$. Assume H_0. Then for (3), the statistics (4) are asymptotically normal (μ_c, σ_c^2) with*

$$\mu_c = \bar{c} \sum_{i=1}^{N} a_N(i) \tag{5}$$

and

$$\sigma_c^2 = \Big[\sum_{i=1}^{N} (c_i - \bar{c})^2 \Big] \int_0^1 [\varphi(u) - \bar{\varphi}]^2\,\mathrm{d}u, \tag{6}$$

or $\sigma_c^2 = \mathsf{var}\, S_c$.

Proof. Rewrite S_c in the following form:

$$S_c = \sum_{i=1}^{N} (c_i - \bar{c}) a_N(R_{Ni}) + \bar{c} \sum_{i=1}^{N} a_N(i). \tag{7}$$

Introduce

$$T_c = \sum_{i=1}^{N} (c_i - \bar{c}) \varphi(U_i) + \bar{c} \sum_{i=1}^{N} a_N(i), \tag{8}$$

where $U_i = F(X_i)$, $1 \le i \le N$. Now drop N in R_{Ni}, and recall that the distribution of $(R_1, \ldots, R_N)$ is independent of $U^{(\cdot)}$. Consequently, by (3.3.1.23), we obtain

$$\begin{aligned} &\mathsf{E}\{(T_c - S_c)^2 \mid U^{(\cdot)} = u^{(\cdot)}\} \qquad (9)\\ &= \mathsf{E}\Big\{\sum_{i=1}^{N}(c_i - \bar{c})\big(a_N(R_i) - \varphi(u^{(R_i)})\big)\Big\}^2 \\ &= \frac{1}{N-1}\sum_{i=1}^{N}(c_i - \bar{c})^2 \sum_{j=1}^{N}[a_N(j) - \varphi(u^{(j)}) - \bar{a}_N + \bar{\varphi}]^2 \\ &\le \frac{1}{N-1}\sum_{i=1}^{N}(c_i - \bar{c})^2 \sum_{j=1}^{N}[a_N(j) - \varphi(u^{(j)})]^2 \\ &= \frac{N}{N-1}\sum_{i=1}^{N}(c_i - \bar{c})^2 \mathsf{E}\{[a_N(R_1) - \varphi(U_1)]^2 \mid U^{(\cdot)} = u^{(\cdot)}\}. \end{aligned}$$

Consequently

$$\mathsf{E}(T_c - S_c)^2 \le \frac{N}{N-1}\sum_{i=1}^{N}(c_i - \bar{c})^2 \mathsf{E}[a_N(R_1) - \varphi(U_1)]^2 \qquad (10)$$

and

$$\begin{aligned} &\mathsf{E}\Big(\frac{T_c - S_c}{\sigma_c}\Big)^2 \qquad (11)\\ &\quad \le \frac{N}{N-1}\Big(\int_0^1 [\varphi(u) - \bar{\varphi}]^2 \,\mathrm{d}u\Big)^{-1} \mathsf{E}[a_N(R_1) - \varphi(U_1)]^2. \end{aligned}$$

On the other hand

$$\frac{\sum_{i=1}^{N}(c_i - \bar{c})^2}{\max_{1\le i\le N}(c_i - \bar{c})^2} \le N \qquad (12)$$

so that (3) entails $N \to \infty$. This fact, in view of Theorem 6.1.4.1 and (11), implies

$$\lim_c \mathsf{E}\Big(\frac{T_c - S_c}{\sigma_c}\Big)^2 = 0, \qquad (13)$$

and a fortiori

$$\lim_c P\Big(\Big|\frac{T_c - S_c}{\sigma_c}\Big| > \varepsilon\Big) = 0, \quad \varepsilon > 0. \qquad (14)$$

Now we know from Theorem in 6.1.2.1 that the random variables T_c are asymptotically normal with parameters given by (5) and (6). Furthermore,

$$\frac{S_c - \mu_c}{\sigma_c} = \frac{T_c - \mu_c}{\sigma_c} + \frac{S_c - T_c}{\sigma_c}, \qquad (15)$$

where the last term converges to 0 in probability according to (14). Thus asymptotic normality $(0,1)$ of $(T_c - \mu_c)/\sigma_c$ implies the same for $(S_c - \mu_c)/\sigma_c$ in view of a well-known lemma (see Cramér (1945), Section 20.6).

Now σ_c^2 given by (6) equals $\mathsf{var}\, T_c$, and (13) implies $\mathsf{var}\, S_c / \mathsf{var}\, T_c \to 1$, since $|(\mathsf{var}\, S_c)^{\frac{1}{2}} - (\mathsf{var}\, T_c)^{\frac{1}{2}}| \leq [\mathsf{E}(T_c - S_c)^2]^{\frac{1}{2}}$. Consequently, we may put $\sigma_c^2 = \mathsf{var}\, S_c$ as well. □

In the *two-sample problem* we consider statistics

$$S_{mn} = \sum_{i=1}^{m} a_{m+n}(R_{m+n,i}) \tag{16}$$

and we are concerned with their limiting properties for

$$\min(m,n) \to \infty. \tag{17}$$

Theorem 2 *Let the scores $a_N^{\varphi}(i)$ be associated with a square integrable function $\varphi(u)$ by* (6.1.4.12) *and assume $\int_0^1 [\varphi(u) - \bar{\varphi}]^2 \,\mathrm{d}u > 0$.*

Then, for $\min(m,n) \to \infty$, the statistics (16) *are asymptotically normal $(\mu_{mn}, \sigma_{mn}^2)$ with*

$$\mu_{mn} = \frac{m}{m+n} \sum_{i=1}^{m+n} a_{m+n}(i) \tag{18}$$

and

$$\sigma_{mn}^2 = \frac{mn}{m+n} \int_0^1 [\varphi(u) - \bar{\varphi}]^2 \,\mathrm{d}u \tag{19}$$

or

$$\sigma_{mn}^2 = \mathsf{var}\, S_{mn}.$$

Proof. The statistic S_{mn} is a particular case of the statistic S_c given by (4) for $N = m+n$ and

$$c_i = \begin{cases} 1, & 1 \leq i \leq m, \\ 0, & m+1 \leq i \leq m+n. \end{cases} \tag{20}$$

For this special vector $c = (1, \ldots, 1, 0, \ldots, 0)$ we have

$$\bar{c} = \frac{m}{m+n}, \tag{21}$$

$$\sum_{i=1}^{m+n} (c_i - \bar{c})^2 = \frac{mn}{m+n} \tag{22}$$

and

$$\min(m,n) \leq \frac{\sum_{i=1}^{m+n} (c_i - \bar{c})^2}{\max_{1 \leq i \leq m+n} (c_i - \bar{c})^2}. \tag{23}$$

Consequently, (17) implies (3) and the theorem follows from Theorem 1. □

Corollary 1 *Under H_0 and $I(f) < \infty$ the statistics*

$$S_c = \sum_{i=1}^{N} c_i a_N(R_{Ni}, f), \quad c \in C, \tag{24}$$

are, for (3), *asymptotically normal* $(0, \sigma_c^2)$ *with*

$$\sigma_c^2 = I(f) \sum_{i=1}^{N} (c_i - \bar{c})^2, \quad c \in C, \tag{25}$$

or $\sigma_c^2 = \operatorname{var} S_c$.

Furthermore, the statistics

$$S_{mn} = \sum_{i=1}^{m} a_{m+n}(R_{m+n,i}, f), \quad m, n \geq 1, \tag{26}$$

are, for $\min(m, n) \to \infty$, *asymptotically normal* $(0, \sigma_{mn}^2)$ *with*

$$\sigma_{mn}^2 = \frac{mn}{m+n} I(f), \quad m, n \geq 1, \tag{27}$$

or $\sigma_{mn}^2 = \operatorname{var} S_{mn}$.

The assertions remain true, if we replace $I(f)$ and $a_N(i, f)$ by $I_1(f)$ and $a_{1N}(i, f)$, respectively.

6.1.6 General simple linear rank statistics for H_0. We shall now investigate the statistics

$$S_c = \sum_{i=1}^{N} c_i a_N(R_{Ni}), \quad c \in C, \tag{1}$$

with $a_N(i)$ being some arbitrary scores.

Assume that for some square integrable function $\varphi(u)$, $\int_0^1 [\varphi(u) - \bar{\varphi}]^2 \,\mathrm{d}u > 0$,

$$\lim_{N \to \infty} \int_0^1 [a_N(1 + [uN]) - \varphi(u)]^2 \,\mathrm{d}u = 0 \tag{2}$$

holds.

Theorem 1 *Under H_0 and* (2) *the statistics* (1) *are for* (6.1.5.3) *asymptotically normal* (μ_c, σ_c^2) *with*

$$\mu_c = \mathsf{E} S_c, \quad c \in C, \tag{3}$$

$$\sigma_c^2 = \Big[\sum_{i=1}^{N} (c_i - \bar{c})^2\Big] \int_0^1 [\varphi(u) - \bar{\varphi}]^2 \,\mathrm{d}u, \quad c \in C, \tag{4}$$

or $\sigma_c^2 = \operatorname{var} S_c$.

Proof. Put

$$S_c^{\varphi} = \sum_{i=1}^{N} c_i a_N^{\varphi}(i), \tag{5}$$

where $a_N^{\varphi}(i)$ is associated with φ by (6.1.4.12). Using the same reasoning as in proving (6.1.5.10), we obtain

$$\begin{aligned} &\mathsf{E}(S_c^{\varphi} - \mathsf{E}S_c^{\varphi} - S_c + \mathsf{E}S_c)^2 \\ &\leq \frac{1}{N-1}\Big[\sum_{i=1}^{N}(c_i - \bar{c})^2\Big]\sum_{j=1}^{N}[a_N^{\varphi}(j) - a_N(j)]^2 \\ &= \frac{N}{N-1}\Big[\sum_{i=1}^{N}(c_i - \bar{c})^2\Big]\int_0^1 \big[a_N^{\varphi}(1+[uN]) - a_N(1+[uN])\big]^2\,\mathrm{d}u. \end{aligned} \tag{6}$$

Now

$$\begin{aligned} &\int_0^1 \big[a_N^{\varphi}(1+[uN]) - a_N(1+[uN])\big]^2\,\mathrm{d}u \\ &\qquad\leq 2\int_0^1 \big[a_N^{\varphi}(1+[uN]) - \varphi(u)\big]^2\,\mathrm{d}u \\ &\qquad\quad +2\int_0^1 \big[a_N(1+[uN]) - \varphi(u)\big]^2\,\mathrm{d}u \end{aligned} \tag{7}$$

so that (6.1.4.18) and (2) entail

$$\lim_{N\to\infty}\int_0^1 \big[a_N^{\varphi}(1+[uN]) - a_N(1+[uN])\big]^2\,\mathrm{d}u = 0. \tag{8}$$

Thus we may conclude, as in the proof of Theorem 6.1.5.1, that $S_c - \mathsf{E}S_c$ has the same limiting distribution as $S_c^{\varphi} - \mathsf{E}S_c^{\varphi}$. □

We know that (2) is satisfied for the scores $a_N^{\varphi}(i)$. Below we give two methods of deriving 'approximate' scores that also satisfy (2).

Lemma 1 *Let the function $\varphi(u)$, $0 < u < 1$, be expressible as a finite sum of square integrable and monotone functions. Put*

$$a_N(i) = \varphi\Big(\frac{i}{N+1}\Big), \quad 1 \leq i \leq N. \tag{9}$$

Then (2) is satisfied.

Proof. Without loss of generality we may assume that $\varphi(u)$ is increasing and $\varphi(0) \geq 0$. Then $\varphi(u)$ is continuous almost everywhere, so that

$$\lim_{N\to\infty} a_N(1+[uN]) = \varphi(u) \quad \text{almost everywhere.} \tag{10}$$

Besides this the monotonicity implies

$$\frac{1}{N}\varphi^2\Big(\frac{i}{N+1}\Big) \le \int_{i/N}^{(i+1)/N} \varphi^2(u)\,\mathrm{d}u, \; 1 \le i \le N-1, \tag{11}$$
$$\frac{1}{N+1}\varphi^2\Big(\frac{N}{N+1}\Big) \le \int_{N/(N+1)}^{1} \varphi^2(u)\,\mathrm{d}u,$$

which entails

$$\int_0^1 \big[a_N(1+[uN])\big]^2\,\mathrm{d}u = \frac{1}{N}\sum_{i=1}^{N}\varphi^2\Big(\frac{i}{N+1}\Big) \tag{12}$$
$$\le \int_{1/N}^{1}\varphi^2(u)\,\mathrm{d}u + \frac{N+1}{N}\int_{N/(N+1)}^{1}\varphi^2(u)\,\mathrm{d}u.$$

Consequently

$$\limsup_{N\to\infty}\int_0^1 \big[a_N(1+[uN])\big]^2\,\mathrm{d}u \le \int_0^1 \varphi^2(u)\,\mathrm{d}u. \tag{13}$$

Now (10) and (13), on account of Theorem 6.1.3.1, imply (2). □

Lemma 2 *Let the function $\varphi(u)$, $0 < u < 1$, be square integrable. Put*

$$a_N(i) = N\int_{(i-1)/N}^{i/N}\varphi(u)\,\mathrm{d}u, \quad 1 \le i \le N. \tag{14}$$

Then (2) *is satisfied.*

Proof. This is a well-known result. However, instead of giving references, we derive it from the foregoing results. We have

$$\int_0^1 \big[a_N(1+[uN]) - \varphi(u)\big]^2\,\mathrm{d}u \le \int_0^1 \big[a_N^{\varphi}(1+[uN]) - \varphi(u)\big]^2\,\mathrm{d}u$$

as a consequence of

$$\int_{(i-1)/N}^{i/N} \big[a_N^{\varphi}(1+[uN]) - \varphi(u)\big]^2\,\mathrm{d}u$$
$$= \int_{(i-1)/N}^{i/N} [a_N^{\varphi}(i) - \varphi(u)]^2\,\mathrm{d}u \ge \int_{(i-1)/N}^{i/N} [a_N(i) - \varphi(u)]^2\,\mathrm{d}u$$
$$= \int_{(i-1)/N}^{i/N} \big[a_N(1+[uN]) - \varphi(u)\big]^2\,\mathrm{d}u.$$

Thus (2) follows from (6.1.4.18). □

Corollary 1 *Under H_0 and $I(f) < \infty$ the statistics*

$$S_c = \sum_{i=1}^{N} c_i \varphi\Big(\frac{R_{Ni}}{N+1}, f\Big), \quad c \in C \tag{15}$$

are for (6.1.5.3) *asymptotically normal* (μ_c, σ_c^2) *with*

$$\mu_c = \bar{c} \sum_{i=1}^{N} \varphi\Big(\frac{i}{N+1}, f\Big) \tag{16}$$

and

$$\sigma_c^2 = I(f) \sum_{i=1}^{N} (c_i - \bar{c})^2, \quad c \in C, \tag{17}$$

or $\sigma_c^2 = \mathsf{var}\, S_c$, *if* $\varphi(u, f)$ *is expressible as a finite sum of monotone and square integrable functions.*

Furthermore the statistics

$$S_{mn} = \sum_{i=1}^{m} \varphi\Big(\frac{R_{m+n,i}}{m+n+1}, f\Big), \quad m, n \geq 1, \tag{18}$$

are for $\min(m, n) \to \infty$, *asymptotically normal with*

$$\mu_{mn} = \frac{m}{m+n} \sum_{i=1}^{m+n} \varphi\Big(\frac{i}{m+n+1}, f\Big) \tag{19}$$

and

$$\sigma_{mn}^2 = \frac{mn}{m+n} I(f) \tag{20}$$

or $\sigma_{mn}^2 = \mathsf{var}\, S_{mn}$, *if* $\varphi(u, f)$ *has the same property.*

The assertions remain true if $I(f)$ *and* $\varphi(u, f)$ *are replaced by* $I_1(f)$ *and* $\varphi_1(u, f)$, *respectively.*

Remark. The same corollary could be formulated for the approximate scores (14).

6.1.7 Rank statistics for H_1.

Put

$$S_N^+ = \sum_{i=1}^{N} a_N(R_i^+) \operatorname{sign} X_i. \tag{1}$$

Theorem 1 *Assume that the scores $a_N(i)$ satisfy*

$$\lim_{N\to\infty}\int_0^1 \left[a_N(1+[uN]) - \varphi^+(u)\right]^2 \mathrm{d}u = 0 \tag{2}$$

for some square integrable φ^+ such that $\int_0^1 [\varphi^+(u)]^2\,\mathrm{d}u > 0$.

Then under H_1 the statistics (1) are asymptotically normal $(0, \sigma_N^2)$ with

$$\sigma_N^2 = \sum_{i=1}^N [a_N(i)]^2 \sim N\int_0^1 [\varphi^+(u)]^2\,\mathrm{d}u. \tag{3}$$

Proof. Consider a density from H_1 and put $U_i^+ = F^+(|X_i|)$, where F^+ denotes the distribution function of $|X_i|$. Introduce the scores

$$a_N^+(i) = \mathsf{E}[\varphi^+(U_1^+) \mid R_1^+ = i] \tag{4}$$

and the statistics

$$T_N = \sum_{i=1}^N \varphi^+(U_i^+)\,\mathrm{sign}\,X_i. \tag{5}$$

Obviously,

$$\begin{aligned}\mathsf{E}(S_N^+ - T_N)^2 &= N\mathsf{E}[a_N(R_1^+) - \varphi^+(U_1^+)]^2 \\ &\le N\{\mathsf{E}[a_N(R_1^+) - a_N^+(R_1^+)]^2 \\ &\qquad + \mathsf{E}[a_N^+(R_1^+) - \varphi^+(U_1^+)]^2\}.\end{aligned}$$

Now both the expectations tend to zero in view of (2) and Theorems 6.1.4.1 and 6.1.4.2. The rest of the proof is immediate. □

Application 1. Putting $a_N(i) = a_N^+(i, f)$, f symmetric about zero, $I(f) < \infty$, we can assert that

$$S_N^+ = \sum_{i=1}^N a_N^+(i, f)\,\mathrm{sign}\,X_i \tag{6}$$

is asymptotically normal $(0, \sigma_N^2)$ with

$$\begin{aligned}\sigma_N^2 &= \sum_{i=1}^N [a_N^+(i,f)]^2 \sim N\int_0^1 [\varphi^+(u,f)]^2\,\mathrm{d}u \\ &= N\int_0^1 \varphi^2(u,f)\,\mathrm{d}u = NI(f).\end{aligned}$$

Recall that

$$\varphi^+(u,f) = \varphi(\tfrac{1}{2} + \tfrac{1}{2}u, f), \quad a_N^+(i,f) = \mathsf{E}[\varphi^+(U_1^+, f) \mid R_1^+ = i].$$

Application 2. Put $a_N(i) = \varphi(\frac{1}{2} + \frac{1}{2}i/(N+1), f)$, where f is symmetric about zero, $I(f) < \infty$, and $\varphi(u, f)$ is a finite sum of monotone square integrable functions. Then the statistics

$$S_N^+ = \sum_{i=1}^{N} \varphi\Big(\frac{1}{2} + \frac{1}{2}\frac{R_i^+}{N+1}, f\Big) \operatorname{sign} X_i$$

are asymptotically normal $(0, \sigma_N^2)$ with

$$\sigma_N^2 = \sum_{i=1}^{N} \varphi^2\Big(\frac{1}{2} + \frac{1}{2}\frac{i}{N+1}, f\Big) \sim NI(f).$$

6.1.8 Simple linear rank statistics for H_2.

We shall investigate the statistics

$$S_N = \sum_{i=1}^{N} a_N(R_{Ni}, f) a_N(Q_{Ni}, g) \tag{1}$$

and the statistics

$$S_N^* = \sum_{i=1}^{N} \varphi\Big(\frac{R_{Ni}}{N+1}, f\Big)\varphi\Big(\frac{Q_{Ni}}{N+1}, g\Big). \tag{2}$$

Theorem 1 *Under H_2 and $I(f) < \infty$ and $I(g) < \infty$, the statistics* (1) *are for $N \to \infty$ asymptotically normal $(0, \sigma_N^2)$ with*

$$\sigma_N^2 = N\,I(f)I(g) \tag{3}$$

or $\sigma_N^2 = \mathsf{var}\, S_N$.

If $\varphi(u, f)$ and $\varphi(u, g)$ are both expressible as a finite sum of monotone and square integrable functions, then the statistics (2) *are also asymptotically normal (μ_N^*, σ_N^{*2}) with*

$$\mu_N^* = \frac{1}{N}\sum_{i=1}^{N} \varphi\Big(\frac{i}{N+1}, f\Big) \sum_{j=1}^{N} \varphi\Big(\frac{i}{N+1}, g\Big)$$

and $\sigma_N^{*2} = NI(f)I(g)$ or $\sigma_N^{*2} = \mathsf{var}\, S_N^*$.

Proof. If we rewrite S_N as follows

$$S_N = \sum_{i=1}^{N} a_N(i, f) a_N(R_{Ni}^0, g), \tag{4}$$

the assertion follows from Corollary 6.1.5.1. For S_N^* the same follows from Corollary 6.1.6.1. The satisfaction of (6.1.5.3) for $c_i = a_N(i, f)$ follows from Theorem 6.1.4.2; for $c_i = \varphi(i/(N+1), f)$ we apply Lemma 6.1.6.1. □

6.1.9 Martingale characterizations. The asymptotic results derived in the preceding subsections revolve around approximations of rank statistics by suitable linear statistics (having independent summands), when hypotheses of invariance hold. These results have been strengthened and unified by Sen and Ghosh (1971, 1972, 1973, 1974), incorporating suitable martingale characterizations of various rank statistics under the null hypotheses. Such martingale properties were implicit to a certain extent in the basic work of Hájek (1961) on permutational central limit theorems, but not exploited adequately to obtain deeper results. The scores $\alpha_N^{\varphi}(i)$, $i = 1, \ldots, N$, defined by (6.1.4.12) play a basic role in this context. While such developments are unifiedly accounted by Sen (1981), we examine here the extent to which the results in the preceding subsections can be simplified by using such inherent martingale structures.

We start with the linear rank statistic $S_c = S_{cN}$ defined by (6.1.5.4) with the scores $a_N^{\varphi}(i)$ defined as in (6.1.4.12) and with $\bar{c} = 0$. First, we consider the following simple lemma.

Lemma 1 *Let $Y_N^{(1)} \leq \ldots \leq Y_N^{(N)}$ be the order statistics of a sample of size N (and use an analogous notation for size $N+1$) from a distribution function G with a finite first moment. Then, for every k, $1 \leq k \leq N$,*

$$\frac{k}{N+1}\mathsf{E}Y_{N+1}^{(k+1)} + \frac{N-k+1}{N+1}\mathsf{E}Y_{N+1}^{(k)} = \mathsf{E}Y_N^{(k)}. \tag{1}$$

The proof follows directly by using (3.1.2.7) and hence is omitted. We also define $T_c = T_{cN}$ as in (6.1.5.8), and note that it involves independent summands, for which Theorem 6.1.2.1 can be readily adopted. Further, we define the ranks R_{Ni} as in Subsection 3.1.1, and let $\mathcal{B}_N = \mathcal{B}(R_{N1}, \ldots, R_{NN})$ be the σ-field generated by $R_N = (R_{N1}, \ldots, R_{NN})$, for $N \geq 1$, so that $\mathcal{B}_N$ is non-decreasing in N. Also, without loss of generality, we let $\bar{\varphi} = \int_0^1 \varphi(u)du = 0$, and this implies that by (6.1.4.12), $\bar{a}_N^{\varphi} = 0$, for all N.

Theorem 1 *For every $N(\geq 1)$, under the hypothesis of invariance, for scores defined by* (6.1.4.12),

$$\mathsf{E}_0\{T_{cN} \mid \mathcal{B}_N\} = S_{cN} \ (a.e.), \tag{2}$$

and hence $\{(S_{cN}, \mathcal{B}_N); N \geq 1\}$ is a zero-mean martingale.

Proof. From (6.1.5.7), (6.1.5.8), and (6.1.4.12), (2) follows as a direct corollary. The martingale property, implicit in (2), can be directly proved as follows. Note that by definition

$$\begin{aligned}\mathsf{E}_0\{S_{cN+1} \mid \mathcal{B}_N\} &= \sum_{i=1}^{N} c_i \mathsf{E}_0\{a_{N+1}^{\varphi}(R_{N+1i}) \mid \mathcal{B}_N\} \\ &\quad + c_{N+1}\mathsf{E}_0\{a_{N+1}^{\varphi}(R_{N+1N+1}) | \mathcal{B}_N\}.\end{aligned} \tag{3}$$

Now, given $\mathcal{B}_N$, R_{N+1N+1} can assume the values $1, \ldots, N+1$ with equal (conditional) probability $(N+1)^{-1}$. Hence,

$$\mathsf{E}_0\{a^{\varphi}_{N+1}(R_{N+1N+1}) \mid \mathcal{B}_N\} = \frac{1}{N+1}\sum_{j=1}^{N+1} a^{\varphi}_{N+1}(j) = \bar{a}^{\varphi}_{N+1} = 0. \quad (4)$$

Also, for any i $(= 1, \ldots, N)$, given $\mathcal{B}_N$, R_{N+1i} can only assume two values R_{Ni} and $R_{Ni}+1$ with respective conditional probabilities $(N+1-R_{Ni})/(N+1)$ and $R_{Ni}/(N+1)$, so that

$$\begin{aligned}\mathsf{E}_0\{a^{\varphi}_{N+1}(R_{N+1i}) \mid \mathcal{B}_N\} &= \frac{R_{Ni}}{N+1} a^{\varphi}_{N+1}(R_{Ni}+1) \\ &\quad + \frac{N - R_{Ni} + 1}{N+1} a^{\varphi}_{N+1}(R_{Ni}).\end{aligned} \quad (5)$$

Therefore, using Lemma 1 along with (4) and (5), it follows from (3) that

$$\mathsf{E}_0\{S_{cN+1} \mid \mathcal{B}_N\} = S_{cN}, \quad \text{a.e., for all} N \geq 1. \quad (6)$$

□

Next we note that by (2),

$$\mathsf{E}_0\{T_{cN} - S_{cN}\}^2 = \mathsf{E}_0\{T^2_{cN}\} - \mathsf{E}_0\{S^2_{cN}\}, \quad (7)$$

so that using Theorem 6.1.4.2, we obtain that

$$\mathsf{E}_0\{T_{cN} - S_{cN}\}^2 / \mathsf{E}_0\{S^2_{cN}\} \to 0, \quad \text{as } N \to \infty, \quad (8)$$

and hence, the central limit theorem whenever applicable to T_{cN} also remains valid for the S_{cN}. The martingale characterization in Theorem 1 also provides access to deeper asymptotic results, such as the law of the iterated logarithm, and weak and strong invariance principles, that may not follow solely from the projection result in (2); we refer to Sen and Ghosh (1972), and shall make more comments on it later on.

Let us consider the general signed rank statistics treated in Subsection 6.1.7. We define S^+_N and T_N as in (6.1.7.1) and (6.1.7.5), and the scores $a^+_N(i)$, $i = 1, \ldots, N$, as in (6.1.7.4). Note that Lemma 1 stated above applies to these scores as well. We define the vector $\operatorname{sign} X$ and R^+ of signs and absolute ranks as in Subsection 3.1.3, and note that under the null hypothesis of symmetry of F around 0, $\operatorname{sign} X$ and R^+ are distributed independently of each other, the former taking on each sign-inversion with the common probability 2^{-N} and the latter each permutation of $\{1, \ldots, N\}$ with equal probability $(N!)^{-1}$. We denote by $\mathcal{B}_N$ the σ-field generated by $(\operatorname{sign} X, R^+)$, for $X_1, \ldots, X_N$, and note that it is non-decreasing in N. Then we have the following.

Theorem 2 *Under the hypothesis of symmetry about* 0 *and with the scores defined by* (6.1.7.4), *we have*

$$\mathsf{E}_0\{T_N \mid \mathcal{B}_N\} = S_N^+ \quad \textit{a.e., for all } N \geq 1, \tag{9}$$

and therefore $\{(S_N^+, \mathcal{B}_N);\ N \geq 1\}$ *is a zero-mean martingale.*

The proof is analogous to that of Theorem 1, and hence, left as an exercise. Here T_N has independent identically distributed summands, and therefore the central limit theorem applies under the finiteness of the second moment of the score function φ. Thus, Theorem 2 extends the asymptotic normality to S_N^+. Again other deeper asymptotic results can be obtained by incorporating the above martingale characterization. We refer to Sen and Ghosh (1971, 1973).

Let us next consider the case of linear rank statistics for testing the hypothesis of independence in a bivariate model. We define S_N as in (6.1.8.1) in a slightly more general form wherein we denote the scores by $a_N(i)$ and $b_N(i)$, $i = 1, \ldots, N$, generated by two possibly different score generating functions φ_1 and φ_2 that are integrable over $(0,1)$. We assume that the scores are defined as in (6.1.4.12), and without loss of generality, we take $\bar{\varphi}_j = \int_0^1 \varphi_j(u)\, \mathrm{d}u = 0$, $j = 1, 2$. Also, we denote by F_1 and F_2 the marginal distribution functions of X and Y, and let

$$T_N = \sum_{i=1}^{N} \varphi_1(F_1(X_i))\varphi_2(F_2(Y_i)). \tag{10}$$

Finally, let R_N and Q_N be the vector of ranks of the X and Y variables among themselves, and let $\mathcal{B}_N$ be the σ-field generated by (R, Q), for $(X_1, Y_1), \ldots, (X_N, Y_N)$; it is non-decreasing in N. Note that under the null hypothesis of stochastic independence of X, Y, the vectors R and Q are independently distributed, each taking on the permutations of $\{1, \ldots, N\}$ with the common probability $(N!)^{-1}$. Then, we have the following.

Theorem 3 *Under the hypothesis of independence, and with the scores defined by* (6.1.4.12),

$$\mathsf{E}_0\{T_N \mid \mathcal{B}_N\} = S_N \quad \textit{a.e., for all } N \geq 1, \tag{11}$$

and $\{(S_N, \mathcal{B}_N);\ N \geq 1\}$ *is a zero-mean martingale.*

The proof virtually follows the lines of that of Theorem 1, and hence is left as an exercise. The statistic T_N has independent identically distributed summands, and hence the central limit theorem applies to it (under the null hypothesis) whenever the score generating functions φ_1, φ_2 are square integrable on $(0,1)$. Therefore, the asymptotic normality of S_N, under the null hypothesis, can be readily deduced from the above projection result, while other deeper asymptotic results can be obtained by incorporating the martingale characterization. We refer to Sen and Ghosh (1974). For linear and signed-rank statistics with scores defined in other ways in the preceding subsections, replacement by scores in (6.1.4.12) does not create any significant difference when conditions like (6.1.7.2) hold. Hence, these martingale characterizations provide even better martingale approximations for such statistics. This implies that the martingale based asymptotics for rank statistics discussed above hold for the general class considered in the preceding subsections.

Besides the linear and signed rank statistics, we have also outlined in Section 3.5 some other statistical functionals that often pertain to the theory of rank tests. Among these, the general U-statistics have an elegant *reverse martingale* property that provides easy access to many asymptotic results (without being confined to the null hypotheses situations). Generalized U-statistics have some directional reverse martingale properties that yield comparable asymptotic results. Of course, in this domain, the classical Hoeffding decomposition yields the projection result that permits easy access to the asymptotic normality. However, for deeper asymptotic results, such reverse martingale properties have been fully exploited in the literature. For a detailed account of these developments, we refer to Chapter 3 of Sen (1981). Finally, we have also considered differentiable statistical functionals in Section 3.5. It follows from Sen (1995) that a reverse martingale (plus a negligible reverse submartingale) representation holds for such functionals under general regularity assumptions.

6.2 RANK STATISTICS OF χ^2-TYPES

6.2.1 Convergence in distribution for random vectors. Let $T_\nu = (T_{\nu 1}, \ldots, T_{\nu k})$, $\nu \geq 1$, and $Z = (Z_1, \ldots, Z_k)$ be k-dimensional random vectors, each having a definite distribution. The notion of convergence of T_ν to Z in distribution may be introduced in several equivalent ways, and all the definitions are important in their own right.

Theorem 1 *The following definitions of convergence of T_ν, $\nu \geq 1$, to Z in distribution are equivalent:*

D1: *$\mathsf{E}h(T_\nu)$ converges to $\mathsf{E}h(Z)$ for every uniformly bounded function h which is continuous on a set C such that $P(Z \in C) = 1$.*

D2: *$h(T_\nu)$ converges to $h(Z)$ in distribution for every function h which is continuous on a set C such that $P(Z \in C) = 1$.*

D3: *$\sum_{j=1}^{k} \lambda_j T_{\nu j}$ converges in distribution to $\sum_{j=1}^{k} \lambda_j Z_j$ for every real vector $(\lambda_1, \ldots, \lambda_k)$.*

D4: *For each Borel subset A such that $P(Z \in$ the boundary of $A) = 0$*

$$\lim_{\nu \to \infty} P(T_\nu \in A) = P(Z \in A) \tag{1}$$

holds.

D5: (1) *holds for each k-dimensional interval A such that $P(Z \in$ the boundary of $A) = 0$.*

Proof. D1 $\Rightarrow$ D2. Given a function h considered in D2, put

$$I_b(z) = \begin{cases} 1, & \text{if } h(z) \le b, \\ 0, & \text{if } h(z) > b. \end{cases}$$

Obviously, each point which is a continuity point of $h(z)$, and is such that $h(z) \neq b$, is also a continuity point of $I_b(z)$. If $P(h(Z) = b) = 0$, both the properties are satisfied with probability 1, so that $I_b(z)$ satisfies the requirement of D1. Thus

$$P(h(T_\nu) \le b) = \mathsf{E} I_b(T_\nu) \to \mathsf{E} I_b(Z) = P(h(Z) \le b),$$

which means that $h(T_\nu)$ converges in distribution to $h(Z)$.

D2 $\Rightarrow$ D3, D4. D4 $\Rightarrow$ D5. Immediate.

D3 $\Rightarrow$ D5. If D3 holds, then obviously

$$\lim_{\nu \to \infty} \mathsf{E} \exp\Big(i \sum_{j=1}^{k} \lambda_j T_{\nu j}\Big) = \mathsf{E} \exp\Big(i \sum_{j=1}^{k} \lambda_j Z_j\Big).$$

Hence the characteristic function of T_ν converges to the characteristic function of Z, from which (1) may be obtained by the well-known inversion formula (see Cramér (1945) Section 10.6) if A is a k-dimensional interval.

D5 $\Rightarrow$ D1. Let $h(z)$ be a continuous function vanishing outside a k-dimensional finite interval A. We may assume that (1) holds without loss of generality. Then for every $\varepsilon > 0$ there exists a partition of A into a system of disjoint subintervals A_j, $j \in J$, such that $P(Z \in$ the boundary of $A_j) = 0$, $j \in J$, and that

$$\sup_{z \in A_j} h(z) - \inf_{z \in A_j} h(z) < \varepsilon, \quad j \in J. \tag{2}$$

Consequently, $h(z)$ may be approximated by a simple function which is constant on the intervals A_j, $j \in J$. On the other hand, D5 implies immediately that D1 holds for simple functions of the above type. Thus, D1 must hold for every continuous h vanishing outside a finite interval.

To prove D1 in general it suffices to note that for the set C considered in D1 and for every $\varepsilon > 0$ there exists a compact subset K such that $K \subset C$ and that $P(Z \in K) > 1 - \varepsilon$, and, if D5 holds,

$$P(T_\nu \in K) > 1 - \varepsilon, \quad \nu \geq 1. \tag{3}$$

Thus we may again utilize (2) with A_j replaced by $A_j \cap K$, and the usual approximation technique. The remaining implications follow immediately from the above ones. □

6.2.2 Statistics for testing H_0 against k samples differing in location or scale. Let $\{s_1, \ldots, s_k\}$ be a partition of $\{1, \ldots, N\}$, and put $n_j = \text{card}\, s_j$ and $N = n_1 + \ldots + n_k$. Moreover, for every $N \geq 1$ let $a_N(i)$ be some scores satisfying (6.1.6.2) for some square integrable $\varphi(u)$ such that $\int_0^1 [\varphi(u) - \bar{\varphi}]^2 \, du > 0$. Let us introduce the statistics

$$S_{Nj} = \sum_{i \in s_j} a_N(R_{Ni}), \quad 1 \leq j \leq k. \tag{1}$$

We shall consider the statistics

$$Q_{n_1 \ldots n_k} = (N-1)\Big\{\sum_{i=1}^{N} [a_N(i) - \bar{a}_N]^2\Big\}^{-1} \sum_{j=1}^{k} \frac{(S_{Nj} - N_j \bar{a}_N)^2}{n_j}, \quad n_1, \ldots, n_k \geq 1. \tag{2}$$

Theorem 1 *Assume that the scores $a_N(i)$ satisfy* (6.1.6.2). *Then under H_0 the statistics* (2) *are, for* $\min(n_1, \ldots, n_k) \to \infty$, *asymptotically χ^2-distributed with $k-1$ degrees of freedom.*

Proof. Assume that the assertion is false. Then there exists a sequence $\{(n_{\nu 1}, \ldots, n_{\nu k})\}$ such that

$$\lim_{\nu \to \infty} \min(n_{\nu 1}, \ldots, n_{\nu k}) = \infty \tag{3}$$

and that the assertion is false for every subsequence of $\{(n_{\nu 1}, \ldots, n_{\nu k})\}$. Thus, passing to a proper subsequence, if necessary, we may assume without loss of generality, that

$$\lim_{\nu \to \infty} \frac{n_{\nu j}}{N_\nu} = \kappa_j^2, \quad 1 \leq j \leq k, \tag{4}$$

where $N_\nu = n_{\nu 1} + \ldots + n_{\nu k}$, and $0 \leq \kappa_j^2 \leq 1$.

If we now succeed in proving that the theorem is true under (3) and (4), the contradiction thus obtained will establish the proof

Denoting $S_{\nu j} = S_{N_\nu j}$, $\bar{a}_{\nu j} = \bar{a}_{N_\nu j}$ and

$$\sigma_\nu^2 = \frac{1}{N_\nu - 1} \sum_{i=1}^{N} [a_{N_\nu}(i) - \bar{a}_{N_\nu}]^2, \tag{5}$$

we conclude from Theorem 6.1.6.1 that the statistics

$$\xi_{\nu j} = \sigma_\nu^{-1}(S_{\nu j} - n_{\nu j}\bar{a}_{\nu j}) n_{\nu j}^{-\frac{1}{2}} \tag{6}$$

are asymptotically normal $(0, 1-\kappa_j^2)$, because $\mathsf{E}S_{\nu j} = n_{\nu j}\bar{a}_{\nu j}$ and

$$\mathsf{var}\, S_{\nu j} = \sigma_\nu^2 n_{\nu j}\Big(1 - \frac{n_{\nu j}}{N_\nu}\Big). \tag{7}$$

Considering the random variables $(Z_1, \dots, Z_k)$ introduced in Subsection 2.4.1, we can also say that $\xi_{\nu j}$ converges in distribution to Z_j and moreover

$$\lim_{\nu\to\infty} \mathsf{var}\, \xi_{\nu j} = \mathsf{var}\, Z_j. \tag{8}$$

On account of (3.3.1.25), we can also show that

$$\lim_{\nu\to\infty} \mathsf{cov}(\xi_{\nu j}, \xi_{\nu g}) = -\kappa_j \kappa_g = \mathsf{cov}(Z_j, Z_g), \quad 1 \le j \ne g \le k. \tag{9}$$

Actually, we only need to note that for

$$c_{ji} = \begin{cases} 1, & \text{if } i \in s_j, \\ 0, & \text{otherwise,} \end{cases}$$

where the sets s_j are disjoint, we have

$$\sum_{i=1}^{N} (c_{ji} - \bar{c}_j)(c_{gi} - \bar{c}_g) = -\frac{n_j n_g}{N}, \quad 1 \le j \ne g \le k.$$

We cannot yet conclude, however, that the vector $(\xi_{\nu 1}, \dots, \xi_{\nu k})$ must converge in distribution to the vector $(Z_1, \dots, Z_k)$. In order to prove this, we have to show, according to D3 of the previous subsection, that $\sum \lambda_j \xi_{\nu j}$ converges in distribution to $\sum \lambda_j Z_j$ for every real vector $(\lambda_1, \dots, \lambda_k)$. We shall distinguish the following two cases.

Case 1. If

$$\mathsf{var}\Big(\sum_{j=1}^{k} \lambda_j Z_j\Big) = \sum_{j=1}^{k} \lambda_j^2 - \Big(\sum_{j=1}^{k} \lambda_j \kappa_j\Big)^2 = 0, \tag{10}$$

then the convergence to the (degenerate) distribution follows from (8) and (9).

Case 2. If

$$\mathsf{var}\Big(\sum_{j=1}^{k}\lambda_j Z_j\Big) = \sum_{j=1}^{k}\lambda_j^2 - \Big(\sum_{j=1}^{k}\lambda_j\kappa_j\Big)^2 > 0, \tag{11}$$

then we put

$$\sum_{j=1}^{k}\lambda_j\xi_{\nu j} = \sigma_\nu^{-1}\sum_{i=1}^{N_\nu} c_{\nu i}[a_N(R_{N_\nu i}) - \bar{a}_{N_\nu}]$$

where

$$c_{\nu i} = \lambda_j n_{\nu j}^{-\frac{1}{2}}, \quad \text{if } i \in s_j.$$

Obviously

$$\begin{aligned}\sum_{i=1}^{N_\nu}(c_{\nu i} - \bar{c}_\nu)^2 &= \sum_{i=1}^{N_\nu} c_{\nu i}^2 - \frac{1}{N_\nu}\Big(\sum_{i=1}^{N_\nu} c_{\nu i}\Big)^2 \\ &= \sum_{j=1}^{k}\lambda_j^2 - \Big(\sum_{j=1}^{k}\lambda_j(n_{\nu j}/N_\nu)^{\frac{1}{2}}\Big)^2 \\ &\to \sum_{j=1}^{k}\lambda_j^2 - \Big(\sum_{j=1}^{k}\lambda_j\kappa_j\Big)^2 > 0.\end{aligned}$$

On the other hand, (3) entails

$$\max_{1\le i\le N}(c_{\nu i} - \bar{c}_\nu)^2 = \max_{1\le j\le k}\Big(\lambda_j n_{\nu j}^{-\frac{1}{2}} - N_\nu^{-1}\sum_{j=1}^{k}\lambda_j n_{\nu j}^{\frac{1}{2}}\Big)^2 \to 0$$

so that

$$\lim_{\nu\to\infty}\frac{\sum_{i=1}^{N_\nu}(c_{\nu i} - \bar{c}_\nu)^2}{\max_{1\le i\le N_\nu}(c_{\nu i} - \bar{c}_\nu)^2} = \infty.$$

Thus, according to Theorem 6.1.6.1 $\sum_{j=1}^{k}\lambda_j\xi_{\nu j}$ is asymptotically normal with $\mu_\nu = 0$ and

$$\sigma_\nu^2 = \mathsf{var}\Big(\sum_{j=1}^{k}\lambda_j\xi_{\nu j}\Big) \to \mathsf{var}\Big(\sum_{j=1}^{k}\lambda_j Z_j\Big),$$

i.e. has asymptotically the same distribution as $\sum\lambda_j Z_j$. Consequently, $(\xi_{\nu 1}, \ldots, \xi_{\nu k})$ converges in distribution to $(Z_1, \ldots, Z_k)$. Now by D2 of Subsection 6.2.1 this means that

$$Q_\nu = \sum_{j=1}^{k}\xi_{\nu j}^2$$

converges in distribution to $Z_1^2 + \ldots + Z_k^2$. The last distribution, however, is the χ^2-distribution with $k-1$ degrees of freedom, in view of Theorem 2.4.1.1. Hence the contradiction. □

Corollary 1 *Under H_0 and $I(f) < \infty$, the statistics*

$$Q_{n_1 \ldots n_k} = (N-1)\Big\{\sum_{i=1}^{N} a_N^2(i,f)\Big\}^{-1} \sum_{j=1}^{k} \Big[\sum_{i \in s_j} a_N(R_{Ni}, f)\Big]^2 n_j^{-1} \tag{12}$$

are, for $\min(n_1, \ldots, n_k) \to \infty$, asymptotically χ^2-distributed with $(k-1)$ degrees of freedom.

If, in addition, $\varphi(u,f)$ is expressible as a sum of monotone and square integrable functions, the same holds for

$$\begin{aligned} Q_{n_1 \ldots n_k} &= (N-1)\Big\{\sum_{i=1}^{N}\Big[\varphi\Big(\frac{i}{N+1}, f\Big) - \bar{\varphi}\Big]^2\Big\}^{-1} \qquad (13) \\ &\cdot \Big\{\sum_{j=1}^{k}\Big[\sum_{i \in s_j} \varphi\Big(\frac{R_{Ni}}{N+1}, f\Big)\Big]^2 n_j^{-1} - N\bar{\varphi}^2\Big\}, \\ \bar{\varphi} &= \frac{1}{N}\sum_{i=1}^{N} \varphi\Big(\frac{i}{N+1}, f\Big). \end{aligned}$$

The assertions remain true for $I_1(f)$, $a_{1N}(i,f)$ and $\varphi_1(u,f)$.

6.2.3 Statistics for H_3 against differences within blocks. The number of observations in each block will be fixed and denoted by k. The scores $a(1), \ldots, a(k)$ will also be fixed and arbitrary but not constant. The number of blocks n will tend to ∞. We consider the statistics

$$S_{ni} = \sum_{j=1}^{n} a(R_{ji}), \quad 1 \le i \le k, \tag{1}$$

which are sums of n independent and equally distributed random variables. It is easy to show that the random vector $(T_{n1}, \ldots, T_{nk})$, where

$$T_{ni} = \Big[\frac{n}{k-1}\sum_{h=1}^{k}\big(a(h) - \bar{a}\big)^2\Big]^{-\frac{1}{2}} (S_{ni} - n\bar{a}) \tag{2}$$

is asymptotically normal with vanishing means, variances

$$\sigma_{ni}^2 = 1 - \frac{1}{k}, \quad 1 \le i \le k, \tag{3}$$

and covariances

$$\sigma_{nig} = -\frac{1}{k}, \quad 1 \leq i,\ g \leq k. \tag{4}$$

Consequently, for $n \to \infty$

$$Q_n = \frac{k-1}{n}\Big[\sum_{j=1}^{k}\big(a(j) - \bar{a}\big)^2\Big]^{-1}\sum_{i=1}^{k}(S_{ni} - n\bar{a})^2 \tag{5}$$

has asymptotically the χ^2-distribution with $k-1$ degrees of freedom.

6.3 STATISTICS OF KOLMOGOROV-SMIRNOV TYPES

6.3.1 Probability distributions in the space of continuous functions. Let $C[0,1]$ be the space of continuous functions $z(t)$, $0 \leq t \leq 1$. With the usual metric,

$$\|z(.) - y(.)\| = \max_t |z(t) - y(t)|, \tag{1}$$

$C[0,1]$ is a separable metric space. Let $\mathcal{C}$ denote the σ-field of Borel subsets of $C[0,1]$, i.e. the smallest σ-field containing all open subsets.

Let $Z(.)$ be a random process whose realizations (sample paths, trajectories) are functions $z(.) \in C[0,1]$. The process may also be considered as a family of random variables $\{Z(t),\ 0 \leq t \leq 1\}$, where $Z(t)$ is the value of the process at the point t. It is easy to see that all random variables are $\mathcal{C}$-measurable.

Denote by $\tau = \{t_1, \ldots, t_n\} \subset [0,1]$ an arbitrary finite subset of $[0,1]$, and by $T = \{\tau\}$ the space of all such subsets.

Let $\mathcal{C}_\tau$ be the sub-σ-field generated by the subfamily $\{Z(t),\ t \in \tau\}$. This means that $\mathcal{C}_\tau$ consists of subsets

$$D = \{z(.) : [z(t_1), \ldots, z(t_n)] \in A_n\}, \tag{2}$$

where A_n is a Borel subset of n-dimensional Euclidean space.

Lemma 1 *$\mathcal{C}_0 = \bigcup_{\tau \in T} \mathcal{C}_\tau$ is a field, and $\mathcal{C}$ is the smallest σ-field over $\mathcal{C}_0$.*

Proof. The first assertion is immediate. To prove the second one, observe that every open subset of $C[0,1]$ is a countable union of open spheres, due to the separability of $C[0,1]$, and that an open sphere $S(y,r)$ with centre at $y = y(.)$ and radius r may be represented as

$$\begin{aligned} S(y,r) &= \{x(.) : y(t) - r < x(t) < y(t) + r,\ t \in [0,1]\} \\ &= \{x(.) : y(t_i) - r < x(t_i) < y(t_i) + r,\ i = 1, 2, \ldots\}, \end{aligned}$$

where the subset $\{t_i\} \subset [0,1]$ is countable and dense in $[0,1]$, but otherwise arbitrary. Thus $S(y,r)$ is a countable intersection of events from $\mathcal{C}_0$, from which the proof is easily concluded. $\square$

Now let a probability distribution P_τ be given on each $\mathcal{C}_\tau$, $\tau \in T$. We say that these probability distributions are *consistent* if $P_{\tau'}$ is a restriction of P_τ whenever $\tau' \subset \tau$, $\tau' \in T$, $\tau \in T$. It is obvious that a consistent family of distributions $\{P_\tau, \tau \in T\}$ defines an additive set function P_0 on $\mathcal{C}_0$. Actually, if $C \in \mathcal{C}_0$, then $C \in \mathcal{C}_\tau$ for some $\tau \in T$, and we may put $P_0(C) = P_\tau(C)$. Due to the consistency $P_0(C)$ is unique, regardless of the choice of τ. Now if we have a finite number of disjoint members of $\mathcal{C}_0$, say $C_1, \ldots, C_m$, then there is a τ such that all these events belong to $\mathcal{C}_\tau$, and

$$P_0\Big(\bigcup_1^m C_i\Big) = \sum_1^m P_0(C_i) \tag{3}$$

follows from the additivity of P_τ.

Unfortunately, (3) does not in general extend to countable unions, i.e. P_0 may not be σ-additive on $\mathcal{C}_0$. In other words, having a monotone sequence of events $D_1 \supset D_2 \supset \ldots$ from $\mathcal{C}_0$ such that

$$\bigcap_1^\infty D_i = \emptyset, \tag{4}$$

we cannot generally assert that

$$\lim_{i\to\infty} P_0(D_i) = 0. \tag{5}$$

If P_0 is σ-additive on $\mathcal{C}_0$, we say that it is a probability distribution on $\mathcal{C}_0$. According to a basic proposition of measure theory there exists a unique extension of a probability distribution P_0 from the field $\mathcal{C}_0$ to the σ-field $\mathcal{C}$. Thus a consistent system of 'marginal' distributions $\{P_\tau, \tau \in T\}$, on the sub-σ-fields $\mathcal{C}_\tau$, generated by finite subfamilies of the random variables $Z(t)$, determines a probability distribution on $\mathcal{C}$ if and only if P_0 is σ-additive on $\mathcal{C}_0$, or, equivalently, if (4) implies (5).

Theorem 1 *P_0 is a probability distribution on $\mathcal{C}_0$ if and only if for every $\varepsilon > 0$*

$$\lim_{\delta\to 0} \inf_{\tau \in T} P_0\big(\max_{\substack{|t-s|<\delta \\ t,s\in\tau}} |Z(t) - Z(s)| < \varepsilon\big) = 1. \tag{6}$$

Proof. *Necessity.* Assume that P_0 is a probability distribution on $\mathcal{C}_0$ and denote by P its extension to $\mathcal{C}$. Then, since all trajectories are continuous,

$$\lim_{\delta\to 0} P\big(\max_{|t-s|<\delta} |Z(t) - Z(s)| < \varepsilon\big) = 1. \tag{7}$$

Furthermore,

$$\inf_{t \in T} P\big(\max_{\substack{|t-s|<\delta \\ t,s \in \tau}} |Z(t) - Z(s)| < \varepsilon\big) \geq P\big(\max_{|t-s|<\delta} |Z(t) - Z(s)| < \varepsilon\big), \tag{8}$$

which together with (7) implies (6).

Sufficiency. Consider a decreasing sequence of events $D_i \in \mathcal{C}_0$, say

$$D_i = \{z(.) : [z(t_{i1}), \ldots, z(t_{in_i})] \in A^0_{n_i}\}, \quad i \geq 1, \tag{9}$$

where $A^0_{n_i} \in \mathcal{A}_{n_i}$, with $\mathcal{A}_n$ denoting the σ-field of Borel subsets of the n-dimensional Euclidean space X_n. We shall assume that

$$\lim_{i \to \infty} P_0(D_i) = \alpha > 0 \tag{10}$$

and show that $\bigcap_1^\infty D_i$ is non-empty under (6).

Denote by $A[0,1]$ the space of *all* functions on $[0,1]$, $A[0,1] = \{y(.)\}$. Without loss of generality we may assume that the finite subsets $\tau_i = \{t_{i1}, \ldots, t_{in_i}\}$ are increasing and that $\tau_\infty = \bigcup_1^\infty \tau_i$ is a dense subset of $[0,1]$. Let $\mathcal{Y}_\infty$ be the smallest σ-field of subsets of $A[0,1]$ containing the events

$$E = \{y(.) : [y(t_1), \ldots, y(t_n)] \in A_n\}, \qquad \{t_1, \ldots, t_n\} \subset \tau_\infty, \quad A_n \in \mathcal{A}_n, \tag{11}$$

and let P^*_∞ be a probability distribution on $\mathcal{Y}_\infty$ such that $P^*_\infty(E) = P_0(D)$, if E and D are given by (11) and (2) with coinciding A_n, or, equivalently, if

$$D = E \cap C[0,1]. \tag{12}$$

Such a probability distribution exists, according to the well-known theorem of Kolmogorov (see Loève (1955), Section 4.3). The existence is proved by showing that P^*_0 defined on the field consisting of the events (11) by $P^*_0(E) = P_0(D)$ is a probability distribution, and then by extending P^*_0 to P^*_∞ on $\mathcal{Y}_\infty$. Now the relation between P^*_∞ and P_0 together with (6) implies that

$$\lim_{\delta \to 0} P^*_\infty\big(\max_{\substack{|t-s|<\delta \\ t,s \in \tau_\infty}} |Y(t) - Y(s)| < \varepsilon\big) = 1. \tag{13}$$

Consequently, we may choose positive numbers $\varepsilon_1, \varepsilon_2, \ldots$ such that

$$\sum_1^\infty \varepsilon_j < \tfrac{1}{2}\alpha \tag{14}$$

and then positive numbers $\delta_1, \delta_2, \ldots$ such that

$$P^*_\infty\Big\{\max_{\substack{|t-s|<\delta_j \\ t,s\in\mathcal{T}_\infty}} |Y(t)-Y(s)| < \frac{1}{j}\Big\} > 1-\varepsilon_j.$$

This means that the event

$$F = \bigcap_{j=1}^{\infty}\Big\{y(.) : \max_{\substack{|t-s|<\delta_j \\ t,s\in\mathcal{T}_\infty}} |y(t)-y(s)| < \frac{1}{j}\Big\} \tag{15}$$

has the probability

$$P^*_\infty(F) > 1 - \tfrac{1}{2}\alpha. \tag{16}$$

On the other hand, putting

$$E_i = \{y(.) : [y(t_{i1}), \ldots, y(t_{in_i})] \in A^0_{n_i}\}, \tag{17}$$

(9) and (10) imply

$$\lim_{i\to\infty} P^*_\infty(E_i) = P^*_\infty\Big(\bigcap_1^\infty E_i\Big) = \alpha > 0. \tag{18}$$

Moreover, accounting for (16), we have

$$P^*_\infty\Big(\bigcap_{i=1}^\infty E_i \cap F\Big) \geq \tfrac{1}{2}\alpha > 0 \tag{19}$$

which means that $\bigcap_{i=1}^\infty E_i \cap F$ is not empty. However, in view of (15), each function from F is uniformly continuous on $\mathcal{T}_\infty$, and hence coincides with some continuous function, which has the same values for $t \in \mathcal{T}_\infty$. This in turn means that the latter function belongs to F, too, since the events from $\mathcal{Y}_\infty$ depend on the coordinates in the points $t \in \mathcal{T}_\infty$ only. Thus $\bigcap_{i=1}^\infty E_i \cap F \cap C[0,1]$ is also not empty, and the same is a fortiori true about

$$\bigcap_{i=1}^\infty D_i = \bigcap_{i=1}^\infty E_i \cap C[0,1] \supset \bigcap_{i=1}^\infty E_i \cap F \cap C[0,1].$$

To review the proof, let us point out that we have used the Kolmogorov theorem for the space of *all* functions, and then showed that (18) under (6) implies not only that $\bigcap_1^\infty E_i$ is not empty, but that it contains at least one continuous function, and hence $\bigcap_1^\infty D_i$ is not empty, too. □

The following useful sufficient condition is due to Kolmogorov.

Theorem 2 *P_0 is a probability distribution on $\mathcal{C}_0$ if*

$$\mathsf{E}_0[Z(t) - Z(s)]^4 \leq M(t-s)^2, \quad 0 \leq t,\ s \leq 1, \tag{20}$$

where E_0 refers to P_0 and to the sub-σ-field $\mathcal{C}_{\{t,s\}}$, and M is a constant independent of t, s.

Proof. Denote by τ_k the subset of points

$$t_i = \frac{i}{2^k}, \quad i = 0, 1, \ldots, 2^k, \tag{21}$$

so that $\tau_\infty = \bigcup_1^\infty \tau_k$ is the subset of dyadic rationals. We know that τ_∞ is dense in $[0,1]$. Since (20) implies that $Z(t)$ is continuous in probability, we have

$$\lim_{j\to\infty} P_0(\max_{\substack{|t-s|<\delta \\ t,s\in\tau_j}} |Z(t) - Z(s)| < \varepsilon) \leq P(\max_{\substack{|t-s|<\delta \\ t,s\in\tau}} |Z(t) - Z(s)| < \varepsilon)$$

for an arbitrary finite set $\tau = (t_1, \ldots, t_n)$, since in τ_j there are points arbitrarily close to $t_1, \ldots, t_n$ for j sufficiently large. Consequently (6) will be implied by

$$\lim_{\delta\to 0} \inf_j P_0(\max_{\substack{|t-s|<\delta \\ t,s\in\tau_j}} |Z(t) - Z(s)| < \varepsilon) = 1. \tag{22}$$

By the Chebyshev inequality for the fourth moment, it follows from (20), that

$$\begin{aligned} &P_0\Big(\max_{1\leq i\leq 2^k} \Big|Z\Big(\frac{i-1}{2^k}\Big) - Z\Big(\frac{i}{2^k}\Big)\Big| < \frac{1}{k^2}\Big) \\ &\geq 1 - \sum_{i=1}^{2^k} P_0\Big(\Big|Z\Big(\frac{i-1}{2^k}\Big) - Z\Big(\frac{i}{2^k}\Big)\Big| \geq \frac{1}{k^2}\Big) \geq 1 - Mk^8 . 2^{-k}. \end{aligned} \tag{23}$$

This further implies that

$$\begin{aligned} &P_0\Big(\max_{1\leq i\leq 2^g} \Big|Z\Big(\frac{i-1}{2^g}\Big) - Z\Big(\frac{i}{2^g}\Big)\Big| < \frac{1}{g^2},\ k \leq g \leq j\Big) \\ &\geq 1 - M\sum_{g=k}^{j} g^8 . 2^{-g} \geq 1 - M\sum_{g=k}^{\infty} g^8 . 2^{-g}. \end{aligned} \tag{24}$$

Take two arbitrary points $t, s \in \tau_j$ such that $t < s$ and

$$|t - s| < 2^{-k}, \quad k \leq j. \tag{25}$$

Let h be the largest integer such that the interval $[t, s]$ and τ_h have just one common point, and denote this point by $i_0/2^h$. Then

$$\begin{aligned} s &= i_0 . 2^{-h} + 2^{-h_1} + \ldots + 2^{-h_a}, \\ t &= i_0 . 2^{-h} - 2^{-g_1} - \ldots - 2^{-g_b}, \end{aligned}$$

where $k \le h < h_1 < \ldots < h_a \le j$ and $h < g_1 < \ldots < g_b \le j$. Denote

$$\begin{aligned} v_i &= i_0 . 2^{-h} - 2^{-g_1} - \ldots - 2^{-g_{b-i}}, \quad 0 \le i \le b, \\ &= i_0 . 2^{-h} + 2^{-h_1} + \ldots + 2^{-h_{i-b}}, \quad b < i \le a + b, \end{aligned}$$

so that $v_0 = t$ and $v_{a+b} = s$. Further

$$|Z(t) - Z(s)| \le \sum_{i=0}^{a+b-1} |Z(v_{i+1}) - Z(v_i)|$$

where

$$|v_{i+1} - v_i| = 2^{-k_i}, \quad j \ge k_i \ge k$$

and each integer g, such that $j \ge g \ge k$, occurs among the numbers $\{k_i\}$ *at most twice*. This entails the following implication:

$$\begin{aligned} &\left[\max_{1 \le i \le 2^g} \left| Z\left(\frac{i-1}{2^g}\right) - Z\left(\frac{i}{2^g}\right) \right| < \frac{1}{g^2}, \ k \le g \le j \right] \qquad (26) \\ &\Rightarrow \left[\max_{\substack{|t-s| \le 2^{-k} \\ t,s \in \tau_j}} |Z(t) - Z(s)| < 2 \sum_{g=k}^{\infty} \frac{1}{g^2} \right]. \end{aligned}$$

From (24) and (26) we conclude that

$$\inf_j P_0\Big(\max_{\substack{|t-s| \le 2^{-k} \\ t,s \in \tau_j}} |Z(t) - Z(s)| < 2 \sum_{g=k}^{\infty} \frac{1}{g^2} \Big) \ge 1 - M \sum_{g=k}^{\infty} g^8 . 2^{-g}.$$

Consequently, for every $\varepsilon > 0$ there exists a k such that

$$2 \sum_{g=k}^{\infty} \frac{1}{g^2} < \varepsilon, \quad M \sum_{g=k}^{\infty} g^8 . 2^{-g} < \varepsilon,$$

so that for $\delta < 2^{-k}$

$$\inf_j P_0(\max_{\substack{|t-s| \le \delta \\ t,s \in \tau_j}} |Z(t) - Z(s)| < \varepsilon) > 1 - \varepsilon,$$

which is equivalent to (22). □

A very simple way of determining a probability distribution on $(C[0,1], \mathcal{C})$ is to distribute the probability among a finite set of trajectories. For example, if we put

$$Z(t) = \sum_{i=1}^{N} c_i a_N(R_i, t),$$

where $a_N(i,t)$ is a continuous function in t for every i, $1 \le i \le N$, then the probability will be distributed among at most $N!$ different trajectories given by

$$z_r(t) = \sum_{i=1}^{N} c_i a_N(r_i, t), \quad r = (r_1, \ldots, r_N) \in R.$$

6.3.2 Convergence in distribution in $C[0,1]$. Each of the definitions D1, D2 and D4 of Subsection 6.2.1 may serve also in $C[0,1]$, and all of them are equivalent. This may be easily proved by following verbally the pattern of the proof of Theorem 6.2.1.1, since the arguments are valid for any separable complete metric space. Recall that $h(z)$, $z \in C[0,1]$, is continuous at a point $z_0 = z_0(.)$, if for every $\varepsilon > 0$ there is a $\delta > 0$ such that

$$[\|z(.) - z_0(.)\| < \delta] \Rightarrow [|h(z) - h(z_0)|] < \varepsilon \tag{1}$$

where $\|.\|$ is given by (6.3.1.1).

Definition D3 must be supplemented by further requirements as follows:

Theorem 1 *Let $Z(.)$ and $T_\nu(.)$, $\nu \ge 1$, be stochastic processes determined by the respective distributions in $(C[0,1], \mathcal{C})$. Then $T_\nu(.)$ converges in distribution to Z* (see D1 or D2 or D4 of Subsection 6.2.1) *if and only if*

(I) *$\sum_{j=1}^{n} \lambda_j T_\nu(t_j)$ converges in distribution to $\sum_{j=1}^{n} \lambda_j Z(t_j)$ for every $n \ge 1$, $0 \le t_1, \ldots, t_n \le 1$, $-\infty < \lambda_1, \ldots, \lambda_n < \infty$, and*

(II) *for every $\varepsilon > 0$*

$$\lim_{\delta \to 0} \liminf_{\nu} P(\max_{|t-s|<\delta} |T_\nu(t) - T_\nu(s)| < \varepsilon) = 1. \tag{2}$$

Proof. *Necessity* follows immediately from D2 since $\max_{|t-s|<\delta} |z(t) - z(s)| = h_\delta(z(.))$ is a continuous functional whose distribution tends to 0 for the Z-process, if $\delta \to 0$.

Sufficiency. Defining convergence in distribution by D4, it suffices to show that for every open set G we have

$$P(Z \in G) \le \liminf_{\nu} P(T_\nu \in G). \tag{3}$$

Then we also have for every closed set F

$$P(Z \in F) \geq \limsup_{\nu} P(T_\nu \in F) \tag{4}$$

so that for an arbitrary set A with $P(Z \in A) = P(Z \in A^0) = P(Z \in \bar{A})$, where A^0 denotes the interior of A, and $\bar{A}$ denotes the closure,

$$P(Z \in A) = P(Z \in A^0) \leq \liminf_{\nu} P(T_\nu \in A^0) \leq \liminf_{\nu} P(T_\nu \in A)$$

and

$$P(Z \in A) = P(Z \in \bar{A}) \geq \limsup_{\nu} P(T_\nu \in \bar{A}) \geq \limsup_{\nu} P(T_\nu \in A),$$

hold, which implies (6.2.1.1).

Now choose an open set G and an arbitrary $\eta > 0$. Let G_ε be a subset of G consisting of points y such that the spheres $S(y, \varepsilon)$ lie within G. Since $G = \bigcup_{\varepsilon > 0} G_\varepsilon$, we may choose ε so that

$$P(Z \in G) - \tfrac{1}{3}\eta < P(Z \in G_\varepsilon). \tag{5}$$

Now put

$$A_k = \{z(.) : \max_{|t-s|<2^{-k}} |z(t) - z(s)| < \tfrac{1}{2}\varepsilon\}$$

and choose k so that

$$P(Z \in C[0,1] - A_k) < \tfrac{1}{3}\eta$$

and also

$$\limsup_{\nu} P(T_\nu \in C[0,1] - A_k) < \tfrac{1}{3}\eta,$$

which is possible in view of (2). Then introduce

$$B_k = \{z(.) : z(t) = z'(t),\ t \in \tau_k,\ \text{for at least one } z'(.) \in (G_\varepsilon \cap A_k)\}$$

where

$$\tau_k = \{i.\, 2^{-k},\ 0 \leq i \leq 2^k\},$$

and note that B_k is open and

$$G \supset B_k \cap A_k.$$

Thus, on utilizing (I) and (5),

$$\begin{aligned} \liminf_{\nu} P(T_\nu \in G) \;&\geq\; \liminf_{\nu} P(T_\nu \in B_k) \\ &\quad - \limsup_{\nu} P(T_\nu \in C[0,1] - A_k) \\ &\geq\; P(Z \in B_k) - \tfrac{1}{3}\eta \\ &\geq\; P(Z \in G_\varepsilon \cap A_k) - \tfrac{1}{3}\eta \\ &\geq\; P(Z \in G_\varepsilon) - P(Z \in C[0,1] - A_k) - \tfrac{1}{3}\eta \\ &\geq\; P(Z \in G_\varepsilon) - \tfrac{2}{3}\eta \geq P(Z \in G) - \eta. \end{aligned} \tag{6}$$

Since η has been chosen arbitrarily, (6) proves (3). $\square$

In what follows we shall establish (I) and (II), and then utilize D2 as a consequence for the following continuous functionals on $C[0,1]$:

$$h^{+}(z(.)) = \max_{0\le t\le 1} z(t), \tag{7}$$

$$h^{\pm}(z(.)) = \max_{0\le t\le 1} |z(t)|, \tag{8}$$

$$h_a^{+}(z(.)) = \max_{a\le t\le 1} \frac{z(t)}{t}, \quad 0<a<1, \tag{9}$$

$$h_a^{\pm}(z(.)) = \max_{a\le t\le 1} \frac{|z(t)|}{t}, \quad 0<a<1, \tag{10}$$

$$h_2(z(.)) = \int_0^1 z^2(t)\,\mathrm{d}t. \tag{11}$$

6.3.3 Weak convergence in the Skorokhod metric. Let now $D[0,1]$ be the space of all real valued functions on $[0,1]$ that are right continuous and have left-hand limits, though the right and left hand limits may not be the same everywhere. That is, they have only *discontinuities of the first kind.* It is clear that the space $C[0,1]$ of all continuous functions on $[0,1]$ considered in Subsection 3.1, is a subspace of $D[0,1]$, and in that vein, we need to replace the *uniform metric* in (6.3.1.1) by a more flexible one (to handle the jump discontinuities); this is due to Skorokhod (1956). We introduce the Skorokhod metric and the topology generated by it (termed the J_1*-topology*) as follows.

Let Λ be the class of all strictly increasing, continuous mappings $\lambda : [0,1] \to [0,1]$, and let x, y both belong to the space $D[0,1]$. Define the Skorokhod distance as

$$\begin{aligned} d(x,y) = \inf_{\varepsilon}\{\varepsilon>0:\ & \text{there exists a } \lambda\in\Lambda, \text{ such that} \\ & \sup_{t\in[0,1]} |\lambda(t)-t|<\varepsilon, \text{ and } \sup_{t\in[0,1]} |x(t)-y(\lambda(t))|<\varepsilon\}. \end{aligned} \tag{1}$$

A necessary and sufficient condition for a sequence $\{f_n\}\in D[0,1]$ to converge to some $f\in D[0,1]$ in the Skorokhod J_1-topology is that there exists a sequence $\{\lambda_n\}\in\Lambda$ such that

$$\lim_{n\to\infty} f_n(\lambda_n(t)) = f(t), \quad \lim_{n\to\infty} \lambda_n(t) = t, \tag{2}$$

both being uniform in $t\in[0,1]$.

Now, we are going to consider the important notions of the *modulus of continuity*, first for $C[0,1]$: If $f\in C[0,1]$, then its modulus of continuity is defined to be

$$\omega_f(\delta) = \sup_{|s-t|<\delta} |f(t)-f(s)|, \quad 0<\delta\le 1. \tag{3}$$

(See Billingsley (1968), formula (8.1).) Note that $\lim_{\delta \to 0} \omega_f(\delta) = 0$ is necessary and sufficient for an arbitrary function f on $[0, 1]$ to lie in $C[0, 1]$. Next, for $f \in D[0, 1]$ and $T_0 \subset [0, 1]$, set

$$\omega_f(T_0) = \sup\{|f(t) - f(s)| : s, t \in T_0\}, \tag{4}$$

and for $f \in D[0, 1]$ its modulus (for $0 < \delta < 1$) is defined as

$$\omega'_f(\delta) = \inf_{\{t_i\}} \max_{0<i\leq r} \omega_f[t_{i-1}, t_i), \tag{5}$$

where the infimum is taken over all finite sets $\{t_i\}$ of points satisfying

$$0 = t_0 < t_1 < \ldots < t_r = 1, \quad t_i - t_{i-1} > \delta, \quad i = 1, 2, \ldots, r.$$

(See Billingsley (1968), formulas (14.6) and (14.7).) Analogously,

$$\lim_{\delta \to 0} \omega'_f(\delta) = 0 \tag{6}$$

if and only if f on $[0, 1]$ belongs to $D[0, 1]$.

Billingsley (1968), formulas (14.9) and (14.11), shows that

$$\omega'_f(\delta) \leq \omega_f(2\delta), \quad \text{for each } 0 < \delta < 1/2, \tag{7}$$

while for $f \in C[0, 1]$,

$$\omega_f(\delta) \leq 2\omega'_f(\delta), \quad \text{for all } \delta > 0. \tag{8}$$

Let $\{X_N(t),\ t \in [0, 1]\}$, $N \geq N_o$ be a sequence of stochastic processes whose trajectories belong to the space $D[0, 1]$. Let $X(t)$, $t \in [0, 1]$, be another stochastic function defined on the space $D[0, 1]$. Then, we say that X_N weakly converges to X (in the Skorokhod J_1-topology on $D[0, 1]$), if for any arbitrary m (≥ 1) and $0 \leq t_1 < \ldots < t_m \leq 1$, $(X_N(t_1), \ldots, X_N(t_m))$ converges in law to $(X(t_1), \ldots, X(t_m))$, and moreover X_N is compact or tight. (For the notion of tightness in $C[0, 1]$ and in $D[0, 1]$ see Billingsley (1968), Sec. 8, and Sec. 15, respectively.) The first criterion is known as the convergence of finite dimensional distributions, and the second one relates to (6) but in a suitable norm to handle the stochastic nature of the processes. Billingsley (1968) contains a number of probability inequalities, named after him, as the Billingsley inequalities, that are quite useful in establishing the desired tightness property. In this context, verification of tightness for processes belonging to the space $C[0, 1]$ is comparatively simpler than for processes belonging to the space $D[0, 1]$. However, the last two inequalities offer some flexibilities to operate with either the $D[0, 1]$ or $C[0, 1]$ space for showing that tightness holds. The convergence of finite dimensional distributions can be handled in a unified manner. In dealing with rank statistics as well as empirical distribution functions, we generally encounter stochastic processes whose trajectories belong to the space $D[0,1]$, and hence, this refinement might be of considerable utility in dealing with their general asymptotics. For rank statistics, in general, we have processes that belong to the space $D[0,1]^p$ of functions that are defined on the unit p-cube $\mathsf{E}^p = [0,1] \times \ldots \times [0,1]$, for some positive integer p. In that case, the modulus of continuity has been further generalized to suit such functions. We refer to Sen (1981) for an extensive account of these results for rank statistics of various types.

6.3.4 The Brownian bridge. A random process $Z(t)$, $0 \le t \le 1$, is called Gaussian if all subsectors $[Z(t_1), \ldots, Z(t_n)]$ are normally distributed. The distribution of a Gaussian process is given by the expectations $\mathsf{E}Z(t)$ of individual observations and by the covariances $\mathsf{cov}\,(Z(t), Z(s))$, $0 \le t \le s \le 1$, of all pairs of observations. Under certain conditions, discussed in general in Subsection 6.3.1, the distribution of a Gaussian process may be realized in $(C[0,1], \mathcal{C})$, and then all sample paths $z(.)$ are continuous functions, and events such as $\max_{0 \le t \le 1} Z(t) <$ const are measurable.

The Gaussian process with zero expectations and covariances

$$\mathsf{cov}\,\big(Z(t), Z(s)\big) = t(1-s), \quad 0 \le t \le s \le 1, \tag{1}$$

will be called a *Brownian bridge.* Since for normally distributed random variables with zero expectations, $\mathsf{E}X^4 = 3[\mathsf{E}X^2]^2$, we have from (1)

$$\mathsf{E}[Z(t) - Z(s)]^4 = 3(t-s)^2(1-|t-s|)^2 \le 3(t-s)^2$$

so that the distribution may be realized in $(C[0,1], \mathcal{C})$ according to Theorem 6.3.1.2. Thus we shall assume the Brownian bridge $Z(.)$ takes its sample paths $z(.)$ in $C[0,1]$, without further mentioning this fact.

The following theorem is well known (see Doob (1949), for example).

Theorem 1 *If $Z(t)$ is a Brownian bridge, then for $\alpha > 0$, $\beta > 0$*

$$P\big(Z(t) < \alpha(1-t) + \beta t,\ 0 \le t \le 1\big) = 1 - \mathrm{e}^{-2\alpha\beta} \tag{2}$$

and

$$\begin{aligned} &P\big(-\alpha(1-t) - \beta t < Z(t) < \alpha(1-t) + \beta t,\ 0 \le t \le 1\big) \\ &\quad = 1 - 2\sum_{m=1}^{\infty} (-1)^{m+1} \mathrm{e}^{-2m^2\alpha\beta}. \end{aligned} \tag{3}$$

Since the process

$$Z^*(t) = (v-u)^{\frac{1}{2}} Z\Big(\frac{t-u}{v-u}\Big) + \frac{v-t}{v-u}x + \frac{t-u}{v-u}y, \quad u \le t \le v,$$

has the same distribution as a Brownian bridge Z' given $Z'(u) = x$ and $Z'(v) = y$, the following generalization of (2) holds for $x < \alpha(1-u) + \beta u$ and $y < \alpha(1-v) + \beta v$:

$$\begin{aligned} &P\big(Z(t) < \alpha(1-t) + \beta t,\, u \le t \le v \mid Z(u) = x,\, Z(v) = y\big) \qquad (4) \\ &= 1 - \exp[-2\big(\alpha(1-u) + \beta u - x\big)\big(\alpha(1-v) + \beta v - y\big)(v-u)^{-1}]. \end{aligned}$$

Furthermore, from the generalization of (3) given in Problems 15 and 16, we may derive for $-\lambda a < x < \lambda a$

$$\begin{aligned} &P\big(-\lambda t < Z(t) < \lambda t,\, a \le t \le 1 \mid Z(a) = x\big) \qquad (5) \\ &= 1 - \sum_{i=1}^{\infty} (\mathrm{e}^{-2A_m/(1-a)} + \mathrm{e}^{-2B_m/(1-a)} \\ &\qquad - \mathrm{e}^{-2C_m/(1-a)} - \mathrm{e}^{-2D_m/(1-a)}) \end{aligned}$$

with $\lambda > 0$ and

$$\begin{aligned} A_m &= (2m-1)^2\lambda^2 a - (2m-1)\lambda x, \\ B_m &= (2m-1)^2\lambda^2 a + (2m-1)\lambda x, \\ C_m &= 4m^2\lambda^2 a - 2m\lambda x, \\ D_m &= 4m^2\lambda^2 a + 2m\lambda x. \end{aligned}$$

From (4) and (5), for $u = a$ and $v = 1$, Theorem 2 easily follows:

Theorem 2 *If $Z(t)$ is a Brownian bridge, then for $a > 0$ and $\lambda > 0$*

$$P\big(Z(t) < \lambda t,\, a \le t \le 1\big) = 2\Phi\Big(\lambda\sqrt{\frac{1}{1-a}}\Big) - 1 \qquad (6)$$

and

$$\begin{aligned} &P\big(-\lambda t < Z(t) < \lambda t,\, a \le t \le 1\big) \qquad (7) \\ &= 4\Phi\Big(\lambda\sqrt{\frac{a}{1-a}}\Big) - 3 \\ &\quad + \sum_{m=1}^{\infty}\Big(\Phi\Big[(2m+1)\lambda\sqrt{\frac{a}{1-a}}\Big] - \Phi\Big[2m\lambda\sqrt{\frac{a}{1-a}}\Big]\Big) \end{aligned}$$

with

$$\Phi(x) = (2\pi)^{\frac{1}{2}} \int_{-\infty}^{x} \exp(-\tfrac{1}{2}y^2)\,\mathrm{d}y.$$

Theorem 3 *If $Z(t)$ is a Brownian bridge, then*

$$P\Big(\int_0^1 Z^2(t)\,\mathrm{d}t < \lambda\Big) = P\Big(\sum_{j=1}^{\infty} \frac{X_j^2}{j^2\pi^2} < \lambda\Big), \tag{8}$$

where $X_1, X_2, \ldots$ are independent standardized normal random variables.

Proof. Put

$$Z_N(t) = 2^{\frac{1}{2}} \sum_{j=1}^{N} X_j \frac{\sin(j\pi t)}{j\pi}$$

and note that

$$2\sum_{j=1}^{\infty} \frac{\sin(j\pi t)\sin(j\pi s)}{(j\pi)^2} = t(1-s), \quad 0 \le t \le s \le 1. \tag{9}$$

Thus the processes $Z_N(.)$ are Gaussian with zero expectations and covariances

$$\mathsf{cov}\,\big(Z_N(t), Z_N(s)\big) = 2\sum_{j=1}^{N} \frac{\sin(j\pi t)\sin(j\pi s)}{(j\pi)^2} \to \mathsf{cov}\,\big(Z(t), Z(s)\big),$$

where $Z(t)$ denotes the Brownian bridge. In order to prove that $Z_N(.)$ converges to $Z(.)$ in distribution in $(C[0,1], \mathcal{C})$, we have only to show that (6.3.2.2) is satisfied. However, (6.3.2.2) will follow from

$$\mathsf{E}[Z_N(t) - Z_N(s)]^4 \le M(t-s)^2, \quad 0 \le t, s \le 1,$$

with M independent of N, in the same manner as (6.3.1.6), and in turn (6.3.1.7), followed from (6.3.1.20).

Now, in fact,

$$\begin{aligned}
\mathsf{E}[Z_N(t) - Z_N(s)]^4 &= 3\big[\,\mathsf{var}\,\big(Z_N(t) - Z_N(s)\big)\big]^2 \\
&\le 3\Big\{2\sum_{j=1}^{N} \Big[\frac{\sin(j\pi t) - \sin(j\pi s)}{j\pi}\Big]^2\Big\}^2 \\
&\le 3\Big\{2\sum_{j=1}^{\infty} \Big[\frac{\sin(j\pi t) - \sin(j\pi s)}{j\pi}\Big]^2\Big\}^2 \\
&= 3(t-s)^2(1-|t-s|)^2 \le 3(t-s)^2.
\end{aligned}$$

Thus $Z_N(.)$ converges in distribution in $(C[0,1], \mathcal{C})$ to $Z(.)$, and, consequently

$$\int_0^1 Z_N^2(t)\,\mathrm{d}t = \sum_{j=1}^{N}\sum_{k=1}^{N} \frac{X_j X_k}{j\pi k\pi} 2\int_0^1 \sin(j\pi t)\sin(k\pi t)\,\mathrm{d}t = \sum_{j=1}^{N} \frac{X_j^2}{j^2\pi^2}$$

converges in distribution to $\int_0^1 Z^2(t)\,\mathrm{d}t$. As clearly $P(\int_0^1 Z^2(t)\,\mathrm{d}t = b) = 0$ for every $b \ge 0$, (8) easily follows. □

6.3.5 Kolmogorov's inequality for dependent summands. Consider an arbitrary vector $(c_1, \dots, c_N)$ and a random vector $D = (D_1, \dots, D_N)$ uniformly distributed over R. In particular, $D = (D_1, \dots, D_N)$ will be the vector of antiranks (see Subsection 3.3.2) under H_0.

Lemma 1 *If $n < N$ and $P(D = r) = 1/N!$, $r \in R$, then*

$$\mathsf{E}\Big(\sum_{i=1}^{n} c_{D_i} - n\bar{c}\Big)^4 = \frac{n(N-n)}{N(N-1)(N-2)(N-3)} \cdot \Big\{(N^2 - 6Nn + 6n^2 + N)\sum_{i=1}^{N}(c_i - \bar{c})^4 + 3(N-n-1)(n-1)\Big[\sum_{i=1}^{N}(c_i - \bar{c})^2\Big]^2\Big\}. \tag{1}$$

Proof. See Isserlis (1931).

Lemma 2 *If $n < N$ and $P(D = r) = 1/N!$, $r \in R$, then*

$$P\Big(\max_{1 \le k \le n} |c_{D_1} + \dots + c_{D_k} - k\bar{c}| > \varepsilon\Big[\sum_{i=1}^{N}(c_i - \bar{c})^2\Big]^{\frac{1}{2}}\Big) \le \frac{n}{N}\left[\frac{\max_{1 \le i \le N}(c_i - \bar{c})^2}{\sum_{i=1}^{N}(c_i - \bar{c})^2} + 3\frac{n}{N}\right]\varepsilon^{-4}\Big(1 - \frac{n}{N}\Big)^{-3}(1 + \varepsilon_N) \tag{2}$$

where $\varepsilon_N \to 0$ as $N \to \infty$.

Proof. Without loss of generality we may assume $\bar{c} = 0$ and $\sum_{i=1}^{N} c_i^2 = 1$ and introduce disjoint events defined as follows:

$$A_k = \{r : \max_{1 \le j \le k-1} |c_{r_1} + \dots + c_{r_j}| \le \varepsilon,\ |c_{r_1} + \dots + c_{r_k}| > \varepsilon\}. \tag{3}$$

Then

$$\mathsf{E}\Big(\sum_{i=1}^{n} c_{D_i}\Big)^4 = \sum_{r \in R}\Big(\sum_{i=1}^{n} c_{r_i}\Big)^4 \frac{1}{N!} \ge \sum_{k=1}^{n}\sum_{r \in A_k}\Big(\sum_{i=1}^{n} c_{r_i}\Big)^4 \frac{1}{N!}.$$

Now, since A_k depends on $r_1, \dots, r_k$ only,

$$\sum_{r \in A_k}\Big(\sum_{i=1}^{n} c_{r_i}\Big)^4 = \sum_{r \in A_k} \mathsf{E}\Big[\Big(\sum_{i=1}^{k} c_{r_i} + \sum_{j=k+1}^{n} c_{D_j}\Big)^4 \mid D_i = r_i, 1 \le i \le k\Big], \quad 1 \le k \le n.$$

As the function x^4 is convex, upon applying the Jensen inequality, we obtain

$$\begin{aligned}
&\mathsf{E}\Big[\Big(\sum_{i=1}^{k} c_{r_i} + \sum_{j=k+1}^{n} c_{D_j}\Big)^4 \mid D_i = r_i,\ 1 \le i \le k\Big] \\
&\ge \Big[\sum_{i=1}^{k} c_{r_i} + \mathsf{E}\Big\{\sum_{j=k+1}^{n} c_{D_j} \mid D_i = r_i,\ 1 \le i \le k\Big\}\Big]^4 \\
&= \Big[\sum_{i=1}^{k} c_{r_i} - \frac{n-k}{N-k}\sum_{i=1}^{k} c_{r_i}\Big]^4 \ge \Big(1 - \frac{n}{N}\Big)^4 \Big(\sum_{i=1}^{k} c_{r_i}\Big)^4.
\end{aligned}$$

Upon combining the above results we obtain

$$\mathsf{E}\Big(\sum_{i=1}^{n} c_{D_i}\Big)^4 \ge \Big(1 - \frac{n}{N}\Big)^4 \sum_{k=1}^{n} \sum_{r \in A_k} \Big(\sum_{i=1}^{k} c_{r_i}\Big)^4 \frac{1}{N!}. \tag{4}$$

However, according to (3), $|c_{r_1} + \ldots + c_{r_k}| > \varepsilon$ for $r \in A_k$, so that (4) implies

$$\begin{aligned}
\mathsf{E}\Big(\sum_{i=1}^{n} c_{D_i}\Big)^4 &\ge \varepsilon^4 \Big(1 - \frac{n}{N}\Big)^4 \sum_{k=1}^{n} P(D \in A_k) \\
&= \varepsilon^4 \Big(1 - \frac{n}{N}\Big)^4 P\big(\max_{1 \le k \le n} |c_{D_1} + \ldots + c_{D_k}| > \varepsilon\big).
\end{aligned}$$

Now the proof is easily concluded by making use of (1). □

We define $c_{Ni} = (c_i - \bar{c}_N)/C_N$, $i = 1, \ldots, N$, where $C_N^2 = \sum_{i=1}^{N} (c_i - \bar{c}_N)^2$. Then note that $\sum_{i=1}^{N} c_{Ni} = 0$ and $\sum_{i=1}^{N} c_{Ni}^2 = 1$. Taking the vector $D_N = (D_1, \ldots, D_N)$ of antiranks, consider the partial sequence

$$S_{Nk} = (N-k)^{-1} \sum_{i=1}^{k} c_{ND_i},\ k = 1, \ldots, N-1, \tag{5}$$

and conventionally, we let $S_{N0} = 0$. Note that D_N takes on each permutation of $\{1, \ldots, N\}$ with the common probability $(N!)^{-1}$. Using this permutational probability law $\mathcal{P}_N$, it is easy to show that

$$\mathsf{E}_{\mathcal{P}_N}\{S_{Nk} | S_{Nj}, j \le k-1\} = S_{Nk-1} \text{ a.e.},\ k : 1 \le k \le N-1. \tag{6}$$

Therefore, the partial sequence $\{S_{Nk};\ k \le N-1\}$ is a zero-mean martingale array. This martingale characterization motivates us to consider the following lemma that extends Lemma 2 to a more general setup (Sen, 1979).

Lemma 3 *Let $q(t)$ be a continuous, non-negative, U-shaped and square integrable function on $(0,1)$. Then, for every $N > n(\geq 1)$,*

$$P\Big\{\max_{k\leq n} q\Big(\frac{k}{N}\Big)\Big|\sum_{i=1}^{k} c_{ND_i}\Big| > 1\Big\} \leq \int_0^{n/N} q^2(t)\,\mathrm{d}t. \tag{7}$$

Proof. Note that by definition,

$$\sum_{i=1}^{k} c_{ND_i} = (N-k)S_{Nk}, \quad k = 0, 1, \ldots, N-1. \tag{8}$$

Moreover, for $n \leq N/2$, by the U-shaped nature of q, $q(\frac{k}{N})(N-k)$ is non-increasing in k. Hence exploiting the martingale characterization in (6), it follows from (8) and the celebrated Chow-Hájek-Rényi inequality for submartingales, we readily arrive at (7) for any $n \leq N/2$. By reflection the proof also follows for the upper tail. □

The inequalities in (2) and (7) are particularly useful for establishing weak convergence properties of suitable rank processes (as will be considered in the next subsection); there we allow n/N to be small, and N to be large. In that case, in (7), by the square integrability of $q(t), t \in (0,1)$, and its U-shaped nature, $\int_0^{n/N} q^2(t)\,dt$ converges to 0 as n/N converges to 0, and that suffices for the purpose. Note further, (7) does not entail any further than the second moment condition; if as in (1) we assume finiteness of fourth or higher order moments, then as in Sen (1970) sharper bounds can be obtained by incorporating the martingale property in (6). However, these refinements are not that essential for our purpose. Rather suitable invariance principles, as will be developed in Section 6.4, provide simpler asymptotic results that are of greater interest.

6.3.6 The basic convergence theorem. For every $t \in [0,1]$ and $N \geq 1$ consider scores $a_N(i,t)$ defined as follows:

$$a_N(i,t) = \begin{cases} 0, & i \leq tN, \\ i - tN, & tN \leq i < tN+1, \\ 1, & tN+1 \leq i. \end{cases} \tag{1}$$

Observe that $a_N(i,t)$ is continuous in t. Consequently, the processes

$$T_c(t) = \Big[\sum_{i=1}^{N}(c_i - \bar{c})^2\Big]^{-\frac{1}{2}} \sum_{i=1}^{N}(c_i - \bar{c})a_N(R_i,t), \quad c \in C, \tag{2}$$

determine probability distributions in $(C[0,1], \mathcal{C})$.

Theorem 1 *Assume that H_0 holds. Then for* $\sum_{i=1}^{N}(c_i-\bar{c})^2/\max_{1\le i\le N}(c_i-\bar{c})^2 \to \infty$ *the processes $T_c(.)$ given by (2) converge in distribution in $(C[0,1],\mathcal{C})$ to the Brownian bridge $Z(t)$.*

Proof. We shall show that the conditions (I) and (II) from Theorem 6.3.2.1 are satisfied.

(I) Put

$$\varphi(u,t)=\begin{cases}0, & \text{if } 0\le u\le t\le 1,\\ 1, & \text{if } 0\le t<u\le 1.\end{cases} \tag{3}$$

Obviously, we have

$$\lim_{N\to\infty}\int[a_N(1+[uN],t)-\varphi(u,t)]^2\,\mathrm{d}u=0,\quad 0\le t\le 1.$$

This implies

$$\begin{aligned}&\lim_c \operatorname{cov}\big(T_c(t),T_c(s)\big)\\ &=\int_0^1\varphi(u,t)\varphi(u,s)\,\mathrm{d}u-\int_0^1\varphi(u,t)\,\mathrm{d}u\int_0^1\varphi(u,s)\,\mathrm{d}u\\ &=t(1-s)=\operatorname{cov}\big(Z(t),Z(s)\big),\quad 0\le t\le s\le 1,\end{aligned}$$

and also ensures the asymptotic normality of $\sum_{j=1}^{n}\lambda_j T_c(t_j)$ with parameters coinciding with the expectation and variance of

$$\sum_{j=1}^{n}\lambda_j Z(t_j),\quad n\ge 1,\ 0\le t_1<\ldots<t_n\le 1,\ -\infty<\lambda_1,\ldots,\lambda_n<\infty.$$

See Theorem 6.1.6.1 and the proof of Theorem 6.2.2.1.

(II) Without loss of generality assume that $\bar{c}=0$ and $\sum_{i=1}^{N}c_i^2=1$. Note that for $\varepsilon>0$

$$\begin{aligned}\Big[\max_{|t-s|<1/N}|T_c(t)-T_c(s)|<\tfrac{1}{3}\varepsilon,\ \max_{|i-j|<\delta N}\Big|T_c\Big(\frac{i}{N}\Big)-T_c\Big(\frac{j}{N}\Big)\Big|<\tfrac{1}{3}\varepsilon\Big]\\ \Rightarrow\big[\max_{|t-s|<\delta}|T_c(t)-T_c(s)|<\varepsilon\big]\end{aligned} \tag{4}$$

and that, in view of the linearity of $T_c(t)$ within the intervals $[(i-1)/N, i/N]$,

$$\max_{|t-s|<1/N}|T_c(t)-T_c(s)|=\max_{1\le j\le N}\Big|T_c\Big(\frac{j-1}{N}\Big)-T_c\Big(\frac{j}{N}\Big)\Big| \tag{5}$$

$$= \max_{1\le j\le N} \Big| \sum_{i=1}^{N} c_i \Big[a_N\Big(R_i, \frac{j-1}{N}\Big) - a_N\Big(R_i, \frac{j}{N}\Big)\Big]\Big|$$
$$= \max_{1\le j\le N} \Big| \sum_{i=1}^{N} c_{D_i} \Big[a_N\Big(i, \frac{j-1}{N}\Big) - a_N\Big(i, \frac{j}{N}\Big)\Big]\Big|$$
$$= \max_{1\le j\le N} |c_{D_j}| = \max_{1\le i\le N} |c_i|.$$

Consequently, since $\max_{1\le i\le N} |c_i| \to 0$, under $\bar{c} = 0$ and $\sum_{i=1}^{N} c_i^2 = 1$, we have

$$\lim_c P(\max_{|t-s|<1/N} |T_c(t) - T_c(s)| < \varepsilon) = 1.$$

Thus, in view of (3), (6.3.2.2) will be proved if we show that for $\varepsilon > 0$

$$\lim_{\delta\to 0} \liminf_c P\Big(\max_{|i-j|<\delta N} \Big| T_c\Big(\frac{i}{N}\Big) - T_c\Big(\frac{j}{N}\Big)\Big| < \varepsilon\Big) = 1 \tag{6}$$

where N corresponds to $c = (c_1, \ldots, c_N)$. Now assume $\delta < 1$ and denote by n_δ the smallest integer such that $n_\delta \ge N\delta$, and by k_δ the greatest integer such that $k_\delta n_\delta \le N$. Obviously

$$\frac{n_\delta}{N} \to \delta \quad \text{and} \quad \delta k_\delta \le 1. \tag{7}$$

Note that for $|i - j| < \delta N \le n_\delta$

$$\begin{aligned} |T_c(i/N) - T_c(j/N)| \;\; \le \;\; & |T_c(i/N) - T(kn_\delta/N)| \\ & + |T_c(j/N) - T_c(k^* n_\delta/N)| \\ & + |T_c(kn_\delta/N) - T_c(k^* n_\delta/N)| \end{aligned}$$

where k (k^*) is the greatest integer $\le i/n_\delta$ (j/n_δ) and $|k^* - k| = 0$ or 1. Consequently,

$$\begin{aligned} & P\Big(\max_{|i-j|<\delta N} \Big| T_c\Big(\frac{i}{N}\Big) - T_c\Big(\frac{j}{N}\Big)\Big| \ge 3\varepsilon\Big) \qquad (8) \\ & \le \sum_{k=0}^{k_\delta} P\Big(\max_{kn_\delta<j\le(k+1)n_\delta} \Big| T_c(\frac{kn_\delta}{N}\Big) - T_c\Big(\frac{j}{N}\Big)\Big| \ge \varepsilon\Big) \\ & \le (k_\delta + 1) \max_{0\le k\le k_\delta} P\Big(\max_{kn_\delta<j\le(k+1)n_\delta} \Big| T_c(\frac{kn_\delta}{N}\Big) - T_c\Big(\frac{j}{N}\Big)\Big| \ge \varepsilon\Big). \end{aligned}$$

Recalling (1), we easily see that

$$T_c\Big(\frac{kn_\delta}{N}\Big) - T_c\Big(\frac{j}{N}\Big) = \sum_{i=kn_\delta+1}^{j} c_{D_i}$$

so that Lemma 6.3.5.2 and (6) entail

$$P\Big(\max_{kn_\delta<j\le(k+1)n_\delta}\Big|T_c\big(\frac{kn_\delta}{N}\big)-T_c\Big(\frac{j}{N}\Big)\Big|>\varepsilon\Big)$$
$$\le\frac{n_\delta}{N}\Big(\max_{1\le i\le N}c_i^2+3\frac{n_\delta}{N}\Big)\varepsilon^{-4}\Big(1-\frac{n_\delta}{N}\Big)^{-3}(1+\varepsilon_N)$$
$$\le\delta(\max_{1\le i\le N}c_i^2+3\delta)\varepsilon^{-4}(1-\delta)^{-3}(1+\varepsilon_N')$$

where $\varepsilon_N'\to 0$ as $N\to\infty$. On substituting this inequality into (7), we obtain

$$P\Big(\max_{|i-j|<\delta N}\Big|T_c\Big(\frac{i}{N}\Big)-T_c\Big(\frac{j}{N}\Big)\Big|\ge 3\varepsilon\Big) \tag{9}$$
$$\le(k_\delta+1)\delta(\max_{1\le i\le N}c_i^2+3\delta)\varepsilon^{-4}(1-\delta)^{-3}(1+\varepsilon_N')$$
$$\le 2(\max_{1\le i\le N}c_i^2+3\delta)\varepsilon^{-4}(1-\delta)^{-3}(1+\varepsilon_N''),$$

where $\varepsilon_N''\to 0$ as $N\to\infty$. Now (5) is an easy consequence of (8) and of the fact that $\max c_i^2$ stands for

$$\max_{1\le i\le N}(c_i-\bar{c})^2/\sum_{i=1}^N(c_i-\bar{c})^2.$$

The proof is thus completed. □

6.3.7 The Kolmogorov-Smirnov statistics.

Define

$$K_c^+=\frac{\max_{1\le k\le N}(k\bar{c}-c_{D_1}-\ldots-c_{D_k})}{[\sum_{i=1}^N(c_i-\bar{c})^2]^{\frac{1}{2}}},\quad c\in C, \tag{1}$$

and

$$K_c^\pm=\frac{\max_{1\le k\le N}|k\bar{c}-c_{D_1}-\ldots-c_{D_k}|}{[\sum_{i=1}^N(c_i-\bar{c})^2]^{\frac{1}{2}}},\quad c\in C, \tag{2}$$

and note that

$$K_c^+=h^+(T_c(.)),$$

and

$$K_c^\pm=h^\pm(T_c(.)),$$

where h^+, $h^\pm$, and $T_c(.)$ are defined by (6.3.2.7), (6.3.2.8), (6.3.6.2), respectively. From Theorem 6.3.4.1 and Theorem 6.3.6.1 we obtain the following

Theorem 1 *Under H_0 we have*

$$\lim_c P(K_c^+ < \lambda) = 1 - \mathrm{e}^{-2\lambda^2}, \quad c \in C, \tag{3}$$

and

$$\lim_c P(K_c^\pm < \lambda) = 1 - 2\sum_{m=1}^{\infty}(-1)^{m+1}\mathrm{e}^{-2m^2\lambda^2}, \quad c \in C, \tag{4}$$

for

$$\frac{\sum_{i=1}^{N}(c_i - \bar{c})^2}{\max_{1\le i\le N}(c_i - \bar{c})^2} \to \infty.$$

In the two-sample problem we put

$$K_{mn}^+ = \left[\frac{m+n}{mn}\right]^{\frac{1}{2}} \max_{0\le k\le m+n}\left(\frac{km}{m+n} - c_{D_1} - \ldots - c_{D_k}\right), \tag{5}$$

and

$$K_{mn}^\pm = \left[\frac{m+n}{mn}\right]^{\frac{1}{2}} \max_{0\le k\le m+n}\left|\frac{km}{m+n} - c_{D_1} - \ldots - c_{D_k}\right| \tag{6}$$

where

$$c_i = 1,\ 1 \le i \le m, \quad \textit{and} \quad c_i = 0,\ m < i \le m+n.$$

Corollary 1 *Under H_0*

$$\lim_{m,n} P(K_{mn}^+ < \lambda) \ = \ 1 - \mathrm{e}^{-2\lambda^2} \tag{7}$$

$$\lim_{m,n} P(K_{mn}^\pm < \lambda) \ = \ 1 - 2\sum_{j=1}^{\infty}(-1)^{j+1}\mathrm{e}^{-2j^2\lambda^2} \tag{8}$$

for $\min(m,n) \to \infty$.

6.3.8 The Rényi statistics.

Put

$$R_{ac}^+ = \frac{\max_{aN\le k\le N} \frac{N}{k}(k\bar{c} - c_{D_1} - \ldots - c_{D_k})}{[\sum_{i=1}^{N}(c_i - \bar{c})^2]^{\frac{1}{2}}}, \quad c \in C, \tag{1}$$

and

$$R_{ac}^\pm = \frac{\max_{aN\le k\le N} \frac{N}{k}|k\bar{c} - c_{D_1} - \ldots - c_{D_k}|}{[\sum_{i=1}^{N}(c_i - \bar{c})^2]^{\frac{1}{2}}}, \quad c \in C. \tag{2}$$

Observe that (6.3.6.4) entails

$$h_a^+(T_c(.)) - \frac{1}{a} \frac{\max_{1 \le i \le N} |c_i - \bar{c}|}{[\sum_{i=1}^{N} (c_i - \bar{c})^2]^{\frac{1}{2}}} \le R_{ac}^+ \le h_a^+(T_c(.))$$

and

$$h_a^\pm(T_c(.)) - \frac{1}{a} \frac{\max_{1 \le i \le N} |c_i - \bar{c}|}{[\sum_{i=1}^{N} (c_i - \bar{c})^2]^{\frac{1}{2}}} \le R_{ac}^\pm \le h_a^\pm(T_c(.)),$$

where h_a^+, $h_a^\pm$ and $T_c(.)$ are given by (6.3.2.9), (6.3.2.10), (6.3.6.2), respectively. From Theorem 6.3.4.2 and Theorem 6.3.6.1 we obtain the following

Theorem 1 *Under H_0 we have*

$$\lim_c P(R_{ac}^+ < \lambda) = 2\Phi\left(\lambda\sqrt{\left(\frac{a}{1-a}\right)}\right) - 1 \tag{3}$$

and

$$\lim_c P(R_{ac}^\pm < \lambda) = 4\Phi\left(\lambda\sqrt{\left(\frac{a}{1-a}\right)}\right) - 3 + \sum_{m=1}^{\infty}\left(\Phi\left[(2m+1)\lambda\sqrt{\left(\frac{a}{1-a}\right)}\right] - \Phi\left[2m\lambda\sqrt{\left(\frac{a}{1-a}\right)}\right]\right) \tag{4}$$

for

$$\frac{\sum_{i=1}^{N} (c_i - \bar{c})^2}{\max_{1 \le i \le N} (c_i - \bar{c})^2} \to \infty.$$

An analogous theorem holds for the two sample versions

$$R_{amn}^+ = \left[\frac{m+n}{mn}\right]^{\frac{1}{2}} \max_{a \le k/(m+n)} \left[\frac{m+n}{k}\left(\frac{km}{m+n} - c_{D_1} - \ldots - c_{D_k}\right)\right]$$

and

$$R_{amn}^\pm = \left[\frac{m+n}{mn}\right]^{\frac{1}{2}} \max_{a \le k/(m+n) \le 1} \left[\frac{m+n}{k}\left|\frac{km}{m+n} - c_{D_1} - \ldots - c_{D_k}\right|\right]$$

where $c_i = 1$, $1 \le i \le m$, and $c_i = 0$, $m < i \le m+n$.

6.3.9 The Cramér-von Mises statistics. Consider the statistics

$$M_c = \Big[\sum_{i=1}^{N}(c_i - \bar{c})^2\Big]^{-1} \sum_{k=1}^{N-1}(k\bar{c} - c_{D_1} - \ldots - c_{D_k})^2 \frac{1}{N} \tag{1}$$

and note that

$$M_c = h_2\big(T_c(.)\big) + \frac{1}{6N}.$$

Thus Theorem 6.3.4.3 and Theorem 6.3.6.1 imply the following

Theorem 1 *Under H_0 we have for*

$$\frac{\sum_{i=1}^{N}(c_i - \bar{c})^2}{\max_{1 \le i \le N}(c_i - \bar{c})^2} \to \infty,$$

$$\lim_c P(M_c < \lambda) \;=\; P\Big(\sum_{j=1}^{\infty} \frac{X_j^2}{j^2\pi^2} < \lambda\Big), \tag{2}$$

where $X_1, X_2, \ldots$ are independent standardized normal random variables.

An analogous theorem holds for

$$M_{mn} = \frac{1}{mn}\sum_{k=1}^{N}\Big(\frac{km}{m+n} - c_{D_1} - \ldots - c_{D_k}\Big)^2$$

where $c_i = 1$, $1 \le i \le m$, and $c_i = 0$, $m < i \le m+n$.

6.4 FUNCTIONAL CENTRAL LIMIT THEOREMS

The results in the preceding section are based on concepts deeper than the asymptotic normality results. In fact, compactness conditions have been invoked to obtain deeper convergence properties. Such results for a general class of rank statistics can be obtained by establishing suitable invariance principles for these statistics. To motivate the results, first we present a brief outline of such invariance principles for linear statistics.

The symbol $\xrightarrow{\mathcal{D}}$, which will be used here many times for brevity, will mean convergence in distribution. If the convergence is in the Skorokhod J_1-topology (see Subsection 6.3.3), it will be expressed explicitely by these words.

6.4.1 Weak invariance principles for martingales. A detailed treatise of invariance principles can be found in Billingsley (1968) and many other contemporary texts. We specialize here to martingales (arrays) which are generally encountered in the study of asymptotic properties of rank tests. For a triangular array $\{(Z_{n,k},\ 0 \leq k \leq k_n); n \geq 1\}$ of random variables (not necessarily independent), we set $Z_{n,0} = 0$, and

$$S_{n,k} = Z_{n,1} + \ldots + Z_{n,k} \text{ for } k = 1, \ldots, k_n, \tag{1}$$

where k_n increases with n. Let $\mathcal{B}_{n,k}$ be the σ-field generated by $S_{n,k}$, for $k \geq 1$ and note that $\mathcal{B}_{n,0}$ is the trivial σ-field, for every $n \geq 1$. We assume that $\mathsf{E}Z_{n,k}^2 < \infty$, for all $k \geq 1$, $n \geq 1$. Further, let

$$\mu_{n,k} = \mathsf{E}\{Z_{n,k}|\mathcal{B}_{n,k-1}\} \text{ and } \sigma_{n,k}^2 = \mathsf{E}\{Z_{n,k}^2|\mathcal{B}_{n,k-1}\} - \mu_{n,k}^2, \tag{2}$$

for $k \geq 1$, $n \geq 1$. Let us now introduce, for $n = 1, 2, \ldots$, a sequence of random functions

$$\{W_n(t) = S_{n,k_n(t)},\ t \in [0,1]\}, \tag{3}$$

where $\{k_n(t), 0 \leq t \leq 1\}$ is a sequence of integer-valued, non-decreasing and right-continuous functions (of t) defined on $[0,1]$ with $k_n(0) = 0$. Then, for every $n(\geq 1)$, W_n belongs to the $D[0,1]$ space, equipped with the Skorokhod J_1-topology, defined in Subsection 6.3.3. Let $W(t)$, $t \in [0,1]$, be a standard Wiener process (that belongs to the space $C[0,1]$ of all continuous functions on $[0,1]$). Then the following weak invariance principle is due to McLeish (1974).

Theorem 1 *Suppose that for each* $t \in [0,1]$,

$$\sum_{k \leq k_n(t)} |\mu_{n,k}| \to 0 \text{ and } \sum_{k \leq k_n(t)} \sigma_{n,k}^2 \to t, \text{ as } n \to \infty, \tag{4}$$

and the conditional Lindeberg condition holds, i.e., for every $\varepsilon > 0$,

$$\sum_{k=1}^{k_n} \mathsf{E}\{Z_{n,k}^2 I(|Z_{n,k}| > \varepsilon)|\mathcal{B}_{n,k-1}\} \to 0, \text{ as } n \to \infty, \tag{5}$$

where all convergences above are in probability. Then W_n *converge in law (in the* J_1*-topology on* $D[0,1]$*) to* W.

A proof of this theorem can be found in most contemporary probability theory texts; it is also given in Sen (1981, Appendix). If the $S_{n,k}$ form a (zero-mean) martingale array, we have by definition $\mu_{n,k} = 0$, for all k. In that case, if we define $s_{n,k}^2 = \mathsf{E}S_{n,k}^2$, for $k \geq 0$, and take, without loss of generality, $k_n = n$, we have

$$k_n(t) = \max\{k : s_{n,k}^2 \leq t s_{n,n}^2\}, \quad t \in [0,1]. \tag{6}$$

Therefore Theorem 1 simplifies to a certain extent if we standardize W_n by letting $W_n(t) = S_{n,k_n(t)}/s_{n,n}$, for $t \in [0,1]$. This weak convergence result has also been strengthened in the following manner (see, Sen (1981), Ch. 2, Th. 2.4.8).

Let $q = \{q(t), 0 < t < 1\}$ be a non-negative, continuous function on $[0,1]$ such that for some a, $0 < a < 1$, $q(t)$ is non-decreasing on $[0,a]$ and bounded away from 0 on $[a,1]$, such that

$$I(q) = \int_0^1 [q(t)]^{-2}\,\mathrm{d}t < \infty. \tag{7}$$

Consider then the metric

$$d_q = \sup_{0 \le t \le 1} \{|x(t) - y(t)|/q(t)\}. \tag{8}$$

Theorem 2 *For a martingale array, when q satisfies (7), and the regularity assumptions in Theorem 1 hold, then W_n converge to W in the d_q-metric as well.*

The last result is of considerable importance for the study of asymptotics for rank tests.

6.4.2 Almost sure invariance principles for martingales. Let $\{Y_i,\ i \ge 1\}$ be a sequence of independent random variables with zero mean and finite variances. Also, let $W(t)$, $t \ge 0$, be a Brownian motion on the non-negative reals. Skorokhod (1956) showed that there exists a sequence $\{T_i,\ i \ge 1\}$ of independent, non-negative random variables, such that for every $n \ge 1$, $W(T_1),\ W(T_1 + T_2) - W(T_1), \dots, W(\sum_{j=1}^{n} T_j) - W(\sum_{j=1}^{n-1})$ have the same joint distribution as do $Y_1, \dots, Y_n$, and $\mathsf{E}(T_j) = \mathsf{var}(Y_j)$, for all $j \ge 1$. This fundamental result is known as the *Skorokhod embedding of Wiener processes*, and it has been extended by Strassen (1964) to a general class of martingales; this basic result is of considerable importance to the theory of rank tests. Let $\{X_n, \mathcal{B}_n; n \ge 1\}$ be a martingale sequence, so that $\{Y_n = X_n - X_{n-1}, \mathcal{B}_n; n \ge 1\}$ forms a martingale-difference sequence. Note that $\mathsf{E}(Y_n|\mathcal{B}_{n-1}) = 0$ (a.e.), and we assume that

$$\mathsf{E}(Y_n^2|\mathcal{B}_{n-1}) \text{ exists a.s. for all } n \ge 1. \tag{1}$$

Let then

$$V_n = \sum_{k=1}^{n} \mathsf{E}(Y_k^2|\mathcal{B}_{k-1}),\ n \ge 1;\ V_0 = 0, \tag{2}$$

so that $\{V_n\}$ is a non-decreasing sequence of non-negative random variables. Let then

$$S_{V_n} = X_n,\ n \ge 1,\ S_0 = X_0 = 0 \text{ with probability 1;} \tag{3}$$

and let

$$S_t = S_{V_n} + Y_{n+1}(t - V_n)/\{V_{n+1} - V_n\}, \text{ for } V_n \le t \le V_{n+1};\ n \ge 0. \quad (4)$$

Defined on this way, S_t, $t \ge 0$, is a continuous process defined on the non-negative reals. Then, we have the following theorem due to Strassen (1964).

Theorem 1 *Let $\{X_n, \mathcal{B}_n; n \ge 1\}$ be a martingale, and define $\{V_n\}$ and S_t as above. Further, let $f(t)$ be a non-negative and non-decreasing function for $t \ge 0$, such that*

$$f(t) \textit{ is } \uparrow \textit{ but } t^{-1}f(t) \textit{ is } \downarrow \textit{ in } t(>0). \quad (5)$$

Suppose further that $V_n \to \infty$ a.s., as $n \to \infty$, and

$$\sum_{n\ge1}[f(V_n)]^{-1}\mathsf{E}\{Y_n^2 I(Y_n^2 > f(V_n))|\mathcal{B}_{n-1}\} < \infty \ a.s. \quad (6)$$

Then there exists a Brownian motion $W(t)$, $t \ge 0$ such that

$$S_t = W(t) + o((log\ t)[tf(t)]^{1/4}) \ a.s., \quad as\ t \to \infty. \quad (7)$$

For other generalizations of the embedding theorems, we refer to Csörgö and Révész (1981).

6.4.3 Invariance principles for empirical processes.

Let $X_1, \ldots, X_n$ be a random sample from a continuous distribution function F. We define the empirical distribution function F_n by

$$F_n(x) = n^{-1}\sum_{i=1}^{n} I(X_i \le x). \quad (1)$$

The corresponding empirical process $V_n(x)$, $-\infty < x < \infty$, is defined by letting

$$V_n(x) = n^{1/2}\{F_n(x) - F(x)\}, \quad -\infty < x < \infty. \quad (2)$$

Let us now consider the reduced model where we set $U_i = F(X_i)$, $i = 1, \ldots, n$, so that $u_1, \ldots, U_n$ is a random sample from the uniform $(0,1)$ distribution function (i.e., $G(u) = P\{U \le u\} = u$, $u \in (0,1)$). Let then

$$G_n(t) = n^{-1}\sum_{i=1}^{n} I(U_i \le t), \ t \in (0,1). \quad (3)$$

The corresponding reduced empirical process $W_n(t)$, $t \in (0,1)$, is defined by letting

$$W_n(t) = n^{1/2}\{G_n(t) - t\}, \quad t \in (0,1). \quad (4)$$

Thus, $W_n(t) = V_n(F^{-1}(t))$, $t \in (0,1)$. Hence, asymptotic properties for W_n are shared by V_n as well. Now, the following two results are easy to verify (see Sen (1981)):

(a) The process $W_n^* = nW_n$, $n \geq 1$ is a martingale (process).

(b) The process W_n, $n \geq 1$ is a reverse martingale (process).

This characterization enables us to incorporate the results in the preceding two subsections to arrive at the following.

Theorem 1 *W_n converges in law (in the Skorokhod J_1-topology on $D[0,1]$) to a standard Brownian bridge W^o. Moreover, defining the D_q metric as in* (6.4.1.8), *the weak convergence of W_n to W^o in the d_q metric holds as well, when q satisfies the condition* (6.4.1.7).

We define a Kiefer process $X^*(s,t), s \in [0,1]$, $t \geq 0$, that is Gaussian with no drift function and covariance kernel $(\min(s_1,s_2) - s_1s_2) \cdot \min(t_1,t_2)$, and note that the following a.s. invariance principle holds.

Theorem 2 *As $n \to \infty$,*

$$\max_{k \leq n} \sup_{0 \leq s \leq 1} |k[G_k(s) - s] - X^*(s,k)| = O((\log n)^2) \ a.s.. \tag{5}$$

There are parallel results for weighted empirical processes, and as these are similar to the ones for rank statistics, to be presented in the next subsection, we omit the duplication.

6.4.4 Invariance principles for rank statistics. By virtue of the martingale characterizations for rank statistics studied in Subsection 6.1.9, and invariance principles considered in the preceding subsections, we are in a position to formulate suitable invariance principles for these rank statistics, under the null hypotheses. We shall show that these results extend to censored rank statistics as well as to multivariate models.

First, we consider a sequence $\{S_{cn},\ 1 \leq n \leq N\}$ of linear rank statistics, defined as in (6.1.5.4) with the scores defined by (6.1.4.12) and with $\bar{c} = 0$. Then under the hypothesis of randomness (i.e., the homogeneity of $F_1, \ldots, F_N$), it follows from Theorem 6.1.9.1 that this sequence is a zero-mean martingale. Note further that $\mathsf{E}_0\{S_{cn}\} = 0$ and $\mathsf{E}_0\{S_{cn}^2\} = C_n^2.A_n^2$, $n \geq 1$, where

$$C_n^2 = \sum_{i=1}^{n}(c_i - \bar{c})^2, \quad A_n^2 = \frac{1}{n-1}\sum_{i=1}^{n}(a_n^{\varphi}(i) - \bar{a}^{\varphi})^2$$

(see Subsections 6.3.5, 4.9.3, with $N = n$, which is in accordance with (3.3.1.23)). Moreover, $S_{c1} = 0$ with probability one, and it is easy to verify that C_n^2 is non-decreasing in n (≥ 1) with $\lim_{n\to\infty} C_n^2 = \infty$, and A_n^2 is also non-decreasing in n $(\geq i)$ with $\lim_{n\to\infty} A_n^2 = A_\varphi^2$. For every $N(\geq 2)$, let us then define a sequence $\{k_N(t), t \in [0,1]\}$ of non-negative numbers (bounded from above by N), by letting

$$k_N(t) = \max\{n : \ C_n^2 A_n^2 \ \leq \ tC_N^2 A_N^2\}, \ t \in [0,1], \tag{1}$$

and note that $k_N(t)$ is non-decreasing in $t \in [0,1]$. Let us then introduce a sequence of stochastic processes $\{W_N(t), \ t \in [0,1]\}$, by letting

$$W_N(t) = \{A_N C_N\}^{-1} S_{ck_N(t)}, \quad t \in [0,1]. \tag{2}$$

Then for every $N(\geq 2)$, W_N belongs to the $D[0,1]$ space equipped with the Skorokhod J_1-topology. Moreover, if we let $\mathcal{B}_{Nt}$ be the σ-field generated by the S_{cj}, $j \leq k_N(t)$, and note that $\mathcal{B}_{Nt}$ is non-decreasing in $t \in [0,1]$, then incorporating the martingale property of the $S_{cn}, n \geq 1$, we arrive at the following.

Lemma 1 *Under the hypothesis of randomness, for every* $N \geq 1$, $\{W_N(t), \mathcal{B}_{Nt}, \ t \in [0,1]\}$ *is a zero-mean martingale (process).*

Side by side, we consider a standard Brownian motion $W(t)$, $t \in [0,1]$. Then for any finite set $\{0 \leq t_1 < \ldots < t_m \leq 1\}$ of distinct points on $[0,1]$, we can use the (pointwise) projection result in Theorem 6.1.9.1, and replace the $S_{ck_N(t)}$ by $T_{ck_N(t)}$, and invoking the independent increment property of the T_n along with the multivariate central limit theorem, we conclude that as $N \to \infty$,

$$\{W_N(t_1), \ldots, W_N(t_m)\} \ \xrightarrow{\mathcal{D}} \ \{W(t_1), \ldots, W(t_m)\}, \tag{3}$$

and this establishes the convergence of the finite dimensional distributions of W_N to W. Moreover, note that the martingale property along with the convergence of finite dimensional distributions imply the compactness or tightness condition (viz., Sen (1981), Ch. 2). Therefore, we arrive at the following.

Theorem 1 *Under the regularity assumptions of Theorem* 6.1.5.1,

$$W_N \ \xrightarrow{\mathcal{D}} \ W, \ \text{as } N \ \to \infty, \tag{4}$$

in the Skorokhod J_1*-topology on the space* $D[0,1]$.

Let us now consider some implications of this basic weak invariance principle for linear rank statistics. A direct consequence of the compactness part of this weak convergence result is the following: for any fixed $t_o \in [0,1]$, as $\delta(> 0)$ converges to 0,

$$\sup\{ \ |W_N(t) - W_N(t_o)| : |t - t_o| < \delta\} \to 0 \tag{5}$$

in probability, where

$$W_N(t_o) \text{ is asymptotically normal } (0, t_o). \tag{6}$$

Hence, if we conceive of any sequence $\{M_N\}$ of integer-valued (possibly random) variables, such that

$$\lim_{N\to\infty} C_N^{-2} C_{M_N}^2 = t_o, \text{ in probability,} \tag{7}$$

where $t_o \in [0,1]$, then under the hypothesis of Theorem 1,

$$S_{cM_N}/\{C_{M_N} A_{M_N} \text{ is asymptotically normal } (0,1). \tag{8}$$

Therefore the asymptotic normality of linear rank statistics holds for random sample sizes as well. This result is useful for setting bounded-width confidence intervals based on such rank statistics (as will be considered in the last chapter). Further, Theorem 1 is itself of basic importance for constructing *sequential rank tests* based on such linear rank statistics; we refer to Sen (1981), Chapter 9 for details.

Let us consider next the case of censored linear rank statistics, treated in Section 4.9. We define a censored linear rank statistic as in (4.9.2.2), and to show its dependence on n^*, we denote it by S_{IIN,n^*}. In this way, we obtain the partial sequence $\{S_{IIN,n}, n \le N\}$, and for notational simplification we drop the subscript II, and write this as $\{S_{N,n},\ n \le N\}$. Then proceeding as in (4.9.3.7)–(4.9.3.12), we conclude that under the null hypothesis (of randomness), for every N, $\{S_{N,n},\ n \le N\}$ is a zero-mean martingale (array). Here we define a sequence $\{k_N(t), t \in [0,1]\}$ of non-negative integers by letting

$$k_N(t) = \max\{n : A_{N,n}^2 \le t A_N^2\}, \quad t \in [0,1], \tag{9}$$

where the $A_{N,n}^2$ and A_N^2 are defined by (4.9.3.10) and (4.9.3.11) respectively; it was shown there that $A_{N,n}^2$ is non-decreasing in $n \le N$. Let us then introduce a sequence of stochastic processes $\{W_N(t),\ t \in [0,1]\}$ by letting

$$W_N(t) = C_N^{-1} A_N^{-1} S_{N,k_N(t)}, \quad t \in [0,1]. \tag{10}$$

Note that by construction, $\{W_N(t), t \in [0,1]\}$ is a zero-mean martingale array (with respect to $t \in [0,1]$), under the null hypothesis. The convergence of finite dimensional distributions of W_N (to those of W, a standard Brownian motion) can be directly proved by invoking the projection result in Theorem 6.1.9.1 after noting that for any given n, $S_{N,n}$ is itself a linear rank statistic. The compactness or tightness property of W_N again follows from its martingale property. Hence, we arrive at the following.

Theorem 2 *Under the hypothesis of randomness, and granted the regularity assumptions of Theorem* 1, *as* $N \to \infty$,

$$W_N \xrightarrow{\mathcal{D}} W, \tag{11}$$

in the Skorokhod J_1*-topology on the space* $D[0,1]$. *The weak convergence result also holds with respect to a* d_q*-metric whenever* $q = \{q(t),\ 0 < t < 1\}$ *satisfies the condition* (6.4.1.7).

Several important results follow directly from Theorem 2. First, as in Subsection 4.9.2, we consider a Type II censoring scheme, where we let $n^* \sim N\alpha$, for some $\alpha \in (0,1]$. Then under the hypothesis of Theorem 2, as $N \to \infty$,

$$A_N^{-1} C_N^{-1} S_{N,n^*} \text{ is asymptotically normal } (0, \nu_\alpha), \tag{12}$$

where

$$\nu_\alpha = \lim_{N\to\infty} A_N^{-2} A_{N,n^*}^2. \tag{13}$$

Thus, a linear rank statistic in the Type II censored case, besides being EDF under the null hypothesis, is asymptotically normal.

Next, we consider as in Subsection 4.9.1, a Type I censored data model. Note that if we let

$$F(T) = P_0\{X \le T\} = \alpha, \quad (0 < \alpha \le 1), \tag{14}$$

then defining $N^* = \sum_{i=1}^{N} I(X_i \le T)$, we obtain by the Glivenko-Cantelli theorem that as $N \to \infty$,

$$N^{-1} N^* \to \alpha \tag{15}$$

in probability. As such, proceeding as in after (5), we find on invoking Theorem 2 that under the regularity assumptions of Theorem 2, as N increases,

$$A_N^{-1} C_N^{-1} S_{N,N^*} \text{ is asymptotically normal } (0,\ \nu_\alpha). \tag{16}$$

where ν_α is defined by (13). Or, equivalently, under the null hypothesis, as $N \to \infty$,

$$A_{N,N^*}^{-1} C_N^{-1} S_{N,N^*} \text{ is asymptotically normal } (0, 1). \tag{17}$$

Thus, besides being CDF under the null hypothesis, a Type I censored linear rank statistic is asymptotically normal; it shares the same asymptotic properties an EDF Type II censored linear rank statistic has.

Let us now consider progressively censored linear rank statistics presented in Subsection 4.9.3. We define the censored rank statistics $\{S_{N,n};\ n \leq N\}$ and W_N as above, and let

$$\begin{aligned} K_N^+ &= \max_{n \leq N}\{A_N^{-1}C_N^{-1}S_{N,n}\} = \sup_{t\in[0,1]}\{W_N(t)\}, \\ K_N &= \max_{n \leq N}\{A_N^{-1}C_N^{-1}|S_{N,n}|\} = \sup_{t\in[0,1]}\{|W_N(t)|\}. \end{aligned} \tag{18}$$

Side by side, we introduce the functionals

$$K^+ = \sup_{t\in[0,1]}\{W(t)\}; \quad K = \sup_{t\in[0,1]}\{|W(t)|\}, \tag{19}$$

where W is a standard Brownian motion on $[0,1]$. Then it follows from Theorem 2 that under the null hypothesis, as $N \to \infty$,

$$K_N^+ \xrightarrow{\mathcal{D}} K^+, \quad K_N \xrightarrow{\mathcal{D}} K. \tag{20}$$

On the other hand, for K^+ and K the distribution theory is well studied (see for example the preceding section), so that they apply to K_N^+ and K_N as well. This shows that the EDF progressively censored linear rank test statistics, studied in Subsection 4.9.3, all have well-known asymptotic null distributions. These results also extend to weighted versions based on weight functions that conform to the class q, considered in Theorem 6.4.3.1.

Let us next consider the Kolmogorov-type inequalities presented in Subsection 6.3.5. We define the S_{Nk} as in (6.3.5.5) and note that the martingale characterization in (6.3.5.6) holds under the null hypothesis. We then define a related sequence $\{U_{Nk}, k \leq N\}$ by letting

$$U_{Nk} = (N-k)S_{Nk}, \quad 1 \leq k \leq N;\ U_{N0} = 0. \tag{21}$$

It is easy to verify that $\{k^{-1}U_{Nk},\ 1 \leq k \leq N\}$ is a reverse martingale, under the null hypothesis (see Sen, 1970), or alternatively, we appeal to the martingale property displayed in (6.3.5.6), and note that the weight function $q = q_N = \{q_N(t) = (1 - [Nt]/N),\ t \in [0,1]\}$ (where $[.]$ stands for the integer part) satisfies the conditions in Theorem 6.4.3.1. Thus, if we define a sequence $\{W_N^o(t),\ t \in [0,1]\}$ by letting

$$W_N^o(t) = U_{N[Nt]}, \quad t \in [0,1], \tag{22}$$

and let W^o be a standard Brownian bridge on $[0,1]$, then we arrive at the following result from Theorem 6.4.3.1.

Theorem 3 *Under the hypothesis of randomness, as $N \to \infty$,*

$$W_N^o \xrightarrow{\mathcal{D}} W^o, \tag{23}$$

in the Skorokhod J_1-topology on the $D[0,1]$ space.

As an immediate corollary to Theorem 3, we claim that for every fixed α $(0 < \alpha \leq 1)$, and $\varepsilon > 0$,

$$\begin{aligned} \lim_{N\to\infty} P_0\{ \sup_{0<t\leq\alpha} |W_N^o(t)| \geq \varepsilon\} &= \\ &= P\{ \sup_{0<t\leq\alpha} |W^o(t)| \geq \varepsilon\} \\ &= P\{ \sup_{0\leq t\leq\alpha/(1-\alpha)} (1+t)^{-1}|W(t)| \geq \varepsilon\} \end{aligned} \tag{24}$$

where W and W^o stand for a standard Brownian motion and Brownian bridge, respectively. At this stage, one may use the basic results of Anderson (1960) to obtain an algebraic expression (although in an infinite series form) for the right hand side of (24). Alternatively, (Sen, 1972) we can bound the right hand side of (24) by

$$\begin{aligned} &P\{ \sup_{0<t\leq\alpha/(1-\alpha)} |W(t)| \geq \varepsilon\} \\ &\leq 2P\{|W(\frac{\alpha}{(1-\alpha)})| \geq \varepsilon\} \\ &\leq 4[1-\Phi(\varepsilon(\alpha^{-1}(1-\alpha))^{1/2})], \end{aligned} \tag{25}$$

where Φ stands for the standard normal distribution function, and whenever $\varepsilon/\sqrt{\alpha}$ is large, the right hand side can be made small. Noting that for all non-negative x, $x[1-\Phi(x)] \leq \varphi(x)$, where φ stands for the standard normal density, we obtain that the right hand side of (25) is bounded from above by

$$\frac{4\sqrt{\alpha}}{(2\pi(1-\alpha)^{1/2})}\exp\{-\epsilon^2(1-\alpha)/2\alpha\}, \tag{26}$$

and for small values of α, (26) converges to zero at a faster rate than the ones obtained in Subsection 6.3.5. Also, the weak convergence results in Subsection 6.3.7 follow directly from the weak convergence result in (23). We preferred to retain the original presentation not only to preserve the flavour of Hájek and Šidák (1967), but also to show how with the progress of time such derivations have been simplified with later developments in the martingale theory and functional limit theorems.

We briefly consider here the multivariate linear rank statistics considered in Subsection 4.10.3. We define the rank-collection matrix $\mathbf{R}_N$ as in (4.10.3.6–7), and denote the reduced one by $\mathbf{R}_N^*$. In this case, the σ-field $\mathcal{B}_N$ is generated by $\mathbf{R}_N$ or equivalently $\mathbf{R}_N^*$. Let us denote the linear rank statistics (vectors) by $\mathbf{T}_{Nj}$, $j = 1,\ldots,p$; see (4.10.3.8) in this respect. Then, letting $\mathcal{B}_{Nj}$ be the σ-field generated by the jth row of $\mathbf{R}_N$, we may easily note that under the null hypothesis,

$$E_0\{\mathbf{T}_{N+1j}|\mathcal{B}_N\} = E_0\{\mathbf{T}_{N+1j}|\mathcal{B}_{Nj}\},\ j = 1,\ldots,p, \tag{27}$$

so that plugging in the univariate results studied in Subsection 6.1.9, we conclude that coordinatewise the martingale property holds for each $\{\mathbf{T}_{Nj}\}$, $j = 1, \ldots, p$. Therefore, as in Subsection 4.10.3, if we denote complete set of linear rank statistics by $\mathbf{T}_N$, we obtain that under the null hypothesis,

$$E_0\{\mathbf{T}_{N+1}|\mathcal{B}_N\} = \mathbf{T}_N \text{ a.e., for all } N \geq 1. \tag{28}$$

This martingale property along with the coordinatewise projection result studied in Subsection 6.1.9 enable us to study weak invariance principles that in turn imply the validity of the multivariate central limit theorem. There is, however, a basic problem for a similar property to hold for the test statistics $\mathcal{L}_N$ considered in Subsection 4.10.3 (see (4.10.3.14)). This is primarily due to the permutation rank-score covariance matrix $\mathbf{V}_N$ defined by (4.10.3.12). The off-diagonal elements of $\mathbf{V}_N$ are random and are not linear rank statistics; they depend on the pairs of ranks in two different rows of $\mathbf{R}_N$. Therefore, the simplification in (26) does not hold here, and as a result, $\mathbf{V}_N$ may not have a (sub-)martingale or reversed (sub-) martingale structure. Since $\mathcal{L}_N$ is a quadratic form in $\mathbf{T}_N$, in the absence of such a nice property for the random $\mathbf{V}_N$, not much can be said about a sub-martingale property for $\mathcal{L}_N$. Hence, the approach based on martingale construction pursued in this section may not generally work out in the multivariate case. Alternative approaches based on coordinatewise martingale structures and projection results along with a.s. convergence properties of $\mathbf{V}_N$ (that works out under quite general regularity assumptions) can be used to derive similar invariance principles for the multivariate linear rank statistics. We refer to Sen (1983) and omit the details here.

The weak invariance principles developed in this subsection extend to signed-rank statistics for testing the hypothesis of symmetry, and also to linear rank statistics for testing the hypothesis of bivariate independence. In both these cases, we have an additional simplification that the time-parameter sequence $\{k_N(t),\ t \in [0,1]\}$ can be taken as $k_N(t) = [Nt]$, for $t \in [0,1]$. There also the strong invariance principles developed in the preceding subsection hold for such rank statistics under the protection of martingale structures that were established in Subsection 6.1.9. Again, we refer to Sen (1981) for a detailed account.

PROBLEMS AND COMPLEMENTS TO CHAPTER 6

Section 6.1

1. An inequality. For arbitrary scores $a(i)$ such that $a(1) \leq \ldots \leq a(N)$, the inequality

$$\mathsf{E}[a(1+[U_1 N]) - a(R_{N1})]^2 \leq \frac{2^{\frac{2}{3}}}{N} \max_{1 \leq i \leq N} |a(i) - \bar{a}| \Big[\sum_{i=1}^{N} (a(i) - \bar{a})^2 \Big]^{\frac{1}{2}} \tag{1}$$

holds. Here $U_1, \dots, U_N$ denotes a sample from the uniform distribution and R_{N1} denotes the rank of U_1. [Hájek (1961).]

2. Necessary and sufficient conditions. Suppose that

$$\frac{\sum_{i=1}^{N_\nu}(a_\nu(i)-\bar{a}_\nu)^2}{\max_{1\le i\le N_\nu}(a_\nu(i)-\bar{a}_\nu)^2} \to \infty$$

and

$$\frac{\sum_{i=1}^{N_\nu}(c_{\nu i}-\bar{c}_\nu)^2}{\max_{1\le i\le N_\nu}(c_{\nu i}-\bar{c}_\nu)^2} \to \infty,$$

for $\nu \to \infty$. Then

$$S_\nu = \sum_{i=1}^{N_\nu} c_{\nu i} a_\nu(R_{\nu i})$$

is asymptotically normal $(\mathsf{E}S_\nu, \mathsf{var}\, S_\nu)$, if and only if for any $\tau > 0$

$$\lim_{\nu\to\infty} \frac{1}{N_\nu} \sum_{|\delta_{\nu ij}|>\tau} \delta_{\nu ij}^2 = 0, \quad \tau > 0, \tag{2}$$

with

$$\delta_{\nu ij} = (c_{\nu i}-\bar{c}_\nu)(a_\nu(j)-\bar{a}_\nu)\Big[\Big(\frac{1}{N_\nu}\Big)\sum_{i=1}^{N_\nu}(c_{\nu i}-\bar{c}_\nu)^2\sum_{j=1}^{N_\nu}(a_\nu(j)-\bar{a}_\nu)^2\Big]^{-\frac{1}{2}}.$$

[Hájek (1961).]

3. Sufficient conditions for general linear statistics. The statistics

$$S_\nu = \sum_{i=1}^{N_\nu} a_\nu(i, R_{\nu i})$$

are asymptotically normal $(\mathsf{E}S_\nu, \mathsf{var}\, S_\nu)$, if (2) holds with

$$\begin{aligned}\delta_{\nu ij} &= [a_\nu(i,j) - a_\nu(.\,,j) - a_\nu(i,.\,) + \bar{a}] \\ &\quad \cdot\Big\{(1/N_\nu)\sum_{i=1}^{N_\nu}\sum_{j=1}^{N_\nu}[a_\nu(i,j) - a_\nu(.\,,j) - a_\nu(i,.\,) + \bar{a}_\nu]^2\Big\}^{-\frac{1}{2}}.\end{aligned}$$

Here $a_\nu(.\,,j)$ and $a_\nu(i,.\,)$ are given by (3.3.1.2) and (3.3.1.3), respectively. [Motoo (1957).]

4. Conditional asymptotic normality. We shall say that the conditional

distribution of T_ν given Z_ν is asymptotically normal $[m_\nu(Z_\nu), s_\nu^2(Z_\nu)]$ if for every $\varepsilon > 0$

$$\lim_{\nu\to\infty} P\Big\{\sup_x \Big|P(T_\nu < x \mid Z_\nu) - \Phi\Big(\frac{x - m_\nu(Z_\nu)}{s_\nu(Z_\nu)}\Big)\Big| > \varepsilon\Big\} = 0. \qquad (3)$$

Prove the following generalization of Theorem 6.1.2.1.

Consider the statistics $T_\nu = \sum_{i=1}^{N_\nu} T_{\nu i}$, $\nu \geq 1$, where, for each ν, the conditional distributions of $T_{\nu i}$ given Z_ν are independent, and let

$$m_{\nu i} = \mathsf{E}(T_{\nu i} \mid Z_\nu), \quad m_\nu = \mathsf{E}(T_\nu \mid Z_\nu), \quad s_{\nu i}^2 = \mathsf{var}(T_{\nu i} \mid Z_\nu)$$

and $s_\nu^2 = \mathsf{var}(T_\nu \mid Z_\nu)$. Assume that for every $\varepsilon > 0$

$$\lim_{\nu\to\infty} P\Big(s_\nu^{-2} \sum_{i=1}^{N_\nu} \int_{|x - m_{\nu i}| > \varepsilon s_\nu} (x - m_{\nu i})^2 \, \mathrm{d}P(T_{\nu i} < x \mid Z_\nu) > \varepsilon\Big) = 0. \quad (4)$$

Then the conditional distribution of T_ν given Z_ν is asymptotically normal $[m_\nu(Z_\nu), s_\nu^2(Z_\nu)]$.

5. A theorem. Let the conditional distribution of T_ν given Z_ν be asymptotically normal (m_ν, s_ν^2).

Then $[T_\nu - m_\nu(Z_\nu)][s_\nu(Z_\nu)]^{-1}$ is asymptotically normal $(0, 1)$.

6. A theorem. Let the conditional distribution of T_ν given Z_ν be asymptotically normal (m_ν, s_ν^2). Assume that $m_\nu = m_\nu(Z_\nu)$ is asymptotically normal (μ, b^2) and that $s_\nu^2 = s_\nu^2(Z_\nu)$ converges in probability to σ^2.

Then T_ν is asymptotically normal $(\mu, \sigma^2 + b^2)$. [Hint: utilize characteristic functions.]

We shall drop the subscript ν in what follows.

7. If S_c is asymptotically normal (μ_c, σ_c^2), then

$$\liminf_c (\mathsf{var}\, S_c/\sigma_c^2) \geq 1$$

and

$$\liminf_c \big[\big(\mathsf{var}\, S_c + (\mathsf{E} S_c)^2\big)/(\sigma_c^2 + \mu_c^2)\big] \geq 1.$$

[Hint: If $F_\nu \to F$ and continuity points, at all $h(x) \geq 0$, then $\liminf \int h(x)\, \mathrm{d}F_\nu \geq \int h(x)\, \mathrm{d}F$. As for the variances, use the relation $\mathsf{var}\, X = \iint (x - y)^2 \, \mathrm{d}F(x)\, \mathrm{d}F(y)$, where $F(x) = P(X \leq x)$.]

8. Complete solution for the asymptotic normality of the statistics

$$S_{mn} = \sum_{i=1}^{m} a_N(R_{Ni}).$$

If

$$\frac{\sum_{i=1}^{N}[a_N(i)-\bar{a}_N]^2}{\max_{1\le i\le N}[a_N(i)-\bar{a}_N]^2}\to\infty,$$

we shall say that the a's satisfy the condition (N). Introduce the condition

$$0<\liminf\frac{m}{N}\le\limsup\frac{m}{N}<1. \tag{5}$$

Put $N=m+n$.

(ia) If (5) holds and the a's satisfy (N), then S_{mn} is asymptotically normal $(\mathsf{E}S_{mn}, \mathsf{var}\, S_{mn})$.

(ib) If (5) holds and the a's do not satisfy (N), then S_{mn} is not asymptotically normal (μ_{mn},σ^2_{mn}) for any choice of μ_{mn} and σ^2_{mn}.

(iia) If $m\to\infty$, $m/N\to 0$, and the functions

$$\gamma_N(u)=\frac{[a_N(1+[uN])-\bar{a}_N]^2}{\int_0^1[a_N(1+[uN])-\bar{a}_N]^2\,\mathrm{d}u}$$

are uniformly integrable, then S_{mn} is asymptotically normal $(\mathsf{E}S_{mn}, \mathsf{var}\, S_{mn})$.

(iib) If the functions $\gamma_N(u)$, introduced above, are not uniformly integrable, then there exists a sequence $\{(m,n)\}$ such that $m\to\infty$, $m/N\to 0$ and that S_{mn} is not asymptotically normal $(\mathsf{E}S_{mn}, \mathsf{var}\, S_{mn})$. However, for that sequence, S_{mn} may be asymptotically normal (μ_{mn},σ^2_{mn}) with $\liminf \mathsf{var}(S_{mn})/\sigma^2_{mn}>1$.

(iic) If the a's do not satisfy (N), then S_{mn} is not asymptotically normal $(\mathsf{E}S_{mn}, \mathsf{var}\, S_{mn})$ for any $m\to\infty$, $m/N\to 0$.

(iiia) If $m<K<\infty$ and the empirical distribution of the a's converges to a normal distribution, then S_{mn} is asymptotically normal $(\mathsf{E}S_{mn}, \mathsf{var}\, S_{mn})$.

(iiib) If $m<K<\infty$ and the empirical distribution of the a's does not converge to a normal distribution, then S_{mn} is not asymptotically normal (μ_{mn},σ^2_{mn}) for any choice of μ_{mn} and σ^2_{mn}.

[Hájek (1960a) and (1961) except for (ib), (iic), (iiia) and (iiib), where additional considerations are necessary.]

9. Necessity of $I(f)<\infty$. Consider the densities

$$f_\beta(x)=\begin{cases}\dfrac{\beta+1}{2}(1-|x|)^\beta, & |x|\le 1\\ 0, & \text{otherwise,}\end{cases}$$

with $0 < \beta < \infty$. Their respective φ-functions are

$$\varphi_\beta(u) = \beta[\text{sign}(2u-1)](1-|2u-1|)^{-1/(\beta+1)}.$$

Note that $\int_{-\infty}^{\infty} |f'_\beta(x)|\,\mathrm{d}x < \infty$ for every $\beta > 0$ and $I(f) < \infty$ for $\beta > 1$. Consider either of the statistics

$$S_{mn}^{\beta} = \sum_{i=1}^{m} \varphi_\beta(R_{Ni}/(N+1)) \quad \text{and} \quad S_{mn}^{\beta} = \sum_{i=1}^{m} a_N(R_{Ni}, f_\beta),$$

and prove the following propositions:

(i) If $\beta > 1$, then S_{mn} is asymptotically normal $(\mathsf{E}S_{mn}, \mathsf{var}\, S_{mn})$ under $\min(m,n) \to \infty$.

(ii) If $\beta = 1$, then S_{mn} is asymptotically normal $(\mathsf{E}S_{mn}, \mathsf{var}\, S_{mn})$ under $m \to \infty$ and (5), but may not be asymptotically normal $(\mathsf{E}S_{mn}, \mathsf{var}\, S_{mn})$ under $m \to \infty$ and $m/N \to 0$. (It is certainly not normal $(\mathsf{E}S_{mn}, \mathsf{var}\, S_{mn})$ for $m < (N/\log N)^{\frac{1}{2}}$.)

(iii) If $0 < \beta < 1$, then S_{mn} is not asymptotically normal $(\mathsf{E}S_{mn}, \mathsf{var}\, S_{mn})$ for any sequence of pairs $\{(m,n)\}$.

10. Asymptotic normality with 'unnatural' parameters. Put $a_N(i) = i$, $1 \le i \le N-1$, $a_N(N) = \frac{1}{2}N^2$, and assume that $m \to \infty$, $m/N \to 0$. Then $S_{mn} = \sum_{i=1}^{m} a_N(R_{Ni})$ is asymptotically normal $(\frac{1}{2}mN, \frac{1}{12}mnN)$, while $\mathsf{var}\, S_{mn}/(\frac{1}{12}mnN) \to \infty$ and $(\mathsf{E}S_{mn} - \frac{1}{2}mN)^2/(\frac{1}{12}mnN) \to \infty$.

11. Consider the following families of scores:

$$\begin{aligned} a_N(i) &= i^k, && 1 \le i \le N,\ k > 0, && (6)\\ a_N(i) &= \log i, && 1 \le i \le N, \\ a_N(i) &= F^{-1}\Big(\frac{i}{N+1}\Big), && 1 \le i \le N, \end{aligned}$$

with F^{-1} being the inverse of a distribution function F such that $\int x^2\,\mathrm{d}F < \infty$.

$$\begin{aligned} a_N(i) &= \frac{1}{\sqrt{i}}, && 1 \le i \le N, && (7)\\ a_N(i) &= F^{-1}\Big(\frac{i}{N+1}\Big), && 1 \le i \le N, \end{aligned}$$

with F being a distribution function attracted by the normal law [i.e. $\int_{|x|>c} \mathrm{d}F = o(1/c^2 \int_{|x|<c} x^2\,\mathrm{d}F)$ for $c \to \infty$] but such that $\int x^2\,\mathrm{d}F(x) = \infty$.

$$\begin{aligned} a_N(i) &= q^i, && 1 \le i \le N,\ q > 1 \text{ or } 0 < q < 1, && (8)\\ a_N(i) &= i^\alpha, && 1 \le i \le N,\ \alpha < -\tfrac{1}{2}, \\ a_N(i) &= F^{-1}\Big(\frac{i}{N+1}\Big), && 1 \le i \le N, \end{aligned}$$

with F not being attracted by the normal law.

Show that the γ-functions introduced in Problem 8 are uniformly integrable under (6) but not under (7) and (8). Further, show that the condition (N) is satisfied under (7) but not under (8).

12. The phrase 'φ is a finite sum of monotone functions' is equivalent to 'φ is monotone or a difference of two non-decreasing functions'. On inspecting Lemma 6.1.6.1, we see that this condition could be replaced by the following more general one: 'φ is continuous almost everywhere, and there exist points $0 = a_0 < a_1 < \ldots < a_k = 1$ such that φ is either bounded or a difference of two non-decreasing functions on each interval $(a_{i=1}, a_i)$, $1 \leq i \leq k$'.

13. Ties. Let $(X_1, \ldots, X_N)$ be a sample from a discontinuous distribution function F, having jumps at points $\{\alpha_h|\}$. Let $A_1, \ldots, A_g$ be groups of indices i such that $X^{(i)} = X^{(j)}$ if and only if i and j both belong to the same group A_k, and let τ_k denote the size of A_k, $1 \leq k \leq g$. Thus $A_1, \ldots, A_g$ is a random partition of $\{1, \ldots, N\}$ corresponding to groups of equal observations. Let $a_N(i)$ be scores satisfying (6.1.6.2) for some square integrable φ. Put $\tau = (\tau_1, \ldots, \tau_g)$ and

$$a_N(i, \tau) = \frac{1}{\tau_k} \sum_{j \in A_k} a_N(j), \quad \text{if } i \in A_k,$$

so that $a_N(i, \tau)$ represent average scores within the groups A_k, $k = k(i)$. Further put $I_h = (F(\alpha_h - 0), F(\alpha_h))$ and

$$\varphi^F(u) = \begin{cases} \varphi(u), & \text{if } u \notin \bigcup_1^\infty I_h, \\ \int_{I_h} \varphi(v)\,\mathrm{d}v \Big/ \int_{I_h} \mathrm{d}v, & \text{if } u \in I_h,\ h \geq 1. \end{cases}$$

Theorem. Assume that $\int_0^1 [\varphi^F(u) - \bar{\varphi}]^2\,\mathrm{d}u > 0$ and that

$$\frac{\sum_{i=1}^N (c_i - \bar{c})^2}{\max_{1 \leq i \leq N} (c_i - \bar{c})^2} \to \infty.$$

Then conditional distribution of the statistic $\bar{S}_c = \sum_{i=1}^N c_i a_N(R_i, \tau)$ given τ is asymptotically normal $[\mathsf{E}(\bar{S}_c \mid \tau), \mathsf{var}(\bar{S}_c \mid \tau)]$. Consequently, the statistic $[\bar{S}_c - \mathsf{E}(\bar{S}_c \mid \tau)][\mathsf{var}(\bar{S}_c \mid \tau)]^{-\frac{1}{2}}$ is asymptotically normal $(0, 1)$.

Section 6.2

14. Ties. Under the conditions of Problem 13, the unconditional as well as the conditional distribution of the statistic

$$\bar{Q}_{n_1\dots n_k} = (N-1)\Big\{\sum_{i=1}^{N}[a_N(i,\tau)-\bar{a}_N]^2\Big\}^{-1}\sum_{j=1}^{k}\frac{(\bar{S}_{Nj}-n_j\bar{a}_N)^2}{n_j},$$

with $\bar{S}_{Nj}=\sum_{i\in s_j}a_N(R_{Ni},\tau)$, is asymptotically χ^2 with $k-1$ degrees of freedom. Apply this to the statistics considered in Subsection 4.4.1.

Section 6.3

15. Let α, β, a, b, be positive numbers, and let $Z(t)$ denote the Brownian bridge. Put

$$\begin{aligned}
A_m &= m^2ab+(m-1)^2\alpha\beta+m(m-1)a\beta+m(m-1)\alpha b,\\
B_m &= (m-1)^2ab+m^2\alpha\beta+m(m-1)a\beta+m(m-1)\alpha b,\\
C_m &= m^2ab+m^2\alpha\beta+m(m-1)a\beta+m(m+1)\alpha b,\\
D_m &= m^2ab+m^2\alpha\beta+m(m+1)a\beta+m(m-1)\alpha b.
\end{aligned}$$

Then

$$\begin{aligned}
&P(-a(1-t)-bt<Z(t)<\alpha(1-t)+\beta t,\ 0\le t\le 1) \\
&\quad = 1-\sum_{m=1}^{\infty}(\mathrm{e}^{-2A_m}+\mathrm{e}^{-2B_m}-\mathrm{e}^{-2C_m}-\mathrm{e}^{-2D_m}).
\end{aligned}\tag{9}$$

16. (Continuation.) Let $|x|\le\alpha(1-u)+\beta u$ and $|y|\le\alpha(1-v)+\beta v$. Put

$$\begin{aligned}
A_m &= [(2m-1)\big(\alpha(1-u)+\beta u\big)-x][(2m-1)\big(\alpha(1-v)+\beta v\big)-y],\\
B_m &= [(2m-1)\big(\alpha(1-u)+\beta u\big)+x][(2m-1)\big(\alpha(1-v)+\beta v\big)+y],\\
C_m &= [2m\big(\alpha(1-u)+\beta u\big)-x][2m\big(\alpha(1-v)+\beta v\big)+y]+xy,\\
D_m &= [2m\big(\alpha(1-u)+\beta u\big)+x][2m\big(\alpha(1-v)+\beta v\big)-y]+xy.
\end{aligned}$$

Then

$$\begin{aligned}
&P(-\alpha(1-t)-\beta t<Z(t)<\alpha(1-t)+\beta t,\\
&\qquad\qquad u\le t\le v\mid Z(u)=x,\ Z(v)=y)\\
&\quad = 1-\sum_{m=1}^{\infty}(\mathrm{e}^{-2A_m/(v-u)}+\mathrm{e}^{-2B_m/(v-u)}\\
&\qquad\qquad -\mathrm{e}^{-2C_m/(v-u)}-\mathrm{e}^{-2D_m/(v-u)}).
\end{aligned}\tag{10}$$

17. An expansion used by Rényi. The right side of (6.3.7.4) may be expressed equivalently by

$$\frac{4}{\pi}\sum_{k=0}^{\infty}(-1)^k(2k+1)^{-1}\exp[-(2k+1)^2\pi^2(1-a)/(8\lambda^2 a)].$$

[Rényi (1953a).]

18. Derive from Theorem 6.3.6.1 the following result due to Kolmogorov (1933): If $F_n(x)$ denotes the empirical distribution function corresponding to a random sample of size n from a continuous distribution function $F(x)$, then

$$\lim_{n\to\infty} P\big(n^{\frac{1}{2}} \max_x \big(F_n(x) - F(x)\big) < \lambda\big) = 1 - \mathrm{e}^{-2\lambda^2}$$

and

$$\lim_{n\to\infty} P\big(n^{\frac{1}{2}} \max_x |F_n(x) - F(x)| < \lambda\big) = 1 - 2\sum_{k=1}^{\infty}(-1)^{k+1}\mathrm{e}^{-2k^2\lambda^2}.$$

[Hint: since m and n may tend to ∞ arbitrarily, we may assume that $m \to \infty$ first and $n \to \infty$ afterwards. Then we utilize the fact that $\max_x |F_m(x) - F(x)|$ converges to 0 in probability when $m \to \infty$.]

19. Let $Z(t)$ be the Brownian bridge and $g(t)$, $0 < t < 1$, be integrable. Then $\int_0^1 g(t)Z(t)\,\mathrm{d}t$ is normal $(0, \sigma^2)$ with

$$\sigma^2 = \int_0^1 [G(t)]^2\,\mathrm{d}t - \Big[\int_0^1 G(t)\,\mathrm{d}t\Big]^2,$$

where $G(t) = \int_0^t g(s)\,\mathrm{d}s$.

20. (a) Observe two samples of equal sizes, $X_1, \dots, X_n$ and $X_{n+1}, \dots, X_{2n}$. Let K^+ be the one-sided Kolmogorov-Smirnov statistic (4.1.3.2), and let I^+ be the first index among the indices k for which $\frac{1}{2}k - c_{D_1} - \dots - c_{D_k}$ attains its maximum. Under H_0 we have

$$\lim_{n\to\infty} P\Big\{K^+ < y, \frac{I^+}{2n} < z\Big\}$$
$$= \Big(\frac{2}{\pi}\Big)^{\frac{1}{2}} \int_0^z \frac{\mathrm{d}v}{[v(1-v)]^{\frac{3}{2}}} \int_0^y u^2 \exp\Big[-\frac{1}{2}\frac{u^2}{v(1-v)}\Big]\,\mathrm{d}u$$

for $0 < y < \infty$, $0 < z < 1$. Further, $\lim_{n\to\infty} P\{(2n)^{-1}I^+ < z\} = z$.

(b) Let $K^{\pm}$ be the two-sided Kolmogorov-Smirnov statistic (4.1.3.3), and let $I^{\pm}$ be the first of the indices k for which $|\frac{1}{2}k - c_{D_1} - \ldots - c_{D_k}|$ attains its maximum. Under H_0 we have

$$\lim_{n\to\infty} P\Big\{K^{\pm} < y, \frac{I^{\pm}}{2n} < z\Big\} = \Big(\frac{8}{\pi}\Big)^{\frac{1}{2}} \int_0^y \mathrm{d}u \int_0^z f(u,v) f(u, 1-v)\, \mathrm{d}v$$

for $0 < y < \infty$, $0 < z < 1$, where

$$f(y,z) = \frac{y}{z^{\frac{3}{2}}} \sum_{j=0}^{\infty} (-1)^j (2j+1) \exp\Big[-\frac{1}{2}(2j+1)^2 \frac{y^2}{z}\Big].$$

[Vincze (1957), where the exact formulas for finite samples may be found, too.]

Chapter 7

Limiting non-null distributions

7.1 CONTIGUITY

The concept of contiguity of probability measures plays a basic role in the development of general asymptotic distribution of rank (as well as other) test statistics, under a suitable sequence of local alternatives. This approach was pioneered in the original text, and hence, we retain the same as much as possible. However, we must update this to a certain extent, and also connect it to some alternative approaches. The concept of contiguity plays a fundamental role in the asymptotic theory of statistical inference. An intimately connected concept is the *locally asymptotically quadratic* (LAQ) family, which we shall also present briefly (for the sake of completeness), and in that context, the classical *Hájek-LeCam-Inagaki* convolution theorem will also be outlined.

7.1.1 Asymptotic methods. Contiguity. The asymptotic approach consists in regarding a given testing problem as a member of a sequence $\{H_\nu, K_\nu\}$, $\nu \geq 1$, of similar testing problems. In this sequence the ν-th testing problem concerns N_ν observations $X_1, \ldots, X_{N_\nu}$ with $N_\nu \to \infty$ as $\nu \to \infty$. As a rule, H_ν depends on ν through N_ν only, i.e. $H_\nu = H(N_\nu)$, whereas K_ν depends on some parameters $d_{\nu i}$, $0 \leq i \leq N_\nu$, in addition. For example, we might assume that $H_\nu = H_0$, H_0 being applied to $N = N_\nu$ observations, and that K_ν consists of a single density q_ν,

$$q_\nu = \prod_{i=1}^{N_\nu} f_0(x_i - d_{\nu i}).$$

Of course, there are infinitely many such sequences, and we try to choose one which resembles the given testing problem as much as possible. First of all it would be desirable to keep the envelope power function $\beta(\alpha, H_\nu, K_\nu)$ independent of ν. Since this is usually difficult or even impossible, we shall be satisfied with the existence of a limit $\beta(\alpha)$:

$$\lim_{\nu\to\infty} \beta(\alpha, H_\nu, K_\nu) = \beta(\alpha), \quad 0 \le \alpha \le 1. \tag{1}$$

As in the previous chapter, we shall also consider indexed sets of testing problems $\{H_d, K_d, d \in D\}$, where the convergence will be equivalent to the convergence to a fixed limit for all sequences selected from the set and satisfying certain requirements. As a rule, K_d will be simple, consisting of a density q_d.

The limiting relation (1) entails that $\beta(\alpha, H_\nu, K_\nu)$ will approximately equal $\beta(\alpha)$ for $\nu \ge \nu_0$. The usefulness of the asymptotic results will depend on whether the problems 'H_ν against K_ν' with $\nu \ge \nu_0$, may occur in practice or not. The value of ν_0 is usually guessed on the basis of numerical calculations for selected ν's and the assumption that the convergence is more or less monotone.

In this book we shall not investigate the somewhat degenerate cases in which

$$\beta(\alpha) = 1, \quad \text{for all } \alpha > 0. \tag{2}$$

We shall even exclude the cases in which

$$\beta(\alpha) \not\to 0 \quad \text{for } \alpha \to 0. \tag{3}$$

However, it may be shown that in problems dealt with in the sequel, (3) implies (2).

The requirement that (3) should not take place finds its theoretical expression in the notion of contiguity, which is due to LeCam (1960). The notion of contiguity is basic for the asymptotic methods of the theory of hypothesis testing.

Consider a sequence $\{p_\nu, q_\nu\}$ of simple hypotheses p_ν and simple alternatives q_ν defined on measure spaces $(\mathcal{X}_\nu, \mathcal{A}_\nu, \mu_\nu)$, $\nu \ge 1$, respectively.

Definition. If for any sequence of events $\{A_\nu\}$, $A_\nu \in \mathcal{A}_\nu$,

$$[P_\nu(A_\nu) \to 0] \Rightarrow [Q_\nu(A_\nu) \to 0] \tag{4}$$

holds, we say that the densities q_ν are *contiguous* to the densities p_ν, where $\mathrm{d}P_\nu = p_\nu \,\mathrm{d}\mu_\nu$, $\mathrm{d}Q_\nu = q_\nu \,\mathrm{d}\mu_\nu$, $\nu \ge 1$.

If H_ν is composite, we say that q_ν is contiguous to H_ν if for each ν the convex hull $\bar{H}_\nu$ of H_ν contains a density p_ν such that (4) holds.

If both H_ν and K_ν are composite, we say that K_ν is contiguous to H_ν if (4) holds for some $p_\nu \in \bar{H}_\nu$ and $q_\nu \in \bar{K}_\nu$.

Contiguity implies that any sequence of random variables converging to zero in P_ν-probability converges to zero in Q_ν-probability, $\nu \to \infty$.

7.1.2 LeCam's first lemma. According to the Neyman-Pearson lemma, for any event A_ν there exists a critical function Ψ_ν such that

$$\Psi_\nu = \begin{cases} 0, & \text{if } q_\nu < k_\nu p_\nu, \\ \xi, & \text{if } q_\nu = k_\nu p_\nu, \\ 1, & \text{if } q_\nu > k_\nu p_\nu, \end{cases} \tag{1}$$

where $0 \le \xi \le 1$, and that

$$\begin{aligned} P_\nu(A_\nu) &= \int \Psi_\nu \,\mathrm{d}P_\nu, \\ Q_\nu(A_\nu) &\le \int \Psi_\nu \,\mathrm{d}Q_\nu. \end{aligned}$$

Thus contiguity will follow if we show that

$$\left[\int \Psi_\nu \,\mathrm{d}P_\nu \to 0\right] \Rightarrow \left[\int \Psi_\nu \,\mathrm{d}Q_\nu \to 0\right] \tag{2}$$

for critical functions of the type (1).

Introduce the likehood ratio $L_\nu = q_\nu / p_\nu$, or more precisely,

$$L_\nu(x_\nu) = \begin{cases} q_\nu(x_\nu)/p_\nu(x_\nu), & \text{if } p_\nu(x_\nu) > 0, \\ 1, & \text{if } p_\nu(x_\nu) = q_\nu(x_\nu) = 0, \\ \infty, & \text{if } p_\nu(x_\nu) = 0 < q_\nu(x_\nu), \end{cases} \tag{3}$$

where x_ν denotes the typical point of the space $\mathcal{X}_\nu$, $\nu \ge 1$.

Let F_ν be the distribution function of L_ν under P_ν:

$$F_\nu(x) = P_\nu(L_\nu \le x), \tag{4}$$

where $L_\nu = L_\nu(X_\nu)$, $\nu \ge 1$.

Lemma 1 *Assume that F_ν given by (4) converges weakly (at continuity points) to a distribution function F such that*

$$\int_0^\infty x \,\mathrm{d}F(x) = 1. \tag{5}$$

Then the densities q_ν are contiguous to the densities p_ν, $\nu \ge 1$.

Proof. Take a sequence of critical functions Ψ_ν of the type (1) and such that

$$\int \Psi_\nu \,\mathrm{d}P_\nu \to 0. \tag{6}$$

Then note that

$$\begin{aligned}
\int \Psi_\nu \,\mathrm{d}Q_\nu &= \int_{\{L_\nu \le y\}} \Psi_\nu \,\mathrm{d}Q_\nu + \int_{\{L_\nu > y\}} \Psi_\nu \,\mathrm{d}Q_\nu \\
&\le y \int \Psi_\nu \,\mathrm{d}P_\nu + \int_{\{L_\nu > y\}} \mathrm{d}Q_\nu \\
&= y \int \Psi_\nu \,\mathrm{d}P_\nu + 1 - \int_{\{L_\nu \le y\}} \mathrm{d}Q_\nu \\
&= y \int \Psi_\nu \,\mathrm{d}P_\nu + 1 - \int_{\{L_\nu \le y\}} L_\nu \,\mathrm{d}P_\nu \\
&= y \int \Psi_\nu \,\mathrm{d}P_\nu + 1 - \int_0^y x \,\mathrm{d}F_\nu.
\end{aligned} \tag{7}$$

Now for any $\varepsilon > 0$ we can find a continuity point y of F such that, in view of (5),

$$1 - \int_0^y x \,\mathrm{d}F < \tfrac{1}{2}\varepsilon.$$

Since $F_\nu \to F$ entails

$$\int_0^y x \,\mathrm{d}F_\nu \to \int_0^y x \,\mathrm{d}F$$

we shall have for some ν_0

$$1 - \int_0^y x \,\mathrm{d}F_\nu < \frac{1}{2}\varepsilon, \quad \nu \ge \nu_0. \tag{8}$$

Furthermore, (6) ensures the existence of ν_1 such that

$$y \int \Psi_\nu \,\mathrm{d}P_\nu < \tfrac{1}{2}\varepsilon, \quad \nu \ge \nu_1. \tag{9}$$

Finally, from (7) through (9) it follows that

$$\int \Psi_\nu \,\mathrm{d}Q_\nu < \varepsilon \quad \text{for } \nu \ge \max(\nu_0, \nu_1).$$

Thus $\int \Psi_\nu \,\mathrm{d}Q_\nu \to 0$, which concludes the proof. □

Remark. Note that contiguity does not entail that the Q_ν are absolutely continuous with respect to the P_ν. The singular part of Q_ν, however, must tend to zero,

$$Q_\nu(p_\nu = 0) \to 0$$

as a consequence of $P_\nu(p_\nu = 0) = 0 \to 0$.

The asymptotic distribution of the likelihoods L_ν will regularly be log-normal. We shall say that a random variable Y is log-normal (μ, σ^2), if

$\log Y$ is normal (μ, σ^2). Now let us establish the condition under which we obtain

$$\mathsf{E}Y = \int_0^\infty x \, \mathrm{d}F = 1$$

for a log-normal random variable Y. We obviously have

$$\begin{aligned} \mathsf{E}Y &= \mathsf{E}\exp(\log Y) \\ &= (2\pi)^{-\frac{1}{2}}\sigma^{-1} \int_{-\infty}^{\infty} \exp[x - \tfrac{1}{2}(x-\mu)^2\sigma^{-2}] \, \mathrm{d}x = \mathrm{e}^{\mu + \frac{1}{2}\sigma^2}. \end{aligned}$$

This equals 1 for

$$\mu = -\tfrac{1}{2}\sigma^2. \tag{10}$$

Thus we have the following

Corollary 1 *If, under P_ν, the ratio L_ν is asymptotically log-normal $(-\frac{1}{2}\sigma^2, \sigma^2)$, then the densities q_ν are contiguous to the densities p_ν.*

7.1.3 LeCam's second lemma. Let $x_\nu = (x_1, \ldots, x_{N_\nu})$ and

$$p_\nu(x_\nu) = \prod_{i=1}^{N_\nu} f_{\nu i}(x_i) \tag{1}$$

and

$$q_\nu(x_\nu) = \prod_{i=1}^{N_\nu} g_{\nu i}(x_i). \tag{2}$$

From (1) and (2) we have

$$\log L_\nu = \sum_{i=1}^{N_\nu} \log[g_{\nu i}(x_i)/f_{\nu i}(x_i)] \tag{3}$$

which makes sense even if $\log L_\nu = \pm\infty$, since on the right side the summands are $< \infty$ with P_ν-probability 1 and are $> -\infty$ with Q_ν-probability 1.

Thus we may regard $\log L_\nu$ as an extended random variable allowed to attain $-\infty$ with positive probability under P_ν. However, asymptotic normality of $\log L_\nu$ is defined in the same way as for an ordinary random variable, i.e. as convergence of $P_\nu(\log L_\nu < x)$ to a normal distribution function in every real point x. Thus asymptotic normality entails $P_\nu(\log L_\nu = -\infty) \to 0$.

In what follows we restrict ourselves to cases in which the summands in (3) are uniformly asymptotically negligible, i.e.

$$\lim_{\nu\to\infty} \max_{1\le i\le N_\nu} P_\nu\Big(\Big|\frac{g_{\nu i}(X_i)}{f_{\nu i}(X_i)} - 1\Big|\Big) = 0. \tag{4}$$

Under this condition necessary and sufficient conditions of asymptotic normality are well known. These conditions are considerably simpler if the summands have finite variance. However, this sometimes fails to be satisfied in (3) within the class of problems considered below. For this reason we instead consider the statistic

$$W_\nu = 2\sum_{i=1}^{N_\nu}\{[g_{\nu i}(X_i)/f_{\nu i}(X_i)]^{\frac{1}{2}} - 1\} \tag{5}$$

which always consists of summands with finite variances, as may be easily seen, and has additional advantages. The following lemma, due to LeCam, shows that asymptotic normality of $\log L_\nu$ may be established by proving asymptotic normality of W_ν.

Lemma 1 *Assume that* (4) *holds and that the statistics* W_ν, $\nu \geq 1$, *are asymptotically normal* $(-\frac{1}{4}\sigma^2, \sigma^2)$ *under* P_ν.

Then the statistics $\log L_\nu$ *satisfy*

$$\lim_{\nu\to\infty} P_\nu(|\log L_\nu - W_\nu + \tfrac{1}{4}\sigma^2| > \varepsilon) = 0, \quad \varepsilon > 0, \tag{6}$$

and are asymptotically normal $(-\frac{1}{2}\sigma^2, \sigma^2)$ *under* P_ν.

Proof. If a function $h(x)$ has a second derivative $h''(x)$, then

$$\begin{aligned} h(x) &= h(x_0) + (x - x_0)h'(x_0) \\ &\quad + \tfrac{1}{2}(x - x_0)^2 \int_0^1 2(1-\lambda)h''[x_0 + \lambda(x - x_0)]\,\mathrm{d}\lambda, \end{aligned} \tag{7}$$

as may be easily seen by integration by parts. Thus, putting

$$T_{\nu i} = 2[g_{\nu i}(X_i)/f_{\nu i}(X_i)]^{\frac{1}{2}} - 2, \tag{8}$$

we obtain

$$\begin{aligned} \log(g_{\nu i}/f_{\nu i}) &= 2\log(1 + \tfrac{1}{2}T_{\nu i}) \\ &= T_{\nu i} - \tfrac{1}{4}T_{\nu i}^2 \int_0^1 [2(1-\lambda)/(1 + \tfrac{1}{2}\lambda T_{\nu i})^2]\,\mathrm{d}\lambda. \end{aligned}$$

Consequently,

$$\log L_\nu = W_\nu - \tfrac{1}{4}\sum_{i=1}^{N_\nu} T_{\nu i}^2 \int_0^1 [2(1-\lambda)/(1 + \tfrac{1}{2}\lambda T_{\nu i})^2]\,\mathrm{d}\lambda. \tag{9}$$

This holds even for $\log L_\nu = -\infty$.

Introduce

$$T_{\nu i}^{\delta} = \begin{cases} T_{\nu i}, & \text{if } |T_{\nu i}| \le \delta, \\ 0, & \text{otherwise.} \end{cases}$$

As is well known (Loève (1955), p. 316), the asymptotic normality $(-\frac{1}{4}\sigma^2, \sigma^2)$ of W_ν implies under (4) that for every $\delta > 0$

$$\sum_{i=1}^{N_\nu} P_\nu(|T_{\nu i}| > \delta) \quad \to \quad 0, \tag{10}$$

$$\sum_{i=1}^{N_\nu} \mathsf{E} T_{\nu i}^{\delta} \quad \to \quad -\tfrac{1}{4}\sigma^2, \tag{11}$$

$$\sum_{i=1}^{N_\nu} \mathsf{var}\, T_{\nu i}^{\delta} \quad \to \quad \sigma^2. \tag{12}$$

Now (10), holding for *every* $\delta > 0$, entails

$$\sum_{i=1}^{N_\nu} T_{\nu i}^2 \int_0^1 [2(1-\lambda)/(1+\tfrac{1}{2}\lambda T_{\nu i})^2]\, \mathrm{d}\lambda \sim \sum_{i=1}^{N_\nu} (T_{\nu i}^{\delta})^2 \tag{13}$$

where $\sim$ denotes that the ratio of both sides tends to 1 in P_ν-probability. Thus, in order to prove (6), it remains to show that

$$\sum_{i=1}^{N_\nu} (T_{\nu i}^{\delta})^2 \to \sigma^2 \tag{14}$$

in P_ν-probability. For this purpose it suffices to prove

$$\sum_{i=1}^{N_\nu} \mathsf{E}(T_{\nu i}^{\delta})^2 \to \sigma^2 \tag{15}$$

and

$$\lim_{\delta \to 0} \limsup_{\nu \to \infty} \sum_{i=1}^{N_\nu} \mathsf{var}(T_{\nu i}^{\delta})^2 = 0, \tag{16}$$

since then (14) will follow by the Chebyshev inequality. Further, in view of (12), (15) is equivalent to

$$\sum_{i=1}^{N_\nu} (\mathsf{E} T_{\nu i}^{\delta})^2 \to 0. \tag{17}$$

We first prove (17). If $\delta > 2$, then $T_{\nu i}^{\delta} \leq T_{\nu i}$, since $T_{\nu i} \geq -2$, in view of (8). Consequently,

$$\begin{aligned} \mathsf{E}T_{\nu i}^{\delta} &\leq \mathsf{E}T_{\nu i} = 2\mathsf{E}\{g_{\nu i}(X_i)/f_{\nu i}(X_i)\}^{\frac{1}{2}} - 2 \\ &\leq 2\{\mathsf{E}[g_{\nu i}(X_i)/f_{\nu i}(X_i)]\}^{\frac{1}{2}} - 2 \\ &= 2\left\{\int_{\{f_{\nu i}>0\}} g_{\nu i}(x)\,\mathrm{d}x\right\}^{\frac{1}{2}} - 2 \leq 0. \end{aligned} \tag{18}$$

Thus, for $\delta > 2$,

$$\sum_{i=1}^{N_\nu} (\mathsf{E}T_{\nu i}^{\delta})^2 \leq \min_{1\leq i\leq N_\nu} \mathsf{E}T_{\nu i}^{\delta} \sum_{i=1}^{N_\nu} \mathsf{E}T_{\nu i}^{\delta},$$

and (17) follows from (11) and from the fact that

$$\min_{1\leq i\leq N} \mathsf{E}T_{\nu i}^{\delta} \to 0,$$

which is an easy consequence of (4). Now it remains to note that the validity of (17) for any $\delta > 2$ entails its validity for any $\delta > 0$, because of (12) and of

$$\sum_{i=1}^{N_\nu} \mathsf{E}(T_{\nu i}^{\delta_1})^2 \leq \sum_{i=1}^{N_\nu} \mathsf{E}(T_{\nu i}^{\delta_2})^2, \quad \delta_1 < \delta_2.$$

As for (16), first note that

$$\sum_{i=1}^{N_\nu} \mathsf{var}[(T_{\nu i}^{\delta})^2] \leq \sum_{i=1}^{N_\nu} \mathsf{E}(T_{\nu i}^{\delta})^4 \leq \delta^2 \sum_{i=1}^{N_\nu} \mathsf{E}(T_{\nu i}^{\delta})^2.$$

Thus, on account of (15),

$$\limsup_{\nu\to\infty} \sum_{i=1}^{N_\nu} \mathsf{var}(T_{\nu i}^{\delta})^2 \leq \delta^2\sigma^2. \tag{19}$$

Consequently, (16) holds.

Finally, asymptotic normality $(-\frac{1}{2}\sigma^2, \sigma^2)$ is an immediate consequence of (6). □

The reader, having finished the above proof, can observe that we did not evade niceties connected with the truncation of summands. However, they are concentrated in the above proof and will not trouble us any more.

LeCam's second lemma, as has been presented here, remains applicable to single as well as multiple parameter models. Moreover, in the case of a product measure, as has been mostly considered here, the observable random variables $X_{\nu i}$ may also be vector valued, a case which arises in multivariate rank tests.

7.1.4 LeCam's third lemma. The limiting distributions of test statistics S_ν *under the alternative* are important from the point of view of the power properties of the respective tests. Unfortunately, their derivations are considerably more difficult than the proofs of limiting distributions under the hypothesis. Nonetheless, in the contiguity case, the difficulties are essentially diminished by the following lemma due to LeCam.

We say that the pair $(S_\nu, \log L_\nu)$ is asymptotically jointly normal $(\mu_1, \mu_2, \sigma_1^2, \sigma_2^2, \sigma_{12})$ if it converges in distribution to a normal vector (Z_1, Z_2) such that $\mathsf{E} Z_i = \mu_i$, $\mathsf{var}\, Z_i = \sigma_i^2$, $i = 1, 2$, and $\mathsf{cov}(Z_1, Z_2) = \sigma_{12}$. (For convergence in distribution in $k \geq 2$ dimensions see the definitions of Subsection 6.2.1.)

Lemma 1 *Assume that the pair* $(S_\nu, \log L_\nu)$ *is under* P_ν *asymptotically jointly normal* $(\mu_1, \mu_2, \sigma_1^2, \sigma_2^2, \sigma_{12})$ *with* $\mu_2 = -\frac{1}{2}\sigma_2^2$.

Proof. Obviously

$$\begin{aligned} Q_\nu(S_\nu \leq x) &= \int_{\{S_\nu \leq x\}} \mathrm{d}Q_\nu \qquad (1) \\ &= \int_{\{S_\nu \leq x\}} L_\nu \,\mathrm{d}P_\nu + Q_\nu(p_\nu = 0,\ S_\nu \leq x) \\ &= \int_\infty^x \int_{-\infty}^\infty \mathrm{e}^v \,\mathrm{d}F_\nu(u, v) + Q_\nu(p_\nu = 0,\ S_\nu \leq x), \end{aligned}$$

where $F_\nu(u, v)$ denotes the distribution function of $(S_\nu, \log L_\nu)$. Now $\mu_2 = -\frac{1}{2}\sigma_2^2$ implies contiguity (Corollary 7.1.2.1), and hence

$$Q_\nu(p_\nu = 0,\ S_\nu \leq x) \to 0, \qquad (2)$$

since $P_\nu(p_\nu = 0,\ S_\nu \leq x) = 0 \to 0$. Furthermore, for any $c > 0$

$$\int_{-\infty}^x \int_{-c}^c \mathrm{e}^v \,\mathrm{d}F_\nu(u, v) \to \int_{-\infty}^x \int_{-c}^c \mathrm{e}^v \,\mathrm{d}\Phi(u, v), \qquad (3)$$

where $\Phi(u, v)$ denotes the two-dimensional normal distribution function with parameters $(\mu_1, \mu_2, \sigma_1^2, \sigma_2^2, \sigma_{12})$. Actually, $F_\nu \to \Phi$ according to our assumption, and the function

$$h(u, v) = \begin{cases} \mathrm{e}^v, & -\infty < u < x,\ -c \leq v \leq c \\ 0, & \text{otherwise,} \end{cases}$$

is uniformly bounded and continuous except on the set $\{(u, v) : v = -c \text{ or } v = c \text{ or } u = x\}$, which obviously has Φ-probability 0. Thus we may apply D1 of Subsection 6.2.1.

Now (1), (2) and (3) will imply

$$Q_\nu(S_\nu \leq x) \to \int_{-\infty}^{x} \int_{\infty}^{\infty} \mathrm{e}^v \,\mathrm{d}\Phi(u,v) \tag{4}$$

if we show that for every ε there exist c_0 and ν_0 such that

$$\int_{-\infty}^{x} \int_{-\infty}^{-c_0} \mathrm{e}^v \,\mathrm{d}F_\nu + \int_{-\infty}^{x} \int_{c_0}^{\infty} \mathrm{e}^v \,\mathrm{d}F_\nu < \varepsilon, \quad \nu \geq \nu_0. \tag{5}$$

In other words we must show that the truncated parts of the integral are uniformly small if c_0 is sufficiently large. However, (5) is an easy consequence of contiguity. Actually, if (5) were not true for some $\varepsilon > 0$, we would have a sequence of pairs (c_j, ν_j) such that

$$\lim_{j\to\infty} c_j = \infty, \quad \lim_{j\to\infty} \nu_j = \infty, \tag{6}$$

and

$$\begin{aligned} &Q_{\nu_j}(\log L_{\nu_j} < -c_j \text{ or } \log L_{\nu_j} > c_j) \\ &= \int_{-\infty}^{\infty} \int_{-\infty}^{-c_j} \mathrm{e}^v \,\mathrm{d}F_{\nu_j} + \int_{-\infty}^{\infty} \int_{c_j}^{\infty} \mathrm{e}^v \,\mathrm{d}F_{\nu_j} \\ &\geq \int_{-\infty}^{x} \int_{-\infty}^{-c_j} \mathrm{e}^v \,\mathrm{d}F_{\nu_j} + \int_{-\infty}^{x} \int_{c_j}^{\infty} \mathrm{e}^v \,\mathrm{d}F_{\nu_j} \geq \varepsilon. \end{aligned}$$

On the other hand, since $\log L_\nu$ has a limit distribution under P_ν,

$$P_{\nu_j}(\log L_{\nu_j} < -c_j \text{ or } \log L_{\nu_j} > c_j) \to 0,$$

because of (6). This contradicts contiguity, and thereby (4) is proved.

Now, by easy computations, we derive that

$$\begin{aligned} &\int_{-\infty}^{x} \int_{-\infty}^{\infty} \mathrm{e}^v \,\mathrm{d}\Phi \\ &= \int_{-\infty}^{x} \int_{-\infty}^{\infty} (\sigma_1\sigma_2 . 2\pi)^{-1}[(1-\rho^2)]^{-\frac{1}{2}} \\ &\quad . \exp\{v - [2(1-\rho^2)]^{-1}[(u-\mu_1)^2\sigma_1^{-2} \\ &\quad -2\rho(u-\mu_1)(v+\tfrac{1}{2}\sigma_2^2)(\sigma_1\sigma_2)^{-1} + (v+\tfrac{1}{2}\sigma_2^2)^2\sigma_2^{-2}]\} \,\mathrm{d}u\,\mathrm{d}v \\ &= \sigma_1^{-1}(2\pi)^{-\frac{1}{2}} \int_{-\infty}^{x} \exp[-\tfrac{1}{2}(u-\mu_1-\sigma_{12})^2\sigma_1^{-2}] \,\mathrm{d}u, \end{aligned} \tag{7}$$

where $\rho = \sigma_{12}(\sigma_1\sigma_2)^{-1}$. Combining (4) and (7), we easily conclude the proof. □

Remark. The above lemma holds even if $\sigma_2^2 = 0$, i.e. if $\log L_\nu$ converges to 0 in probability.

LeCam's third lemma plays a central role in the study of general asymptotics for rank statistics under contiguous alternatives. The following result due to Behnen and Neuhaus (1975) deserves mention in this context.

Lemma 2 *Let $\{Q_\nu\}$ be contiguous to $\{P_\nu\}$, and let $S_\nu = \sum_{i=1}^{\nu} d_{\nu i} Z_{\nu i}$ where the $Z_{\nu i}$ are independent under P_ν with 0 means and unit variances, and the constants $d_{\nu i}$ satisfy the conditions that*

$$\sum_{i \le \nu} d_{\nu i}^2 = 1; \quad \max_{1 \le i \le \nu} |d_{\nu i}| \to 0, \quad \textit{as } \nu \to \infty. \tag{8}$$

Then whenever the $Z_{\nu i}^2$ are uniformly integrable under P_ν, there exists a sequence $\{a_\nu\}$ of real numbers, such that

$$S_\nu, \ \textit{under } P_\nu, \quad \textit{and} \quad S_\nu - a_\nu, \ \textit{under } Q_\nu, \tag{9}$$

are asymptotically normal (0,1).

This is a special case of LeCam's third lemma, and the proof is omitted.

We may further note that, in the context of a multivariate model, generally we have a vector of rank statistics (viz., the linear model and multivariate models treated in Section 4.10). As such, if we have a vector $\mathbf{T}_\nu$ of statistics, and our intention is to study the asymptotic joint normality of $\mathbf{T}_\nu$ under contiguous alternatives, we could take advantage of the classical Cramér-Wold characterization of multinormality, and formulate the following result.

Lemma 3 *Let under P_ν, for every $\mathbf{c}$, the pair $(\mathbf{c}'\mathbf{T}_\nu, \log L_\nu)$ be asymptotically jointly normal $(\mathbf{c}'\mu_1, \mu_2; \mathbf{c}'\mathbf{\Sigma}_{11}\mathbf{c}, \sigma_2^2, \mathbf{c}'\sigma_{12})$ with $\mu_2 = -\frac{1}{2}\sigma_2^2$. Then, under Q_ν,*

$$\mathbf{T}_\nu \ \textit{is asymptotically multivariate normal} \ (\mu_1 + \sigma_{12}, \Sigma_{11}). \tag{10}$$

Note that μ_1 and σ_{12} are vectors. Moreover in the above characterization, the restriction on μ_2 and σ_2^2 is solely related to the log-likelihood ratio statistic, and has nothing to do with the statistic $\mathbf{T}_\nu$, but the centring of $\mathbf{T}_\nu$ depends on the vector σ_{12}.

In the case of product probability measures involving a triangular array $\mathbf{X}_\nu = \{X_{\nu i}; i \le \nu\}, \nu \ge 1$, as has been considered here, we might have a stochastic process Y_ν that is a mapping of $\mathbf{X}_\nu$ into the space $D[0,1]$, or more generally, $D[0,1]^p$, for some $p \ge 1$; we refer to Section 6.4 for such processes with special reference to rank processes. We denote by P_ν^* and

Q^*_ν the probability measures induced by P_ν and Q_ν respectively. Then the following result (viz., Theorem 4.3.4 of Sen (1981)) plays a basic role in the asymptotics for rank statistics and processes; here even the rowwise independence of the $X_{\nu i}$ is not that crucial.

Lemma 4 *If $\{Q_\nu\}$ is contiguous to $\{P_\nu\}$, and Y_ν is tight under the P_ν-measure, then it remains tight under the Q_ν-measure as well, so that $\{Q^*_\nu\}$ is contiguous to $\{P^*_\nu\}$.*

We refer to Sen (1981) for a derivation and some allied discussion.

7.1.5 LAQ family and the convolution theorem. We consider here a rather parametric formulation wherein a family of probability measures $\{P_{\theta,n} : \theta \in \boldsymbol{\Theta}\}$ on some measure space $(\mathcal{X}_n, \mathcal{A}_n)$, $n \geq 1$ is conceived; here θ is a parameter belonging to some p-dimensional open set $\boldsymbol{\Theta}$. Then, following LeCam (1960), we consider a bounded p-dimensional set B and a sequence δ_n of real numbers, such that

$$\{P_{\theta+\delta_n \mathbf{t}_n,n}\} \text{ is contiguous to } \{P_{\theta,n}\}, \text{ for all } \mathbf{t}_n \in B. \tag{1}$$

We denote the log-likelihood ratio statistic by

$$\Lambda(\theta + \delta_n \mathbf{t}_n; \theta) = \log\{\,\mathrm{d}P_{\theta+\delta_n \mathbf{t}_n,n} / \,\mathrm{d}P_{\theta,n}\}. \tag{2}$$

Then, the family $\mathcal{P}_n = \{P_{\gamma,n}; \gamma \in \boldsymbol{\Theta}\}$ is termed locally asymptotically quadratic (LAQ) at $\boldsymbol{\Theta}$ if there exist a vector $\mathbf{S}_n$ and an a.s. non-negative definite matrix $\mathbf{K}_n$, both possibly random, such that for every $\mathbf{t}_n \in B$,

$$\Lambda(\theta + \delta_n \mathbf{t}_n; \theta) - \mathbf{t}_n' \mathbf{S}_n + \frac{1}{2} \mathbf{t}_n' \mathbf{K}_n \mathbf{t}_n \to 0, \tag{3}$$

in $P_{\theta,n}$-probability. In particular, if $\mathbf{K}_n$ can be taken as non-random, then the LAQ family reduces to the well-known LAN (*locally asymptotically normal*) family. LeCam's second lemma considered in Subsection 7.1.3 essentially relates to the LAN condition. In that sense, the LAQ condition is more general that the LAN one. Moreover, if we consider an intermediate setup where $\mathbf{K}_n$, though random, has a limiting distribution that does not depend on $\mathbf{t}_n$ ($\in B$), then we have a LAMN (*locally asymptotically mixed normal*) family. In a general setup, log-likelihood ratio statistics may relate to the LAMN family rather than the LAN one. Finally, a density $f(x, \gamma)$ with respect to a measure μ is said to be DQM (*differentiable in quadratic mean*) at θ if there exists a vector $\mathbf{V}(x)$, such that as $\gamma \to \theta$,

$$\int \left| \sqrt{f(x,\gamma)} - \sqrt{f(x,\theta)} - (\gamma - \theta)' \mathbf{V}(x) \right|^2 \mathrm{d}\mu(x) \tag{4}$$
$$= o(||\gamma - \theta||).$$

It follows from the above definitions that DQM $\Rightarrow$ LAQ. The DQM condition of LeCam (1960) is also implicit in the derivations of the main results in the preceding subsections.

In the above setup, we shall present the Hájek-LeCam-Inagaki theorem that pertains to *regular estimators*. Since we shall consider rank estimators in Chapter 9, this result will be of considerable interest in that context. We refrain from a derivation, and refer to LeCam and Yang (1990). For a p-dimensional parameter θ, let $\hat{\theta}_n$ be an estimator, based on a sample of size n, such that as $n \to \infty$,

$$\sqrt{n}(\hat{\theta}_n - \theta) \text{ is asymptotically } p\text{-variate normal } (\mathbf{0}, \mathbf{\Gamma}), \tag{5}$$

where $\mathbf{\Gamma}$ is the inverse of the Fisher information matrix (with respect to θ) that is assumed to exist and satisfy certain other regularity assumptions. Then $\hat{\theta}_n$ is termed a FOE (*first-order efficient*) or BAN (*best asymptotically normal*) estimator of θ. The existence of such BAN estimators has been studied in detail for various families of densities with special emphasis on the LAQ family.

Theorem 1 *Consider the LAMN family* $\mathcal{F}_{\theta,n} = \{P_{\theta+\delta_n\tau,n};\ \tau$ *being a p-dimensional vector*$\}$, *and assume that under* $P_{\theta,n}$, *the matrix* $\mathbf{K}_n$ *converges in probability to some non-random* $\mathbf{K}$. *For a given non-random matrix* $\mathbf{A}$, *consider the parameter* $\mathbf{A}\tau$, *and let* $\mathbf{T}_n$ *be an estimator of* $\mathbf{A}\tau$, *such that*

$$\mathbf{T}_n - \mathbf{A}\tau H(\cdot), \tag{6}$$

in distribution under $P_{\theta+\delta_n\tau,n}$, *where* $H(\cdot)$ *does not depend on* τ. *Then* H *is the distribution of the random vector* $\mathbf{A}\mathbf{K}^{-1/2}\mathbf{Z}+\mathbf{U}$, *where* $\mathbf{Z}$ *and* $\mathbf{U}$ *are stochastically independent, and* $\mathbf{Z}$ *is asymptotically normal* $(\mathbf{0},\mathbf{I})$, *where* $\mathbf{I}$ *is the identity matrix.*

Note that if $\hat{\xi}_n$ is a BAN estimator of $\mathbf{A}\tau$ (in the sense of (5)), then $\hat{\xi}_n \xrightarrow{\mathcal{D}} \mathbf{A}\mathbf{K}^{-1/2}\mathbf{Z}$, so that

$$\mathbf{T}_n \xrightarrow{\mathcal{D}} \hat{\xi}_n + \mathbf{U}_n, \tag{7}$$

where $\hat{\xi}_n$ and $\mathbf{U}_n$ are asymptotically independent. Note that in general $\mathbf{U}_n$ need not be asymptotically normal, but it is so whenever $\mathbf{T}_n$ is asymptotically normal. Finally, the convolution result (7) implies that the difference of the asymptotic dispersion matrix of $\mathbf{T}_n$ and of ξ_n is positive semidefinite.

7.1.6 Contiguity and Hellinger distance. The Bhattacharya-Kakutani-Hellinger (BKH) distance $H(P,Q)$ between two probability measures P and Q with densities $p = \mathrm{d}P/\,\mathrm{d}\mu$, $q = \mathrm{d}Q/\,\mathrm{d}\mu$, with respect to a common σ-finite measure μ dominating $P+Q$, is defined by

$$H(P,Q) = \left\{ \int (\sqrt{p} - \sqrt{q})^2 \,\mathrm{d}\mu \right\}^{1/2}. \tag{1}$$

We also define the *total variation* distance between $\{P_\nu\}$ and $\{Q_\nu\}$ by

$$||P_\nu - Q_\nu|| = \sup_{A \in \mathcal{A}} |P_\nu(A) - Q_\nu(A)|. \tag{2}$$

Then it is easy to verify that

$$H^2(P_\nu, Q_\nu) \leq 2||P_\nu - Q_\nu|| \leq 2H(P_\nu, Q_\nu). \tag{3}$$

Oosterhoff and van Zwet (1979) studied the relationship between contiguity and BKH distance when $\{P_\nu = \prod_{i=1}^{\nu} P_{\nu i}\}$ and $\{Q_\nu = \prod_{i=1}^{\nu} Q_{\nu i}\}$ are product measures; their main result can be stated as follows.

Theorem 1 *$\{Q_\nu\}$ is contiguous to $\{P_\nu\}$ if and only if*

$$\limsup_{\nu\to\infty} \sum_{i=1}^{\nu} H^2(P_{\nu i}, Q_{\nu i}) < \infty, \tag{4}$$

and for every $\{c_\nu\}$ with $\lim_{\nu\to\infty} c_\nu = \infty$,

$$\lim_{\nu\to\infty} \sum_{i=1}^{\nu} Q_{\nu i}(\frac{q_{\nu i}(X_{\nu i})}{p_{\nu i}(X_{\nu i})} \geq c_\nu) = 0. \tag{5}$$

$\{Q_\nu\}$ and $\{P_\nu\}$ are mutually contiguous one to another if and only if (4) *and* (5) *hold, and moreover*

$$\lim_{\nu\to\infty} \sum_{i=1}^{\nu} P_{\nu i}(\frac{p_{\nu i}(X_{\nu i})}{q_{\nu i}(X_{\nu i})} \geq c_\nu) = 0. \tag{6}$$

As we shall see later on, for rank statistics verification of contiguity does not entail a serious problem, and hence we shall not enter into any detailed discussion on the role of the BKH distance in the theory of rank tests.

7.2 SIMPLE LINEAR RANK STATISTICS

7.2.1 Location alternatives for H_0. We shall consider alternatives

$$q_d = \prod_{i=1}^{N} f_0(x_i - d_i), \tag{1}$$

where f_0 is a known density with $I(f_0) < \infty$, and $d = (d_1, \ldots, d_N)$ is an arbitrary vector. Recall that the vector d runs through the set of all real vectors of all finite dimensions, and that the asymptotic statements concern sequences $\{d_\nu = (d_{\nu_1}, \ldots, d_{\nu N_\nu})\}$ selected from this set. However, to simplify the notation, we shall drop the index ν. First of all, we shall establish conditions under which sequences of such alternatives are contiguous with respect to corresponding sequences of the hypotheses $H_0 = H_{0N}$. For this purpose we associate with each q_d the following density

$$p_d = \prod_{i=1}^{N} f_o(x_i - \bar{d}). \tag{2}$$

Obviously $p_d \in H_{0N}$, where N is the dimension of d, and it depends on d only through $\bar{d} = N^{-1} \sum d_i$. If we show that under certain conditions the densities q_d are contiguous with respect to the densities p_d, then, a fortiori, they will be contiguous with respect to the hypotheses H_{0N}. The densities p_d have been chosen so as to be least favourable for H_{0N} against q_d, in an asymptotic sense.

According to LeCam's first lemma, the densities q_d are contiguous to the densities p_d, if $\log L_d$, where $L_d = q_d/p_d$, is asymptotically normal $(-\frac{1}{2}\sigma^2, \sigma^2)$. Moreover, on account of LeCam's second lemma, under (7.1.3.4), the asymptotic normality of $\log L_d$ may be derived from the asymptotic normality of

$$W_d = 2\sum_{i=1}^{N} \{[f_0(X_i - d_i)/f_0(X_i - \bar{d})]^{\frac{1}{2}} - 1\}. \tag{3}$$

Since $f_0(x)$ has finite Fisher information, and consequently is absolutely continuous, (7.1.3.4) in our case amounts to

$$\max_{1 \le i \le N} (d_i - \bar{d})^2 \to 0. \tag{4}$$

Let us assume, in addition, that

$$I(f_0) \sum_{i=1}^{N} (d_i - \bar{d})^2 \to b^2, \quad 0 < b^2 < \infty, \tag{5}$$

and first prove two lemmas. Recall that the dependence of N and d_i on ν is not indicated in (4) and (5), and that (4) and (5) entail $N \to \infty$.

Lemma 1 *Let* E *denote the expectation with respect to* p_d. *Then, under* (4) *and* (5),

$$\mathsf{E}W_d \to -\tfrac{1}{4}b^2. \tag{6}$$

Proof. Denote $s(x) = [f_0(x)]^{\frac{1}{2}}$ and observe that

$$s'(x) = \tfrac{1}{2} f_0'(x)[f_0(x)]^{-\frac{1}{2}}, \tag{7}$$

and

$$I(f_0) = 4\int_{-\infty}^{\infty} [s'(x)]^2 \, \mathrm{d}x. \tag{8}$$

Further let $(X_1, \ldots, X_N)$ be governed by p_d, and put $Y_i = X_i - \bar{d}$. Then, obviously,

$$\begin{aligned} \mathsf{E}W_d &= 2\sum_{i=1}^{N} \mathsf{E}\Big[\frac{s(Y_i - d_i + \bar{d})}{s(Y_i)} - 1\Big] \qquad (9) \\ &= -\sum_{i=1}^{N}(d_i - \bar{d})^2 \int_{\infty}^{\infty} \Big[\frac{s(x - d_i + \bar{d}) - s(x)}{d_i - \bar{d}}\Big]^2 \, \mathrm{d}x. \end{aligned}$$

Thus, on account of (4) and (5), in order to prove (6) it suffices to show that

$$\begin{aligned} \lim_{h\to 0}\int_{-\infty}^{\infty} \Big[\frac{s(x-h) - s(x)}{h}\Big]^2 \mathrm{d}x &= \int_{-\infty}^{\infty} [s'(x)]^2 \, \mathrm{d}x \qquad (10) \\ &= \frac{1}{4}\int_{-\infty}^{\infty} [f_0'(x)/f_0(x)]^2 f_0(x) \, \mathrm{d}x. \end{aligned}$$

Now, since $s(x)$ is, under $I(f_0) < \infty$, also absolutely continuous, we have

$$\lim_{h\to 0} \frac{s(x-h) - s(x)}{h} = -s'(x) \tag{11}$$

almost everywhere. Furthermore,

$$\begin{aligned} \{[s(x-h) - s(x)]/h\}^2 &= \Big[(1/h)\int_0^h s'(x-t) \, \mathrm{d}t\Big]^2 \\ &\le (1/h)\int_0^h [s'(x-t)]^2 \, \mathrm{d}t, \end{aligned}$$

so that

$$\begin{aligned} &\int_{-\infty}^{\infty} \{[s(x-h) - s(x)]/h\}^2 \, \mathrm{d}x \qquad (12) \\ &\le (1/h)\int_0^h \int_{-\infty}^{\infty} [s'(x-t)]^2 \, \mathrm{d}x \, \mathrm{d}t = \int_{-\infty}^{\infty} [s'(x)]^2 \, \mathrm{d}x, \end{aligned}$$

since the interchange of the integration order is justified for non-negative integrands. Now (10) may be concluded from (11) and (12) by means of Theorem 3.4.2.1. □

Introducing the statistics

$$T_d = -\sum_{i=1}^{N}(d_i - \bar{d})\frac{f_0'(X_i - \bar{d})}{f_0(X_i - \bar{d})}, \tag{13}$$

we have the following

Lemma 2 *Let* var *denote the variance with respect to* p_d. *Then, under* (4) *and* (5),

$$\mathsf{var}(W_d - T_d) \to 0. \tag{14}$$

Proof. Using $s(x)$ and Y_i in the above meaning and expressing $(W_d - T_d)$ as N independent summands, we have

$$\begin{aligned}
\mathsf{var}(W_d - T_d) &= \qquad\qquad (15)\\
&= 4\sum_{i=1}^{N} \mathsf{var}\left[\frac{s(Y_i - d_i + \bar{d})}{s(Y_i)} - 1 + \tfrac{1}{2}(d_i - \bar{d})\frac{f_0'(Y_i)}{f_0(Y_i)}\right]\\
&\le 4\sum_{i=1}^{N} \mathsf{E}\left[\frac{s(Y_i - d_i + \bar{d})}{s(Y_i)} - 1 + \tfrac{1}{2}(d_i - \bar{d})\frac{f_0'(Y_i)}{f_0(Y_i)}\right]^2\\
&= 4\sum_{i=1}^{N}(d_i - \bar{d})^2 \int_{-\infty}^{\infty}\left[\frac{s(x - d_i + \bar{d}) - s(x)}{d_i - \bar{d}} + s'(x)\right]^2 \mathrm{d}x.
\end{aligned}$$

Now (4), (11) and (12), in view of Theorem 6.1.3.1, imply that the last integral converges to 0 uniformly in $i = 1, \ldots, N$. □

Theorem 1 *Under* (4) *and* (5),

$$(\log L_d - T_d + \tfrac{1}{2}b^2) \to 0 \tag{16}$$

in P_d*-probability. Hence,* $\log L_d$ *is asymptotically normal* $(-\frac{1}{2}b^2, b^2)$ *and the densities* q_d *are contiguous to the densities* p_d.

Proof. From (6) and (14) it follows that

$$\mathsf{E}(W_d - T_d + \tfrac{1}{4}b^2)^2 \to 0,$$

which, together with LeCam's second lemma, entails (16). Thus it remains to show that T_d is asymptotically normal $(0, b^2)$. This follows, however, from Theorem 6.1.2.1. □

7.2.2 Scale alternatives for H_0. Here we shall consider alternatives

$$q_d = \prod_{i=1}^{N} \exp(-d_i) f_0[x_i \exp(-d_i)] \tag{1}$$

where f_0 is a known density with $I_1(f_0) < \infty$. The densities q_d will be compared with densities

$$p_d = \prod_{i=1}^{N} \exp(-\bar{d}) f_0[x_i \exp(-\bar{d})], \tag{2}$$

which belong to H_{0N}. The condition (7.2.1.4) will work even here, and (7.2.1.5) will be replaced by

$$I_1(f_0)\sum_{i=1}^{N}(d_i-\bar{d})^2 \to b^2, \quad 0<b^2<\infty. \tag{3}$$

Theorem 1 *Put $L_d = q_d/p_d$, q_d and p_d given by* (1) *and* (2), *and introduce*

$$T_d = -\sum_{i=1}^{N}(d_i-\bar{d})\Big\{1+X_i\exp(-\bar{d})\frac{f_0'[X_i\exp(-\bar{d})]}{f_0[X_i\exp(-\bar{d})]}\Big\}. \tag{4}$$

Then under (7.2.1.4) *and* (3),

$$(\log L_d - T_d + \tfrac{1}{2}b^2) \to 0 \tag{5}$$

in P_d-probability. Hence, $\log L_d$ *is asymptotically normal* $(-\frac{1}{2}b^2, b^2)$ *and the densities q_d are contiguous to the densities p_d.*

Proof. The proof is analogous to the proof of Theorem 7.2.1.1. Introducing $Y_i = X_i\exp(-\bar{d})$ and $s(x) = [f_0(x)]^{\frac{1}{2}}$, we have

$$\begin{aligned}\mathsf{E}W_d &= 2\sum_{i=1}^{N}\mathsf{E}\Big[\frac{s[Y_i\exp(-d_i+\bar{d})]}{\exp(\frac{1}{2}d_i-\frac{1}{2}\bar{d})s(Y_i)}-1\Big]\\ &= -\sum_{i=1}^{N}(d_i-\bar{d})^2\\ &\quad\cdot\int_{-\infty}^{\infty}\Big\{\frac{\exp(-\frac{1}{2}d_i+\frac{1}{2}\bar{d})s[x\exp(-d_i+\bar{d})]-s(x)}{d_i-\bar{d}}\Big\}^2\,\mathrm{d}x\end{aligned}$$

and

$$\begin{aligned}\mathsf{var}(W_d-T_d) &\le 4\sum_{i=1}^{N}(d_i-\bar{d})^2\\ &\quad\cdot\int_{-\infty}^{\infty}\Big\{\frac{\exp(-\frac{1}{2}d_i+\frac{1}{2}\bar{d})s[x\exp(-d_i+\bar{d})]-s(x)}{d_i-\bar{d}}\\ &\qquad+\tfrac{1}{2}s(x)+x\,s'(x)\Big\}^2\,\mathrm{d}x.\end{aligned}$$

Now the conclusions analogous to (7.2.1.6) and (7.2.1.14) follow from

$$\lim_{h\to 0}\frac{1}{h}[\mathrm{e}^{-\frac{1}{2}h}s(x\mathrm{e}^{-h})-s(x)] = -\frac{1}{2}s(x)-x\,s'(x)$$

and

$$\int_{-\infty}^{\infty} \{(1/h)[\mathrm{e}^{-\frac{1}{2}h} s(x\mathrm{e}^{-h}) - s(x)]\}^2 \,\mathrm{d}x$$
$$\leq (1/h) \int_0^h \int_{-\infty}^{\infty} [\tfrac{1}{2}\mathrm{e}^{-\frac{1}{2}t} s(x\mathrm{e}^{-t}) + x\mathrm{e}^{-\frac{3}{2}t} s'(x\mathrm{e}^{-t})]^2 \,\mathrm{d}x\,\mathrm{d}t$$
$$= \int_{-\infty}^{\infty} [\frac{1}{2} s(x) + x\, s'(x)]^2 \,\mathrm{d}x.$$

The rest is immediate. □

7.2.3 Rank statistics for H_0 against two samples. The results of this subsection will be presented without proofs, since they are special cases of the results of the next subsection. Consider again some approximative scores $a_N(i)$ converging in quadratic mean to a square integrable function φ:

$$\lim_{N\to\infty} \int_0^1 [a_N(1 + [uN]) - \varphi(u)]^2 \,\mathrm{d}u = 0. \tag{1}$$

Throughout the sequel we shall assume

$$\int_0^1 [\varphi(u) - \bar{\varphi}]^2 \,\mathrm{d}u > 0$$

without further mention. We shall be interested in the limiting distributions of the statistics

$$S_{mn} = \sum_{i=1}^{m} a_{m+n}(R_i) - m\bar{a}_{m+n} \tag{2}$$

under q_d. As we know from Theorem 6.1.6.1, S_{mn} is under H_0 asymptotically normal $(0, \sigma_{mn}^2)$ with $\sigma_{mn}^2 = mn(m+n)^{-1} \int_0^1 [\varphi(u) - \bar{\varphi}]^2 \,\mathrm{d}u$ or $\sigma_{mn}^2 = \mathsf{var}\, S_{mn}$.

Theorem 1 *Let q_d be given by (7.2.1.1) and assume that (7.2.1.4) and (7.2.1.5) hold. Then the statistics S_{mn} given by (2), where the scores satisfy (1), are, for $\min(m,n) \to \infty$, under q_d asymptotically normal $(\mu_{dmn}, \sigma_{mn}^2)$ with*

$$\mu_{dmn} = \Big[\sum_{i=1}^{m} (d_i - \bar{d})\Big] \int_0^1 \varphi(u)\varphi(u, f_0) \,\mathrm{d}u \tag{3}$$

and

$$\sigma_{mn}^2 = \frac{mn}{m+n} \int_0^1 [\varphi(u) - \bar{\varphi}]^2 \,\mathrm{d}u. \tag{4}$$

If $d_i = \Delta$ *for* $1 \le i \le m$, *and* $d_i = 0$ *for* $m < i \le m+n$, *then*

$$\mu_{\Delta mn} = \Delta \frac{mn}{m+n} \int_0^1 \varphi(u)\varphi(u, f_0)\,du. \tag{5}$$

The assertions remain true if we replace (7.2.1.1), (7.2.1.5) *and* $\varphi(u, f_0)$ *by* (7.2.2.1), (7.2.2.3) *and* $\varphi_1(u, f_0)$, *respectively.*

Remark. Note that for $d_i = \Delta$, $1 \le i \le m$, and $d_i = 0$ otherwise, assumption (7.2.1.5) becomes

$$I(f_0)\Delta^2 \frac{mn}{m+n} \to b^2, \quad 0 < b^2 < \infty. \tag{6}$$

Application 1. Scores corresponding to some density: $a_N(i) = a_N(i, f)$, $I(f) < \infty$. Then

$$S_{mn} = \sum_{i=1}^{m} a_{m+n}(R_i, f), \tag{7}$$

and

$$\mu_{\Delta mn} = \Delta \frac{mn}{m+n} \int_0^1 \varphi(u, f)\varphi(u, f_0)\,du, \tag{8}$$

$$\sigma^2_{mn} = \frac{mn}{m+n} I(f). \tag{9}$$

The same with $a_{1N}(i, f)$, $\varphi_1(u, f)$ and $I_1(f)$.

Application 2. Approximate scores corresponding to some density f such that $\varphi(u, f)$ is a sum of monotone and square integrable functions. If, in addition, $\varphi(u, f)$ is skew symmetric, then

$$S_{mn} = \sum_{i=1}^{m} \varphi\Big(\frac{R_i}{m+n+1}, f\Big), \tag{10}$$

with (8) and (9) still applicable. If $\varphi(u, f)$ is not symmetric, we have to subtract the expectation under H_0, i.e. to put

$$S_{mn} = \sum_{i=1}^{m} \varphi\Big(\frac{R_i}{m+n+1}, f\Big) - \frac{m}{N} \sum_{i=1}^{N} \varphi\Big(\frac{i}{m+n+1}, f\Big).$$

However, since $\int_0^1 \varphi(u, f)\,du = 0$, the correction is asymptotically negligible, as may be shown.

7.2.4 Rank statistics for H_0 against regression. Now we shall investigate the limiting distribution under q_d of the statistics

$$S_c = \sum_{i=1}^{N} (c_i - \bar{c}) a_N(R_i). \tag{1}$$

Theorem 1 *Let q_d be given by (7.2.1.1) and assume that (7.2.1.4) and (7.2.1.5) hold. Then, under q_d the statistics S_c given by (1), where the scores satisfy (7.2.3.1), are for $\sum_{i=1}^{N}(c_i - \bar{c})^2 / \max_{1\le i\le N}(c_i - \bar{c})^2 \to \infty$ asymptotically normal (μ_{dc}, σ_c^2) with*

$$\mu_{dc} = \Big[\sum_{i=1}^{N}(c_i - \bar{c})(d_i - \bar{d})\Big] \int_0^1 \varphi(u)\varphi(u, f_0)\, du \tag{2}$$

and

$$\sigma_c^2 = \Big[\sum_{i=1}^{N}(c_i - \bar{c})^2\Big] \int_0^1 [\varphi(u) - \bar{\varphi}]^2\, du. \tag{3}$$

The assertions remain true if we replace (7.2.1.1), (7.2.1.5) and $\varphi(u, f_0)$ by (7.2.2.1), (7.2.2.3) and $\varphi_1(u, f)$, respectively.

Proof. Without loss of generality, we may assume that

$$\sum_{i=1}^{N}(c_i - \bar{c})^2 = 1 \tag{4}$$

and

$$\sum_{i=1}^{N}(c_i - \bar{c})(d_i - \bar{d}) \to b_{12}. \tag{5}$$

Note that under (4), $\sum(c_i - \bar{c})^2/\max(c_i - \bar{c})^2 \to \infty$ is equivalent to

$$\max_{1\le i\le N}(c_i - \bar{c})^2 \to 0. \tag{6}$$

Furthermore, if S_c^{φ} is given by

$$S_c^{\varphi} = \sum_{i=1}^{N}(c_i - \bar{c}) a_N^{\varphi}(R_i), \tag{7}$$

with the scores a_N^{φ} as in (6.1.4.12), then $(S_c - S_c^{\varphi})(\operatorname{var} S_c^{\varphi})^{-\frac{1}{2}} \to 0$ in probability under H_0 (see the proof of Theorem 6.1.6.1). We shall denote

this briefly by $S_c \sim S_c^\varphi$. Furthermore, inspecting the proof of Theorem 1 in 6.1.5.1, we see that $S_c^\varphi \sim T_c$, T_c given by

$$T_c = \sum_{i=1}^{N}(c_i - \bar{c})\varphi(U_i), \tag{8}$$

where $U_i = F_d(X_i)$, $F_d(x) = P_d\,(X_i \le x)$ with P_d given by (7.2.1.2). Thus $S_c \sim T_c$, and S_c may be replaced by T_c in considerations concerning the limiting distribution.

On the other hand, we know from Theorem 7.2.1.1 that $\log L_d \sim (T_d - \frac{1}{2}b^2)$, T_d given by (7.2.1.13), or equivalently, by

$$T_d = \sum_{i=1}^{N}(d_i - \bar{d})\varphi(U_i, f_0). \tag{9}$$

Thus

$$(S_c, \log L_d) \sim (T_c, T_d - \tfrac{1}{2}b^2), \tag{10}$$

where it should be noted that T_c and T_d differ not only in their regression constants but also in their φ-functions. Consequently, if we show that (T_c, T_d) is under P_d asymptotically jointly normal with $\mu_1 = \mu_2 = 0$, variances $\sigma_1^2 = \int_0^1 [\varphi(u) - \bar{\varphi}]^2\,du$ and $\sigma_2^2 = b^2$, and covariance $\sigma_{12} = b_{12}\int_0^1 \varphi(u)\varphi(u, f_0)\,du$, we can conclude that $(S_c, \log L_d)$ is asymptotically jointly normal with the same parameters except for $\mu_2 = -\frac{1}{2}b^2$, and the theorem will follow immediately from LeCam's third lemma.

Now, since the U_i's are independent and uniformly distributed under P_d, we have $\mathsf{E}T_c = \mathsf{E}T_d = 0$, and in view of (7.2.1.5) and (5),

$$\begin{aligned}
\operatorname{var} T_c &= \sum_{i=1}^{N}(c_i - \bar{c})^2 \int_0^1 [\varphi(u) - \bar{\varphi}]^2\,du, \\
\operatorname{var} T_d &= \sum_{i=1}^{N}(d_i - \bar{d})^2 \int_0^1 \varphi^2(u, f_0)\,du \to b^2, \\
\operatorname{cov}(T_c, T_d) &= \Big[\sum_{i=1}^{N}(c_i - \bar{c})(d_i - \bar{d})\Big] \int_0^1 \varphi(u)\varphi(u, f_0)\,du \\
&\to b_{12} \int_0^1 \varphi(u)\varphi(u, f_0)\,du.
\end{aligned}$$

Thus the limiting parameters have the required values. In view of D3 of Subsection 6.2.1, it remains to show, for all real λ_1 and λ_2, that $\lambda_1 T_c + \lambda_2 T_d$ is either asymptotically normal $(0, \sigma_{cd}^2)$ with $\sigma_{cd}^2 = \operatorname{var}(\lambda_1 T_c + \lambda_2 T_d)$, or

$\mathsf{var}(\lambda_1 T_c + \lambda_2 T_d) \to 0$. Write

$$\lambda_1 T_c + \lambda_2 T_d = \sum_{i=1}^{N} [\lambda_1 (c_i - \bar{c})\varphi(U_i) + \lambda_2 (d_i - \bar{d})\varphi(U_i, f_0)],$$

and assume that

$$\mathsf{var}(\lambda_1 T_c + \lambda_2 T_d) \to v^2 > 0. \tag{11}$$

Now put

$$\begin{aligned} Z_{1i} &= \lambda_1 (c_i - \bar{c})[\varphi(U_i) - \bar{\varphi}], \\ Z_{2i} &= \lambda_2 (d_i - \bar{d})\varphi(U_i, f_0), \\ Z_i &= Z_{1i} + Z_{2i}, \\ Z_i(\delta) &= \begin{cases} Z_i, & \text{if } |Z_i| > \delta, \\ 0, & \text{if } |Z_i| \le \delta, \end{cases} \end{aligned}$$

and define $Z_{1i}(\delta)$ and $Z_{2i}(\delta)$ similarly. In view of (11) the Lindeberg condition for $\lambda_1 T_c + \lambda_2 T_d$ may be expressed as follows:

$$\sum_{i=1}^{N} \mathsf{E}[Z_i(\delta)]^2 \to 0, \quad \delta > 0. \tag{12}$$

However, we obviously have

$$[Z_i(\delta)]^2 \le 4[Z_{1i}(\tfrac{1}{2}\delta)]^2 + 4[Z_{2i}(\tfrac{1}{2}\delta)]^2,$$

so that (12) follows from

$$\sum_{i=1}^{N} \mathsf{E}[Z_{1i}(\delta)]^2 \to 0, \quad \delta > 0, \tag{13}$$

and

$$\sum_{i=1}^{N} \mathsf{E}[Z_{2i}(\delta)]^2 \to 0, \quad \delta > 0. \tag{14}$$

Finally, observe that (13) and (14) are equivalent to the Lindeberg condition for $\lambda_1 T_c$ and $\lambda_2 T_d$, respectively, in view of (4) and (7.2.1.5). Moreover, from (6) and (7.2.1.4) it follows that this condition is satisfied in both cases (see the proof of Theorem 6.1.2.1). This concludes the proof for q_d given by (7.2.1.1). If q_d were given by (7.2.2.1), we would proceed quite similarly. □

Remark. If (7.2.1.5) is replaced by

$$\left[I(f_0) < \infty, \ \sum_{i=1}^{N} (d_i - \bar{d})^2 \to 0\right],$$

the L_ν converges in P_ν-probability to 1, i.e. the distribution of $\log L_\nu$ converges to the degenerate normal distribution $(0,0)$. This could be proved as follows: First, we note that T_d given by (7.2.1.13) satisfies $\mathsf{E}T_d^2 \to 0$, and that considerations employed in the proofs of Lemmas 7.2.1.1 and 7.2.1.2 yield the relation $\mathsf{E}(W_d - T_d)^2 \to 0$. Thus $\mathsf{E}W_d^2 \to 0$, and consequently, the distribution of W_d converges to the degenerate normal distribution $(0,0)$. Thus it remains to show that LeCam's second lemma extends to this case, too. However, the degenerate convergence of the distribution of W_d entails that (7.1.3.10), (7.1.3.11) and (7.1.3.12) hold with $\sigma^2 = 0$ (Loève (1955), p. 317). The rest of the proof needs no change, and we obtain that $(\log L_\nu - W_\nu) \to 0$ in probability, and hence $\log L_\nu \to 0$ in probability. Consequently the theorem remains valid even for $b = 0$ and $I(f_0) < \infty$, and every statistic has the same limiting distribution, if any, under q_d as under p_d. Furthermore, the theorem remains valid even if we replace (7.2.1.5) by $I(f_0) < \infty$ and

$$\sum_{i=1}^{N}(d_i - \bar{d})^2 \le b^2 < \infty. \tag{15}$$

For, if it were not valid, there would exist a sequence $\{d_\nu\}$ satisfying (15) and such that the theorem would not be valid for any of its subsequences. However, since we can draw a subsequence $\{d_j\} \subset \{d_\nu\}$ such that

$$\sum_{i=1}^{N}(d_{ji} - \bar{d}_j)^2 \to b_1^2 \le b^2,$$

this would contradict the theorem if $b_1^2 > 0$, and the above extension of the theorem, if $b_1^2 = 0$.

The results of the present section and related results were adapted and generalized from Hájek (1962).

Theorem 7.2.4.1 remains applicable with minor modifications to both Type I and II censoring schemes, considered in Section 4.9. The situation is simpler for Type II censoring, as there the asymptotic normality result of Section 6.4 holds, under the P_ν-measure, with a modification of the score function based on the fact that for $N \to \infty$,

$$\begin{aligned} n^*/N \to \alpha \quad &\Rightarrow \quad a_N^*(n^*) \to \varphi^*(\alpha); \\ \varphi^*(\alpha) \quad &= \quad (1-\alpha)^{-1}\int_\alpha^1 \varphi(u)\,du, \; 0 < \alpha < 1. \end{aligned} \tag{16}$$

Therefore, verification of LeCam's third lemma (based on the likelihood criterion for Type II censoring outlined in Section 4.9) is quite easy, so that for contiguous alternatives the asymptotic normality result follows with the following changes: (i) in (2), the integral on the right hand side is to be replaced by $\int_0^\alpha \varphi(u)\varphi(u,f_o)\,du + \varphi^*(\alpha)\int_\alpha^1 \varphi(u,f_o)\,du$, and (ii) in (3) the integral on the right hand side is to be replaced by $\int_0^\alpha [\varphi(u) - \bar{\varphi}]^2\,du + (1-\alpha)[\varphi^*(\alpha) - \bar{\varphi}]^2$. The case of Type I is slightly more complex. In Section 6.4, we have obtained the asymptotic normality result (under a conditional model that yields the parallel result for the unconditional null distribution) through a random change of time argument. As such, if we make use of Lemma 7.1.4.4, we obtain that contiguity holds, so that the non-null distribution theory under such alternatives is the same as in Type II censoring.

The case of progressive censoring schemes can be handled with additional manipulations. We define the sequence $\{W_N\}$ of stochastic processes defined on the space $D[0,1]$ as in (6.4.4.10), and appeal to Theorem 6.4.4.2 that characterizes the weak convergence to a Wiener process, under the $\{P_N\}$-measure. We incorporate the tightness property in this theorem in Lemma 7.1.4.4, and conclude that the contiguity of $\{Q_N\}$ to $\{P_N\}$ holds. The convergence of the finite dimensional distributions under $\{Q_N\}$ can then be directly established and this would then lead to the weak convergence of $\{W_N\}$, under $\{Q_N\}$, to a drifted Wiener process, with the drift function $\omega(t)$, $t \in (0,1)$, defined by

$$\omega(t) = \frac{\int_0^1 \tilde{\varphi}_t(u)\varphi(u,f_o)\,du}{\{\int_0^1 [\varphi(u) - \bar{\varphi}]^2\,du\}^{1/2}}, \quad t \in [0,1]; \tag{17}$$

we use the same notation as in Theorem 7.2.4.1, and let

$$\tilde{\varphi}_t(u) = \varphi(u) \text{ or } \varphi^*(\nu^{-1}(t)) \text{ according as } u < \text{ or } \geq \nu^{-1}(t), \tag{18}$$

with $\varphi^*(t)$ defined by (16), and as in (6.4.4.13), we let

$$\begin{aligned} \nu(t) \;=\; \sup\Big\{u : \int_0^u [\varphi(s) - \bar{\varphi}]^2\,ds + (1-u)[\varphi^*(u) - \bar{\varphi}]^2 \\ \leq t\int_0^1 [\varphi(s) - \bar{\varphi}]^2\,ds\Big\}, \quad t \in [0,1]. \end{aligned} \tag{19}$$

In passing, we may note that typically $\nu(t)$ is not equal to t on $[0,1]$, and $\omega(t)$ is not a linear function of t on $[0,1]$. Hence, the classical Anderson (1960) type probability results on the boundary crossing of a Wiener process on a compact interval are not applicable for such non-linear boundaries. This makes the asymptotic non-null distribution theory of the rank statistics in progressive censoring schemes under contiguous alternatives rather complex in nature. However, this will be comparable to the results that will be presented in Subsection 7.4.7.

7.2.5 Rank statistics for H_1. Consider

$$q_{N\Delta} = \prod_{i=1}^{N} f_0(x_i - \Delta), \tag{1}$$

where f_0 is a known density *symmetric about zero*, and Δ satisfies

$$N\Delta^2 \to b^2, \quad 0 < b^2 < \infty. \tag{2}$$

The density q_Δ will be associated with the density

$$p = \prod_{i=1}^{N} f_0(x_i), \tag{3}$$

which obviously belongs to H_1. Our aim is to establish the limiting distribution of

$$S_N^+ = \sum_{i=1}^{N} a_N(R_i^+) \operatorname{sign} X_i \tag{4}$$

under q_Δ.

Theorem 1 *Let $q_{N\Delta}$ be given by (1), where f_0 is symmetric about zero, $I(f_0) < \infty$, and (N, Δ) satisfy (2). Further assume that the functions $a_N(1+[uN])$, $0 < u < 1$, converge in quadratic mean to a square integrable function $\varphi^+(u)$. Then the statistics (4) are under $q_{N\Delta}$ asymptotically normal (μ_N, σ_N^2) with*

$$\mu_N = \Delta N \int_0^1 \varphi^+(u)\varphi^+(u, f_0)\, du \tag{5}$$

and

$$\sigma_N^2 = \sum_{i=1}^{N} [a_N(i)]^2 \sim N \int_0^1 [\varphi^+(u)]^2\, du. \tag{6}$$

Remark. Recall that $\varphi^+(u, f_0) = \varphi(\frac{1}{2} + \frac{1}{2}u, f_0)$.

Proof. Following the pattern of the proof of Theorem 7.2.1.1, we can show that

$$\log L_{N\Delta} \sim \Delta \sum_{i=1}^{N} \varphi^+(U_i^+, f_0) \operatorname{sign} X_i - \tfrac{1}{2} b^2 \tag{7}$$

where $U_i^+ = F^+(|X_i|)$. The rest follows from $S_N^+ \sim T_N$, T_N given by (6.1.7.5). $\square$

Application 1. Put $a_N(i) = a_N^+(i, f)$, f symmetric about zero, $I(f) < \infty$. Then

$$S_N^+ = \sum_{i=1}^{N} a_N^+(i, f) \operatorname{sign} X_i \tag{8}$$

is under $q_{N\Delta}$ asymptotically normal (μ_N, σ_N^2) with

$$\mu_N = \Delta N \int_0^1 \varphi^+(u, f)\varphi^+(u, f_0)\, du, \tag{9}$$

$$\sigma_N^2 = N \int_0^1 [\varphi^+(u, f)]^2\, du. \tag{10}$$

Note that

$$\int_0^1 \varphi^+(u, f)\varphi^+(u, f_0)\, du = \int_0^1 \varphi(u, f)\varphi(u, f_0)\, du.$$

Application 2. Put $a_N(i) = \varphi(\frac{1}{2} + \frac{1}{2}i/(N+1), f)$, where f is symmetric about zero, $I(f) < \infty$, and $\varphi(u, f)$ is a finite sum of monotone and square integrable functions. Then the statistics

$$S_N^+ = \sum_{i=1}^{N} \varphi\Big(\frac{1}{2} + \frac{R_i^+}{N+1}, f\Big) \operatorname{sign} X_i \tag{11}$$

are asymptotically normal (μ_N, σ_N^2) with μ_N and σ_N^2 given by (9) and (10) respectively.

7.2.6 Rank statistics for H_2. Similarly, as in Subsection 3.4.11, we will work with the following model for two-dimensional random variables (X, Y). Let the sequence of their densities be

$$q_{N\Delta} = \prod_{i=1}^{N} h_\Delta(x_i, y_i), \tag{1}$$

with

$$h_\Delta(x, y) = \int_{-\infty}^{\infty} f_0(x - \Delta z) g_0(y - \Delta z)\, dM(z), \tag{2}$$

where f_0, g_0 are some densities having finite Fisher information, and $M(z)$ is a distribution function of some random variable Z with a finite and positive variance.

Theorem 1 *Suppose*

$$N\Delta^4 I(f_0)I(g_0) \to b^2, \quad 0 < b^2 < \infty. \tag{3}$$

We will deal with the statistics

$$S_N = \sum_{i=1}^{N} a_N(R_{Ni})b_N(Q_{Ni}), \tag{4}$$

based on approximate scores a_N, b_N, *i.e.*

$$\lim_{N\to\infty} \int_0^1 [a_N(1+[uN] - \varphi(u))]^2 \,\mathrm{d}u = 0, \tag{5}$$

$$\lim_{N\to\infty} \int_0^1 [b_N(1+[uN] - \psi(u))]^2 \,\mathrm{d}u = 0, \tag{6}$$

with some square integrable functions φ, ψ.

Under these assumptions, the statistics S_N *are for* $N \to \infty$ *asymptotically normal* (μ_N, σ_N^2) *with*

$$\mu_N = N\Delta^2 \int_0^1 \varphi(u)\varphi(u, f_0)\,\mathrm{d}u \int_0^1 \psi(u)\varphi(u, g_0)\,\mathrm{d}u \tag{7}$$

and

$$\sigma_N^2 = N \int_0^1 [\varphi(u) - \bar{\varphi}]^2 \,\mathrm{d}u \int_0^1 [\psi(u) - \bar{\psi}]^2 \,\mathrm{d}u. \tag{8}$$

The idea of the proof is similar to those above. Moreover, this is a special case of Behnen's (1971) theorem, which holds even for much more general alternatives of positive quadrant dependence. Further asymptotic results for possibly non-contiguous alternatives are contained e.g. in Ruymgaart, Shorack and van Zwet (1972), and in Ruymgaart (1974).

7.3 FAMILIES OF SIMPLE LINEAR RANK STATISTICS

7.3.1 Finite families (vectors). First we are going to derive a theorem analogous to Theorem 6.2.2.1. Again considering

$$S_{Nj} = \sum_{i\in s_j} a_N(R_{Ni}), \quad j = 1, \ldots, k \tag{1}$$

where $\{s_1, \ldots, s_k\}$ is a partition of $\{1, \ldots, N\}$, $n_j = \operatorname{card} s_j$, $N = n_1 + \ldots + n_k$, and introducing

$$Q_{n_1\ldots n_k} = (N-1)\Big\{\sum_{i=1}^{N}[a_N(i) - \bar{a}_N]^2\Big\}^{-1} \sum_{j=1}^{k} \frac{(S_{Nj} - n_j\bar{a}_N)^2}{n_j} \tag{2}$$

we have the following

Theorem 1 *Assume that the scores $a_N(i)$ satisfy (7.2.3.1). Let q_d be given by (7.2.1.1) and let (7.2.1.4) and (7.2.1.5) hold.*

Then, under q_d, the statistics $Q_{n_1 \dots n_k}$ have for $\min(n_1, \dots, n_k) \to \infty$ asymptotically the non-central χ^2-distribution with $k-1$ degrees of freedom and the non-centrality parameter

$$\delta = \Big\{\sum_{j=1}^{k} \Big[\Big(\sum_{i\in s_j}(d_i - \bar{d})\Big)^2 n_j^{-1}\Big]\Big\}\Big[\int_0^1 \varphi(u)\varphi(u, f_0)\,\mathrm{d}u\Big]^2 \cdot\Big\{\int_0^1 [\varphi(u) - \bar{\varphi}]^2\,\mathrm{d}u\Big\}^{-1}. \tag{3}$$

Remark 1. If $d_i = \Delta_j$, $i \in s_j$, then (7.2.1.5) amounts to

$$I(f_0)\sum_{j=1}^{k} n_j(\Delta_j - \bar{\Delta})^2 \to b^2, \quad 0 < b^2 < \infty, \tag{4}$$

where

$$\bar{\Delta} = \frac{1}{N}\sum_{j=1}^{k} n_j\Delta_j,$$

and

$$\delta = \Big\{\sum_{j=1}^{k} n_j(\Delta_j - \bar{\Delta})^2\Big\}\Big[\int_0^1 \varphi(u)\varphi(u, f_0)\,\mathrm{d}u\Big]^2 \cdot\Big\{\int_0^1 [\varphi(u) - \bar{\varphi}]^2\,\mathrm{d}u\Big\}^{-1}. \tag{5}$$

Remark 2. The theorem holds with q_d given by (7.2.2.1), if (7.2.1.5) and $\varphi(u, f_0)$ are replaced by (7.2.2.3) and $\varphi_1(u, f_0)$, respectively.

Proof. The proof follows the pattern of the proof of Theorem 6.2.2.1, except that the limiting distribution of $\sum \lambda_j\xi_{\nu j}$ should be concluded from Theorem 7.2.4.1 instead of Theorem 6.1.6.1. Thus we obtain that $(\xi_{\nu 1}, \dots, \xi_{\nu k})$ are asymptotically normal with expectations (see (7.2.4.2))

$$\mu_{\nu j} = n_{\nu j}^{-\frac{1}{2}}\sum_{i\in s_j}(d_i - \bar{d})\int_0^1 \varphi(u)\varphi(u, f_0)\,\mathrm{d}u\Big\{\int_0^1 [\varphi(u) - \bar{\varphi}]^2\,\mathrm{d}u\Big\}^{-\frac{1}{2}}$$

and the same variances and covariances as in the proof of Theorem 6.2.2.1. The rest follows from Theorem 2.4.1.1. □

Corollary 1 *Let q_d be given by (7.2.1.1) with $d_i = \Delta_j$, $i \in s_j$, and assume that (4) is satisfied. Then, under q_d, and $I(f) < \infty$, the statistics*

$$Q_{n_1 \ldots n_k} = (N-1)\Big[\sum_{i=1}^{N} a_N^2(i,f)\Big]^{-1} \sum_{j=1}^{k} \Big[\sum_{i \in s_j} a_N(R_{Ni}, f)\Big]^2 n_j^{-1} \quad (6)$$

$$\sim [I(f)]^{-1} \sum_{j=1}^{k} \Big[\sum_{i \in s_j} a_N(R_{Ni}, f)\Big]^2 n_j^{-1}$$

have for $\min(n_1, \ldots, n_k) \to \infty$ asymptotically the non-central χ^2-distribution with $k-1$ degrees of freedom and the non-centrality parameter

$$\delta = \Big[\sum_{j=1}^{k} n_j(\Delta_j - \bar{\Delta})^2\Big]\Big\{\int_0^1 \varphi(u,f)\varphi(u,f_0)\,\mathrm{d}u\Big\}^2 [I(f)]^{-1}. \quad (7)$$

If, in addition, $\varphi(u,f)$ is expressible as a sum of monotone and square integrable functions, the same holds for

$$Q_{n_1 \ldots n_k} = (N-1)\Big\{\sum_{i=1}^{N}\Big[\varphi\Big(\frac{i}{N+1}, f\Big) - \tilde{\varphi}\Big]^2\Big\}^{-1} \quad (8)$$

$$\cdot \sum_{j=1}^{k}\Big\{\sum_{i \in s_j}\Big[\varphi\Big(\frac{R_{Ni}}{N+1}, f\Big)\Big]^2 n_j^{-1} - N\tilde{\varphi}^2\Big\}$$

$$\sim [I(f)]^{-1} \sum_{j=1}^{k}\Big\{\sum_{i \in s_j}\Big[\varphi\Big(\frac{R_{Ni}}{N+1}, f\Big)\Big]^2 n_j^{-1} - N\tilde{\varphi}^2\Big\}$$

with

$$\tilde{\varphi} = \frac{1}{N}\sum_{i=1}^{N} \varphi\Big(\frac{i}{N+1}, f\Big).$$

The assertion remains true with (7.2.1.1) and (7.2.1.5) replaced by (7.2.2.1) and (7.2.2.3), respectively, and with $I(f)$, $a_N(i,f)$, $\varphi(u,f)$ and $\varphi(u,f_0)$ replaced by $I_1(f)$, $a_{1N}(i,f)$, $\varphi_1(u,f)$ and $\varphi_1(u,f_0)$, respectively.

To conclude, let us sketch the respective model and results for random blocks. We consider the density

$$q_\Delta = \prod_{j=1}^{n}\prod_{i=1}^{k} f_0(x_{ji} - \Delta_i - \alpha_j) \quad (9)$$

where $\alpha_1, \ldots, \alpha_n$ are arbitrary and

$$n\sum_{i=1}^{k}(\Delta_i - \bar{\Delta})^2 \to b^2, \quad 0 < b^2 < \infty, \quad (10)$$

where

$$\bar{\Delta} = k^{-1} \sum_{j=1}^{k} \Delta_j.$$

Then, under q_Δ, the statistics

$$Q_n = \frac{k-1}{n} \Big\{ \sum_{j=1}^{k} [a(j) - \bar{a}]^2 \Big\}^{-1} \Big\{ \sum_{i=1}^{k} \Big[\sum_{j=1}^{n} a(R_{ji}) \Big]^2 - n^2 k \bar{a}^2 \Big\} \tag{11}$$

have for $n \to \infty$ asymptotically the non-central χ^2-distribution with $k-1$ degrees of freedom and the non-centrality parameter

$$\delta = \frac{n}{k-1} \Big[\sum_{i=1}^{k} (\Delta_i - \bar{\Delta})^2 \Big] \Big[\sum_{j=1}^{k} a(j) a_k(j, f_0) \Big]^2 \tag{12}$$
$$\cdot \Big\{ \sum_{j=1}^{k} [a(j) - \bar{a}]^2 \Big\}^{-1}.$$

The proof is easy and is left for the reader as an exercise.

7.3.2 Continuous families (processes). Refer to Subsection 6.3.6 for the definition of the following symbols: $a_N(i,t)$, $T_c(t)$, $\varphi(u,t)$, and rewrite the relation following the definition of $\varphi(u,t)$:

$$\lim_{N \to \infty} \int_0^1 [a_N(1 + [uN], t) - \varphi(u,t)]^2 \, du = 0, \quad 0 \le t \le 1. \tag{1}$$

As we know from Subsection 6.3.6, under H_0, the stochastic process $T_c(t)$ converges in distribution in $(C[0,1], \mathcal{C})$ to the Brownian bridge.

Theorem 1 *Let q_d be given by (7.2.1.1), and assume that (7.2.1.4) and (7.2.1.5) hold. Let $\sum_{i=1}^{N} (c_i - \bar{c})^2 / \max_{1 \le i \le N} (c_i - \bar{c})^2 \to \infty$ hold. Put*

$$\mu_{dc}(t) = \Big[\sum_{i=1}^{N} (c_i - \bar{c})^2 \Big]^{-\frac{1}{2}} \Big[\sum_{i=1}^{N} (c_i - \bar{c})(d_i - \bar{d}) \Big] f_0\big(F_0^{-1}(t)\big). \tag{2}$$

Then, under q_d, the process $T_c(t) - \mu_{dc}(t)$ converges in distribution in $(C[0,1], \mathcal{C})$ to the Brownian bridge $Z(t)$.

Proof. For a fixed $t \in [0,1]$, Theorem 7.2.4.1 implies, in view of (1), that $T_c(t)$ is asymptotically normal $(\mu_{dc}(t), \sigma^2(t))$ with

$$\mu_{dc}(t) = \Big[\sum_{i=1}^{N} (c_i - \bar{c})^2 \Big]^{-\frac{1}{2}} \Big[\sum_{i=1}^{N} (c_i - \bar{c})(d_i - \bar{d}) \Big] \tag{3}$$

$$
\begin{aligned}
&\cdot \int_0^1 \varphi(u,t)\varphi(u,f_0)\,\mathrm{d}u, \\
\sigma^2(t) \;&=\; \int_0^1 [\varphi(u,t) - \bar{\varphi}_t]^2\,\mathrm{d}u. \qquad (4)
\end{aligned}
$$

Now since $\varphi(u,t) = 1$ for $t \leq u \leq 1$ and vanishes otherwise, we have (see the proof of Lemma 2.2.4.6)

$$\int_0^1 \varphi(u,t)\varphi(u,f_0)\,\mathrm{d}u = \int_t^1 \varphi(u,f_0)\,\mathrm{d}u = f_0\big(F_0^{-1}(t)\big) \tag{5}$$

and

$$\int_0^1 [\varphi(u,t) - \bar{\varphi}_t]^2\,\mathrm{d}u = t(1-t). \tag{6}$$

From (3) and (5) we obtain (2), and (6) equals the variance of the Brownian bridge at the point t.

We also easily show that under H_0

$$\mathsf{var}\Big(\sum_{j=1}^n \lambda_j T_c(t_j)\Big) \to \mathsf{var}\Big(\sum_{j=1}^n \lambda_j Z(t_j)\Big),$$

where $Z(t)$ is the Brownian bridge, and $n \geq 1$ and $t_j \in [0,1]$ are arbitrary. Thus the convergence of the vectors

$$[T_c(t_1) - \mu_{dc}(t_1), \ldots, T_c(t_n) - \mu_{dc}(t_n)] \to [Z(t_1), \ldots, Z(t_n)]$$

in distribution will be proved if we show that

$$\mathsf{var}\Big(\sum_{j=1}^n \lambda_j T_c(t_j)\Big) \to v^2 > 0$$

implies that

$$\sum_{j=1}^n \lambda_j [T_c(t_j) - \mu_{dc}(t_j)]$$

is asymptotically normal $(0, v^2)$ under q_d. This follows, however, from Theorem 7.2.4.1 applied to

$$\sum_{j=1}^n \lambda_j T_c(t_j) = \Big[\sum_{i=1}^N (c_i - \bar{c})^2\Big]^{-\frac{1}{2}} \sum_{i=1}^N (c_i - \bar{c}) a_N(R_i),$$

where

$$a_N(t) = \sum_{j=1}^n \lambda_j a_N(i, t_j).$$

It remains to verify condition (II) of Theorem 6.3.2.1. We know that this condition is satisfied under H_0 (see the proof of Theorem 6.3.6.1) and that the densities q_d are contiguous to the densities $p_d \in H_0$ (we do not indicate the dependence of H_0 on N). This implies, however, that condition (II) is satisfied under q_d, too. Actually, contiguity implies that for every sequence $\{d_\nu\}$ satisfying (7.2.1.4) and (7.2.1.5) and for every $\varepsilon > 0$ there exist $\alpha > 0$ and ν_0 such that

$$[P_\nu(A_\nu) < \alpha,\, \nu \geq \nu_0] \Rightarrow [Q_\nu(A_\nu) < \varepsilon], \tag{7}$$

where $\mathrm{d}P_\nu = p_{d_\nu}\,\mathrm{d}x$ and $\mathrm{d}Q_\nu = q_{d_\nu}\,\mathrm{d}x$. Now, inspecting (6.3.2.2), we can see the following: under H_0 (II) means that for every sequence $\{c_\nu\}$ satisfying $\sum(c_{\nu i} - \bar{c}_\nu)^2/\max(c_{\nu i} - \bar{c}_\nu)^2 \to \infty$ and for every $\varepsilon > 0$ and $\beta > 0$ there exist $\delta > 0$ and ν_0 such that

$$P_\nu\big(\max_{|t-s|<\delta} |T_\nu(t) - T_\nu(s)| \geq \varepsilon\big) < \beta, \quad \nu \geq \nu_0, \tag{8}$$

where $T_\nu = T_{c_\nu}$. Putting

$$A_\nu = \{\max_{|t-s|<\delta} |T_\nu(t) - T_\nu(s)| \geq \varepsilon\} \tag{9}$$

we can see that (7) entails that (8) holds with P_ν replaced by Q_ν, i.e. that (6.3.2.2) is satisfied under q_d, too. This concludes the proof. □

Application. If $c_i = 1$ and $d_i = \Delta$ for $1 \leq i \leq m$, and $c_i = d_i = 0$ otherwise, then (2) specializes to

$$\mu_{mn\Delta}(t) = \Delta\Big[\frac{mn}{m+n}\Big]^{\frac{1}{2}} f_0\big(F_0^{-1}(t)\big). \tag{10}$$

Remark. If q_d is given by (7.2.2.1), then (10) is replaced by

$$\mu_{mn\Delta}(t) = \Delta\Big[\frac{mn}{m+n}\Big]^{\frac{1}{2}} F_0^{-1}(t) f_0\big(F_0^{-1}(t)\big). \tag{11}$$

7.4 ASYMPTOTIC POWER

Here we reformulate the previous results in terms of the power of tests.

7.4.1 One-sided S-tests. Consider the rank statistic S_c given by (7.2.4.1) and the test with critical region

$$\{S_c \geq k_{1-\alpha}(\mathsf{var}\, S_c)^{\frac{1}{2}}\}, \tag{1}$$

where $k_{1-\alpha}$ denotes the $(1-\alpha)$-quantile of the standardized normal distribution. Under the conditions of Theorem 6.1.6.1 we have

$$P\big(S_c \geq k_{1-\alpha}(\mathsf{var}\, S_c)^{\frac{1}{2}}\big) \to \alpha \tag{2}$$

for $P \in H_0$. Now let

$$\rho_1 = \Big[\int_0^1 \varphi(u)\varphi(u, f_0)\,\mathrm{d}u\Big] \cdot \Big\{\int_0^1 \varphi^2(u, f_0)\,\mathrm{d}u \int_0^1 [\varphi(u)-\bar{\varphi}]^2\,\mathrm{d}u\Big\}^{-\frac{1}{2}} \tag{3}$$

and assume that

$$\Big[\sum_{i=1}^{N}(c_i-\bar{c})(d_i-\bar{d})\Big]\Big\{\sum_{i=1}^{N}(c_i-\bar{c})^2\sum_{i=1}^{N}(d_i-\bar{d})^2\Big\}^{-\frac{1}{2}} \to \rho_2. \tag{4}$$

Then, under the conditions of Theorem 7.2.4.1 and under (4),

$$Q_d\big(S_c \geq k_{1-\alpha}(\mathsf{var}\, S_c)^{\frac{1}{2}}\big) \to [1-\Phi(k_{1-\alpha}-\rho_1\rho_2 b)], \tag{5}$$

with $b > 0$ being determined by (7.2.1.5) and Φ denoting the standardized normal distribution function.

The limit $1-\Phi(k_{1-\alpha}-\rho_1\rho_2 b)$ appearing in (5) will be called the *asymptotic power* of the one-sided S_c-test at level α. It applies to all particular forms of S_c.

If $c_i = 1$ and $d_i = \Delta$ for $1 \leq i \leq m < N$ and $c_i = d_i = 0$ otherwise, then $\rho_2 = 1$ for $\Delta > 0$ and $\rho_2 = -1$ for $\Delta < 0$. We may also consider the scale alternative (7.2.2.1) and define b by (7.2.2.3).

7.4.2 Two-sided S-tests. Under the notation of the previous subsection we have

$$P\big(|S_c| \geq k_{1-\frac{1}{2}\alpha}(\mathsf{var}\, S_c)^{\frac{1}{2}}\big) \to \alpha, \quad P \in H_0, \tag{1}$$

and

$$Q_d\big(|S_c| \geq k_{1-\frac{1}{2}\alpha}(\mathsf{var}\, S_c)^{\frac{1}{2}}\big) \to 1-\Phi(k_{1-\frac{1}{2}\alpha}-\rho_1\rho_2 b)+\Phi(k_{\frac{1}{2}\alpha}-\rho_1\rho_2 b). \tag{2}$$

The last expression is called the asymptotic power of the two-sided S_c-test at level α. We again assume that the assumptions of Theorem 7.2.4.1 and condition (7.4.1.4) are satisfied. Note that $\rho_2 = 1$ for S_{mn} and for $d_i = \Delta > 0$, $1 \leq i \leq m$, and $d_i = 0$ otherwise.

7.4.3 Q-tests. Let the statistic Q be given by (7.3.1.2), and $q_d = q_\Delta$, where

$$q_\Delta = \prod_{j=1}^{k} \prod_{i \in s_j} f_0(x_i - \Delta_j).$$

Let $\chi^2_{1-\alpha,k-1}$ be the $(1-\alpha)$-quantile of the χ^2-distribution with $k-1$ degrees of freedom, and let $F_{k-1}(.\,,\delta)$ be the distribution function of the non-central χ^2 with the non-centrality parameter δ and with $k-1$ degrees of freedom. Thus

$$F_{k-1}(\chi^2_{1-\alpha,k-1}, 0) = 1-\alpha.$$

Then, for $\min(n_1, \dots, n_k) \to \infty$, we have

$$P(Q_{n_1 \dots n_k} \geq \chi^2_{1-\alpha,k-1}) \to \alpha, \quad P \in H_0. \tag{1}$$

If, in addition, $I(f_0) \sum n_j(\Delta_j - \bar{\Delta})^2 \to b^2$, then

$$Q_\Delta(Q_{n_1 \dots n_k} \geq \chi^2_{1-\alpha,k-1}) \to [1 - F_{k-1}(\chi^2_{1-\alpha,k-1}, \rho_1^2 b^2)], \tag{2}$$

where ρ_1 is given by (7.4.1.3). The limit in (2) will be called the asymptotic power of the Q-test at level α.

Corresponding results hold for the scale alternative

$$q_\Delta = \prod_{j=1}^{k} \prod_{i \in s_j} \exp(-\Delta_j) f_0\big(x_i \exp(-\Delta_j)\big)$$

with $I_1(f_0) \sum n_j(\Delta_j - \bar{\Delta})^2 \to b^2$ and $\varphi_1(u, f_0)$ instead of $\varphi(u, f_0)$ in (7.4.1.3). For proofs see Theorem 6.2.2.1 and Theorem 7.3.1.1.

7.4.4 Tests of Kolmogorov-Smirnov types. Let $T_c(t)$ be defined by (6.3.6.2) and consider the statistic

$$K_c^+ = \max_{0 \leq t \leq 1} T_c(t). \tag{1}$$

In accordance with Theorem 6.3.7.1, we have

$$P\big(K_c^+ \geq (-\tfrac{1}{2}\log\alpha)^{\frac{1}{2}}\big) \to \alpha, \quad P \in H_0. \tag{2}$$

Moreover, under the conditions of Theorem 7.3.2.1,

$$\begin{aligned} &Q_d\big(K_c^+ \geq (-\tfrac{1}{2}\log\alpha)^{\frac{1}{2}}\big) \\ &\to P\Big[\max_{0 \leq t \leq 1} \{Z(t) + b\rho_2[I(f_0)]^{-\frac{1}{2}} f_0\big(F_0^{-1}(t)\big)\} \geq (-\tfrac{1}{2}\log\alpha)^{\frac{1}{2}}\Big], \end{aligned} \tag{3}$$

where $Z(t)$ denotes the Brownian bridge (see Subsection 6.3.4), ρ_2 is given by (7.4.1.4) and b by (7.2.1.5). The limit in (3) will be called the asymptotic power of the one-sided Kolmogorov-Smirnov test at level α. For K_{mn}^+ and for $d_i = \Delta > 0$, $1 \leq i \leq m$, and $d_i = 0$ otherwise, we have $\rho_2 = 1$.

It is left for the reader to derive, on the basis of Theorem 7.3.2.1, the asymptotic power for $K_c^\pm$, R_{ac}^+, $R_{ac}^\pm$ and M_c, introduced in Subsections 6.3.7, 6.3.8 and 6.3.9, respectively.

7.4.5 Local behaviour of the asymptotic power (one-sided tests). In the previous subsection we have established the asymptotic power of the one-sided Kolmogorov-Smirnov test in terms of the probability of an event involving the Brownian bridge and the density f_0. The evaluation of this probability, however, is a rather complex task, for the solution of which no straightforward methods are available. This fact makes a direct comparison of the asymptotic power of the K^+-test with the asymptotic power of S-tests very difficult.

In order to arrive at an approximate solution of this problem, let us regard the asymptotic powers appearing in (7.4.1.5) and (7.4.4.3) as functions of b and compare the derivatives at the point $b = 0$. From (7.4.1.5) it follows that

$$[1 - \Phi(k_{1-\alpha} - \rho_1\rho_2 b) - \alpha] \sim b\rho_1\rho_2(2\pi)^{-\frac{1}{2}} \exp(-\tfrac{1}{2}k_{1-\alpha}^2), \tag{1}$$

where $\sim$ denotes that the ratio of the two sides tends to 1 as $b \to 0$. As regards (7.4.4.3), put

$$\begin{aligned} B(\alpha, f_0, b) &= P\{\max_{0\le t\le 1}[Z(t) + b\rho_2[I(f_0)]^{-\frac{1}{2}} f_0(F_0^{-1}(t))] \\ &\ge (-\tfrac{1}{2}\log\alpha)^{\frac{1}{2}}\} \end{aligned} \tag{2}$$

and

$$\psi(\alpha, u) = 2\Phi\{(-\tfrac{1}{2}\log\alpha)^{\frac{1}{2}}(2u-1)/[u(1-u)]^{\frac{1}{2}}\} - 1, \tag{3}$$

where Φ is the standardized normal distribution function.

Theorem 1 *In the above notation*

$$\begin{aligned} &[B(\alpha, f_0, b) - \alpha] \\ &\quad\sim 2b\rho_2\alpha(-\tfrac{1}{2}\log\alpha)^{\frac{1}{2}}[I(f_0)]^{-\frac{1}{2}} \int_0^1 \varphi(u, f_0)\psi(\alpha, u)\,\mathrm{d}u \end{aligned} \tag{4}$$

holds for $b \to 0$.

Proof. To simplify the notation, let us assume that $\rho_2 = 1$. Let $(C[0,1], \mathcal{C})$ be the space of continuous functions on $[0,1]$ with the σ-field of Borel subsets, as was introduced in Subsection 6.3.1. Denote by Q_b the probability distribution corresponding to the process

$$Z_b(t) = Z(t) + b[I(f_0)]^{-\frac{1}{2}} f_0(F_0^{-1}(t)), \quad b \ge 0, \tag{5}$$

where $Z(t)$ is the Brownian bridge. Obviously, Q_0 coincides with the distribution of the Brownian bridge. Put $Q_0 = P$. Recall that $[\mathrm{d}/\,\mathrm{d}t]f_0(F_0^{-1}(t)) = \varphi(t, f_0)$, and $\int_0^1 \varphi^2(u, f_0)\,\mathrm{d}u = I(f_0)$. It may be shown that, for $I(f_0) <$

∞, Q_b is absolutely continuous with respect to P, and the respective likelihood ratio $h_b = \mathrm{d}Q_b/\mathrm{d}P$ equals (see Problem 13)

$$h_b = \exp(b\xi - \tfrac{1}{2}b^2) \tag{6}$$

where

$$\xi = -[I(f_0)]^{-\frac{1}{2}} \int_0^1 \varphi(t, f_0)\,\mathrm{d}Z(t). \tag{7}$$

This stochastic integral is to be understood in the sense of Doob (1953, Chap. IX, §2) and Problems 8 and 9.

Putting

$$A = \{z(.) : \max_{0\le t\le 1} z(t) \ge (-\tfrac{1}{2}\log\alpha)^{\frac{1}{2}}\}, \tag{8}$$

we can write

$$\begin{aligned} B(\alpha, f_0, b) &= Q_b(A) = \int_A \mathrm{d}Q_b = \int_A h_b\,\mathrm{d}P \\ &= \int_A \exp(b\xi - \tfrac{1}{2}b^2)\,\mathrm{d}P. \end{aligned} \tag{9}$$

Note that

$$\left|\frac{\exp(b\xi - \frac{1}{2}b^2) - 1}{b}\right| \le \mathrm{e}^{\xi} + 2, \quad 0 \le b \le 1.$$

Moreover, $\mathsf{E}\mathrm{e}^{\xi} < \infty$, since ξ is normally distributed. Consequently, we may differentiate under the integral sign in (9), and get

$$\frac{\partial}{\partial b} B(\alpha, f_0, b)\Big|_{b=0} = \int_A \xi\,\mathrm{d}P. \tag{10}$$

It remains to show that $b\int_A \xi\,\mathrm{d}P$ coincides with the right side of (4). As the first step in this direction, let us establish the conditional distribution of $Z(t)$, given $A' = C[0,1] - A$. Note that the subprocesses $\{Z(s), 0 \le s \le t\}$ and $\{Z(s), t \le s \le 1\}$ are conditionally independent, given $Z(t) = x$. Furthermore, putting $\lambda = (-\frac{1}{2}\log\alpha)^{\frac{1}{2}}$, (6.3.4.4) implies that

$$\begin{aligned} &P(A' \mid Z(t) = x) \\ &\quad = (1 - \exp[-2\lambda(\lambda - x)/t])(1 - \exp[-2\lambda(\lambda - x)/(1-t)]), \\ &\qquad -\infty < x \le \lambda. \end{aligned} \tag{11}$$

Consequently, denoting

$$\begin{aligned} f_t(x, \lambda) &= [2\pi t(1-t)]^{-\frac{1}{2}} \exp\big(-\tfrac{1}{2}x^2/[t(1-t)]\big) \\ &\quad \cdot\big(1 - \exp[-2\lambda(\lambda - x)/t]\big)\big(1 - \exp[-2\lambda(\lambda - x)/(1-t)]\big), \\ &\qquad -\infty < x \le \lambda, \end{aligned} \tag{12}$$

we can write

$$\int_A Z(t)\,\mathrm{d}P = -\int_{A'} Z(t)\,\mathrm{d}P = -\int_{-\infty}^{\lambda} x f_t(x,\lambda)\,\mathrm{d}x. \tag{13}$$

By easy but somewhat lengthy computations, from (12) and (13) we obtain

$$\frac{\partial}{\partial t}\int_A Z(t)\,\mathrm{d}P = -2\alpha(-\tfrac{1}{2}\log\alpha)^{\frac{1}{2}}\psi(\alpha,t) \tag{14}$$

with $(-\frac{1}{2}\log\alpha)^{\frac{1}{2}} = \lambda$ and ψ being given by (3).

On comparing (7), (10) and (14), we see that the proof will be concluded if we show that

$$\begin{aligned}\int_A \xi\,\mathrm{d}P &= -[I(f_0)]^{-\frac{1}{2}}\int_A\Big[\int_0^1 \varphi(t,f_0)\,\mathrm{d}Z(t)\Big]\,\mathrm{d}P \\ &= -[I(f_0)]^{-\frac{1}{2}}\int_0^1 \varphi(t,f_0)\,\mathrm{d}_t\Big[\int_A Z(t)\,\mathrm{d}P\Big].\end{aligned} \tag{15}$$

If $\varphi(u,f_0)$ is an indicator of an interval (a,b), then

$$\xi = [I(f_0)]^{-\frac{1}{2}}[Z(a)-Z(b)],$$

and (15) obviously holds. This obviously extends to any linear combinations of such indicators. Next, let us choose an $\varepsilon > 0$ and approximate a general $\varphi(t)$, square integrable and satisfying $\varphi(0)=\varphi(1)=0$, by a linear combination $\bar{\varphi}$ of indicators of disjoint intervals such that

$$\int_0^1 [\varphi-\bar{\varphi}]^2\,\mathrm{d}t < \varepsilon^2.$$

Then, denoting

$$\bar{\xi} = -[I(f_0)]^{-\frac{1}{2}}\int_0^1 \bar{\varphi}\,\mathrm{d}Z(t),$$

we have

$$\begin{aligned}\Big|\int_A \xi\,\mathrm{d}P - \int_A \bar{\xi}\,\mathrm{d}P\Big| &\le \int |\xi-\bar{\xi}|\,\mathrm{d}P \le \Big[\int (\xi-\bar{\xi})^2\,\mathrm{d}P\Big]^{\frac{1}{2}} \\ &\le [I(f_0)]^{-\frac{1}{2}}\Big[\int_0^1 (\varphi-\bar{\varphi})^2\,\mathrm{d}t\Big]^{\frac{1}{2}} < \varepsilon[I(f_0)]^{-\frac{1}{2}}.\end{aligned}$$

On the other hand, since $|\psi(\alpha,u)| \le 1$,

$$\Big| -\int_0^1 \varphi(t)\,\mathrm{d}_t\int_A Z(t)\,\mathrm{d}P + \int_0^1 \bar{\varphi}(t)\,\mathrm{d}_t\int_A Z(t)\,\mathrm{d}P\Big|$$

$$= \left| \int_0^1 (\varphi(t) - \bar{\varphi}(t))\alpha(-\tfrac{1}{2}\log\alpha)^{\frac{1}{2}}\psi(\alpha, t)\,\mathrm{d}t \right|$$
$$\leq \alpha(-\tfrac{1}{2}\log\alpha)^{\frac{1}{2}} \left[\int_0^1 (\varphi - \bar{\varphi})^2\,\mathrm{d}t \int_0^1 \psi^2(\alpha, t)\,\mathrm{d}t \right]^{\frac{1}{2}}$$
$$\leq \alpha(-\tfrac{1}{2}\log\alpha)^{\frac{1}{2}} \left[\int_0^1 \psi^2(\alpha, t)\,\mathrm{d}t \right]^{\frac{1}{2}} \varepsilon = \alpha(-\tfrac{1}{2}\log\alpha)^{\frac{1}{2}}\varepsilon.$$

Consequently, (15) holds generally. □

7.4.6 Local behaviour of asymptotic power (multi-sided tests). Let us return to the asymptotic power of the two-sided S-test, given by (7.4.2.2). We easily verify that the first derivative of this asymptotic power with respect to b at the point $b = 0$ equals 0. Consequently, we have to characterize the local behaviour by the second derivative. This yields the following relation:

$$[1 - \Phi(k_{1-\frac{1}{2}\alpha} - \rho_1\rho_2 b) + \Phi(k_{\frac{1}{2}\alpha} - \rho_1\rho_2 b) - \alpha] \qquad (1)$$
$$\sim b^2\rho_1^2\rho_2^2(2\pi)^{-\frac{1}{2}}k_{1-\frac{1}{2}\alpha}\exp(-\tfrac{1}{2}k^2_{1-\frac{1}{2}\alpha}).$$

A similar situation arises with Q-tests, considered in Subsection 7.4.3.

Theorem 1 *Under the notation of Subsection* 7.4.3,

$$[1 - F_{k-1}(\chi^2_{1-\alpha,k-1}, \rho_1^2 b^2) - \alpha] \sim \rho_1^2 b^2 \cdot 2^{-\frac{1}{2}(k+1)} \qquad (2)$$
$$\cdot [\Gamma(\tfrac{1}{2}(k+1))]^{-1}(\chi^2_{1-\alpha,k-1})^{\frac{1}{2}(k-1)}\exp(-\tfrac{1}{2}\chi^2_{1-\alpha,k-1})$$

holds for $b \to 0$.

Proof. Let $Z_1, \ldots, Z_{k-1}$ be independent standardized normal random variables. For simplicity, assume $\rho_1^2 = 1$. Then the statistic

$$T = Z_1^2 + \ldots + Z_{k-2}^2 + (Z_{k-1} - b)^2 \qquad (3)$$

has the distribution function $F_{k-1}(x, b^2)$ considered above. On the other hand

$$P(T < x \mid Z_1^2 + \ldots + Z_{k-2}^2 = y) = P\big((Z - b)^2 < x - y\big) \qquad (4)$$
$$= \Phi\big(b + \sqrt{(x-y)}\big) - \Phi\big(b - \sqrt{(x-y)}\big), \quad y < x,$$

and

$$\frac{\partial}{\partial b^2}[\Phi\big(b + \sqrt{(x-y)}\big) - \Phi\big(b - \sqrt{(x-y)}\big)]\Big|_{b=0} \qquad (5)$$
$$= -[\sqrt{(x-y)}](2\pi)^{-\frac{1}{2}}\exp\big(-\tfrac{1}{2}(x-y)\big), \quad y < x.$$

Next, denote the density of χ^2 with $k-2$ degrees of freedom by $g_{k-2}(y)$, and recall that

$$g_{k-2}(y) = 2^{-\frac{1}{2}(k-2)}[\Gamma(\tfrac{1}{2}(k-2))]^{-1} y^{\frac{1}{2}(k-4)} \mathrm{e}^{-\frac{1}{2}y}. \tag{6}$$

Utilizing the above facts, we derive by easy but somewhat lengthy computations

$$\begin{aligned} &\frac{\partial}{\partial b^2} F_{k-1}(x, b^2) \qquad (7)\\ &= \frac{\partial}{\partial b^2} \int_{-\infty}^{x} P(T < x \mid Z_1^2 + \ldots + Z_{k-2}^2 = y) g_{k-2}(y)\,\mathrm{d}y \\ &= -2^{-\frac{1}{2}(k-2)}[\Gamma(\tfrac{1}{2}(k-2))]^{-1}(2\pi)^{-\frac{1}{2}}\mathrm{e}^{-\frac{1}{2}x} \int_{-\infty}^{x} (x-y)^{\frac{1}{2}} y^{\frac{1}{2}(k-4)}\,\mathrm{d}y \\ &= -2^{-\frac{1}{2}(k+1)}[\Gamma(\tfrac{1}{2}(k+1))]^{-1} x^{\frac{1}{2}(k-1)} \mathrm{e}^{-\frac{1}{2}x}. \end{aligned}$$

This concludes the proof. □

For $k = 2$ and $\rho_2^2 = 1$, the right side of (2) coincides with the right side of (1).

7.4.7 Progressive censoring. Let us consider the progressive censoring scheme rank statistics presented in Subsection 7.2.4. We briefly comment here on the asymptotic power for contiguous alternatives. These alternatives are the same as in the case of regression ones, treated in Subsection 7.2.4, but because of the non-linear structure of the test statistics, we have a somewhat different formulation (that resembles the structure dealt with in Subsection 7.4.4). We define for every $t \in (0,1)$, $\tilde{\varphi}_t(u)$, $u \in (0,1)$ as in (7.2.4.18) and $\nu(t)$ as in (7.2.4.19). Let then $\varphi(u, f_0)$, $u \in (0,1)$, be the usual score function, and $I(f_0)$ be the Fisher information. Further, let

$$\rho(t) = \frac{\int_0^1 \tilde{\varphi}_t(u)\varphi(u, f_0)\,\mathrm{d}u}{\{\nu(t) I(f_0)\}^{1/2}}, \quad t \in (0,1). \tag{1}$$

Finally, we note that $\omega(t)$, defined in (7.2.4.17), can also be written as $\rho(t)\sqrt{I(f_0)}$, so that the drift function is proportional to $\rho(t)$, $t \in (0,1)$. At this stage, we note that typically $\rho(t)$ is not linear in $t \in (0,1)$, so we encounter a drifted Wiener process with a non-linear drift function, and hence, the asymptotic power situation is quite comparable to the Kolmogorov-Smirnov type tests considered in the two previous subsections. For futher details, we refer to Chapter 11 of Sen (1981).

7.5 NON-CONTIGUOUS ALTERNATIVES

7.5.1 A historical note. Prior to the pioneering work of Hájek (1962), contiguity was not incorporated for the study of the asymptotic power of nonparametric statistics. Nevertheless, sparked by the early development of local alternatives by Pitman (1948), there has been considerable work on the asymptotic power of rank tests for local alternatives that need not be contiguous. Hoeffding's (1948a) treatment of U-statistics opened the doors for the study of asymptotic power for general alternatives for a large class of tests. Dwass (1956, 1957) incorporated the Hoeffding methodology for the study of asymptotic power of rank tests that can be well approximated by test statistics based on (generalized) U-statistics. The most significant result concerning non-contiguous alternatives, for the two-sample model, is due to Chernoff and Savage (1958). During the 1960s, the classical Chernoff-Savage theorem went through an evolution wherein their stringent regularity conditions on the score functions were replaced by less restrictive ones, and the two-sample model was extended to multi-sample (as well as single sample) ones. A very detailed treatise of the Chernoff-Savage representation of linear rank statistics, covering some multivariate cases as well, is due to Puri and Sen (1971). Pyke and Shorack (1968 a, b) employed the then novel concept of weak convergence of empirical processes (in the one and two-sample case) to derive asymptotic normality results for (signed) linear rank statistics in a more fashionable way; the work of Govindarajulu, LeCam and Raghavachari (1967) also deserves mention in this context. The most general result in this direction is due to Hájek (1968) who considered (as in his 1962 paper) general linear rank statistics that are appropriate for linear models as well, and provided an anatomical picture of the asymptotic distribution theory of linear rank statistics for general alternatives that cover the contiguous case as well. During the 1970s and afterwards, Hájek's theorems have been extended to more complex situations by a host of researchers, including some of his former students and colleagues in Prague. A detailed treatment of these developments may be found in Puri and Sen (1985). With our primary emphasis on contiguous alternatives, it is of considerable interest to us to depict side by side the case of non-contiguous alternatives, so that we can have a clear picture of the underlying regularity conditions on the score functions, as well as parent distributions (densities).

Hájek's approach rests on three major steps:

(i) the variance inequality for linear rank statistics;

(ii) the projection approximation for linear rank statistics;

(iii) the asymptotic normality of linear rank statistics.

We shall therefore outline these developments in the next three subsections, and in the concluding one, compare the contiguous case with the non-contiguous case.

7.5.2 Variance inequality for linear rank statistics. As in earlier sections, consider a linear rank statistic

$$S_c = \sum_{i=1}^{N} (c_i - \bar{c}) a_N(R_i). \tag{1}$$

(According to the terminology of Subsection 3.3.1, the statistic (1) should be termed precisely a *simple* linear rank statistic. However, as was done usually in the previous sections, and will be done in the sequel, the term 'simple' will be omitted for brevity.) Assume also without loss of generality that (i) $\bar{c} = 0$, (ii) $a_N(1) \leq \cdots \leq a_N(N)$ (with at least one inequality sign holding), and (iii) $\bar{a}_N = 0$. On the other hand, here we do not restrict ourselves to the underlying distribution functions $F_1, \ldots, F_N$ being identical or contiguous.

Theorem 1 *Assume that the scores $a_N(i)$ are non-decreasing, and the distribution functions $F_1, \ldots, F_N$ are all continuous (a.e.). Then*

$$\mathsf{var}(S_c) \leq 21\{\max_{1 \leq i \leq N} c_i^2\}\Big\{\sum_{i=1}^{N} a_N^2(i)\Big\}. \tag{2}$$

A key to the proof of the above theorem is the following lemma also due to Hájek (1968).

Lemma 1 *Let $X_1, \ldots, X_N$ be independent random variables, and let $\alpha(x_1, \ldots, x_N)$ and $\beta(x_1, \ldots, x_N)$ be two real functions that are non-decreasing in each argument x_i. Then*

$$\bar{\alpha}(x_1, \ldots, x_k) = \mathsf{E}[\alpha(X_1, \ldots, X_N) | X_i = x_i, i = 1, \ldots, k] \tag{3}$$

is also a non-decreasing function in each x_i, $i = 1, \ldots, k$, and

$$\mathsf{cov}\,\{\alpha(X_1, \ldots, X_N),\ \beta(X_1, \ldots, X_N)\} \geq 0, \tag{4}$$

whenever the covariance exists.

The lemma can be easily proved by the method of induction using the identity:

$$\begin{aligned} \mathsf{cov}\{\alpha(X_1, \ldots, X_N), \beta(X_1, \ldots, X_N)\} &= \mathsf{cov}\{\bar{\alpha}(X_1), \bar{\beta}(X_1)\} \\ &+ \int \mathsf{cov}\{\alpha(x, X_2, \ldots, X_N), \beta(x, X_2, \ldots, X_N)\}\, \mathrm{d}F_1(x). \end{aligned} \tag{5}$$

We relegate the proof of the theorem as an exercise at the end of this chapter. In passing, we may note that by definition $R_i = \sum_{j=1}^{N} I(X_j \leq X_i)$ is non-decreasing in X_i while non-increasing in $X_j, j \neq i$, for $i = 1, \ldots, N$. Hájek (1968) incorporated this fact in a very clever way to articulate the proof of the main theorem through a set of lemmas that we pose as exercises at the end.

7.5.3 The Hájek projection approximation. Led by the elegant projection results on U-statistics by Hoeffding (1948a), a similar result for linear rank statistics was developed by Hájek (1968); for a more general result in this direction we refer to van Zwet (1984). Towards this basic result, we consider first the following.

Lemma 1 *Let $X_1, \ldots, X_N$ be independent random variables, and let $T_N = T(X_1, \ldots, X_N)$ be a statistic such that $\mathsf{E}T_N^2 < \infty$. Let*

$$\hat{T}_{Ni} = \mathsf{E}(T_N|X_i) - \mathsf{E}(T_N), i = 1, \ldots, N; \quad (1)$$

$$\hat{T}_N = \sum_{i=1}^{N} \hat{T}_{Ni} + \mathsf{E}T_N. \quad (2)$$

Also, let $L_N = \sum_{i=1}^{N} l_{Ni}(X_i)$ be a linear statistic with independent square integrable summands. Then

$$E(T_N - \hat{T}_N)^2 = \mathsf{var}(T_N) - \mathsf{var}(\hat{T}_N), \quad (3)$$

and

$$\mathsf{E}(T_N - L_N)^2 = \mathsf{E}(T_N - \hat{T}_N)^2 + \mathsf{E}(\hat{T}_N - L_N)^2. \quad (4)$$

The proof is fairly straightforward, and hence is left as an exercise. Hájek (1968) made use of this lemma along with the results in the previous subsection to formulate the following basic result for linear rank statistics when the scores $a_N(k), k = 1, \ldots, N$, are generated by a score generating function φ that has a bounded second derivative inside $(0, 1)$.

Theorem 1 *If $\varphi(u)$ possesses a bounded second derivative $\varphi''(u)$ for all $u \in (0, 1)$, and $T_N = \sum_{i=1}^{N} c_i\varphi(R_{Ni}/(N+1))$, with the c_i as known constants, then there exists a constant $K(\varphi)$ such that*

$$E(T_N - \hat{T}_N)^2 \leq K(\varphi)N^{-1}\sum_{i=1}^{N}(c_i - \bar{c})^2, \quad (5)$$

where $\hat{T}_N$ is defined as in (2).

Note that the boundedness of the second derivative of φ implies the uniform continuity and boundedness of φ inside $(0, 1)$, and hence the scores can also be taken in a more general form. The proof of the theorem is relegated as an exercise.

The assumed boundedness condition enables us to make use of the variance inequality in the preceding subsection, as well as the projection result in Lemma 1. It may be recalled that for the projection $\hat{T}_N$, though the variance satisfies some upper bounds, it could be close to 0 in some pathological cases. Therefore, in the asymptotic normality results in Subsection 7.5.4, a mild condition is imposed to avert this limiting degeneracy.

Theorem 2 *Suppose that in addition to the regularity assumptions in Theorem 1, for every $\varepsilon > 0$, there exists a $K = K_\varepsilon(< \infty)$, such that*

$$\mathsf{var}(T_N) > K_\varepsilon\{\max_{1\le i\le N} (c_i - \bar{c})^2\}. \tag{6}$$

Then

$$\sup_{-\infty<x<\infty} |P\{T_N - \mathsf{E}T_N < x(\mathsf{var}(T_N))^{1/2}\} - \Phi(x)| < \varepsilon, \tag{7}$$

for all sufficiently large N, where $\Phi(x)$ stands for the standard normal distribution function. In both (6) and (7), it is possible to replace $\mathsf{var}(T_N)$ *by* $(\mathsf{var}\,\hat{T}_N)$, *where by definition*

$$\mathsf{var}(\hat{T}_N) = \sum_{i=1}^{N} \mathsf{var}(\hat{T}_{Ni}). \tag{8}$$

The proof is left as an exercise. Both these theorems provide the basis for the general result that is presented in the next subsection.

7.5.4 Asymptotic normality of linear rank statistics. The basic approach to extending the previous theorem to general scores is to make use of the following elegant polynomial approximation to absolutely continuous score functions that can be expressed as the difference of two monotone score functions.

Lemma 1 *Let $\varphi(u) = \varphi_1(u) - \varphi_2(u)$, $u \in (0,1)$, where both the φ_j are absolutely continuous, non-decreasing and square integrable inside $(0,1)$. Then, for every $\eta > 0$, there exists a decomposition*

$$\varphi(u) = \psi(u) + \tilde{\varphi}_1(u) - \tilde{\varphi}_2(u), \quad u \in (0,1), \tag{1}$$

where $\psi(\cdot)$ is a polynomial, and $\tilde{\varphi}_1(\cdot)$ and $\tilde{\varphi}_2(\cdot)$ are non-decreasing, such that

$$\int_0^1 \tilde{\varphi}_1^2(u)\,\mathrm{d}u + \int_0^1 \tilde{\varphi}_2^2(u)\,\mathrm{d}u < \eta. \tag{2}$$

The lemma is based on the Weierstrass theorem, and the monotonicity and square integrability of the φ_j provide an easy truncation at the two ends to prove the desired result; we pose this as an exercise.

Theorem 7.5.3.2 and the above lemma pave the way for the following general result on the asymptotic normality of linear rank statistics under general alternatives.

Theorem 1 *Suppose that the score function φ satisfies the regularity conditions in Lemma 1, and further that for every $\eta > 0$, there exists an $N_\eta(< \infty)$ such that*

$$\mathsf{var}(T_N) > N\eta\{\max_{1\le i\le N}(c_i - \bar{c})^2\} \quad \textit{for } N \ge N_\eta. \tag{3}$$

Then the conclusions of Theorem 7.5.3.2 hold for linear rank statistics based on general (possibly unbounded) scores as well.

The theorem is due to Hájek (1968). A somewhat more general theorem (permitting also discontinuous score-generating functions, i.e. dropping the assumption of their absolute continuity) was given by Dupač and Hájek (1969).

The theorem extends directly to various censored cases discussed earlier in this chapter, as well as to signed rank statistics, and multivariate cases that were considered earlier. It should be noted however that in a general setup where the X_i are not necessarily identically distributed, the centring of linear rank statistics (i.e., $\mathsf{E}T_N$) deserves some simplifications (that were not so evident from the Hájek (1968) theorems). Hoeffding (1973) studied this problem under some additional mild regularity assumptions.

7.5.5 Integration with contiguity based approaches. Note that Hájek (1968) put down side by side the simplifications that could be made when the underlying distribution functions $F_1, \ldots, F_N$ are close to each other (though the contiguity might not hold). These results exhibit a clearcut picture of the interplay of regularity assumptions on the densities f_i, the constants c_i and the score function φ in the asymptotic normality results for related rank statistics. The following result is due to Hájek (1968).

Theorem 1 *Suppose that the regularity assumptions of Theorem 7.5.4.1 hold, and assume further that for every $\varepsilon > 0$ and $\eta > 0$, there exist $N^* = N_{\varepsilon\eta}$ and $\delta^* = \delta_{\varepsilon\eta}$, such that*

$$\text{(a)} \qquad N \max_{1\le i\le N} (c_i - \bar{c})^2 < \eta^{-1} \sum_{i=1}^{N} (c_i - \bar{c})^2 \ \textit{for } N \ge N^*, \tag{1}$$

and

$$\text{(b)} \qquad \max_{1\le i\ne j\le N} \sup_{-\infty<x\infty} |F_i(x) - F_j(x)| < \delta^* \tag{2}$$

hold. Then in the asymptotic normality result for T_N, the variance can be replaced by its counterpart prevailing in the null case.

Note that here though the alternative is of local type, contiguity is not enforced; however, condition (a) is more restrictive than the Noether condition that as $N \to \infty$,

$$\frac{\max_{1 \le i \le N} (c_i - \bar{c})^2}{\sum_{i=1}^{N} (c_i - \bar{c})^2} \to 0, \tag{3}$$

as has been assumed for the contiguous case. Moreover, for both Theorems 7.5.4.1 and 7.5.5.1, the regularity assumptions on the score functions are more stringent than the square integrability condition in the contiguous case; but the finite Fisher information condition in the contiguous case is not necessary in the other cases. This picture leaves us with a variety of other local alternatives where the finite Fisher information clause can be eliminated with slightly more restrictive conditions on the c_i and/or the score function. The original Chernoff-Savage (1958) theorem for the two-sample model relates to the c_i that are either 0 or 1 (and hence bounded) so that condition (a) in the above theorem holds, and therefore, contiguity can be avoided with more stringent regularity conditions on the score functions. The Chernoff-Savage score functions have been relaxed considerably by Hájek (1968), and in that way, the passage from their alternatives to contiguous ones for multisample models is clear.

PROBLEMS AND COMPLEMENTS TO CHAPTER 7

Section 7.1

1. Let F_ν, defined by (7.1.2.4), converge weakly to F. Then $\int_0^\infty x \, \mathrm{d}F(x) \le 1$.

2. A generalization of LeCam's first lemma. Assume that F_ν given by (7.1.2.4) converge weakly to F, and that $P_\nu(A_\nu) \to 0$. Then

$$\limsup_{\nu \to \infty} Q_\nu(A_\nu) \le 1 - \int_0^\infty x \, \mathrm{d}F. \tag{1}$$

3. Show that (7.1.3.4) entails

$$\lim_{\nu \to \infty} \max_{1 \le i \le N_\nu} P_\nu\Big(\Big|\Big[\frac{g_{\nu i}(X_i)}{f_{\nu i}(X_i)}\Big]^\alpha - 1\Big| > \varepsilon\Big) = 0, \quad \alpha, \varepsilon > 0.$$

4. Suppose that for some $t \in (0, 1]$ the statistic

$$\sum_{i=1}^{N_\nu} \{[g_{\nu i}(X_i)/f_{\nu i}(X_i)]^t - 1\}$$

is asymptotically normal $(\frac{1}{2}t(t-1)\sigma^2, t^2\sigma^2)$. Then, under (7.1.3.4), the same result holds for every $t \in (0,1]$ (cf. LeCam).

5. Under (7.1.3.4), the following conditions are equivalent: (a) $\log L_\nu$ is asymptotically normal $(-\frac{1}{2}\sigma^2, \sigma^2)$; (b) for some $t \in (0,1]$ (and hence for all)

$$\sum_{i=1}^{N_\nu} \{[g_{\nu i}(X_i)/f_{\nu i}(X_i)]^t - 1\}$$

is asymptotically normal $(\frac{1}{2}t(t-1)\sigma^2, t^2\sigma^2)$ (cf. LeCam).

Section 7.2

6. If a density $f(x)$ has finite Fisher information, then $s(x) = [f(x)]^{\frac{1}{2}}$ is absolutely continuous. On the other hand, absolute continuity of $f(x)$ does not imply the same result for $s(x)$. [Hint: see the definition of absolute continuity in Subsection 2.2.1, and utilize $\int_{-\infty}^{\infty} |f'(x)/f^{\frac{1}{2}}(x)|\,\mathrm{d}x < \infty$.]

Definition. Let J be an open interval containing 0. A family of densities $d(x,\Theta)$, $\Theta \in J$ will be said to satisfy condition A_2, if
(i) $d(x,\Theta)$ is absolutely continuous in Θ for almost every x;
(ii) the limit

$$\dot{d}(x,0) = \lim_{\Theta\to 0} \frac{1}{\Theta}[d(x,\Theta) - d(x,0)] \tag{2}$$

exists a.e.
(iii)

$$\lim_{\Theta\to 0} \int_{-\infty}^{\infty} \frac{\dot{d}(x,\Theta)]^2}{d(x,\Theta)}\,\mathrm{d}x = \int_{-\infty}^{\infty} \frac{[\dot{d}(x,0)]^2}{d(x,0)}\,\mathrm{d}x < \infty$$

holds, with $\dot{d}(x,\Theta)$ denoting the partial derivative with respect to Θ. [Cf. condition A_1 in Subsection 3.4.8.]

7. A modification of Theorem 7.2.4.1. Put

$$q_\Theta = \prod_{i=1}^{N} d(x_i, \Theta_i)$$

and assume that $\max_{1\le i\le N} \Theta_i^2 \to 0$ and $I(d)\sum_{i=1}^N \Theta_i^2 \to b^2$, $0 < b^2 < \infty$, hold, with

$$I(d) = \int_{-\infty}^{\infty} \frac{[\dot{d}(x,0)]^2}{d(x,0)}\,\mathrm{d}x.$$

Then, under condition A_2, the statistics S_c given by (7.2.4.1) are, for

$$\sum_{i=1}^{N} (c_i - \bar{c})^2 / \max_{1\le i\le N} (c_i - \bar{c})^2 \to \infty,$$

asymptotically normal $(\mu_{\Theta_c}, \sigma_c^2)$ with

$$\mu_{\Theta_c} = \Big[\sum_{i=1}^{N}(c_i - \bar{c})(\Theta_i - \bar{\Theta})\Big] \int_0^1 \varphi(u)\varphi(u, d, 0)\,\mathrm{d}u$$

and

$$\sigma_c^2 = \Big[\sum_{i=1}^{N}(c_i - \bar{c})^2\Big] \int_0^1 [\varphi(u) - \bar{\varphi}]^2\,\mathrm{d}u,$$

where $\varphi(u, d, 0)$ is defined by (2.2.4.4).

Section 7.4

Definition. The Gaussian process $\{Z(t),\ 0 \le t \le 1\}$ with zero expectations and covariances $\mathsf{cov}(Z(t), Z(s)) = \min(t, s)$ will be called a *Brownian process.*

8. Stochastic integral. If $Z(t)$, $0 \le t \le 1$, is a Brownian process and $h(t)$ a square integrable function, then the stochastic integral $\int_0^1 h(t)\,\mathrm{d}Z(t)$ may be defined as follows: First, if $h(t)$ is a step function constant in each interval $(t_{i-1}, t_i]$, $1 \le i \le n$, $t_0 = 0$, $t_n = 1$, we put

$$\int_0^1 h(t)\,\mathrm{d}Z(t) = \sum_{i=1}^{n} h(t_i)[Z(t_i) - Z(t_{i-1})].$$

Generally, we select a sequence of functions $\{h_N(t)\}$, such that

$$\lim_{N\to\infty} \int_0^1 [h_N(t) - h(t)]^2\,\mathrm{d}t = 0,$$

and that each function $h_N(t)$ is a step function. Since

$$\begin{aligned} &\mathsf{E}\Big[\int_0^1 h_N(t)\,\mathrm{d}Z(t) - \int_0^1 h_{N+k}(t)\,\mathrm{d}Z(t)\Big]^2 \\ &= \int_0^1 [h_N(t) - h_{N+k}(t)]^2\,\mathrm{d}t \to 0, \quad \text{for } N \to \infty. \end{aligned} \tag{3}$$

uniformly in $k \ge 0$, there exists a limit in the mean,

$$Y = \text{l.i.m.} \int_0^1 h_N(t)\,\mathrm{d}Z(t).$$

Next, we show that Y does not depend on the choice of the particular sequence $\{h_N\}$, and define $\int_0^1 h(t)\,\mathrm{d}Z(t) = Y$.

9. Continuation. Since in (3) the inequality $\leq$ performs the same service as the equality, the procedure is also applicable if $Z(t)$ has a covariance $\sigma(t,s)$ which is dominated by the covariance $\min(t,s)$ of the Brownian process, i.e. such that $\min(t,s)-\sigma(t,s)$ is a covariance, too. Show that this case occurs if $\sigma(t,s)$ denotes the conditional covariances of the Brownian process given $v=0$, with v being some finite linear combination $\sum_{k=1}^{n}\lambda_k Z(t_k)$ or a limit in the mean of such linear combinations.

10. Continuation. The Brownian bridge may be regarded as a conditional Brownian process given $Z(1)=0$. Thus, for square integrable $h(t)$, the integral $\int_0^1 h(t)\,\mathrm{d}Z(t)$ is well defined for a Brownian bridge, too.

11. Absolute continuity and Radon-Nikodym derivatives. Let Q_b, $b \geq 0$, denote the distribution of the Gaussian process $Z(t)$ with a fixed covariance $\sigma(t,s)$ and expectations $\mathsf{E}Z(t) = b\psi(t)$, $t \in T$. Put $P = Q_0$, and denote by $\mathbf{M} = \{\xi\}$ the space of all finite linear combinations $\sum_{k=1}^{n}\lambda_k Z(t_k)$ and their limits in the mean.

Then Q_b is absolutely continuous with respect to P, if and only if there exists a random variable $\xi \in \mathbf{M}$ such that

$$\mathsf{E}[Z(t)\xi] = \psi(t), \quad t \in T, \tag{4}$$

where E refers to P. If (4) holds, then

$$\frac{\mathrm{d}Q_b}{\mathrm{d}P} = \exp(b\xi - \tfrac{1}{2}b^2 \operatorname{var}\xi), \quad -\infty < b < \infty. \tag{5}$$

[Hájek (1960b).]

12. Let $Z(t)$ denote a Brownian bridge and consider the set $\mathbf{M}$ of all random variables

$$\xi = \int_0^1 h(t)\,\mathrm{d}Z(t) \tag{6}$$

where

$$\int_0^1 h^2(t)\,\mathrm{d}t < \infty \quad \text{and} \quad \int_0^1 h(t)\,\mathrm{d}t = 0. \tag{7}$$

Prove that $Z(t) \in \mathbf{M}$, $0 \leq t \leq 1$, and that $\mathbf{M}$ is linear and closed with respect to the convergence in the mean.

13. Apply Problem 11 to the Brownian bridge considered in Problem 12 and show that absolute continuity occurs if and only if $\psi(0) = \psi(1) = 0$ and ψ is absolutely continuous with square integrable $\psi'(t)$. Prove (7.4.5.6).

Section 7.5

14. Prove Theorem 7.5.2.1 with the help of Lemma 7.5.2.1. [Hájek (1968).]

15. Prove Lemma 7.5.3.1. [Hájek (1968).]

16. Prove Theorem 7.5.3.1. [Hájek (1968).]

17. Prove Theorem 7.5.3.2. [Hájek (1968).]

18. Prove Lemma 7.5.4.1. [Hájek (1968).]

Chapter 8

Asymptotic optimality and efficiency

8.1 ASYMPTOTICALLY OPTIMUM TESTS

8.1.1 Introduction. Consider testing H_0 against some simple alternative q. Assume that for some $p^0 \in H_0$ the likehood ratio q/p^0 is a function of the ranks $r = (r_1, \dots, r_N)$ only, i.e., that

$$q/p^0 = h(r). \tag{1}$$

Then the test Ψ^0 defined by

$$\Psi^0 = \begin{cases} 1, & \text{if } h(r) > k \\ 0, & \text{if } h(r) < k \end{cases} \tag{2}$$

and arbitrarily otherwise, is most powerful at level $\alpha = \int \Psi^0 \,\mathrm{d}P^0$ for p^0 against q. Further, since $\Psi^0 = \Psi^0(R)$ and the distribution of R does not depend on the choice of $P \in H_0$ the distribution of Ψ^0 is invariant over H_0, too, and a fortiori

$$\int \Psi^0 \,\mathrm{d}P = \alpha \quad \text{for all } P \in H_0. \tag{3}$$

Thus p^0 is least favourable for H_0 against q, in the sense of Subsection 2.3.5, and Ψ^0 is most powerful for H_0 against q, according to Lemma 2.3.5.1 If we put $K = \{q\}$, where q are all the densities such that

$$q = p\,h(r), \quad p \in H_0, \tag{4}$$

then Ψ^0 is most powerful for H_0 against K, too. This follows easily from the fact that $\int \Psi^0 \,\mathrm{d}Q$ does not depend on the choice of $Q \in K$.

More generally, if $K = \{q_d, d \in W\}$ and

$$q_d/p^0 = h_d(r), \quad d \in W, \tag{5}$$

for some density $p^0 \in H_0$, then the maximin most powerful test for H_0 against K may be found within the family of rank tests. In such a situation, giving the term sufficiency a broader meaning, we could say that the sub-σ-field generated by ranks is sufficient for distinguishing between H_0 and K.

Unfortunately, relation (5) does not hold for any problem of practical importance. We may still hope, however, to have more luck with an appropriate asymptotic version of this relation.

Denote by $\|p - q\|$ the L_1-distance of two probability densities:

$$\|p - q\| = \int |p - q| \, \mathrm{d}\nu, \tag{6}$$

where ν is some σ-finite measure with respect to which the densities are defined. If P and Q are the respective probability distributions, then $\|p-q\|$ equals twice the variation of $P - Q$:

$$\|p - q\| = 2 \sup_{A \in \mathcal{A}} |P(A) - Q(A)|. \tag{7}$$

If $\mathcal{B}$ is a sub-σ-field of $\mathcal{A}, \mathcal{B} \subset \mathcal{A}$, $\bar{P}$ and $\bar{Q}$ are restrictions of P and Q on $\mathcal{B}$, respectively, and $\bar{p}$ and $\bar{q}$ are $\mathcal{B}$-measurable densities corresponding to $\bar{P}$ and $\bar{Q}$, respectively, then (7) entails

$$\|\bar{p} - \bar{q}\| \leq \|p - q\|. \tag{8}$$

Finally, if Ψ is a critical function, then

$$\left| \int \Psi \, \mathrm{d}P - \int \Psi \, \mathrm{d}Q \right| \leq \|p - q\|, \tag{9}$$

which suggests that $\|p - q\|$ might be a useful concept in asymptotic problems of the theory of hypotheses testing.

Let us now consider a sequence of testing problems 'H_ν against K_ν', $\nu \geq 1$, with H_ν being H_0 applied to N_ν observations and $K_\nu = \{q_{\nu d}, d \in W_\nu\}$. We shall assume that $W_\nu = W_{\nu 0} \times W_{\nu 1}$, $d = (a, b)$, $a \in W_{\nu 0}$, $b \in W_{\nu 1}$.

Theorem 1 *Let us consider testing H_ν against K_ν and denote by $\beta(\alpha, H_\nu, K_\nu)$ the power of the maximin most powerful test, and by $\bar{\beta}(\alpha, H_\nu, K_\nu)$ the power of the maximin most powerful rank test. Assume that there are densities $p^0_{\nu a} \in H_\nu$ and rank statistics $h_{\nu b} = h_{\nu b}(r_\nu)$ such that the functions*

$$q^0_{\nu d} = p^0_{\nu a} h_{\nu b}, \quad \nu \geq 1, \; d = (a, b) \in W_\nu \tag{10}$$

are densities and satisfy

$$\lim_{\nu\to\infty} \sup_{d\in W_\nu} \|q_{\nu d} - q^0_{\nu d}\| = 0. \tag{11}$$

Then

$$\lim_{\nu\to\infty} [\beta(\alpha, H_\nu, K_\nu) - \bar{\beta}(\alpha, H_\nu, K_\nu)] = 0, \quad 0 \le \alpha \le 1. \tag{12}$$

Proof. Let Ψ_ν be a critical function of size $\le \alpha$, and let us denote by $\bar{\psi}_{\nu a}$ its conditional expectation under $P^0_{\nu a}$ given R_ν:

$$\bar{\psi}_{\nu a}(r_\nu) = N! \int_{\{R_\nu = r_\nu\}} \Psi_\nu \, \mathrm{d}P^0_{\nu a}. \tag{13}$$

Since $\int \Psi_\nu \, \mathrm{d}P_\nu \le \alpha$, $P_\nu \in H_\nu$, we also have $\int \bar{\psi}_{\nu a} \, \mathrm{d}P^0_{\nu a} \le \alpha$. Moreover, since the distribution of $\bar{\psi}_{\nu a} = \bar{\psi}_{\nu a}(R_\nu)$ under P_ν does not depend on the choice of $P_\nu \in H_\nu$, all critical rank functions $\bar{\psi}_{\nu a}$, $a \in W_{\nu 0}$, have the correct size:

$$\int \bar{\psi}_{\nu a} \, \mathrm{d}P_\nu \le \alpha, \quad P_\nu \in H_\nu. \tag{14}$$

On the other hand, according to (9),

$$\int \Psi_\nu \, \mathrm{d}Q_{\nu d} \le \int \Psi_\nu \, \mathrm{d}Q^0_{\nu d} + \|q_{\nu d} - q^0_{\nu d}\| \tag{15}$$

holds. Furthermore, since the distribution of $\bar{\psi}_{\nu a}(R_\nu) h_{\nu b}(R_\nu)$ under P_ν does not depend on the choice of $P_\nu \in H_\nu$, we can write, in view of (10),

$$\begin{aligned}\int \Psi_\nu \, \mathrm{d}Q^0_{\nu d} &= \int \Psi_\nu h_{\nu b} \, \mathrm{d}P^0_{\nu a} = \int \bar{\psi}_{\nu a} h_{\nu b} \, \mathrm{d}P^0_{\nu a} \\ &= \int \bar{\psi}_{\nu a} h_{\nu b} \, \mathrm{d}P^0_{\nu a'} = \int \bar{\psi}_{\nu a} \, \mathrm{d}Q^0_{\nu d'}, \\ &\qquad\qquad d = (a, b),\ d' = (a', b).\end{aligned} \tag{16}$$

Thus

$$\begin{aligned}\inf_{d\in W_\nu} \int \Psi_\nu \, \mathrm{d}Q^0_{\nu d} &= \inf_{\substack{a,a'\in W_{\nu 0}\\ b\in W_{\nu 1}}} \int \bar{\psi}_{\nu a} h_{\nu b} \, \mathrm{d}P_{\nu a'} \le \inf_{\substack{a'\in W_{\nu 0}\\ b\in W_{\nu 1}}} \int \bar{\psi}_{\nu a} h_{\nu b} \, \mathrm{d}P_{\nu a'} \\ &= \inf_{d'\in W_\nu} \int \bar{\psi}_{\nu a} \, \mathrm{d}Q^0_{\nu d'}, \quad a \in W_{\nu 0}.\end{aligned} \tag{17}$$

Consequently, for every $a \in W_{\nu 0}$, $\bar{\psi}_{\nu a}$ is not inferior to Ψ_ν in testing H_ν against $\{q^0_{\nu d}, d \in W_\nu\}$. Now, on combining

$$\int \bar{\psi}_{\nu a} \, \mathrm{d}Q^0_{\nu d} \le \int \bar{\psi}_{\nu a} \, \mathrm{d}Q_{\nu d} + \|q_{\nu d} - q^0_{\nu d}\| \tag{18}$$

with (15), and (17), we obtain

$$\inf_{d\in W_\nu} \int \Psi_\nu \,\mathrm{d}Q_{\nu d} \le \inf_{d\in W_\nu} \int \bar{\psi}_{\nu a} \,\mathrm{d}Q_{\nu d} + 2 \sup_{d\in W_\nu} \|q_{\nu d} - q^0_{\nu d}\|, \quad a \in W_{\nu 0}, \tag{19}$$

which, in view of (11), concludes the proof. □

Remark. Similar situations may be considered with H_ν denoting some other hypothesis and R_ν replaced by some other statistic. For example, H_ν may denote H_1 (or H_2) with the role of R_ν played by $(R_\nu^+, \operatorname{sign} X_\nu)$ (or (R_ν, Q_ν)).

Definition. If (10) and (11) are satisfied, we say that the vector of ranks R_ν is *asymptotically sufficient* for distinguishing between H_ν and K_ν.

8.1.2 Asymptotic sufficiency of the vector of ranks.

Let us consider the alternatives

$$q_{\nu d} = \prod_{i=1}^{N_\nu} f_0(x_i - d_{\nu i}), \quad d_\nu \in W_\nu, \tag{1}$$

or

$$q_{\nu d} = \prod_{i=1}^{N_\nu} \exp(-d_{\nu i}) f_0(x_i \exp(-d_{\nu i})), \quad d_\nu \in W_\nu \tag{2}$$

where W_ν denotes a subset of an N_ν-dimensional vector. Let H_ν denote H_0, applied to N_ν observations. For simplicity, we sometimes write d instead of d_ν.

Theorem 1 *Let us assume that*

$$I(f_0) \sum_{i=1}^{N_\nu} (d_{\nu i} - \bar{d}_\nu)^2 < K < \infty, \quad d_\nu \in W_\nu, \ \nu \ge 1, \tag{3}$$

and

$$\lim_{\nu\to\infty} \sup_{d\in W_\nu} \max_{1\le i\le N_\nu} (d_{\nu i} - \bar{d}_\nu)^2 = 0 \tag{4}$$

hold. Put

$$\bar{d}_\nu = (1/N_\nu) \sum_{i=1}^{N_\nu} d_{\nu i}, \quad d_\nu^0 = (d_{\nu 1} - \bar{d}_\nu, \ldots, d_{\nu N_\nu} - \bar{d}_\nu),$$

and

$$p^0_{\nu\bar{d}} = \prod_{i=1}^{N_\nu} f_0(x_i - \bar{d}_\nu), \quad \nu \geq 1,\ d_\nu \in W_\nu, \tag{5}$$

and consider the alternatives (1).

Then there exist rank statistics $h_{\nu d^0}$ *such that* (8.1.1.11) *holds for densities* $q^0_{\nu d}$ *given by*

$$q^0_{\nu d} = p^0_{\nu\bar{d}} h_{\nu d^0}, \quad \nu \geq 1,\ d \in W_\nu. \tag{6}$$

The same holds for the alternatives (2), *if* $I(f_0)$ *is replaced by* $I_1(f_0)$ *in* (3), *and if*

$$p^0_{\nu\bar{d}} = \exp(-N_\nu \bar{d}_\nu) \prod_{i=1}^{N_\nu} f_0(x_i \exp(-\bar{d}_\nu)).$$

Remark. In the sense of Definition 8.1.1, we could also say that the vector of ranks $R_\nu = (R_{\nu 1}, \ldots, R_{\nu N_\nu})$ is asymptotically sufficient for H_ν against $\{q_{\nu d}, d \in W_\nu\}$. By Theorem 8.1.1.1, the loss of power due to restricting ourselves to rank tests tends to zero as $\nu \to \infty$. Here the role of a and b is played by $\bar{d}$ and d^0, respectively. Obviously, d may be decomposed into and composed from $\bar{d}$ and d^0, $d = (\bar{d}, d^0)$.

Proof. Let us introduce

$$S_{\nu d^0} = \sum_{i=1}^{N_\nu} (d_{\nu i} - \bar{d}_\nu) a_N(R_{\nu i}, f_0) \tag{7}$$

and

$$\sigma^2_{\nu d^0} = I(f_0) \sum_{i=1}^{N_\nu} (d_{\nu i} - \bar{d}_\nu)^2. \tag{8}$$

We shall show that

$$[\log(q_{\nu d}/p_{\nu\bar{d}}) - S_{\nu d^0} + \tfrac{1}{2}\sigma^2_{\nu d^0}] \to 0 \tag{9}$$

in $P_{\nu\bar{d}}$-probability, and that this convergence is uniform with respect to $d \in W_\nu$.

From $\mathsf{E}(q_{\nu d}/p_{\nu\bar{d}}) \leq 1$ we may conclude that for every $\varepsilon > 0$ there exists an M such that $P_{\nu\bar{d}}(q_{\nu d}/p_{\nu\bar{d}} > M) < \varepsilon$ holds for every $\nu \geq 1$. Further, (3) entails that $\exp(S_{\nu d^0} - \frac{1}{2}\sigma^2_{\nu d^0})$ is bounded away from 0 in $P_{\nu\bar{d}}$-probability. Consequently, the uniform fulfilment of (9) is equivalent to

$$\lim_{\nu\to\infty} \sup_{d \in W_\nu} P_{\nu\bar{d}}\{|q_{\nu d}/p_{\nu\bar{d}} - \exp(S_{\nu d^0} - \tfrac{1}{2}\sigma^2_{\nu d^0})| > \varepsilon\} = 0, \quad \varepsilon > 0. \tag{10}$$

If (10) were not true, there would exist an $\varepsilon_0 > 0$ and a sequence $\{d_j\}$, $d_j \in W_{\nu_j}$, $\nu_j \to \infty$, such that

$$P_{\nu_j \bar{d}}\{|q_{\nu_j d}/p_{\nu_j \bar{d}} - \exp(S_{\nu_j d^0} - \tfrac{1}{2}\sigma^2_{\nu_j d^0})| > \varepsilon_0\} > \varepsilon_0. \tag{11}$$

On the other hand, in view of (3) and (4), the sequence $\{d_j\}$ contains a subsequence $\{d_h\} \subset \{d_j\}$ such that

$$I(f_0) \sum_{i=1}^{N_h} (d_{hi} - \bar{d}_h)^2 \to b^2, \quad 0 \le b^2 < \infty, \tag{12}$$

and

$$\max_{1 \le i \le N_h} (d_{hi} - \bar{d}_h)^2 \to 0. \tag{13}$$

Consequently, if $b^2 > 0$, Theorems 7.2.1.1 and 7.2.4.1 and formula (6.1.5.14) entail

$$P_{\nu_h \bar{d}}\{|q_{\nu_h d}/p_{\nu_h \bar{d}} - \exp(S_{\nu_h d^0} - \tfrac{1}{2}\sigma^2_{\nu_h d^0})| > \varepsilon_0\} \to 0 \tag{14}$$

which contradicts (11), since $\{\nu_h\} \subset \{\nu_j\}$. Noting that for $b^2 = 0$, (14) is justified by Remark in 7.2.4, we see that the proof of (10) is finished.

Now we shall put

$$\begin{aligned} h_{\nu d^0} &= B_{\nu d^0} \exp(S_{\nu d^0} - \tfrac{1}{2}\sigma^2_{\nu d^0}), \quad \text{if } |S_{\nu d^0}| \le C_\nu \\ &= 0, \quad \text{if } |S_{\nu d^0}| > C_\nu \end{aligned} \tag{15}$$

with

$$B_{\nu d^0} = \left[\int_{-C_\nu}^{C_\nu} \exp(x - \tfrac{1}{2}\sigma^2_{\nu d^0})\, \mathrm{d}P_\nu(S_{\nu d^0} \le x) \right]^{-1}, \tag{16}$$

where the constants C_ν will be approximately determined in a moment. Before doing that, let us note that $h_{\nu d^0}$ is a rank statistic depending on d^0_ν only, because it is a function of the statistic $S_{\nu d^0}$ possessing the same property. Furthermore, let us observe that $h_{\nu d^0}$ inserted into the right side of (6) provides a density, i.e. $\int q^0_{\nu d}\, \mathrm{d}x = 1$, $x = (x_1, \ldots, x_{N_\nu})$. Now we may proceed to the choice of the constants C_ν, which will be a somewhat lengthy procedure. We shall put

$$C_\nu = k, \quad \nu_k \le \nu < \nu_{k+1}, \tag{17}$$

for a strictly increasing sequence $\{\nu_k, k \ge 1\}$ to be chosen below.

It may be easily shown that

$$\|p - q\| = 2 \int_{\{q<p\}} (1 - q/p)\, \mathrm{d}P. \tag{18}$$

Consequently for $q^0_{\nu d}$ given by (6),

$$\begin{aligned} \|q_{\nu d} - q^0_{\nu d}\| &= 2\int_{\{q<q^0\}} (1 - q_{\nu d}/q^0_{\nu d})\,\mathrm{d}Q^0_{\nu d} \\ &= 2\int_{\{q/p<h\}} (h_{\nu d^0} - q_{\nu d}/p_{\nu \bar{d}})\,\mathrm{d}P_{\nu \bar{d}} \\ &= 2\int_{\{q/p<h\}} [h_{\nu d^0} - \exp(S_{\nu d^0} - \tfrac{1}{2}\sigma^2_{\nu d^0})]\,\mathrm{d}P_{\nu \bar{d}} \\ &\quad +2\int_{\{q/p<h\}} [\exp(S_{\nu d^0} - \tfrac{1}{2}\sigma^2_{\nu d^0}) - q_{\nu d}/p_{\nu \bar{d}}]\,\mathrm{d}P_{\nu \bar{d}}. \end{aligned} \tag{19}$$

The first of the last two integrals is bounded by $|1 - B^{-1}_{\nu d^0}|$ and the second by $\varepsilon_\nu + \alpha_{\nu d}\exp(C_\nu)$, with

$$\alpha_{\nu d} = P_{\nu \bar{d}}(\exp(S_{\nu d^0} - \tfrac{1}{2}\sigma^2_{\nu d^0}) - q_{\nu d}/p_{\nu \bar{d}} > \varepsilon_\nu). \tag{20}$$

Consequently,

$$\|q_{\nu d} - q^0_{\nu d}\| \le 2|1 - B^{-1}_{\nu d^0}| + 2\varepsilon_\nu + 2\alpha_{\nu d}\exp(C_\nu). \tag{21}$$

Our next aim is to explore the limiting behaviour of $B_{\nu d^0}$. First we need to show that for every natural number k

$$\begin{aligned} \lim_{\nu\to\infty} \sup_{d\in W_\nu} \Bigg| &\int_{-k}^{k} \exp(x - \tfrac{1}{2}\sigma^2_{\nu d^0})\,\mathrm{d}P_\nu(S_{\nu d^0} \le x) \\ &- \int_{-k}^{k} \exp(x - \tfrac{1}{2}\sigma^2_{\nu d^0})\,\mathrm{d}\Phi(x/\sigma_{\nu d^0}) \Bigg| = 0, \end{aligned} \tag{22}$$

with Φ denoting the standardized normal distribution function. This may be proved by reasoning similar to that used in the proof of (10), i.e. by assuming the opposite and then drawing a contradictory subsequence. Here the contradiction has to be derived either from the asymptotic normality of $S_{\nu d^0}$, which is ensured, by Corollary 6.1.5.1, for any sequence $\{d_h\}$ such that

$$I(f_0)\sum_{i=1}^{N_h}(d_{hi} - \bar{d}_h)^2 \to b^2,$$

$0 < b^2 < \infty$, and $\max(d_{hi} - \bar{d}_h)^2 \to 0$, or from the degenerate asymptotic normality of $S_{\nu d^0}$ for $b = 0$, which is trivial.

Further, note that

$$\int_{-\infty}^{\infty} \exp(x - \tfrac{1}{2}\sigma^2)\,\mathrm{d}\Phi(x/\sigma) = 1, \quad \sigma \ge 0, \tag{23}$$

and that

$$\lim_{k \to \infty} u_k = 0 \tag{24}$$

for

$$u_k = \sup_{\sigma^2 \leq K} \left| \int_{-k}^{k} \exp(x - \tfrac{1}{2}\sigma^2)\, \mathrm{d}\Phi(x/\sigma) - 1 \right|, \tag{25}$$

where K is the same as in (3). This may be proved by easy algebraic transformations. Now we are ready to define the subsequence $\{\nu_k\}$ appearing in (17). Having defined $\nu_1, \ldots, \nu_{k-1}$, let ν_k be some integer satisfying the following requirements: (i) $\nu_k > \nu_{k-1}$; (ii) $|B_{\nu d^0}^{-1} - 1| < 2u_k$ for $\nu \geq \nu_k$ and $d \in W_\nu$, with $B_{\nu d^0}$ defined by (16) and (17); (iii) $\alpha_{\nu d} \exp(k) < 1/k$ for $\nu \geq \nu_k$ and $d \in W_\nu$, with $\alpha_{\nu d}$ defined by (20), where $\varepsilon_\nu = 1/k$. The possibility of fulfilling (ii) is apparent from (22) and (23), and the possibility of fulfilling (iii) follows from (10). With this definition of ν_k (21) entails

$$\|q_{\nu d} - q_{\nu d}^0\| \leq 4u_k + 2/k + 2/k, \quad \nu_k \leq \nu < \nu_{k+1},\ d \in W_\nu \tag{26}$$

which, on account of (24), implies (8.1.1.11).

We have simultaneously proved that, with the above definition of $\{\nu_k\}$,

$$\lim_{\nu \to \infty} B_{\nu d^0} = 1 \tag{27}$$

uniformly in $d \in W_\nu$.

The case of alternatives (2) could be treated similarly with the help of the last part of Corollary 6.1.5.1 and Theorem 7.2.2.1. □

Theorem 2 *Let H_ν denote H_1 applied to N_ν observations, $N_\nu \to \infty$. Put*

$$q_\nu = \prod_{i=1}^{N_\nu} f_0(x_i - \Delta_\nu),$$

where

$$N_\nu \Delta_\nu^2 \leq K < \infty,$$

and f_0 is some density symmetric about zero and such that $I(f_0) < \infty$.

Then there exist rank statistics $h_\nu = h_\nu(R_\nu^+, \operatorname{sign} X_\nu)$ such that $\|q_\nu - q_\nu^0\| \to 0$ holds for densities q_ν^0 given by

$$q_\nu^0 = h_\nu p_\nu^0,$$

where

$$p_\nu^0 = \prod_{i=1}^{N_\nu} f_0(x_i).$$

Proof. We omit this proof, since it is similar to the proof of Theorem 1. As a matter of fact, the present proof would be even simpler, since the alternatives are simple for every fixed $\nu \geq 1$. See Theorems 6.1.7.1 and 7.2.5.1. □

8.1.3 Asymptotically optimum one-sided tests. For problems treated in the previous subsection we have shown that asymptotically optimum test may be found among the rank tests. Now we are going to establish these optimum tests explicitly for some important particular cases. Let us stress that these tests will be asymptotically optimum not only within the class of rank tests, but even within the class of all possible tests. Their optimality is preserved even for some subhypotheses of $H_0(H_1)$, provided these subhypotheses include the densities $p^0_{\nu\bar{d}}$ (p^0_ν). It is also worth mentioning that in contrast to possible 'parametric' solutions for testing the sub-hypotheses against $q_{\nu d}$, the rank tests do not involve any estimation problems (estimation of location and scale in partial samples, for example, as in the case of the t-test). On the other hand, if the subhypotheses do not include the densities $p^0_{\nu\bar{d}}$, then rank tests may be quite useless.

Definition. A test Ψ_ν will be called asymptotically maximin most powerful for testing H_ν against K_ν at level α if

$$\limsup_{\nu\to\infty} \sup_{P_\nu \in H_\nu} \int \Psi_\nu \, \mathrm{d}P_\nu \le \alpha \tag{1}$$

and

$$\lim_{\nu\to\infty} \left[\beta(\alpha, H_\nu, K_\nu) - \inf_{Q_\nu \in K_\nu} \int \Psi_\nu \, \mathrm{d}Q_\nu\right] = 0. \tag{2}$$

For simplicity, let us again delete the subscript ν in what follows, writing

$$\sum_{i=1}^{N} (d_i - \bar{d}) \to b^2 \quad \text{for} \quad \lim_{\nu\to\infty} \sum_{i=1}^{N_\nu} (d_{\nu i} - \bar{d}_\nu)^2 = b^2 \quad \text{etc.}$$

We say that a test is based on a statistic S if the hypothesis is accepted for $S < c$, and rejected for $S > c$, whereas for $S = c$ randomization is permitted. By a proper choice of c and of the randomization, the size of test based on S may attain any value from $[0, 1]$. If the distribution of S does not depend on the choice of $P \in H$ and is asymptotically normal $(\mathsf{E}S, \mathsf{var}\, S)$, then (1) will be satisfied if $\Psi = 1$ for $S \ge \mathsf{E}S + k_{1-\alpha}(\mathsf{var}\, S)^{\frac{1}{2}}$ and $\Psi = 0$ otherwise. Let us begin with the easiest case, where the alternative is simple.

Theorem 1 *Consider testing H_0 against*

$$q = \prod_{i=1}^{N} f_0(x_i - d_i). \tag{3}$$

Then, for

$$I(f_0) \sum_{i=1}^{N} (d_i - \bar{d})^2 \to b^2, \quad 0 < b^2 < \infty \tag{4}$$

and

$$\max_{i \le i \le N} (d_i - \bar{d})^2 \to 0 \tag{5}$$

the following relation holds:

$$\beta(\alpha, H_0, q) \to 1 - \Phi(k_{1-\alpha} - b). \tag{6}$$

The maximum power is asymptotically reached by the test based on the statistic

$$S = \sum_{i=1}^{N} d_i a_N(R_i, f_0). \tag{7}$$

The same holds if $I(f_0)$, (3) *and* (7) *are replaced by* $I_1(f_0)$,

$$q = \exp\Big(-\sum_{i=1}^{N} d_i\Big) \prod_{i=1}^{N} f_0(x_i \exp(-d_i)), \tag{8}$$

and

$$S = \sum_{i=1}^{N} d_i a_{1N}(R_i, f_0), \tag{9}$$

respectively.

Remark. The scores $a_N(i, f_0)$ could be replaced by any other scores $a_N(i)$ such that

$$\int_0^1 [a_N(1 + [uN]) - \varphi(u, f_0)]^2 \, du \to 0,$$

and similarly for $a_{1N}(i, f_0)$.

Proof. First, it is clear that

$$\beta(\alpha, H_0, q) \le \beta(\alpha, p^0, q), \tag{10}$$

where

$$p^0 = \prod_{i=1}^{N} f_0(x_i - \bar{d}). \tag{11}$$

On the other hand, from LeCam's third lemma and from Theorem 7.2.1.1 it follows that $\log(q/p^0)$ is asymptotically normal $(-\frac{1}{2}b^2, b^2)$ under p^0, and is asymptotically normal $(\frac{1}{2}b^2, b^2)$ under q. Consequently, a test based on q/p^0 has the following asymptotic power:

$$\beta(\alpha, p^0, q) \to 1 - \Phi(k_{1-\alpha} - b). \tag{12}$$

On the other hand, this asymptotic power belongs to the test based on S, according to Subsection 7.4.1. Consequently

$$\liminf \beta(\alpha, H_0, q) \geq 1 - \Phi(k_{1-\alpha} - b). \tag{13}$$

The rest follows upon combining (10), (12) and (13). The proof for q given by (8) is quite similar. □

Corollary 1 *Consider testing H_0 against*

$$q = \prod_{i=1}^{m} f_0(x_i - \Delta) \prod_{i=m+1}^{m+n} f_0(x_i). \tag{14}$$

Then, for

$$I(f_0)\Delta^2 \frac{mn}{m+n} \to b^2 \tag{15}$$

and

$$\min(m, n) \to \infty \tag{16}$$

the following relation holds:

$$\beta(\alpha, H_0, q) \to 1 - \Phi(k_{1-\alpha} - b). \tag{17}$$

The maximum power is asymptotically reached by the test based on the statistic

$$S = \sum_{i=1}^{m} a_N(R_i, f_0). \tag{18}$$

The same holds if $I(f_0)$, (14) and (18) are replaced by $I_1(f_0)$,

$$q = \exp(-m\Delta) \prod_{i=1}^{m} f_0(x_i \exp(-\Delta)) \prod_{i=m+1}^{m+n} f_0(x_i) \tag{19}$$

and

$$S = \sum_{i=1}^{m} a_{1N}(R_i, f_0), \tag{20}$$

respectively.

8.1.4 Asymptotically optimum multi-sided tests. Let us first reformulate and extend the definitions of Subsection 6.2.1 in terms of distribution functions. We say that a sequence of k-variate distribution functions $\{F_\nu\}$ converges *completely* to F if for every continuous and uniformly bounded function $h(z) = h(z_1, \ldots, z_k)$

$$\lim_{\nu\to\infty} \int h \,\mathrm{d}F_\nu = \int h \,\mathrm{d}F \tag{1}$$

holds. We say that F_ν converges *weakly* if (1) holds for every continuous function vanishing outside a compact (i.e. closed and bounded) set. Complete convergence corresponds essentially to D1. Both convergences, weak and complete, extend to all functions satisfying $\mathrm{d}F \geq 0$ and $\int \mathrm{d}F < \infty$. Equivalent versions of the above definitions, corresponding to D2–D5 of Subsection 6.2.1, are possible here, too.

It is easy to show that the following lemma holds.

Lemma 1 *A sequence $\{F_\nu\}$ converging weakly converges completely if and only if either of the following two conditions holds:*

(i) *For every $\varepsilon > 0$ there exists a bounded set K_ε such that*

$$\int_{E-K_\varepsilon} \mathrm{d}F_\nu < \varepsilon, \quad \textit{for all } \nu \geq 1, \tag{2}$$

with E denoting the whole Euclidean space.

(ii)

$$\lim_{\nu\to\infty} \int \mathrm{d}F_\nu = \int \mathrm{d}F. \tag{3}$$

In what follows $\mathrm{d}G = \psi \,\mathrm{d}F$ denotes that $\int_A \mathrm{d}G = \int_A \psi \,\mathrm{d}F$, $A \in \mathcal{A}$.

Lemma 2 *Let $F_\nu \to F$ completely and $G_\nu \to G$ weakly. Further, assume that $\mathrm{d}G_\nu = \psi_\nu \,\mathrm{d}F_\nu$, $0 \leq \psi_\nu \leq 1$, $\nu \geq 1$.*

Then $G_\nu \to G$ completely and $\mathrm{d}G = \psi \,\mathrm{d}F$, $0 \leq \psi \leq 1$.

Proof. Denote the respective Euclidean space by E. For any nonnegative, continuous function $h(z)$, $z \in E$, vanishing outside of a compact set,

$$\int h \,\mathrm{d}G_\nu \leq \int h \,\mathrm{d}F_\nu \tag{4}$$

holds. Consequently,

$$\int h \,\mathrm{d}G \leq \int h \,\mathrm{d}F,$$

which entails $\mathrm{d}G = \psi \,\mathrm{d}F$, $0 \leq \psi \leq 1$.

Moreover, since $F_\nu \to F$ completely, there exists for every $\varepsilon > 0$ a bounded set K_ε such that (2) holds. Consequently, in view of $0 \leq \psi_\nu \leq 1$, we also have

$$\int_{E-K_\varepsilon} \mathrm{d}G_\nu < \varepsilon, \quad \nu \geq 1, \tag{5}$$

which together with weak convergence entails complete convergence. See Lemma 1.

□

Lemma 3 *Let $h_{\nu\Theta}$ and h_Θ, $\nu \geq 1$, $\Theta \in M$, be non-negative measurable functions such that*

$$\lim_{\nu\to\infty} \sup_{\Theta\in M} \sup_{z\in K} |h_{\nu\Theta}(z) - h_\Theta(z)| = 0 \tag{6}$$

holds for every compact subset K of a Euclidean space E. Further, assume that the functions h_Θ, $\Theta \in M$, are uniformly bounded and uniformly continuous on every compact K.

Let $F_\nu \to F$ completely. Suppose that

$$1 = \int h_{\nu\Theta}\,\mathrm{d}F_\nu = \int h_\Theta\,\mathrm{d}F, \quad \nu \geq 1, \ \Theta \in M, \tag{7}$$

and that for every $\varepsilon > 0$ there exists a compact K_ε such that

$$\sup_{\Theta\in M} \int_{E-K_\varepsilon} h_\Theta\,\mathrm{d}F < \varepsilon. \tag{8}$$

Then

$$\lim_{\nu\to\infty} \int h_{\nu\Theta}\,\mathrm{d}G_\nu = \int h_\Theta\,\mathrm{d}G \tag{9}$$

holds uniformly in $\Theta \in M$ for every sequence G_ν such that $G_\nu \to G$ weakly and such that $dG_\nu = \psi_\nu\,\mathrm{d}F_\nu$, $0 \leq \psi_\nu \leq 1$.

Proof. From the assumption that the functions h_Θ, $\Theta \in M$, are uniformly bounded and uniformly continuous on every compact K, it follows that

$$\lim_{\nu\to\infty} \int_K h_\Theta\,\mathrm{d}G_\nu = \int_K h_\Theta\,\mathrm{d}G \tag{10}$$

holds uniformly in $\Theta \in M$ for every compact K the boundary of which has G-measure zero. On the other hand, (6) entails

$$\lim_{\nu\to\infty} \int_K (h_{\nu\Theta} - h_\Theta)\,\mathrm{d}G_\nu = 0 \tag{11}$$

uniformly in $\Theta \in M$ for every compact K. Consequently,

$$\lim_{\nu\to\infty} \int_K h_{\nu\Theta}\,\mathrm{d}G_\nu = \int_K h_\Theta\,\mathrm{d}G \tag{12}$$

uniformly in $\Theta \in M$ for every compact K with zero boundary. Thus it remains to show that the truncated parts of the integrals appearing in (9) are uniformly negligible.

From (8) and Lemma 2 it follows that

$$\int_{E-K_\varepsilon} h_\Theta \,\mathrm{d}G \leq \int_{E-K_\varepsilon} h_\Theta \,\mathrm{d}F < \varepsilon. \tag{13}$$

On the other hand

$$\begin{aligned}\int_{E-K_\varepsilon} h_{\nu\Theta} \,\mathrm{d}G_\nu &\leq \int_{E-K_\varepsilon} h_{\nu\Theta} \,\mathrm{d}F_\nu = 1 - \int_{K_\varepsilon} h_{\nu\Theta} \,\mathrm{d}F_\nu \\ &\leq 1 + \left| \int_{K_\varepsilon} (h_{\nu\Theta} - h_\Theta) \,\mathrm{d}F_\nu \right| - \int_{K_\varepsilon} h_\Theta \,\mathrm{d}F_\nu \\ &\quad \to 1 - \int_{K_\varepsilon} h_\Theta \,\mathrm{d}F = \int_{E-K_\varepsilon} h_\Theta \,\mathrm{d}F < \varepsilon.\end{aligned} \tag{14}$$

□

Theorem 1 *Consider testing H_0 against $K = \{q_\Delta, \Delta \in D\}$ with*

$$q_\Delta = \prod_{j=1}^{k} \prod_{i \in s_j} f_0(x_i - \Delta_j), \quad \Delta \in D, \tag{15}$$

where $s_1, \ldots, s_k$ is a partition of $\{1, \ldots, N\}$, $n_j = \operatorname{card} s_j$, $N = n_1 + \ldots + n_k$, and $D = D(b^2, n_1, \ldots, n_k)$ denotes the set of all vectors $\Delta = (\Delta_1, \ldots, \Delta_k)$ such that

$$I(f_0) \sum_{j=1}^{k} n_j (\Delta_j - \bar{\Delta})^2 = b^2, \quad 0 < b^2 < \infty \tag{16}$$

with

$$\bar{\Delta} = (1/N) \sum_{j=1}^{k} n_j \Delta_j.$$

Assume that

$$\min(n_1, \ldots, n_k) \to \infty. \tag{17}$$

Then

$$\beta(\alpha, H_0, K) \to 1 - F_{k-1}(\chi^2_{1-\alpha,k-1}, b^2) \tag{18}$$

with the right-hand side having the same meaning as in Subsection 7.4.3.

The power on the right-hand side of (18) is reached asymptotically by the test based on the statistic

$$Q = [I(f_0)]^{-1} \sum_{j=1}^{k} \Big[\sum_{i \in s_j} a_N(R_{Ni}, f_0) \Big]^2 / n_j. \tag{19}$$

Proof. The fact that the asymptotic power of the Q-test equals

$$1 - F_{k-1}(\chi^2_{1-\alpha,k-1}, b^2)$$

follows immediately from Corollary 7.3.1.1. Consequently, it remains to prove (18). This will be accomplished by piecing together Lemma 2, the proof of Theorem 8.1.2.1, Theorem 2.4.2.3 and Problem 4 to Section 8.1. To avoid confusion, let us resume denoting the members of sequences by ν. Thus (17) may be rewritten as $\lim_{\nu\to\infty} \min(n_{\nu 1}, \ldots, n_{\nu k}) = \infty$, for example.

Suppose that (18) does not hold. Passing to a subsequence, if necessary, we may then assume that

$$\lim_{\nu\to\infty} \beta(\alpha, H_{0\nu}, K_\nu) > 1 - F_{k-1}(\chi^2_{1-\alpha,k-1}, b^2) \tag{20}$$

and that

$$\lim_{\nu\to\infty} n_{\nu j}/N_\nu = \kappa_j^2, \quad 1 \leq j \leq k. \tag{21}$$

Introduce random variables

$$\xi_{\nu j} = [n_{\nu j} I(f_0)]^{-\frac{1}{2}} \sum_{i \in s_{\nu j}} a_{N_\nu}(R_{\nu i}, f_0) \tag{22}$$

and note that they are under H_0 asymptotically normal with variances $1-\kappa_j^2$ and covariances $-\kappa_j\kappa_g$, $1 \leq j \neq g \leq k$, and with vanishing means. This distribution was considered in Subsection 2.4.1.

Now observe that we are dealing with a special case of the situation considered in Theorem 8.1.2.1. In the present case

$$d_{\nu i} = \Delta_{\nu j}, \quad i \in s_{\nu j}, \quad 1 \leq j \leq k, \quad \bar{d}_\nu = \bar{\Delta}_\nu,$$
$$\sum_{i=1}^{n} (d_{\nu i} - \bar{d}_\nu)^2 = \sum_{j=1}^{k} n_{\nu j}(\Delta_{\nu j} - \bar{\Delta}_\nu)^2,$$

and $d \in W_\nu$ is equivalent to $\Delta \in D_\nu$. Furthermore, (16) and (17) entail (8.1.2.3) and (8.1.2.4), so that the assumptions of Theorem 8.1.2.1 are satisfied.

In the course of the proof of Theorem 8.1.2.1, we have established the statistics (8.1.2.15), which in our special case are all functions of $(\xi_{\nu 1}, \ldots, \xi_{\nu k})$. Putting

$$\Theta_j = [n_{\nu j} I(f_0)]^{\frac{1}{2}} (\Delta_{\nu j} - \bar{\Delta}_\nu), \tag{23}$$
$$\kappa_{\nu j} = [n_{\nu j}/N_\nu]^{\frac{1}{2}}, \tag{24}$$
$$V_\nu = \left\{\Theta : \sum_{j=1}^{k} \Theta_j^2 = b^2, \sum_{j=1}^{k} \kappa_{\nu j}\Theta_j = 0\right\}, \tag{25}$$

and writing $h_{\nu\Theta}$ and $B_{\nu\Theta}$ instead of $h_{\nu d^0}$ and $B_{\nu d^0}$, we can present (8.1.2.15) in the following form:

$$h_{\nu\Theta} = \begin{cases} B_{\nu\Theta} \exp\Big(\sum_{j=1}^{k} \Theta_j \xi_{\nu j} - \frac{1}{2} b^2\Big), & \text{if } \Big|\sum_{j=1}^{k} \Theta_j \xi_{\nu j}\Big| < C_\nu \\ 0, & \text{otherwise,} \end{cases} \tag{26}$$

with $\Theta \in V_\nu$.

Now (20) entails the existence of a critical function Ψ_ν of size $\leq \alpha$ such that

$$\lim_{\nu\to\infty} \left[\inf_{\Theta\in V_\nu} \int \Psi_\nu \, \mathrm{d}Q_{\nu\Theta}\right] > 1 - F_{k-1}(\chi^2_{1-\alpha,k-1}, b^2) \tag{27}$$

where, for Θ given by (23), $Q_{\nu\Theta}$ is identical with $Q_{\nu\Delta}$, which is determined by (15) with $n_j = n_{\nu j}$. From (8.1.1.11), which holds under the conditions of Theorem 8.1.2.1, it follows that also

$$\lim_{\nu\to\infty} \left[\inf_{\Theta\in V_\nu} \int \Psi_\nu \, \mathrm{d}Q^0_{\nu\Theta}\right] > 1 - F_{k-1}(\chi^2_{1-\alpha,k-1}, b^2) \tag{28}$$

where

$$q^0_{\nu\Theta} = h_{\nu\Theta} p_{\nu\bar{\Delta}}, \quad p_{\nu\bar{\Delta}} = \prod_{i=1}^{N_\nu} f_0(x_i - \bar{\Delta}_\nu).$$

Introduce

$$\psi_\nu(\xi_{\nu 1}, \ldots, \xi_{\nu k}) = \mathsf{E}(\Psi_\nu \mid \xi_{\nu 1}, \ldots, \xi_{\nu k}) \tag{29}$$

with the conditional expectation referring to density $p_{\nu 0} = \prod_{i=1}^{N_\nu} f_0(x_i)$. By the same argument as used in (8.1.1.17), we can show that

$$\inf_{\Theta\in V_\nu} \int \Psi_\nu \, \mathrm{d}Q^0_{\nu\Theta} = \inf_{\Theta\in V_\nu} \int \psi_\nu \, \mathrm{d}Q^0_{\nu\Theta} \tag{30}$$

which, in view of (28), entails

$$\liminf_{\nu\to\infty} \left[\inf_{\Theta\in V_\nu} \int \psi_\nu \, \mathrm{d}Q^0_{\nu\Theta}\right] > 1 - F_{k-1}(\chi^2_{1-\alpha,k-1}, b^2). \tag{31}$$

Carrying over the integration into the space of $(\xi_1, \ldots, \xi_k)$-values, we can write

$$\begin{aligned} \int \psi_\nu \, \mathrm{d}Q^0_{\nu d} &= \int \psi_\nu h_{\nu\Theta} \, \mathrm{d}P_{\nu\bar{\Delta}} \\ &= \int \psi_\nu(z_1, \ldots, z_k) h_{\nu\Theta}(z_1, \ldots, z_k) \, \mathrm{d}F_\nu, \quad \Theta \in V_\nu, \end{aligned} \tag{32}$$

with F_ν denoting the distribution function of $(\xi_1, \ldots, \xi_k)$ and $h_{\nu\Theta}$ is determined by (26) with $z_j = \xi_{\nu j}$.

Note that $\sum z_j \kappa_{\nu j} = 0$ with F_ν-probability 1, so that $h_{\nu\Theta} = h_{\nu\Theta'}$ with F_ν-probability 1, if $\Theta_j - \Theta_j' = c\kappa_{\nu j}$. Consequently

$$\inf_{\Theta \in V_\nu} \int \psi_\nu h_{\nu\Theta} \, \mathrm{d}F_\nu = \inf_{\sum \Theta_j^2 = b^2} \int \psi_\nu h_{\nu\Theta} \, \mathrm{d}F_\nu. \tag{33}$$

From the proof of Theorem 8.1.2.1, namely from the properties of the constants $B_{\nu d^0}$ and C_ν, it follows that

$$h_{\nu\Theta} \to h_\Theta = \exp\Big(\sum_{j=1}^k \Theta_j z_j - \tfrac{1}{2} b^2\Big), \quad \sum_{j=1}^k \Theta_j^2 = b^2 \tag{34}$$

and that the convergence satisfies condition (6). Further, recall that $F_\nu \to F$ completely, F being the k-dimensional normal distribution function with vanishing means and variances $1 - \kappa_j^2$ and covariances $-\kappa_j \kappa_g$, $1 \le j \ne g \le k$, $\sum\limits_{j=1}^k \kappa_j^2 = 1$. Denoting by F_Θ the k-dimensional normal distribution function with the same variances-covariances and with mean values $\Theta_1, \ldots, \Theta_k$, $\sum \Theta_j^2 = b^2$, $\sum \Theta_j \kappa_j = 0$, we can easily show that $\mathrm{d}F_\Theta = h_\Theta \, \mathrm{d}F$, with h_Θ given by (34) (see Problem 4 to Section 8.1). Generally, if

$$\sum_{j=1}^k \Theta_j \kappa_j = 0$$

is not necessarily satisfied, $h_\Theta \, \mathrm{d}F = \mathrm{d}F_{\Theta'}$ holds with

$$\Theta_j' = \Theta_j - \kappa_j \sum_{i=1}^k \Theta_i \kappa_i, \quad 1 \le j \le k,$$

(note that $\sum\limits_{j=1}^k \kappa_j Z_j = 0$ with probability 1 under F, so that $h_\Theta = h_{\Theta'}$ with probability 1). This entails that conditions (7) and (8) are satisfied for $M = \{\sum \Theta_j^2 = b^2\}$. Finally, passing to a subsequence, if necessary, we may assume that the sequence $\{G_\nu\}$, $\mathrm{d}G_\nu = \psi_\nu \, \mathrm{d}F_\nu$ converges weakly to G. Then, by Lemma 8.1.4.2, $\mathrm{d}G = \psi \, \mathrm{d}F$, with ψ being a critical function of size $\le \alpha$, if the functions ψ_ν have the same property. Further, by Lemma 8.1.4.3,

$$\inf_{\sum \Theta_j^2 = b^2} \int \psi_\nu h_{\nu\Theta} \, \mathrm{d}F_\nu \to \inf_{\sum \Theta_j^2 = b^2} \int \psi h_\Theta \, \mathrm{d}F. \tag{35}$$

Now from Theorem 2.4.2.3 it follows that

$$\inf_{\sum \Theta_j^2 = b^2} \int \psi h_\Theta \, \mathrm{d}F = \inf_{\substack{\sum \Theta_j^2 = b^2 \\ \sum \Theta_j \kappa_j = 0}} \int \psi \, \mathrm{d}F_\Theta \tag{36}$$

$$= 1 - F_{k-1}(\chi^2_{1-\alpha,k-1}, b^2).$$

On combining (31), (32), (35) and (36), we easily come to a contradiction. □

For $k = 2$ we obtain the following

Corollary 1 *Consider testing H_0 against $K = \{q_1, q_2\}$ with*

$$q_t = \prod_{i=1}^{m} f_0(x_i + (-1)^t \Delta) \prod_{i=m+1}^{m+n} f_0(x_i), \quad t = 1, 2. \tag{37}$$

Then, for

$$I(f_0) \frac{mn}{m+n} \Delta^2 = b^2, \quad 0 < b^2 < \infty, \tag{38}$$

and

$$\min(m, n) \to \infty, \tag{39}$$

the following relation holds:

$$\beta(\alpha, H_0, K) \to 1 - \Phi(k_{1-\frac{1}{2}\alpha} - b) + \Phi(k_{\frac{1}{2}\alpha} - b). \tag{40}$$

The power on the right hand side of (40) *is asymptotically reached by the test based on the statistic $|S|$, with S given by* (8.1.3.18).

Proof. It is sufficient to note that

$$F_1(\chi^2_{1-\alpha,1}, b^2) = \Phi(k_{1-\frac{1}{2}\alpha} - b) - \Phi(k_{\frac{1}{2}\alpha} - b).$$

□

8.2 ASYMPTOTIC EFFICIENCY OF TESTS

8.2.1 One-sided tests. Complete information on the limiting properties of a test is provided by its asymptotic power. Comparing this asymptotic power with that of the asymptotically most powerful test, we obtain a picture of the 'asymptotic efficiency' of the test. If all statistics under consideration have asymptotically normal distributions under both the hypothesis and the alternative, the notion of 'asymptotic efficiency' may be defined by one number, independent of the level of significance and admitting a suggestive interpretation in k-sample problems, $k \geq 2$.

Assume that the asymptotically most powerful test for H against q may be based on a statistic S_0 which is asymptotically normal $(0, \sigma_0^2)$ under H_0, and which is asymptotically normal (μ_0, σ_0^2) under q. Further, consider another test based on a statistic S, which is asymptotically normal $(0, \sigma^2)$ under H_0, and which is asymptotically normal (μ, σ^2) under q. Then the asymptotic powers of the S_0-test and of the S-test equal

$$1 - \Phi(k_{1-\alpha} - \mu_0\sigma_0^{-1}) \quad \text{and} \quad 1 - \Phi(k_{1-\alpha} - \mu\sigma^{-1}), \tag{1}$$

respectively. This suggests the following

Definition. The number

$$e = (\mu\sigma_0/\mu_0\sigma)^2 \tag{2}$$

will be called the *asymptotic efficiency* of the S-test. (If we wish to stress its distinction from other definitions of efficiency, it is called the *Pitman asymptotic efficiency.*)

In this definition we assume tacitly $\mu \geq 0$. If $\mu < 0$, then the S-test is worse than the test for which $\Psi \equiv \alpha$ constantly, disregarding the observations, and the notion of 'efficiency' makes no sense.

We shall consider the alternatives

$$q = \prod_{i=1}^{N} f_0(x_i - d_i), \tag{3}$$

or

$$q = \exp\Big(-\sum_{i=1}^{N} d_i\Big) \prod_{i=1}^{N} f_0[x_i \exp(-d_i)], \tag{4}$$

and the statistics

$$S_c = \sum_{i=1}^{N} (c_i - \bar{c}) a_N(R_i). \tag{5}$$

We shall assume that the scores $a_N(i)$ are related to some square integrable function $\varphi(u)$ as follows:

$$\lim_{N\to\infty} \int_0^1 \big(a_N(1 + [uN]) - \varphi(u)\big)^2 \,\mathrm{d}u = 0. \tag{6}$$

Furthermore, we shall assume that

$$\max_{1\leq i\leq N} (d_i - \bar{d})^2 \ \to\ 0, \tag{7}$$

$$I(f_0) \sum_{i=1}^{N} (d_i - \bar{d})^2 \ \to\ b^2, \quad 0 < b^2 < \infty, \tag{8}$$

$$\sum_{i=1}^{N} (c_i - \bar{c})^2 / \max_{1\leq i\leq N} (c_i - \bar{c})^2 \ \to\ \infty, \tag{9}$$

and

$$\sum_{i=1}^{N}(c_i - \bar{c})(d_i - \bar{d})\Big[\sum_{i=1}^{N}(c_i - \bar{c})^2 \sum_{i=1}^{N}(d_i - \bar{d})^2\Big]^{-\frac{1}{2}} \to \rho_2. \tag{10}$$

Finally, we shall denote

$$\int_0^1 \varphi(u)\varphi(u, f_0)\,\mathrm{d}u\Big[\int_0^1 \varphi^2(u, f_0)\,\mathrm{d}u \cdot \int_0^1 \left(\varphi(u) - \bar{\varphi}\right)^2 \mathrm{d}u\Big]^{-\frac{1}{2}} = \rho_1. \tag{11}$$

Theorem 1 *Consider testing H_0 against q given by (3). Under conditions (6) through (10) the (Pitman) asymptotic efficiency of the S_c-test equals*

$$e = \rho_1^2\rho_2^2, \tag{12}$$

with ρ_1 and ρ_2 defined by (11) and (10), respectively.

The same holds if (3), $I(f_0)$ in (8), and $\varphi(u, f_0)$ in (11) are replaced by (4), $I_1(f_0)$ and $\varphi_1(u, f_0)$, respectively.

Proof. According to (8.1.3.6) the asymptotically most powerful test has the asymptotic power $1 - \Phi(k_{1-\alpha} - b)$, whereas the S_c-test yields, according to (7.4.1.5), the asymptotic power $1 - \Phi(k_{1-\alpha} - \rho_1\rho_2 b)$. The rest is immediate. □

Remark 1. We again tacitly assume $\rho_1\rho_2 \geq 0$. Typically we have $\rho_1 \geq 0$ and $\rho_2 \geq 0$.

Remark 2. The product $\rho_1\rho_2$ also equals the limiting correlation coefficient of S_c and S given by (8.1.3.7). Thus the asymptotic efficiency equals the limiting square correlation coefficient of an asymptotically best test statistic and the test statistic under consideration.

Remark 3. (Pitman's interpretation of asymptotic efficiency.) In some situations the number $100(1 - e)$ may be regarded as the percentage of observations 'wasted' by making use of the S-test instead of the asymptotically most powerful test. This may be illustrated by the two-sample problem, in which $d_i = \Delta$, $1 \leq i \leq m$, and $d_i = 0$, $m < i \leq m + n$. Then the S-test distinguishes H_0 from

$$q = \prod_{i=1}^{m} f_0(x_i - \Delta) \prod_{i=m+1}^{m+n} f_0(x_i) \tag{13}$$

with the same asymptotic power as the asymptotically most powerful test based on $S_0 = \sum_{i=1}^{m'} a_N(i, f_0)$ distinguishes H_0 from

$$q' = \prod_{i=1}^{m'} f_0(x_i - \Delta) \prod_{i=m'+1}^{m'+n'} f_0(x_i),$$

where

$$m' = [em] \quad \text{and} \quad n' = [en].$$

Actually, if $\min(m, n) \to \infty$ and $\Delta^2 mn(m+n)^{-1} \to b^2$, the former asymptotic power equals $1 - \Phi(k_{1-\alpha} - \rho_1\rho_2 b)$, and the latter equals $1 - \Phi(k_{1-\alpha} - e^{\frac{1}{2}} b)$, so that equality is obtained for e given by (12). Since, for $\min(m, n) \to \infty$, $m'/m \to e$, $n'/n \to e$, the interpretation follows.

Remark 4. Similarly, we could show that the S-test distinguishes H_0 from (3) asymptotically with the same power as the most powerful test distinguishes H_0 from

$$q' = \prod_{i=1}^{N} f_0(x_i - e^{\frac{1}{2}} d_i).$$

This yields another interpretation for e: The effect of making use of the S-test instead of the best test causes asymptotically the same decrement of the power as the replacement of the regression constants d_i by constants $e^{\frac{1}{2}} d_i$, $1 \leq i \leq N$.

We conclude the present section by reformulating the above results for the two-sample problem. Consider the alternatives (13) or

$$q = \exp(-m\Delta) \prod_{i=1}^{m} f_0(x_i e^{-\Delta}) \prod_{i=m+1}^{m+n} f_0(x_i) \tag{14}$$

and statistics

$$S_{mn} = \sum_{i=1}^{m} a_N(R_i). \tag{15}$$

Furthermore, assume that

$$I(f_0)\Delta^2 mn(m+n)^{-1} \to b^2, \quad 0 < b^2 < \infty, \tag{16}$$

and

$$\min(m, n) \to \infty. \tag{17}$$

Corollary 1 *Consider testing H_0 against q given by (13). Under conditions (6), (16), (17) and $\Delta > 0$, the asymptotic efficiency of the S_{mn}-test equals*

$$e = \rho_1^2, \tag{18}$$

with ρ_1 defined by (11).

The same holds if (13), $I(f_0)$ in (16), and $\varphi(u, f_0)$ in (11) are replaced by (14), $I_1(f_0)$ and $\varphi_1(u, f_0)$, respectively.

8.2.2 Multi-sided tests. Denote by $F_{k-1}(.\,,\delta^2)$ the distribution function of non-central χ^2 with $k-1$ degrees of freedom and with the coefficient of non-centrality δ^2. Assume that the asymptotically maximin most powerful test for H against K may be based on a statistic Q_0, which is asymptotically χ^2-distributed with $k-1$ degrees of freedom under H_0, and which is asymptotically distributed according to $F_{k-1}(.\,,\delta_0^2)$ under K. Further, consider another test based on a statistic Q, which is also asymptotically χ^2-distributed with $k-1$ degrees of freedom under H_0, but which is $F_{k-1}(.\,,\delta^2)$-distributed under K. Then the asymptotic powers of the Q_0-test and of the Q-test equal

$$1 - F_{k-1}(\chi^2_{1-\alpha,k-1}, \delta_0^2) \quad \text{and} \quad 1 - F_{k-1}(\chi^2_{1-\alpha,k-1}, \delta^2), \tag{1}$$

respectively.

Definition. The number

$$e = \delta^2/\delta_0^2 \tag{2}$$

will be called the asymptotic efficiency of the Q-test.

We shall consider the alternatives $K = \{q_\Delta, \Delta \in D\}$ with

$$q_\Delta = \prod_{j=1}^{k} \prod_{i \in s_j} f_0(x_i - \Delta_j), \tag{3}$$

where $s_1, \ldots, s_k$ is a partition of $\{1, \ldots, N\}$, $n_j = \operatorname{card} s_j$, $N = n_1 + \ldots + n_k$, and $D = D(b^2, n_1, \ldots, n_k)$ denotes the set of all vectors $\Delta = (\Delta_1, \ldots, \Delta_k)$ such that

$$I(f_0) \sum_{j=1}^{k} n_j (\Delta_j - \bar{\Delta})^2 = b^2, \tag{4}$$

with $\bar{\Delta}(1/N) \sum_{j=1}^{k} n_j \Delta_j$. Further, we shall consider tests of H_0 against K based on statistics

$$Q = (N-1)\Big[\sum_{i=1}^{N} \big(a_N(i) - \bar{a}_N\big)^2\Big]^{-1} \cdot \sum_{j=1}^{k} \Big[\sum_{i \in s_j} a_N(R_{Ni}) - n_j \bar{a}_N\Big]^2 n_j^{-1}. \tag{5}$$

Theorem 1 *Assume that* (8.2.1.6) *and* $\min(n_1, \ldots, n_k) \to \infty$ *hold. Then the asymptotic efficiency of the Q-test, with Q given by* (5), *equals*

$$e = \rho_1^2 \tag{6}$$

with ρ_1 given by (8.2.1.11).

Proof. The proof follows from Theorem 8.1.4.1 and Corollary 7.3.1.1 immediately. □

For $k = 2$ we have $K = \{q_1, q_2\}$ with q_t given by (8.1.4.37), $t = 1, 2$, and the Q-test coincides with the $|S_{mn} - \mathsf{E}S_{mn}|$-test.

Remark. (Pitman's interpretation.) Similarly as in Remark 8.2.1.3, we may regard $100(1 - e)$ as the percentage of observations 'wasted' by making use of the Q-test instead of the asymptotically most powerful test. This means that the Q-test distinguishes H_0 from K, given by (3), with the same asymptotic power as the maximin most powerful test distinguishes H_0 from $K' = \{q'_\Delta, \Delta \in D\}$,

$$q'_\Delta = \prod_{j=1}^{k} \prod_{i \in s'_j} f_0(x_i - \Delta),$$

where $n'_j = \operatorname{card} s'_j$ satisfy

$$n'_j = [en_j], \quad 1 \leq j \leq k,$$

and D is the same as in K.

8.2.3 Local asymptotic efficiency. If a test is based on a statistic which has asymptotic distribution different from normal or chi-square, a simple determination of the asymptotic efficiency is not possible. We may define the asymptotic efficiency e along the lines of Remark 8.2.1.3 and Remark 8.2.2, or alternatively along the lines of Remark 8.2.1.4. In the former case, e is defined so that the test under consideration reveals the shift Δ in $m + n$ observations with the same asymptotic power as the most powerful test in $m' + n'$ observations with $m' = [em]$, $n' = [en]$, and similarly for multi-sided tests. We omit the details. Note that the resulting e will depend on both the significance level and the limiting 'distance' of K from H_0. Denoting the distance by b, we shall have $e = e(\alpha, b)$. Most often b is defined by

$$I(f_0) \sum_{i=1}^{N} (d_i - \bar{d})^2 \to b^2. \tag{1}$$

Definition. The limit

$$e(\alpha) = \lim_{b \downarrow 0} e(\alpha, b), \tag{2}$$

if it exists, will be called the *local asymptotic efficiency* of the test under consideration.

Theorem 1 *Under conditions of Theorem* 8.1.3.1, *the local asymptotic efficiency of the test based on one-sided Kolmogorov-Smirnov statistic* K_c^+, *given by* (6.3.7.1), *equals*

$$e(\alpha) = 4\pi\rho_2^2\alpha^2(-\log\alpha)\exp(k_{1-\alpha}^2) \cdot [I(f_0)]^{-1}\left[\int_0^1 \varphi(u, f_0)\psi(\alpha, u)\, du\right]^2, \tag{3}$$

where ρ_2 *and* $\psi(\alpha, u)$ *are defined by* (8.2.1.10) *and* (7.4.5.3), *respectively.*

Furthermore,

$$\lim_{\alpha\to 0} e(\alpha) = \rho_2^2[I(f_0)]^{-1}\left[\int_0^1 \varphi(u, f_0)\operatorname{sign}(2u-1)\, du\right]^2 \tag{4}$$

holds.

Proof. (3) follows from (7.4.5.1) and Theorem 7.4.5.1. Further, (4) is equivalent to

$$\lim_{\alpha\to 0}[4\pi\alpha^2(-\log\alpha)\exp(k_{1-\alpha}^2)] = 1.$$

This follows, however, from the well-known fact that (see Feller (1950), Chap. VII, §1)

$$\lim_{x\to\infty}[(1-\Phi(x))^2 2\pi x^2\exp(x^2)] = 1. \tag{5}$$

Actually, it suffices to put

$$\alpha = 1 - \Phi(k_{1-\alpha})$$

and note that (5) implies

$$\lim_{\alpha\to 0}[-2\log\alpha]k_{1-\alpha}^{-2} = 1.$$

□

(Cf. Anděl (1967), and for related results also Klotz (1967).)

Remark 1. (4) shows that, for $\alpha \to 0$, the local asymptotic power of the K_c^+-test approaches the asymptotic power of the S_c-test with S_c being defined by

$$S_c = \sum_{i=1}^{N}(c_i - \bar{c})\operatorname{sign}\left(R_i - \tfrac{1}{2}(N+1)\right).$$

If $c_i = 1$, $1 \le i \le m$, and $c_i = 0$, $m < i \le m + n = N$, then the S_c-test coincides with the median test, and the K_c^+-test coincides with the ordinary one-sided Kolmogorov-Smirnov test.

Remark 2. The convergence of $e(\alpha)$ to the right side of (4) is very slow, and, consequently, useless for the usual values of α, which satisfy $0.001 \leq \alpha \leq 0.05$.

The local asymptotic efficiency $e(\alpha)$ may serve as a first approximation to the exact asymptotic efficiency $e(\alpha, b)$:

$$e(\alpha) \simeq e(\alpha, b), \quad b \geq 0. \tag{6}$$

In terms of asymptotic powers, (6) means that we assume

$$B(\alpha, f_0, b) \simeq 1 - \Phi\big(k_{1-\alpha} - [e(\alpha)]^{\frac{1}{2}} b\big), \quad b \geq 0, \tag{7}$$

where $B(\alpha, f_0, b)$ denotes the asymptotic power of the Kolmogorov-Smirnov test. The following example suggests that the approximation (7) is reasonably exact for $[e(\alpha)]^{\frac{1}{2}} b \leq 2$, and that it tends to underestimate the actual asymptotic power.

Example. Let us take $f_0 = \frac{1}{2} e^{-|x|}$, and assume that $\rho_2 = 1$. Then (3) yields (see Problem 14 to Section 8.2)

$$e(\alpha) = 4\pi(-\log \alpha) \exp(k_{1-\alpha}^2)[\alpha - 2\Phi(-(-2\log\alpha)^{\frac{1}{2}})]^2. \tag{8}$$

On the other hand, utilizing (7.4.5.2) and Problem 16 to Section 6.3, we could show that $B(\alpha, f_0, b)$ equals

$$\begin{aligned} B(\alpha, f_0, b) \; &= \; 1 - \Phi\big((-2\log\alpha)^{\frac{1}{2}} - b\big) \\ &\quad - \exp\big(4b(-\tfrac{1}{2}\log\alpha)^{\frac{1}{2}}\big)\Phi\big(-(-2\log\alpha)^{\frac{1}{2}} - b\big) \\ &\quad + 2\exp\big((-2\log\alpha)^{\frac{1}{2}}\big)[b - (-\tfrac{1}{2}\log\alpha)^{\frac{1}{2}}]\Phi(-b). \end{aligned} \tag{9}$$

In Table 2 we display the exact values of $B(\alpha, f_0, b)$, their approximation by (7), and the envelope asymptotic power function $\beta(\alpha, b) = 1 - \Phi(k_{1-\alpha} - b)$.

8.2.4. Bibliographical notes. The notion of asymptotic sufficiency, introduced in Subsection 8.1.1, and some ideas of the proof of Theorem 8.1.2.1 are due to LeCam (1960). The results of Subsection 8.1.3 come essentially from Hájek (1962), those in 8.1.4 are proved for the first time in Hájek and Šidák (1967).

As for our Remark 8.2.1.3, Pitman's definition of the asymptotic efficiency was introduced by Pitman (1948), and later generalized by Noether (1955) and by other authors. Older surveys and expositions on this notion are contained e.g. in Fraser (1957a), Noether (1958) and Stuart (1958). An analogue of our Remark 8.2.1.2 was proved by van Eeden (1963), and with its aid many efficiencies of tests for symmetry and for the two-sample problems have been calculated. For the relation of Pitman's efficiency to the derivative of the power function cf. Stuart (1954).

Table 2

b	$\alpha = 0.05$			$\alpha = 0.01$			$\alpha = 0.001$		
	(A)	(B)	(C)	(A)	(B)	(C)	(A)	(B)	(C)
0.0	0.050	0.050	0.050	0.010	0.010	0.010	0.001	0.001	0.001
0.5	.111	.112	.126	.029	.030	.034	.004	.005	.005
1.0	.212	.220	.259	.072	.075	.092	.014	.015	.018
1.5	.353	.373	.442	.152	.163	.204	.038	.046	.056
2.0	.519	.554	.639	.275	.302	.372	.092	.105	.138
2.5	.681	.726	.804	.434	.481	.569	.187	.217	.278
3.0	.814	.858	.912	.605	.664	.750	.327	.380	.464
3.5	.906	.938	.968	.758	.817	.880	.497	.569	.659
4.0	.959	.975	.991	.871	.916	.953	.668	.744	.818
4.5	.985	.994	.998	.941	.969	.985	.809	.872	.921
5.0	.995	.999	1.000	.977	.991	.996	.905	.949	.972

In Table 2 the columns (A) denote $1 - \Phi(k_{1-\alpha} - [e(\alpha)]^{\frac{1}{2}} b)$,
the columns (B) denote $B(\alpha, f_0, b)$,
the columns (C) denote $1 - \Phi(k_{1-\alpha} - b)$.

Other notions of efficiency, but related to Pitman's definition, have been investigated e.g. by Hoeffding and Rosenblatt (1955), who formulated its general analogue for families of tests and composite alternatives, by Witting (1960), who established correction terms for finite samples for Pitman's efficiency, further by Hodges and Lehmann (1956), Blyth (1958) (the latter two dealing also with non-contiguous alternatives), etc.

The notion of local asymptotic efficiency and the relevant results in Subsection 8.2.3 were given for the first time in book form by Hájek and Šidák (1967).

For tests of independence the problem is more complicated, since first a suitable general model for alternatives must be fixed. We have used here throughout the book the model of Subsection 3.4.11. Another model, considered in Problem 4 to Section 8.2, may be discussed by the same method as the regression alternatives against H_0. Konijn (1956) makes use of the model $X = \lambda_1 U + \lambda_2 V$, $Y = \lambda_3 U + \lambda_4 V$, with U and V independent. Farlie (1961) specifies the two-dimensional distribution function $H(x, y) = F(x)G(y)[1 + \Delta A(x)B(y)]$, where $\Delta \geq 0$ is the parameter being tested. Both of them generalize the notion of Pitman's efficiency and obtain for normal densities the value $9/\pi^2$ for the Spearman and Kendall correlation coefficients, the value $4/\pi^2$ for the quadrant test, and the value 1 for the Fisher-Yates coefficient. See also Noether (1958) and Elandt (1962), where the power function of the quadrant test is studied.

Small sample powers and efficiencies of the one-sample Wilcoxon test and of the Fraser test are studied by Klotz (1963). Walsh (1946) and Dixon (1953) studied the same for the sign test, van der Laan (1964) for the van der Waerden test, Sukhatme (1960) for the Kamat test, Teichroew (1955) and Klotz (1964) for the Fisher-Yates-Terry-Hoeffding test. Cf. also Hemelrijk (1960) and Dixon (1954).

Some concrete numerical results on the asymptotic efficiencies of rank tests are given in the Problems and Complements to this chapter, with indications of the places where the relevant results are mentioned or proved for the first time.

There is a vast newer literature on the asymptotic efficiencies of rank tests. We must also mention that, in addition to Pitman's efficiency, some other notions of asymptotic efficiency, based on different ideas, have been introduced and investigated. Among them, probably the best known and studied is the notion of Bahadur's efficiency, which will be dealt with in Section 8.3 of this book.

A new survey book on asymptotic efficiency is Nikitin (1995).

8.3 BAHADUR EFFICIENCY

8.3.1 Exact Bahadur efficiency. Asymptotic efficiency of tests defined in Section 8.2 gives a comparison of tests locally, i.e. for alternatives approaching the hypothesis. However, there are many other ways of comparing tests, see e.g. a survey in Singh (1984), or also Nikitin (1995).

Among the non-local measures of asymptotic efficiency the most popular one is the Bahadur efficiency. It works for fixed alternatives (and fixed hypotheses) and its main idea is to compare the rates how fast the attained significance levels of tests approach zero when the sample sizes tend to infinity.

We will present here basic definitions and theorems for some important special cases, as they were elaborated by Bahadur (1960, 1967, 1971). In principle, we are now turning back to the end of Subsection 2.3.1.

Let $X = (X_1, X_2, \dots)$ be an infinite sequence of independent and identically distributed random variables (possibly vector or abstract) with their distributions depending on a parameter Θ in a set $\boldsymbol{\Theta}$. Consider testing a null hypothesis that Θ lies in a subset $\boldsymbol{\Theta}_0$ of $\boldsymbol{\Theta}$. For each $n = 1, 2, \dots$, let $T_n(X)$ be a test statistic that depends only on $(X_1, \dots, X_n)$, and whose large values form the relevant critical regions. If T_n has the distribution function F_n under the null hypothesis, i.e.

$$P_\Theta(T_n \le z) = F_n(z) \quad \text{for all } \Theta \in \boldsymbol{\Theta}_0, \ -\infty < z < \infty,$$

(where $F_n(z)$ does not depend on Θ), then the *level attained by* T_n is defined to be the random variable

$$L_n(X) = 1 - F_n\big(T_n(X)\big).$$

If in a given case we observe $(x_1, \dots, x_n)$, then $L_n(x_1, \dots, x_n)$ is the probability of getting a larger value of T_n than the observed value $T_n(x_1, \dots, x_n)$ if the null hypothesis is true. Under the alternative hypothesis we expect generally that $L_n(X)$ will be small.

The level attained is also popularly known as the *p-value*.

(At the end of 2.3.1 and in some other publications, the level actually attained was defined to be $\ell(X)$ with $\ell(x) = P\big(T_n(X) \geq T(x)\big)$ while here we define $\ell_n(x) = P\big(T_n(X) > T(x)\big)$. Thus the difference lies only in the inclusion or no inclusion of $P\big(T_n(X) = T(x)\big)$ which is rather a matter of taste, and usually its probability tends to 0 asymptotically. Moreover, the discussion in 2.3.1 was more 'philosophical', while the present 'level attained by T_n' is a focal point in the formulation of Bahadur efficiency.)

For finite samples in our area of interest an exact handling of L_n is usually intractable, because F_n is, as a rule, discontinuous for rank tests. However, when speaking about asymptotics, typically under the null hypothesis, L_n is asymptotically uniformly distributed over $(0, 1)$, and for the alternative hypothesis $L_n \to 0$ exponentially fast (with probability one). We will say that for the alternative characterized by Θ the sequence $\{T_n\}$ has *exact slope* $c(\Theta)$ if

$$\lim_{n\to\infty} n^{-1} \log L_n(X) = -c(\Theta)/2 \quad [P_\Theta \text{ a.s.}]. \tag{1}$$

Bahadur's exact slope, for a reason immediately to be seen, can be taken as a measure of asymptotic efficiency of the test at hand. *The relative Bahadur efficiency* of two tests is defined as the ratio of relevant exact Bahadur slopes. (Here we use Bahadur's (1971) original definition of the exact slope. However, in some publications Bahadur's exact slope is defined as $-c(\Theta)/2$; of course, this does not bring any change into *relative* efficiencies.)

Now for given α, $0 < \alpha < 1$, and given $x = (x_1, x_2, \ldots)$, let $N = N(\alpha, x)$ be the smallest integer m such that $L_n(x) < \alpha$ for all $n \geq m$, and let $N = \infty$ if no such m exists, i.e. N is the smallest sample size for which $\{T_n\}$ becomes and remains significant at the level α. The following theorem shows that, for small α, the size N is approximately inversely proportional to the exact slope.

Theorem 1 *If* (1) *holds with* $0 < c(\Theta) < \infty$, *then,*

$$\lim_{\alpha\to 0} N(\alpha, x)/2 \log \alpha^{-1} = 1/c(\Theta) \quad [P_\Theta \text{ } a.s.]. \tag{2}$$

Proof. Fix some Θ such that $0 < c(\Theta) < \infty$ and fix some $x = (x_1, x_2, \ldots)$ such that $n^{-1} \log L_n(X) \to -c(\Theta)/2$. Then $L_n > 0$ for all sufficiently large n and $L_n \to 0$ as $n \to \infty$. It follows that $N < \infty$ for every $\alpha > 0$ and that $N \to \infty$ through a subsequence of integers as $\alpha \to 0$. Hence $2 \leq N < \infty$ for all sufficiently small α, say for $\alpha < \alpha_1$. For $\alpha < \alpha_1$, we have, by definition, $L_N < \alpha \leq L_{N-1}$. Thus $N^{-1} \log L_N < N^{-1} \log \alpha \leq (N-1)N^{-1}(N-1)^{-1} \log L_{N-1}$. It follows from the present choice of x that $N^{-1} \log \alpha \to -c(\Theta)/2$ as $\alpha \to 0$. □

Let us compare two sequences of test statistic $\{T_n^{(1)}\}$ and $\{T_n^{(2)}\}$ and positive finite exact slopes $c_1(\Theta)$ and $c_2(\Theta)$, respectively, under a non-null Θ alternative. It then follows from Theorem 1 that, defining $N_i(\alpha, x)$ the size required to make $T_n^{(i)}$ significant at level α, we have $\frac{N_2(\alpha,x)}{N_1(\alpha,x)} \to \frac{c_1(\Theta)}{c_2(\Theta)}$ with P_Θ probability 1. Consequently, $c_1(\Theta)/c_2(\Theta)$ is a measure of the asymptotic efficiency (exact Bahadur efficiency) of $T_n^{(1)}$ relative to $T_n^{(2)}$ when the Θ alternative obtains.

A useful method of finding the exact slope is often given by the following theorem.

Theorem 2 *Suppose that*

$$\lim_{n\to\infty} n^{-1/2} T_n(X) = \tau(\Theta) \quad [P_\Theta \ a.s.] \tag{3}$$

for each $\Theta \in \boldsymbol{\Theta} - \boldsymbol{\Theta}_0$, *where* $-\infty < \tau(\Theta) < \infty$, *and that*

$$\lim_{n\to\infty} n^{-1} \log[1 - F_n(n^{1/2}t)] = -v(t) \tag{4}$$

for each t *in an open interval* I, *where* v *is a continuous function on* I, *and* $\{\tau(\Theta) : \Theta \in \boldsymbol{\Theta} - \boldsymbol{\Theta}_0\} \subset I$. *Then* (1) *holds with* $c(\Theta) = 2v\big(\tau(\Theta)\big)$ *for each* $\Theta \in \boldsymbol{\Theta} - \boldsymbol{\Theta}_0$.

Proof. Fix some $\Theta \in \boldsymbol{\Theta} - \boldsymbol{\Theta}_0$, and fix some $x = (x_1, \ldots, x_n, \ldots)$ such that $n^{-1/2}T_n(x) \to \tau$ for $n \to \infty$. Let $\varepsilon > 0$ be so small that $\tau + \varepsilon \in I$, $\tau - \varepsilon \in I$. Since $F_n(t)$ is non-decreasing in t it follows from the definition of L_n that $n^{1/2}(\tau - \varepsilon) < T_n < n^{1/2}(\tau + \varepsilon)$ implies $1 - F_n\big(n^{1/2}(\tau - \varepsilon)\big) \geq L_n \geq 1 - F_n\big(n^{1/2}(\tau + \varepsilon)\big)$; hence the latter inequality holds for all sufficiently large n. Assumption (4) now implies that $\limsup_{n\to\infty} n^{-1} \log L_n \leq -v(\tau - \varepsilon)$ and similarly $\liminf_{n\to\infty} n^{-1} \log L_n \geq -v(\tau + \varepsilon)$. Since $v(t)$ is continuous and ε is arbitrary this implies that $\lim_{n\to\infty} n^{-1} \log L_n = -v(\tau)$. □

Theorem 2 shows therefore that the standard method for evaluating the Bahadur slope usually involves two ingredients: a strong convergence result expressed by (3) and a large deviation result expressed by (4), the latter being under the null hypothesis.

Bahadur's monograph (1971) as well as works of other authors contain a number of examples based on Theorem 2. While almost always the strong convergence result, corresponding to (3), is easy to verify, the second part, corresponding to (4), i.e. the large deviation result, is, as a rule, much more difficult to establish. It can also be remarked that historically, in former times, there were only very few results on Bahadur efficiency, namely for the reason that large deviation theory was still insufficiently developed but, later on, this link to Bahadur efficiency served as a strong impetus for the development of large deviation theory.

The following two examples illustrate how to find Bahadur slopes in specific situations.

Example 1. (Bahadur (1971), p. 31) Let X be the real line, and $\boldsymbol{\Theta}$ be the set of all continuous probability distribution functions $\Theta(x)$ on X, and let $P_\Theta(B)$ denote the probability measure on X determined by the distribution function Θ. The null hypothesis is that $\Theta = \Theta_0$ where Θ_0 is a given continuous distribution function.

For each n let $F_n(t) = F_n(t|x_1, \ldots, x_n)$ be the empirical distribution function based on $\{x_1, \ldots, x_n\}$, and let T_n be the Kolmogorov statistic, i.e. $T_n(x) = n^{1/2} \sup\{|F_n(t) - \Theta_0(t)| : -\infty < t < \infty\}$. It follows from the Glivenko-Cantelli theorem that (3) holds for T_n, with $\tau(\Theta) = \sup\{|\Theta(t) - \Theta_0(t)| : -\infty < t < \infty\}$ where $0 < \tau(\Theta) < 1$ for $\Theta \neq \Theta_0$.

Further, it follows from Bahadur (1971); Ex. 5.3, that T_n satisfies (4) where v is defined as follows. Let

$$\begin{aligned} f(a,y) &= \begin{cases} (a+y)\log\left(\frac{a+y}{y}\right) + & \\ \quad +(1-a-y)\log\left(\frac{1-a-y}{1-y}\right) & \text{for } 0 \le y \le 1-a, \\ \infty & \text{for } y > 1-a \end{cases} \\ g(a) &= \inf\{f(a,y) : 0 \le y \le 1\}. \end{aligned}$$

In (4) we put $v(t) = g(t)$, and, consequently, the exact slope of T_n is $c(\Theta) = 2g\big(\tau(\Theta)\big)$.

Example 2. The formulas for the two-sample rank test problem are still more difficult. The following case and its solution was adapted from Kremer (1983). Let $X_1, \ldots, X_m$, and $Y_1, \ldots, Y_n$ be two samples of $N = m+n$ independent random variables where the X_i's have a continuous distribution function F, while the Y_j's have a continuous distribution function G. We wish to test the two-sample problem of randomness $F = G$, against alternatives $F \neq G$. To this end, use a linear rank statistic

$$T_N = N^{-1/2} \sum_{i=1}^{m} a_N(R_i),$$

where $R_1, \ldots, R_N$ are the ranks of the combined sample $X_1, \ldots, X_m$, $Y_1, \ldots, Y_n$, and the scores $a_N(i)$, $i = 1, \ldots, N$, satisfy

$$\lim_{N\to\infty} \int_0^1 |a_N(1 + [uN]) - \varphi(u)|\, du = 0,$$

with some non-constant integrable function $\varphi(u)$ (for simplicity supposed to be non-decreasing). We will present basic asymptotic results for m, $n \to \infty$, where $m/N \to s$ ($s \in (0,1)$ fixed).

The verification of condition (3) concerning the limit $\tau(\Theta)$ in Theorem 2 is usually not difficult, so suppose it was done.

However, the derivation of the large deviation statement (4) is much more complicated: First define $t_0(\varphi, s) = s \int_0^1 \varphi(u)\, du$, $t_1(\varphi, s) = \int_{1-s}^1 \varphi(u)\, du$. Now (4) holds for $t < t_1(\varphi, s)$ and for $v(t)$ defined

$$v(t) = \begin{cases} 0, & \text{for } t \leq t_0(\varphi, s), \\ at + s \log z - \displaystyle\int_0^1 \log\big((1-s) + s e^{a\varphi(u)}\big)\, du & \\ \qquad \text{for } t \in \big(t_0(\varphi, s), t_1(\varphi, s)\big), & \end{cases}$$

where a, $z \geq 0$ are the unique solutions of the integral equations

$$\int_0^1 s\varphi(u) \frac{z e^{a\varphi(u)}}{(1-s) + sz e^{a\varphi(u)}}\, du = t,$$
$$\int_0^1 \frac{z e^{a\varphi(u)}}{(1-s) + sz e^{a\varphi(u)}}\, du = 1.$$

This statement can be proved in two stages. First, we assume that $a_N(1 + [uN]) = \varphi(u)$ is a step function which reduces (4) to a large deviation statement for a multinomially distributed statistic. Second, we utilize the assumption that the step function $a_N(1 + [uN]) \to \varphi(u)$.

Consequently, the exact slope of T_N equals $2v\big(\tau(\Theta)\big)$.

Concerning the literature on these topics, in addition to Bahadur's works quoted above, we refer to Kremer (1983) (a survey paper), Woodworth (1970) (general class of rank tests including linear rank tests for the two-sample and independence problem), Hoadley (1965) (Wilcoxon test), Stone (1967, 1968) (Wilcoxon and normal scores test), and Nikitin (1995).

As a complement to our considerations, we will now present a useful non-asymptotic property of $L_n(X)$ in the null case, namely its relation to the uniform distribution. We will now work with left-continuous distribution functions.

Theorem 3 *For each* $\Theta \in \boldsymbol{\Theta}_0$ *and each* n,

$$P_\Theta\big(L_n(X) \leq u\big) \leq u \quad \text{for } 0 \leq u \leq 1.$$

Proof. Let us have some $\Theta \in \boldsymbol{\Theta}_0$, and some statistic $T = T_n$, but we will omit Θ and n in the sequel since they are fixed. If F, the distribution function of T, is continuous, then L is uniformly distributed over $[0, 1]$ and $P(L \leq u) = u$ for $0 \leq u \leq 1$. In the general case, let U be a random variable distributed uniformly over $(0, 1)$ and independent of X, and put $T^* = T^*(X, U) = T(X) + \alpha U$, $\alpha > 0$ being a constant. (This corresponds essentially to randomization of the outcome X, cf. also Problems to Section

2.3.) Then the distribution function F^* of T^* is continuous, hence $F^*(T^*)$ is uniform on $[0, 1]$. Now, for any t, $F^*(t) = P(T + \alpha U < t) \geq P(T < t - \alpha) = F(t - \alpha)$; this shows $F^*(T^*) \geq F(T^* - \alpha) \geq F(T - \alpha)$ since $T^* \geq T$ and F is non-decreasing. Therefore $P(1 - F(T - \alpha) < t) \leq t$ for $t \geq 0$. Now, let us have a decreasing sequence of positive constants $\alpha_1, \alpha_2, \ldots$, such that $\alpha_k \to 0$, and let $A_k(t)$ be the event $1 - F(T - \alpha_k) < t$. Then $P(A_k(t)) \leq t$ for each k. Since F is nondecreasing and left-continuous, we have $A_k(t) \subset A_{k+1}(t)$ for each k and $\bigcup_{k=1}^{\infty} A_k(t)$ is the event $1 - F(T)(\equiv L) < t$. Consequently, $P(L < t) = \lim_{k \to \infty} P(A_k(t)) \leq t$. Since t is arbitrary, we get easily now $P(L \leq u) \leq u$ for $0 \leq u \leq 1$. □

8.3.2 Kullback-Leibler information. We need to make a small digression now.

Let X be the sample space of points x (single observation), and $\mathcal{A}$ a δ-field of sets of X. If p and q are some probability measures on $\mathcal{A}$, the Kullback-Leibler information $K(q, p)$ is defined as follows: If q is absolutely continuous with respect to p on $\mathcal{A}$, there exists an $\mathcal{A}$-measurable function $r(x)$, $0 \leq r(x) < \infty$, such that

$$\frac{\mathrm{d}q}{\mathrm{d}p} = r(x) \quad \text{on } \mathcal{A},$$

and the Kullback-Leibler information is defined to be

$$K(q, p) = \int_{\mathsf{X}} [\log r(x)] \, \mathrm{d}q.$$

If q is not absolutely continuous with respect to p, let $K(q, p) = \infty$.

It is easy to see that K is well defined, $0 \leq K \leq \infty$, and $K = 0$ if and only if $p(B) = q(B)$ for all $B \in \mathcal{A}$. In the case of absolute continuity, an equivalent definition is

$$K(q, p) = \int_{\mathsf{X}} r(x) \log r(x) \, \mathrm{d}p \quad \text{with } 0 \log 0 = 0.$$

The Kullback-Leibler information has an important role in asymptotic statistics because $\sqrt{2K(q, p)}$, $\sqrt{2K(p, q)}$, $\sqrt{K(q, p) + K(p, q)}$ may serve as indices of statistical distance between p and q. There is a vast literature on it but here we will not enter into that.

Instead, we will show how Kullback-Leibler information can be utilized to establish a bound on the asymptotic behaviour of the significance level, and thus a bound on the exact Bahadur slope. We emphasize that we now need the structure of T_n, namely that T_n depends only on a finite number $X_1, \ldots, X_n$ of independent and identically distributed random variables. Related to this, let $\mathsf{X}^{(n)}$ be the space of n-variate vectors

$x^{(n)} = (x_1, \ldots, x_n)$, $\mathcal{A}^{(n)} = \mathcal{A} \times \ldots \times \mathcal{A}$ (n-times), $p^{(n)} = p \times \ldots \times p$ (n-times), etc.

For Θ and Θ_0 in $\mathbf{\Theta}$, let $K(\Theta, \Theta_0)$ be defined with the measures P_Θ and P_{Θ_0}, as above, and let

$$J(\Theta) = \inf\{K(\Theta, \Theta_0) : \Theta_0 \in \mathbf{\Theta}_0\}. \tag{1}$$

Then $0 \leq J(\Theta) \leq \infty$ for all Θ, and $J(\Theta) = 0$ for $\Theta \in \mathbf{\Theta}_0$. The following theorem implies that the exact slope of any sequence $\{T_n\}$ cannot exceed $2J(\Theta)$ for any fixed Θ.

Theorem 1 *For each* $\Theta \in \mathbf{\Theta}$,

$$\liminf_{n\to\infty} n^{-1} \log L_n(X_1, \ldots, X_n) \geq -J(\Theta) \quad [P_\Theta\text{-}a.s..] \tag{2}$$

Proof. For $J = \infty$ this inequality holds automatically, so let us choose and fix points for which $J(\Theta) < \infty$. Let $\varepsilon > 0$ be a constant. It follows from (1) that there exists a $\Theta_0 \in \mathbf{\Theta}_0$ such that

$$0 \leq K(\Theta, \Theta_0) < J(\Theta) + \varepsilon < \infty. \tag{3}$$

With Θ and Θ_0 fixed, let us write K and J in place of $K(\Theta, \Theta_0)$ and $J(\Theta)$, respectively. Since $K < \infty$, P_Θ is absolutely continuous with respect to P_{Θ_0}, so we can write $P_\Theta(A) = \int_A r(x)\, \mathrm{d}P_{\Theta_0}$ for $A \in \mathcal{A}$. If we set $r_n(x^{(n)}) = \prod_{i=1}^{n} r(x_i)$ we have

$$\lim_{n\to\infty} n^{-1} \log r_n(x^{(n)}) = K(\Theta, \Theta_0) \quad [P_\Theta \text{ a.s.}] \tag{4}$$

and $\mathrm{d}P_\Theta^{(n)} = r_n\, \mathrm{d}P_{\Theta_0}^{(n)}$ for $(X^{(n)}, \mathcal{A}^{(n)})$. For each n let A_n be the event that $L_n < \exp(-n[K + 2\varepsilon])$ and B_n the event that $r_n < \exp(n[K + \varepsilon])$. Then

$$\begin{aligned} P_\Theta(A_n \cap B_n) &= \int_{A_n \cap B_n} \mathrm{d}P_\Theta^{(n)} = \int_{A_n \cap B_n} r_n\, \mathrm{d}P_{\Theta_0}^{(n)} \\ &\leq \mathrm{e}^{n(K+\varepsilon)} \int_{A_n \cap B_n} \mathrm{d}P_{\Theta_0}^{(n)} \leq \mathrm{e}^{n(K+\varepsilon)} \int_{A_n} \mathrm{d}P_{\Theta_0}^{(n)} \\ &= \mathrm{e}^{n(K+\varepsilon)} P_{\Theta_0}^{(n)}\big(L_n < \exp(-n[K + 2\varepsilon])\big) \leq \mathrm{e}^{-n\varepsilon} \end{aligned} \tag{5}$$

by Theorem 8.3.1.3. It follows from (5) that $\sum_{n=1}^{\infty} P_\Theta(A_n \cap B_n) < \infty$. Hence by (4) and by the definitions of A_n and B_n, we see for the fixed chosen Θ that with probability one $L_n(X) \geq \exp[-n(K + 2\varepsilon)]$ for all sufficiently large n. Thus the left-hand side of (2) is not less than $-K - 2\varepsilon$ [P_Θ-a.s.]. It now follows by (3) that

$$\liminf_{n\to\infty} n^{-1} \log L_n(X) \geq -J(\Theta) - 3\varepsilon \quad [P_\Theta\text{-a.s.}].$$

Since ε is arbitrary Theorem 1 follows. □

8.3.3 Best exact slope of two-sample rank statistics. We will now investigate the two-sample problem of randomness in a setup similar to that in Example 2, in 8.3.1. Thus we have two samples $X_1, \dots, X_m$ and $Y_1, \dots, Y_n$ of independent random variables from continuous distribution functions F, and G, respectively. We are testing randomness $F = G$ against $F \neq G$. Again, let $N = m + n$, and let $R_1, \dots,\ R_N$ be the ranks in the pooled sample $X_1, \dots, X_m,\ Y_1, \dots, Y_n$. We will show that the best exact slope is attainable by rank statistics. As usual, our basic tools will be a law of large numbers for simple linear statistics and a large deviation theorem that are taken from Hájek (1974).

Let $m,\ n \to \infty$ so that $m/N \to s$ for some fixed $0 < s < 1$. Set

$$H(x) = sF(x) + (1 - s)G(x), \quad -\infty < x < \infty.$$

Introduce new densities

$$\bar{f}(u) = \frac{\mathrm{d}}{\mathrm{d}u} F\big(H^{-1}(u)\big), \quad \bar{g}(u) = \frac{\mathrm{d}}{\mathrm{d}u} G\big(H^{-1}(u)\big), \quad 0 < n < 1.$$

Obviously, if $f = F'$ and $g = G'$ correspond to X and Y, then $\bar{f}$ and $\bar{g}$ correspond to $H(X)$ and $H(Y)$, respectively. Since $sF\big(H^{-1}(u)\big) + (1 - s)G\big(H^{-1}(u)\big) = u$, we have

$$s\bar{f}(u) + (1 - s)\bar{g}(u) = 1, \quad 0 < u < 1. \tag{1}$$

Now consider the simple linear rank statistic

$$S_N = \sum_{i=1}^{m} a_N(R_i) \tag{2}$$

where the scores $a_N(i)$, $1 \leq i \leq N < \infty$, satisfy

$$\lim_{N \to \infty} \int_0^1 |a_N(1 + [uN]) - \varphi(u)| \,\mathrm{d}u = 0 \tag{3}$$

where $\varphi(u)$, $0 < u < 1$, is some non-constant integrable function and $[uN]$ denotes the integral part of uN.

The statistics S_N satisfy the following law of large numbers:

Theorem 1 *Assume that the functions $a_N(1 + [uN])$ have uniformly bounded variation on closed subintervals of $(0, 1)$ and (3) holds. Then*

$$N^{-1} S_N \to s \int_0^1 \varphi(u)\bar{f}(u)\,\mathrm{d}u \quad \textit{with probability } 1.$$

Proof. For every $\delta > 0$ we can find $K > 0$ such that the truncated scores

$$\bar{a}_N(i) = \begin{cases} a_N(i), & \text{if } |a_N(i)| \leq K, \\ 0 & \text{otherwise} \end{cases}$$

satisfy

$$N^{-1} \sum_{i=1}^{N} |a_N(R_i) - \bar{a}_N(R_i)| < \delta \tag{4}$$

for all $N > N_0$ and $R_1, \ldots, R_N$. The similarly defined

$$\bar{\varphi}(u) = \begin{cases} \varphi(u), & \text{if } |\varphi(u)| \leq K, \\ 0 & \text{otherwise} \end{cases}$$

will satisfy

$$\int_0^1 |\varphi(u) - \bar{\varphi}(u)| \, \mathrm{d}u < \delta.$$

By (1) it follows that $s\bar{f}(u) \leq 1$, so that we have also

$$\left| s \int_0^1 \varphi(u)\bar{f}(u) \, \mathrm{d}u - s \int_0^1 \bar{\varphi}(u)\bar{f}(u) \, \mathrm{d}u \right| < \delta. \tag{5}$$

By (4) and (5) it follows that it is sufficient to prove the theorem for the case when φ and $a_N(.)$ have bounded variations over all $(0, 1)$.

Denoting by F_m the empirical distribution function corresponding to $X_1, \ldots, X_m$, and by H_N the empirical distribution function corresponding to $X_1, \ldots, X_m, Y_1, \ldots, Y_n$, we can write

$$\begin{aligned} N^{-1}S_N &= mN^{-1} \int_0^1 a_N(1 + [uN]) \, \mathrm{d}F_m\big(H_N^{-1}(u)\big) \qquad (6) \\ &= mN^{-1} \int_0^1 a_N(1 + [uN]) \, \mathrm{d}\bar{F}_m(u) \\ &= mN^{-1} \int_0^1 \varphi\bar{f} \, \mathrm{d}u + mN^{-1} \int_0^1 (a_N - \varphi)\bar{f} \, \mathrm{d}u \\ &\quad + mN^{-1} \int_0^1 a_N \, \mathrm{d}(\bar{F}_m - \bar{F}), \end{aligned}$$

where $\bar{F}(u) = F\big(H^{-1}(u)\big)$, $\bar{F}_m(u) = F_m\big(H_N^{-1}(u)\big)$.

Consequently, by (3) and $mN^{-1} = s$ we get

$$\left| mN^{-1} \int_0^1 (a_N - \varphi)\bar{f} \, \mathrm{d}u \right| \to 0 \quad \text{for } N \to \infty. \tag{7}$$

Moreover, by the Glivenko-Cantelli theorem

$$H_N(x) \to H(x), \quad F_m(y) \to F(y), \; \bar{F}_m(u) \to \bar{F}(u) \quad \text{for } N \to \infty,$$

the convergence being uniform. Thus, if the variations of $a_N(.)$ are bounded by V, we have

$$\begin{aligned} \left|mN^{-1}\int_0^1 a_N \,\mathrm{d}(\bar{F}_m - \bar{F})\right| &= mN^{-1}\left|\int_0^1 (\bar{F}_m - \bar{F})\,\mathrm{d}a_N(.)\right| \qquad (8)\\ &\leq \max_{0<u<1} |\bar{F}_m(u) - \bar{F}(u)|V \to 0. \end{aligned}$$

Substituting relations (7) and (8) into (6), the theorem is proved. □

As we said before, our second goal is to show the necessary large deviation result. Such a result was contained Woodworth (1970), which has the following consequence: for any ϱ satisfying

$$s\int_0^1 \varphi\,\mathrm{d}u < \varrho < \sup_A \left\{\int_A \varphi\,\mathrm{d}u;\ \int_A \mathrm{d}u = s\right\}$$

we have, under the hypothesis of randomness, that

$$\lim_{N\to\infty} N^{-1}\log P(N^{-1}S_N > \varrho) = -v(s,\varrho)$$

where

$$\begin{aligned} v(s,\varrho) &= a\varrho + (1-s)\log z - \int_0^1 \log(\mathrm{e}^{a\varphi(u)} + z)\,\mathrm{d}u \qquad (9)\\ &\quad -s\log s - (1-s)\log(1-s), \end{aligned}$$

with a, z being the unique solution of the following integral equations

$$\int_0^1 [1 + z\mathrm{e}^{-a\varphi(u)}]^{-1}\,\mathrm{d}u = s, \qquad (10)$$

$$\int_0^1 \varphi(u)[1 + z\mathrm{e}^{-a\varphi(u)}]^{-1}\,\mathrm{d}u = \varrho. \qquad (11)$$

The above complicated result simplifies substantially if we apply it to our special case. Let us put

$$\varphi(u) = \log\frac{\bar{f}(u)}{\bar{g}(u)}, \qquad (12)$$

and

$$\varrho = s\int_0^1 \bar{f}(u)\log\frac{\bar{f}(u)}{\bar{g}(u)}\,\mathrm{d}u = sK(\bar{f},\bar{g}), \qquad (13)$$

where K denotes the Kullback-Leibler information (cf. Subsection 8.3.2). By Theorem 1 clearly $N^{-1}S_N \to \varrho$ under the alternative (f,g) and also under $(\bar{f},\bar{g})$. Introduce now

$$\bar{J}(f,g,s) = s\int_0^1 \bar{f}\log\bar{f}\,\mathrm{d}u + (1-s)\int_0^1 \bar{g}\log\bar{g}\,\mathrm{d}u. \qquad (14)$$

Theorem 2 *Let* (3) *be satisfied for* $\varphi(u)$ *defined by* (12). *Then, under the hypothesis of randomness, we have*

$$\lim_{N\to\infty} N^{-1}\log P\big(N^{-1}S_N > sK(\bar{f},\bar{g})\big) = -\bar{J}(f,g,s). \tag{15}$$

Proof. For φ and ϱ defined by (12) and (13), the equations (10) and (11) become

$$\int_0^1 [1 + z(\bar{g}/\bar{f})^a]^{-1}\,\mathrm{d}u \;=\; s, \tag{16}$$

$$\int_0^1 \log(\bar{f}/\bar{g})[1 + z(\bar{g}/\bar{f})^a]^{-1}\,\mathrm{d}u \;=\; sK(\bar{f},\bar{g}). \tag{17}$$

Now, their solution is $a = 1$, $z = (1-s)/s$. Substituting the present ϱ, φ, a, z into (1), we get for (9)

$$\begin{aligned} v(s,\varrho) \;&=\; sK(\bar{f},\bar{g}) + (1-s)\log[(1-s)/s] \\ &\quad - \int_0^1 \log[\bar{f}/\bar{g} + (1-s)/s]\,\mathrm{d}u - s\log s - (1-s)\log(1-s) \\ &=\; sK(\bar{f},\bar{g}) + \int_0^1 \log\bar{g}\,\mathrm{d}u = \bar{J}(f,g,s), \end{aligned}$$

where the last equality follows by (1). Theorem 2 is thus proved. □

Putting $L_N(t) = P(N^{-1}S_N \geq t)$ we can get the exact slope of the S_N-test for the (f,g,s) alternative (cf. Subsection 8.3.1 and Bahadur (1967)).

Corollary 1 *Under the conditions of Theorem* 2, *and under the* (f,g,s) *alternative described in and before Theorem* 1,

$$\lim_{N\to\infty} N^{-1}\log L_N(N^{-1}S_N) = -\bar{J}(f,g,s) \tag{18}$$

with probability 1. *Hence Bahadur's exact slope defined in* (8.3.1.1) *equals* $2\bar{J}(f,g,s)$.

The proof follows on considering $N^{-1}\log L_N(t)$ and putting together the preceding derivations.

Using an adaptation of Theorem 8.3.2.1 for the present two-sample problem and again the equality (8.3.1.1), we can get an upper bound for the exact slope for any statistic whatsoever. Next, if we insert the least favourable distributions, we see that the best possible exact slope for testing randomness is $2\bar{J}(f,g,s)$ for any statistic, but this equals the exact slope given above for rank statistics. Thus the best possible slope for the present problem is attainable by means of rank statistics (Hájek (1974) termed this property '*the asymptotic sufficiency in the Bahadur sense*').

8.3.4 Approximate Bahadur efficiency. The exact Bahadur efficiency (cf. Subsection 8.3.1) starts with the exact distribution function $P_\Theta(T_n \le z) = F_n(z)$ and with the exact level attained $L_n(X) = 1 - F_n\big(T_n(X)\big)$. Supposing that $F_n(z) \to F(z)$ for all z, this suggests we approximate $L_n(X)$ by

$$L_n^a(X) = 1 - F\big(T_n(X)\big).$$

If then

$$\lim_{n\to\infty} n^{-1} \log L_n^a(X) = -c^a(\Theta)/2 \quad [P_\Theta \text{ a.s.}]$$

for some $c^a(\Theta)$, than this $c^a(\Theta)$ is called the *approximate slope* of the test based on T_n. Similarly as in 8.3.1, for two test statistics $T_n^{(1)}$, $T_n^{(2)}$ with approximate slopes $c_1^a(\Theta)$, $c_2^a(\Theta)$, the approximate Bahadur efficiency of $T_n^{(1)}$ relative to $T_n^{(2)}$ is defined by $c_1^a(\Theta)/c_2^a(\Theta)$ for a Θ alternative. (Consult Bahadur's papers and Kremer (1983).)

However, we must recall here a known fact that the approximate Bahadur efficiency may differ rather much for general Θ alternatives from the exact Bahadur efficiency, except the case of local alternatives where they coincide. (For details cf. Kremer (1983).)

8.3.5 Relation between Bahadur efficiency and Pitman efficiency. Wieand (1976) presented some conditions under which the Pitman and Bahadur approaches to efficiency coincide. In principle, the main additional condition is that $n^{-1/2}T_n \to \tau(\Theta)$ in probability at a specified rate, namely

$$P_\Theta\big(|n^{-1/2}T_n - \tau(\Theta)| < \varepsilon\tau(\Theta)\big) > 1 - \delta$$

for all sufficiently large n. (Compare with (8.3.1.3) and Theorem 8.3.1.2.)

The result is that, under these conditions, the limiting Bahadur efficiency of $T_n^{(1)}$ with respect to $T_n^{(2)}$ for $\Theta \to \Theta_0$ is equal to the limiting Pitman efficiency for these statistics for the significance level $\alpha \to 0$. The precise statement of the pertaining theorem and its proof are rather lengthy and would lead us to somewhat remote notions. Some examples are given by Wieand (1976). We could emphasize the interesting fact that here we take limits with respect to different parameters, namely for Bahadur efficiency the limit for $\Theta \to \Theta_0$, for Pitman efficiency the limit for $\alpha \to 0$ but, nevertheless, the result is identical.

8.4 HODGES-LEHMANN DEFICIENCY

This section will be quite short; its aim is only to provide the reader with some basic information on this part of the rank statistics, namely on principal achievements published since the first edition of the book. Though it is an interesting area, a thorough exposition (including theorems, proofs, etc.) would require much space and many details and niceties of a highly technical character.

8.4.1 Asymptotic expansions for distributions of rank statistics. Let us begin with a general definition. Suppose we are given a sequence of statistics T_N, $N \geq 1$, where N usually denotes the sample size. We say that the distribution function F_N of T_N has an *asymptotic expansion* valid to $(r+1)$ terms if there exist functions $A_0, \ldots, A_r$ such that

$$\left| F_N(x) - A_0(x) - \sum_{j=1}^{r} \frac{A_j(x)}{N^{j/2}} \right| = o(N^{-r/2}). \tag{1}$$

If $\sup_x$ of the left-hand side of (1) is again $o(N^{-r/2})$, we say that the *expansion is uniformly valid* to $(r+1)$ terms. The asymptotic expansion itself is then $A_0(x) + \sum_{j=1}^{r} \frac{A_j(x)}{N^{j/2}}$. An expansion valid to one term is simply an ordinary limit theorem.

Higher order terms are of interest mainly for the following three reasons.

(1) They improve the basic approximation A_0 which leads, in turn, to improving numerical approximations of the critical values and of the power. As experience shows even taking the term with $A_1(x)$ of the expansion frequently improves the basic approximation A_0 strikingly. (For examples cf. Fix and Hodges (1955), Thompson, Govindarajulu and Doksum (1967).)

(2) The higher order terms give some insight into the regions of unreliability of first order results. For instance, when the limits A_0 is normal the higher order terms A_1 and A_2 are typically corrections for skewness and kurtosis.

(3) The expansions can be used to discriminate between procedures equivalent to the first order. Namely, this is a starting point for the definition and investigations of the so called Hodges and Lehmann deficiency (cf. the next Subsection 8.4.2).

Asymptotic expansions mostly used are the Edgeworth series (or the Gram-Charlier series of type A). For a survey see Bickel (1974).

In order to give the reader at least a rough idea of how these approximations may look, we will present here only the following very simple examples.

If the distribution of a rank statistic T_N is symmetric about $\mathsf{E}T_N$, we may take the approximation

$$P(T_N \leq t_j) \doteq \Phi(u) + \tfrac{1}{24}\gamma_{2N}\varphi'''(u) \tag{2}$$

where $\varphi'''(u)$ is the third derivative of $\varphi(u) = (2\pi)^{-1/2}\exp(-\frac{1}{2}u^2)$, γ_{2N} is the coefficient of excess (kurtosis),

$$\gamma_{2N} = \frac{\mathsf{E}(T_N - \mathsf{E}T_N)^4}{(\operatorname{var} T_N)^2} - 3, \tag{3}$$

and $u = (t_j - \mathsf{E}T_N)(\operatorname{var} T_N)^{-1/2}$, or, including the corrections for continuity from Subsection 5.4.1, $u = (t_j - \mathsf{E}T_N + \frac{1}{2}\lambda)(\operatorname{var} T_N - \frac{1}{12}\lambda^2)^{-1/2}$. Actually, (2) represents the initial part of the Edgeworth series up to the third term, since for symmetric distributions $\mathsf{E}(T_N - \mathsf{E}T_N)^3 = 0$ so that the second term with φ'' vanishes. In spite of the simplicity of (2), it often gives very good results for approximations.

Another (more specialized) example for a one-sample rank test of symmetry is as follows. Let $X_1, \ldots, X_N$ be a random sample, and let our aim be testing the symmetry of the basic distribution. As usual, the relevant ranks are $R_1^+, \ldots, R_N^+$ (cf. Subsection 3.1.3 or 4.5.1). Let the relevant scores be $a_N(i) = \mathsf{E}(J(U_{Ni}))$ where $U_{N1} < \ldots < U_{NN}$ are the order statistics of a sample of size N from the uniform distribution on $(0,1)$, and J is a continuously differentiable and non-constant function on $(0,1)$. Put

$$T_N^+ = \frac{1}{\sigma_N} \sum_{i=1}^{N} a_N(i) \operatorname{sign} X_{R_i^+}, \tag{4}$$

where $\sigma_N^2 = \sum_{i=1}^{N} a_N^2(i)$. (E.g. many usual rank statistics for testing symmetry, including the sign, the Wilcoxon one-sample, and normal scores statistics, can be put in this form.) It can be shown under certain conditions and under the null hypothesis of symmetry, that the distribution of T_N^+ has the asymptotic expansion

$$F_N(x) = \Phi(x) + \frac{\int_0^1 J^4(t)\,\mathrm{d}t}{12N(\int_0^1 J^2(t)\,\mathrm{d}t)^2}\varphi(x)H_3(x) + o(N^{-1}) \tag{5}$$

up to three terms, where H_3 is the Hermite polynomial of degree 3.

For a general and thorough exposition on asymptotic expansions for one-sample rank tests we refer to Bickel (1974), Albers (1974), and Albers, Bickel and van Zwet (1976). The latter paper establishes the asymptotic expansions up to order N^{-1} not only for the null distributions of the one-sample linear rank statistics but also for their powers under contiguous location alternatives.

Now, a similar remark on two-sample rank statistics: suppose that $X_1, \ldots, X_m$ is the first sample with density f, and $Y_1 = X_{m+1}, \ldots, Y_n = X_{m+n}$ is the second sample with density g. Let $N = m + n$, the ranks R_i

be defined as usual. (cf. Subsection 3.1.1), and our wish is to test $f = g$. For this purpose use the rank statistics

$$T_N = -\sum_{i=1}^{m} a_N(R_i)/\tau_N^2, \tag{6}$$

where $a_N(j)$ are some specified scores,

$$\tau_N^2 = \sum_{j=1}^{N}(a_N(j) - \bar{a}_N)^2 mnN^{-1}(N-1)^{-1}$$

and $\bar{a}_N = N^{-1}\sum\limits_{j=1}^{N} a_N(j)$.

Suppose again that $a_N(j)$ are defined as above. Under some assumptions, the distribution of T_N has the asymptotic expansion

$$\begin{aligned} F_N(x) &= \Phi(x) - \varphi(x)\Big[\frac{K_{3N}}{6}H_2(x) + \frac{K_{4N}}{24}H_3(x) \\ &\quad + \frac{K_{3N}^2}{72}H_5(x)\Big] + o(N^{-1}), \end{aligned} \tag{7}$$

where K_{jN} are the cumulants of T_N, and $H_j(x)$ are Hermite polynomials of the j-th degree.

As a very specialized example, let U_{mn} be the Mann-Whitney statistic (a form of the two-sample Wilcoxon statistic, cf. Subsection 4.1.1) for sample sizes m, n of the same order, but $m \le n$. Then the asymptotic expansion up to three terms for the distribution function of U_{mn} is

$$\begin{aligned} P(U_{mn} \le u) &= \Phi(x) - \frac{m^2 + n^2 + mn + m + n}{20mn(m+n+1)}\varphi'''(x) \\ &\quad + o(m^{-1}), \end{aligned} \tag{8}$$

where $x = (u - \frac{1}{2}mn)(\frac{1}{12}mn(m+n+1))^{-1/2}$, or, including the corrections for continuity,

$$x = (u - \tfrac{1}{2}mn + \tfrac{1}{2})\big(\tfrac{1}{12}mn(m+n+1) - \tfrac{1}{12}\big)^{-1/2}.$$

Formula (8) corresponds essentially to formula (2) or formula (7). This result may be found in an old paper by Fix and Hodges (1955) where also higher terms up to $\varphi^{(7)}$ of the asymptotic expansion are established.

The basic paper on asymptotic expansions for two-sample rank tests is Bickel and van Zwet (1978) which is essentially a companion paper to Albers, Bickel and van Zwet (1976), and deals almost completely with analogous topics; only, of course, the two-sample case is considerably more difficult, which requires some additional developments.

Some further literature on asymptotic expansions are: David, Kendall and Stuart (1951), Glasser and Winter (1961), Fellingham and Stoker (1964), Ruymgaart (1974), Prášková (1976), Albers (1978), van Zwet (1982), Does (1983, 1984), and Schneller (1989).

Till now we have concentrated essentially only on linear rank statistics. Asymptotic expansions for the distributions of the Kolmogorov-Smirnov statistics (and similar ones) are a topic of many other papers, but they are based on entirely different ideas related mainly to the first passage problem for Markov chains and Wiener processes. See e.g. Kemperman (1959), and Borovkov (1962). The results of Korolyuk (1955b) are not correct, as was pointed out by Gihman, Gnedenko and Smirnov (1956) and Darling (1960).

8.4.2 Hodges-Lehmann's concept of deficiency. Roughly and intuitively speaking, the concept of deficiency is a refinement to higher order terms of the concept of efficiency. Namely, deficiency is crudely defined as the difference (or its limit) in sample sizes required to reach equal power for the same alternative.

Let us have two sequences of test statistics T_N and T'_N for the same hypothesis against the same (possibly contiguous) alternatives at the same fixed level α. Let $\pi_N(\theta_N)$ and $\pi'_N(\theta_N)$ denote the powers of the corresponding tests against the same sequence of contiguous alternatives parametrized by a parameter θ. If T_N is more powerful than T'_N we are looking for a number $k_N = N + d_N$ such that $\pi_N(\theta_N) = \pi'_{k_N}(\theta_N)$. Here k_N and d_N are treated as continuous variables, the power π'_N being defined for all real N by linear interpolation between consecutive integers. The quantity d_N was called the *deficiency* of T'_N with respect to T_N by Hodges and Lehmann (1970), who introduced this concept and initiated its study. Unfortunately, d_N is often analytically intractable and we can only analyse its asymptotic behaviour for $N \to \infty$.

We should distinguish between two sharply different cases. Generally, for $N \to \infty$ let the ratio N/k_N tend to a limit e, the asymptotic relative efficiency of T'_N with respect to T_N. Now, the first (and simpler) case is when $0 < e < 1$. Then we have $d_N \sim (e^{-1} - 1)N$ and further asymptotic information about d_N is not particularly interesting. In the second case suppose that the asymptotic relative efficiency is $e = 1$. Then the asymptotic behaviour of d_N may be anything from $o(1)$ to $o(N)$, and this is exactly the case where the behaviour of d_N may provide a finer comparison between the tests. Especially interesting is the case when d_N tends to a finite limit, the so called *asymptotic deficiency* of T'_N with respect to T_N.

However, the evaluation (even asymptotic) of d_N is a more delicate matter than showing that $e = 1$. This is usually done by comparison of the corresponding two asymptotic expansions.

Again, there is not enough space to go into details and niceties of this sophisticated problem. We refer to basic papers by Hodges and Lehmann (1970), Bickel (1974), Albers (1974, 1975), Albers, Bickel and van Zwet (1976), and Bickel and van Zwet (1978). Here we will display only a couple of examples. For testing symmetry against a sequence of contiguous alternatives, the deficiency of both the Fraser normal scores test and the van der Waerden test with respect to the most powerful parametric test based on $\bar{X}$ for normal alternatives tends to ∞ at the rate of $\frac{1}{2}\log\log N$. For logistic alternatives the deficiency of the Wilcoxon one-sample (signed rank) test with respect to the most powerful parametric test tends to a finite limit. Qualitatively similar results hold true also for the two-sample tests against a sequence of contiguous alternatives. Namely, the deficiency of both the Fisher-Yates-Terry-Hoeffding normal scores test and the van der Waerden test with respect to the most powerful test based on the difference of the sample means for normal alternatives tends to ∞ at the rate of $\log\log N$. For logistic alternatives the deficiency of the Wilcoxon two-sample rank test with respect to the most powerful parametric test tends to a finite limit (other than before).

One remarkable phenomenon should be noticed. In the one-sample location problem the underlying distribution is always supposed to be symmetric, which entails that the asymptotic expansions for the hypothesis and for contiguous location alternatives do not contain the term of order $N^{-1/2}$ for any parametric or rank tests. If we restrict our attention to sequences of tests T_N and T'_N with asymptotic relative efficiency 1, the leading terms of the expansions coincide, so that these expansions have the form $\pi_N = c_0 + c_{2N}N^{-1} + o(N^{-1})$, $\pi'_N = c_0 + c'_{2N}N^{-1} + o(N^{-1})$. Thus the deficiency d_N is of the order $N(\pi_N - \pi'_N) = c_{2N} - c'_{2N} + o(1)$. In the two-sample problem, however, the underlying distributions are not required to be symmetric, consequently, the asymptotic expansions do in general contain the term of order $N^{-1/2}$. It is then remarkable that for a number of important two-sample tests, like for the most powerful test, the locally most powerful test, the locally most powerful rank test and its approximate scores analogue, the term of order $N^{-1/2}$ in the expansion for contiguous location alternatives is in fact the same for each of these four tests. This implies that again d_N is of the order $N(\pi_N - \pi'_N) = c_{2N} - c'_{2N} + o(1)$, so that the results for d_N are qualitatively the same as for the one-sample case. For details see Bickel and van Zwet (1978), who say that this phenomenon is similar in nature to the phenomenon 'first order efficiency implies second order efficiency' proved by Pfanzagl (1979) (though for the parametric, but possibly asymmetric, one-sample problem).

8.5 ADAPTIVE RANK TESTS

Throughout the book, except the present section, we have supposed that the distributions, the relevant scores and rank tests for basic observations are known. However, in practice this is usually not the case, and we know either nothing about them, or only have a partial little information. Thus, having the observations, we often need first to estimate their distribution or scores, or at least to find out to which broad type they belong, and only then can we select and apply some rank test which is reasonably 'good' for this case. Such tests are called *adaptive rank tests.*

There are numerous procedures, developments, and papers on this topic, so that a detailed description of them would require a separate book in itself. Therefore we will restrict ourselves here only to basic ideas and some important contributions. Some older surveys are Hogg (1974), Behnen (1983), and Hušková (1983, 1984).

8.5.1 Restrictive adaptive tests. These tests are based rather on intuitive grounds and simple formulas, and they originated in principle from earlier descriptive statistics. For their closer inspection mainly Monte Carlo methods are used. The basic steps of such tests are as follows:

(1) Fix a reasonable family $\mathcal{F}$ of distributions or densities whose members can be supposed to govern the observations.

(2) Choose a decision rule selecting, on the basis of observations, the density $f_0 \in \mathcal{F}$ which seems to be the closest (in a certain sense) in $\mathcal{F}$ to the actual distribution.

(3) By this decision rule select the appropriate $f_0 \in \mathcal{F}$, and perform the optimal rank test for it.

The restrictive tests are simpler than the non-restrictive ones, but they tend to be optimal, or at least close to 'reasonable' tests only if the true density is close in some sense to the family $\mathcal{F}$.

First, the *one-sample problem* for testing symmetry of the basic density will be considered.

Generally, we will work with the family of densities

$$\mathcal{F}(f) = \{g;\, g(x) = \lambda f(\lambda x - u),\ -\infty < u < \infty,\ \lambda > 0\}$$

where f is some given density.

Now, one idea for choosing the decision rule is motivated by the behaviour of the tails of distribution (e.g. Hájek (1970), Randles and Hogg (1973), Jones (1979)). In such a case the family $\mathcal{F}$ contains densities ranging from the light-tailed (like uniform) to heavy-tailed (like double-exponential or Cauchy).

Randles and Hogg (1973) proposed the three-member family $\mathcal{F}(f_1)$, $\mathcal{F}(f_2), \mathcal{F}(f_3)$, where f_1 is the double exponential density (heavy-tailed), f_2 is logistic (medium-tailed), and f_3 is uniform (light-tailed). If $X^{(1)} \leq X^{(2)} \leq \ldots \leq X^{(N)}$ are the ordered statistics of the sample $X_1, X_2, \ldots, X_N$, define the following measure of tail weight

$$
\begin{aligned}
Q &= (X^{(N)} - X^{(1)})N\Big\{2\sum_{i=1}^{N}|X^{(i)} - M|\Big\}^{-1} \quad \text{for } N \leq 20, \qquad (1)\\
Q &= (\bar{U}_{0.05} - \bar{L}_{0.05})(\bar{U}_{0.5} - \bar{L}_{0.5})^{-1} \quad \text{for } N > 20,
\end{aligned}
$$

where M is the sample median, and $\bar{U}_\beta(\bar{L}_\beta)$ is the average of the largest (smallest) $N\beta$ order statistics (fractional items are used if $N\beta$ is not an integer).

The decision rule is as follows: choose

$$
\begin{aligned}
&\mathcal{F}(f_1) \quad \text{for } Q > 2.96 - 5.5/N, \qquad (2)\\
&\mathcal{F}(f_2) \quad \text{for } 2.96 - 5.5/N \geq Q \geq 2.08 - 2/N,\\
&\mathcal{F}(f_3) \quad \text{for } 2.08 - 2/N > Q,
\end{aligned}
$$

and for $\mathcal{F}(f_1)$ apply the sign test, for $\mathcal{F}(f_2)$ the one-sample Wilcoxon signed-rank test, for $\mathcal{F}(f_3)$ the authors propose a modified Wilcoxon test eliminating about one-half of the observations with smallest $|X_i|$. The tail weight Q is essentially the range of the sample divided by the mean deviation from the sample median, which gives a reasonable motivation for the rule (2). In particular, for $N \to \infty$ we have in probability $Q \to 3.3$ for $f \in \mathcal{F}(f_1)$, $Q \to 2.6$ for $f \in \mathcal{F}(f_2)$, $Q \to 1.96$ for $f \in \mathcal{F}(f_3)$.

Hájek (1970) proposed another procedure based on tail behaviour for the family $\mathcal{F}(f_1), \ldots, F(f_K)$, where f_i are distinct symmetric densities. The decision rule is defined by choosing that $\mathcal{F}(f_i)$ for which the quantile function corresponding to f_i is close to the sample quantile function. The procedure is quick but no further properties have been studied.

Jones (1979) introduced the continuous infinite family $\mathcal{F} = \{f_\lambda, \lambda \text{ real}\}$, where f_λ is defined by

$$F_\lambda^{-1}(u) = \big(u^\lambda - (1-u)^\lambda\big)/\lambda,$$

so that $\varphi(u, f_\lambda) = (\lambda - 1)(u^{\lambda-2} - (1-u)^{\lambda-2})(u^{\lambda-1} + (1-u)^{\lambda-1})^{-2}$. This family contains densities ranging from light-tailed ones (for $\lambda > 0$) to heavy tailed ones (for $\lambda < 0$). In particular, for $\lambda = 1$ and $\lambda = 2$, f_λ is uniform; for $\lambda = 0.135$, f_λ is approximately normal; for $\lambda = 0$, f_λ is logistic. The author proposed to estimate λ by means of the ordered absolute values of the observations, namely

$$
\begin{aligned}
\hat{\lambda} = (\log 2)^{-1} \log\{(|X|^{(N-2M+1)} - |X|^{(N-4M+1)})\,\cdot \\
\cdot(|X|^{(N-M+1)} - |X|^{(N-2M+1)})^{-1}\}
\end{aligned}
$$

where M is chosen in some proper way reflecting the behaviour of the tail. The score φ-function in the suitable one-sample signed-rank statistic is taken as $\varphi(u, f_\lambda)$.

There are more papers dealing with this problem, among them also those based on different ideas than the tail behaviour, but we cannot go into the details (see e.g. Hájek (1970), Albers (1979), etc.).

Second, consider the *two-sample problem* that two independent symmetric densities having the same shape are identical. The restricted adaptive tests are again based on analogous principles.

Randles and Hogg (1973) propose the following procedure similar to that described above. Let Q_X and Q_Y be the Q statistics, as in (1), computed from the two respective samples, $X_1, \ldots, X_m$ and $Y_1, \ldots, Y_n$, where $N = m + n$. Define $\bar{Q} = (mQ_X + nQ_Y)/N$, and $\bar{N} = (m^2 + n^2)/N$. Then the decision rule is as (2), where Q and N are replaced by $\bar{Q}$ and $\bar{N}$, respectively, and for $\mathcal{F}(f_1)$ the two-sample median test is performed, for $\mathcal{F}(f_2)$ the two-sample Wilcoxon test, and for $\mathcal{F}(f_3)$ the authors thought first of the Haga test (cf. Subsection 4.1.2) but it gives too much weight on a few extreme order statistics, so that they recommend their newly modified Wilcoxon test eliminating about one-half of the middle observations.

The two-sample problem for testing the equality of location, but for asymmetric densities, was dealt with by Hogg, Fisher and Randles (1975). Their procedure is essentially similar. Let us suppose that the null hypothesis of equality of locations is true. Then we may pool both samples $X_1, \ldots, X_m$ and $X_{m+1}, \ldots, X_N$ into one sample, and by means of it define $\bar{U}_\beta(\bar{L}_\beta)$ as in (1), and $\bar{M}_{0.5}$ as the average of the middle 50 percent of the pooled sample. Then consider

$$Q_1 = (\bar{U}_{0.05} - \bar{M}_{0.5})(\bar{M}_{0.5} - \bar{L}_{0.05})^{-1}, \tag{3}$$
$$Q_2 = (\bar{U}_{0.05} - \bar{L}_{0.05})(\bar{U}_{0.5} - \bar{L}_{0.5})^{-1}. \tag{4}$$

Q_1 essentially indicates the skewness: Q_1 large indicates that probably the right tail of the density is longer than the left, i.e. skewness to the right (e.g. the right exponential density has $Q_1 = 4.5$), etc. Q_2 essentially indicates the tail weight as described in (1) and (2). After computing Q_1, Q_2, the authors recommend selecting one of the following five rank tests: If $Q_2 > 7$, we use a heavy-tailed model and the median test. If $2 < Q_2 < 7$, we use a medium-tailed model, and we might use the two-sample Wilcoxon test, provided Q_1 is not too extreme. If $Q_2 \leq 2$, $Q_1 \leq 2$, this is light-tailed symmetric model, and we use the modified Wilcoxon test where about 50% of the middle observations are discarded. If the data indicate skewness to the right, say $Q_1 \geq 2$ (or skewness to the left, say $Q_1 \leq \frac{1}{2}$), we use a modified Wilcoxon test, where about 50% of the largest (of the smallest, respectively) observations are discarded.

Ruberg (1986) proposed a rather elaborate procedure of this kind. First, by means of Q_1 and Q_2 he found certain 'truncation' points. Then, for light-tailed distributions he used a modified Wilcoxon test, where the middle part of the sample between the truncation points has scores 0, while the left-hand and right-hand parts have modified Wilcoxon scores; for heavy-tailed distributions the middle part has modified Wilcoxon scores, while the left-hand and right-hand parts have constant scores. (For details and Monte Carlo results, of how to select the best test, cf. Ruberg (1986).)

Hüsler (1987) tried to improve the power function of the latter procedures by the following modification. Let Q_{1X}, Q_{2X} (resp. Q_{1Y}, Q_{2Y}) be defined as in (3) and (4) but only for the X-sample (Y-sample, respectively). The decision rule is as above, after formula (3) and (4), only in place of Q_i we use now $Q_i^* = \frac{m}{N}Q_{iX} + \frac{n}{N}Q_{iY}$, $i = 1, 2$.

A survey of similar procedures is given by Hogg and Lenth (1984).

Other ideas for the present problem may be found e.g. in Hájek (1970).

8.5.2 Non-restrictive adaptive tests. This type of adaptive test involves the following steps:

(1) Prechoose the type of the test statistic.

(2) Estimate either the density f, or rather the score function $\varphi(.\,, f)$ (the latter is more straightforward, and much more often used) from the data at hand. Namely, this is done by means of the order statistics.

(3) Construct the rank test generated by this estimate of the score function $\varphi(.\,, f)$, and decide by the result of this test.

This type of test may be considered more modern in spirit. Such tests are generally asymptotically optimal for a broad class of densities, but their use involves, as a rule, tedious computations, and the convergence rates for the asymptotic results may be quite slow. (Of course, each such test generated by an estimate of $\varphi(.\,, f)$ is essentially 'new', so we must also compute critical values for it.)

One of the first attempts to estimate the optimal $\varphi(.\,, f)$ is due to Hájek (1962), cf. also Hájek and Šidák (1967), VII.1.5:

Theorem 1 *Let $\mathbf{Z} = (Z_1, \ldots, Z_N)$ be a random sample from the density f, and let $Z^{(1)}, \ldots, Z^{(N)}$ be its order statistics.*

Put $m_N = [N^{3/4}\varepsilon_N^{-2}]$, $n_N = [N^{1/4}\varepsilon_N^3]$, where $\{\varepsilon_N\}$ is some sequence of positive numbers such that

$$\varepsilon_N \to 0, \quad N^{1/4}\varepsilon_N^3 \to \infty.$$

Furthermore, put

$$h_{Nj} = [jN/(n_N + 1)], \quad 1 \le j \le n_N,$$

where $[x]$ always denotes the largest integer not exceeding x. Now take

$$\tilde{\varphi}_N(u, \mathbf{Z}) = \begin{cases} 2\dfrac{m_N n_N}{N+1}\{[Z^{(h_{Nj}+m_n)} - Z^{(h_{Nj}-m_N)}]^{-1} \\ -[Z^{(h_{Nj+1}+m_N)} - Z^{(h_{Nj+1}-m_N)}]^{-1}\}, \\ \qquad h_{Nj}/N \le u < h_{Nj+1}/N,\ 1 \le j \le n_N, \\ 0, \qquad \textit{otherwise}. \end{cases} \tag{1}$$

Then $\tilde{\varphi}_N(u, \mathbf{Z})$ is a consistent estimate of $\varphi(u, f)$, i.e.

$$\lim_{N\to\infty} P\Big\{ \int_0^1 [\tilde{\varphi}_N(u, \mathbf{Z}) - \varphi(u, f)]^2 \, du > \varepsilon \Big\} = 0 \tag{2}$$

for every $\varepsilon > 0$ and the density f satisfying $I(f) < \infty$.

Remarks. Note that h_{Nj} and h_{Nj+1} should be read as $h_{N,j}$ and $h_{N,j+1}$, respectively. Since the proof is somewhat lengthy, we will omit it (it can be found e.g. in Hájek and Šidák (1967), VII.1.5). It suffices to say that the motivation for the theorem comes from the following two facts:
(1) $Z^{[Nu]} \to F^{-1}(u)$ in probability as $N \to \infty$, $0 < u < 1$,
(2) $\lim_{r\downarrow 0, s\downarrow 0} 2rs\{\frac{1}{F^{-1}(u+r)-F^{-1}(u-r)} - \frac{1}{F^{-1}(u+s+r)-F^{-1}(u+s-r)}\}$
$= \varphi(u, f)$, $0 < u < 1$.

Further, Hájek and Šidák (1967), VII.1.6, introduce the statistic $\hat{S}$, analogous to S in (8.1.3.7) but where $a_N(R_i, f_0)$ are replaced by $\hat{a}_N(R_i)$ generated by the estimate $\tilde{\varphi}_N(u, \mathbf{Z})$ from (1). They show that $\hat{S}$ (after proper normalization) is asymptotically normal $(0, 1)$, and yields an asymptotically most powerful test for the problem in Theorem 8.1.3.1 for any f with $I(f) < \infty$. A related assertion holds also for the k-sample problem and Theorem 8.1.4.1. (In proofs one must be cautious since $\hat{S}$ is not a rank statistic; it depends also on $\mathbf{Z}^{(\cdot)}$ through $\tilde{\varphi}_N(u, \mathbf{Z})$.)

Unfortunately, $\tilde{\varphi}_N(u, \mathbf{Z})$ constructed in Theorem 1 converges to $\varphi(u, f)$ very slowly, which makes its practical usefulness doubtful. Therefore, different authors proposed other estimates of $\varphi(u, f)$ hoping that their convergence might be more rapid.

Other, more modern, devices of estimating $\varphi(u, f)$ employ, e.g. the ideas of *Fourier analysis.*

Observe that if the density f possesses finite Fisher information $I(f)$, then for testing the null hypothesis of randomness against local (or contiguous) regression alternatives, a locally most powerful (or asymptotically efficient) rank test, based on a linear rank statistic, corresponds to a score generating function

$$\varphi(u, f) = -f'(F^{-1}(u))/f(F^{-1}(u)), \quad u \in (0, 1). \tag{3}$$

We will sometimes write more concisely φ_f instead of $\varphi(u, f)$. For testing the hypothesis of symmetry (about the origin) against a shift in location, a similar situation holds. Moreover, if we let $\bar{\varphi}_f = \int_0^1 \varphi_f(u)\, du$ and $\|\varphi_f\|^2 = \int_0^1 \varphi_f^2(u)\, du$, then note that $I(f) < \infty$ implies that $\bar{\varphi}_f = 0$ and

$$\|\varphi_f\|^2 = \int_0^1 \left(f'(F^{-1}(u))/f(F^{-1}(u))\right)^2 du = I(f) < \infty, \tag{4}$$

so that $\varphi_f \in L_2(0,1)$. This makes it, at least intuitively, appealing to incorporate a suitable *orthonormal system*, say $\{(P_k(u),\, u \in [0,1]);\, k \geq 0\}$, in a *Fourier series* representation:

$$\varphi_f(u) \sim \sum_{k\geq 0} \gamma_k P_k(u), \quad u \in [0,1], \tag{5}$$

where the γ_k are the Fourier coefficients, (conventionally) we let $P_0(u) \equiv 1$, and

$$\int_0^1 P_k(u)P_q(u)\,\mathrm{d}u = \delta_{kq} = \begin{cases} 1, & k = q, \\ 0, & k \neq q, \end{cases} \quad k, q \geq 0. \tag{6}$$

Beran (1974) applied a *trigonometric system* for the location model, and suggested the estimate $\hat{\varphi}_N$, of $\varphi(.\,, f)$ through the estimates $\hat{c}_k$ of the Fourier coefficients c_k belonging to $\varphi(.\,, f)$, as shown in Theorem 2.

Theorem 2 *Let* $\mathbf{Z}$ *be as in Theorem* 1, *and let* $T_N(\mathbf{Z}, g)$ *be the functional on* $L_2(0,1)$ *given by*

$$\begin{aligned} &T_N(\mathbf{Z}, g) \\ &= (2N\theta_N)^{-1} \sum_{\nu=1}^{N} \Big\{ g\big((N-1)^{-1} \sum_{j=1, j\neq\nu}^{N} v(Z_\nu - Z_j + \theta_N)\big) \\ &\quad - g\big((N-1)^{-1} \sum_{j=1, j\neq\nu}^{N} v(Z_\nu - Z_j - \theta_N)\big) \Big\} \end{aligned} \tag{7}$$

where $v(x) = 1$ *for* $x \geq 0$, $v(x) = 0$ *for* $x < 0$, $\theta_N = bN^{-1/2}$ *for some* $b \neq 0$. *If* $\varphi(u, f)$ *has the Fourier expansion* $\sum_{|k|=1}^{\infty} c_k \exp(2\pi iku)$, *then for an estimate of* $\varphi(u, f)$ *we can take*

$$\hat{\varphi}_N(u, \mathbf{Z}) = \sum_{|k|=1}^{M_N} \hat{c}_k \exp(2\pi iku) \ \textit{with}\ \hat{c}_k = T_N\big(\mathbf{Z}, \exp(-2\pi iku)\big). \tag{8}$$

In the regular case, and if $M_N \to \infty$, $M_N^{7/2} N^{-1} \to 0$, *then* $\hat{\varphi}_N(u, \mathbf{Z})$ *is again a consistent estimate in the sense of Theorem* 1.

Remark. A motivation comes from the fact that $T_N(\mathbf{Z}, g)$ is an estimate of the functional

$$T(g) = \int_0^1 \varphi(u, f) g(u)\,\mathrm{d}u = \int_{-\infty}^{\infty} \frac{\mathrm{d}g(F(x))}{\mathrm{d}x}\,\mathrm{d}F(x), \tag{9}$$

obtained by replacing the theoretical distribution by the empirical one and replacing the derivative by a difference.

If we consider the *two-sample location model* (cf. Subsection 3.4.4) with the samples $\mathbf{X} = (X_1, \ldots, X_m)$ and $\mathbf{Y} = (Y_1, \ldots, Y_n)$ and the densities $f(x - \Delta)$ and $f(x)$, respectively, define the estimators

$$\tilde{\varphi}_N^*(u) = N^{-1}\big(m\tilde{\varphi}_m(u, \mathbf{X}) + n\tilde{\varphi}_n(u, \mathbf{Y})\big), \quad 0 < u < 1, \tag{10}$$

and

$$\hat{\varphi}_N^*(u) = N^{-1}\big(m\hat{\varphi}_m(u, \mathbf{X}) + n\hat{\varphi}_n(u, \mathbf{Y})\big), \quad 0 < u < 1, \tag{11}$$

with $N = m+n$. The respective adaptive rank test statistics in this model are then given by the usual rank test statistic, but in which the relevant scores are generated by (10) or (11) in place of $\varphi(u, f)$. If $\min(m, n) \to \infty$, these adaptive tests provide the asymptotically most powerful tests of H_0 against contiguous alternatives.

For the *one-sample location model* (cf. Subsection 3.4.9) the procedures can be easily modified, see Beran (1974).

On the other hand, Hušková and Sen (1985, 1986) incorporated the following *Legendre polynomial system*, covering both the location and regression model,

$$P_k(u) = \sqrt{2k+1}(-1)^k(k!)^{-1}\{(\mathrm{d}^k/\,\mathrm{d}u^k)[u(1-u)]^k\}, \qquad u \in [0, 1];\ k \geq 0. \tag{12}$$

Note that $P_0(.) \equiv 1$ and $P_1(u)$ $(= \sqrt{3}(2u-1))$ corresponds to the classical Wilcoxon score generating function, which coincides with $\varphi_f(u)$ when f is a logistic density. This adds a distinct advantage for this system over the trigonometric one. Further note that for every $k \geq 0$, $P_{2k}(u) = P_{2k}(1 - u)$ is symmetric about $u = 1/2$ and $P_{2k+1}(u) = -P_{2k+1}(1-u)$ is skew-symmetric. Also, by (5) and (12),

$$I(f) = \langle \varphi_f, \varphi_f \rangle = \sum_{k \geq 0} \gamma_k^2 < \infty, \tag{13}$$

and for every $k \geq 0$,

$$\langle \varphi_f, P_k \rangle = \int_0^1 \varphi_f(u) P_k(u)\,\mathrm{d}u = \gamma_k, \tag{14}$$

and these provide us with a way to estimate the γ_k and $I(f)$ in a convenient nonparametric manner.

Based on the convergence in (13), we claim that for every $\varepsilon > 0$, there exists a positive integer $M = M_\varepsilon$, such that

$$\sum_{k>m} \gamma_k^2 \leq \varepsilon, \quad \text{for all } m \geq M_\varepsilon. \tag{15}$$

Therefore, we are tempted to approximate $\varphi_f(.)$ by

$$\varphi_{f,M}(u) = \sum_{k \leq M} \gamma_k P_k(u), \quad u \in [0,1], \tag{16}$$

and we proceed to estimate only the set $\{\gamma_k, k \leq M\}$ by suitable rank methods, denote these estimates by $\hat{\gamma}_{k,N}$, $k \leq M$, and consider the estimated score generating function:

$$\hat{\varphi}_{f,M;N}(u) = \sum_{k \leq M} \hat{\gamma}_{k,N} P_k(u), \quad u \in (0,1). \tag{17}$$

While this simple prescription works out well in many situations, we may bear in mind that the γ_k, by (14), are dependent on the unknown f, so that $M = M_\varepsilon$ may also depend on f. As a result, the approximation in (16), based on (15), cannot possibly be uniform in all f belonging to the contemplated class with $I(f) < \infty$. This can be achieved either by confining to a subclass of finite $I(f)$ densities for which such a uniformity result holds, or using a sequential setup where $M_\varepsilon = M_{\varepsilon,N}$ is allowed to depend on the sample size N (i.e., $\varepsilon = \varepsilon_N \downarrow 0$ as $N \to \infty$), so that as $N \to \infty$, such a regularity condition holds for the entire class of densities with finite $I(f)$. Hušková and Sen (1985, 1986) considered such a sequential procedure, and studied various properties of their proposed estimates.

We may also note that with the γ_k estimated from the same sample, the estimated $\hat{\varphi}_{f,M;N}(.)$ is an adaptive score function, and as a result, a rank statistic based on $\hat{\varphi}_{f,M;N}$ is an adaptive rank statistic. Since such adaptive (linear or signed) rank statistics are to be used either as suitable rank tests for testing a null hypothesis of randomness (or symmetry) against regression (or shift) alternatives, we need an additional regularity assumption that the $\hat{\gamma}_{k,N}$ are translation (or regression) invariant. For our testing problems, we have taken $F(x) = F_0(x - \theta)$ for the location model, and $F_0(x - \mathbf{t}'\beta)$ for the linear model, where the form of F_0 is assumed to be free from the parameter θ (or β).

In this setup, using (14), we write

$$\begin{aligned} \gamma_k &= \int_0^1 P_k(u)\varphi_f(u)\,\mathrm{d}u \\ &= \int_{-\infty}^{\infty} P_k(F_0(x))\{-f_0'(x)/f_0(x)\}\,\mathrm{d}F_0(x) \\ &= \int_{-\infty}^{\infty} f_0(x)\,\mathrm{d}P_k(F_0(x)) \\ &= \int_{-\infty}^{\infty} P_k'(F_0(x))f_0^2(x)\,\mathrm{d}x, \quad k \geq 0, \end{aligned} \tag{18}$$

where by (12)

$$\begin{aligned} P_k'(u) &= \frac{\mathrm{d}}{\mathrm{d}u} P_k(u) \\ &= \frac{(-1)^k}{k!}\sqrt{2k+1}\frac{\mathrm{d}}{\mathrm{d}u}\{(\mathrm{d}^k/\,\mathrm{d}u^k)[u(1-u)]^k\} \\ &= \frac{(-1)^k}{k!}\sqrt{2k+1}\{(\mathrm{d}^{k+1}/\,\mathrm{d}u^{k+1})[u(1-u)]^k\} \\ &= \text{a polynomial of degree } (k-1), \quad k \geq 1, \end{aligned} \tag{19}$$

while $P_0'(u) \equiv 0$, by definition. As a result, if we define the coefficients $\xi_p = \int u^p f_0(F_0^{-1}(u))\,\mathrm{d}u$, for $p \geq 0$, then the γ_k can be expressed (recursively) in terms of the ξ_p, $p \leq k-1$. But this process needs the estimation of $f_0(F_0^{-1}(u))$ (as was pursued in the original text), which in turn, may need a much larger sample size to achieve a reasonable degree of consistency. Hušková and Sen (1985, 1986) avoided these complications by appealing to an R-(rank)estimate, of the γ_k. We present the case of the simple regression model, and similar solutions exist for other models as well. Here we have $F_i(x) = F_0(x - \theta - \beta t_i)$, $i \geq 1$. Based on $X_1, \ldots, X_N$, we define the ranks $R_1, \ldots, R_N$ as before. Then note that the ranks are invariant under any translation of the X_i's, and hence, without loss of generality, we may take $\theta = 0$. Consider then the linear rank statistic

$$L_{Nk} = \sum_{i=1}^{N}(t_i - \bar{t}_N)P_k\Big(\frac{R_i}{N+1}\Big), \quad k \geq 1. \tag{20}$$

Next, consider the aligned observations

$$X_i(b) = X_i - bt_i, \quad i = 1, \ldots, N, \quad -\infty < b < \infty. \tag{21}$$

Let $R_i(b)$ be the rank of $X_i(b)$ among $X_1(b), \ldots, X_N(b)$, for $i = 1, \ldots, N$, and let

$$L_{Nk}(b) = \sum_{i=1}^{N}(t_i - \bar{t}_N)P_k\Big(\frac{R_i(b)}{N+1}\Big). \tag{22}$$

For $b = \beta$, $L_{Nk}(\beta)$ has the same distribution as L_{Nk}, under the null hypotheses $\beta = 0$. Therefore $L_{Nk}(\beta)$ is distribution-free, and has mean 0. Take $k = 1$, and equate $L_{N1}(b) = 0$; this gives the classical Wilcoxon-score R-estimate of β, denoted by $\hat{\beta}_{N,1}$. Take an arbitrary (fixed) $a > 0$, and let

$$\hat{\gamma}_{k,N} = \frac{1}{2a\sqrt{N}}\{L_{Nk}(\hat{\beta}_{N,1} - \tfrac{1}{\sqrt{N}}a) - L_{Nk}(\hat{\beta}_{N,2} + \frac{1}{\sqrt{N}}a)\}, \tag{23}$$

for $k \geq 1$. Asymptotic properties of these invariant estimators are based on the asymptotic linearity of linear rank statistics in the regression parameter that will be presented in detail in Section 9.2. In addition, here the $P_k(.)$ have bounded second derivatives (as they are polynomials on [0,1]), and hence these $\hat{\gamma}_{k,N}$ will have better convergence rates than in the general case where the score function is only taken to be square integrable. Hušková and Sen (1985, 1986) have established these convergence properties, and cited other references as well. We omit these details here.

In passing, we may remark that if F_0 is symmetric about 0 (which we generally assume for testing the hypothesis of symmetry), then by the symmetric nature of $P_{2k}(u)$, $k \geq 0$, we have $\gamma_{2k} = 0$, for every $k \geq 0$. Therefore in this case, in the Fourier series representation in (5), we may take

$$\varphi_f(u) \sim \sum_{k\geq 0} \gamma_{2k+1} P_{2k+1}(u), \quad u \in (0,1). \tag{24}$$

As such, in (17) we would only have the odd-order estimated Fourier coefficients $\hat{\gamma}_{2k+1,N}$, $k \geq 0$, so that for any chosen M (≥ 1), there is about 50% computational savings. Also, for the particular case of the double-exponential density $f_0(x) = \frac{1}{2}\mathrm{e}^{-|x|}$, $-\infty < x < \infty$, we have $\varphi_f(u) = \mathrm{sign}(u - \frac{1}{2})$, and there is a jump-discontinuity at $u = \frac{1}{2}$. Further, $I(f_0) = 1$. In this case using the representation in (5) that involves bounded and absolutely continuous $P_k(u)$ may therefore require M to be large. On the other hand, for this f_0, the sign-test is known to be optimal. For a logistic f_0, we have $\varphi_f(u) = 2u - 1$, so that $P_1(u)$ alone suffices in (5), i.e., $\gamma_k = 0$, for all $k \geq 2$. For a Cauchy density $f_0(x) = \pi^{-1}(1 + x^2)^{-1}$, $-\infty < x < \infty$, because of the arctan transformation that reduces f_0 to a uniform $(0,1)$ density, a trigonometric system, such as that considered by Beran (1974) may be more convenient. Finally, for a normal f_0, we may note that $\gamma_1^2 = 3/\pi \sim 0.955$, so that $P_1(.)$ alone explains about 95% of the information, while $\gamma_2 = \gamma_4 = \ldots = 0$. Further, note that $(\mathrm{d}/\mathrm{d}u)P_{2k+1}(u)$ is itself a polynomial of degree $2k$, so that by partial integration of (14), it can be shown that including $P_1(.)$, $P_3(.)$ and $P_5(.)$ only in (5) may provide an almost full recovery of the information $I(f)$, so that from this point, in practice we may prescribe $M = 5$. From a theoretical perspective, however, we need M to be large to cover all f's.

Let us now add a remark on the *two-sample problem* with samples $X_1, \ldots, X_m$ and $Y_1, \ldots, Y_n$, where F and G are distribution functions (f and g densities) of the two samples, respectively, and $m + n = N$, $\lambda_N = m/N \to \lambda \in (0,1)$ as $N \to \infty$. Throughout the book we suppose that F and G have the same functional form, and that they differ only by some parameter (mainly in location, or in scale). More recently, a number of contributions appeared on adaptive rank tests of the null hypothesis $H_0 : F = G$ against a much more general alternative $K : F \leq G, F \neq G$, where both F and G are unknown. The start of the approach to such problems is the same as in our Subsection 8.3.3. Put

$H(x) = \lambda F(x) + (1-\lambda)G(x)$, and define the densities

$$\bar{f}(u) = \frac{\mathrm{d}}{\mathrm{d}u} F\big(H^{-1}(u)\big), \quad \bar{g}(u) = \frac{\mathrm{d}}{\mathrm{d}u} G\big(H^{-1}(u)\big), \quad 0 < u < 1. \tag{25}$$

Clearly $\bar{f}$ ($\bar{g}$) corresponds to $H(X)$ ($H(Y)$, respectively). It can be seen that, for fixed F, G, the optimal score function is $\log \bar{f}(u) - \log \bar{g}(u)$, but for F close to G (i.e. F, G close to the null hypothesis H_0) the last function is close to $\bar{f}(u) - \bar{g}(u)$. Therefore usually a simpler problem of estimating an approximately optimal score function $\varphi = \bar{f} - \bar{g}$ is dealt with. Towards this end, the basic principle is usually to begin with $\hat{F}_m$, $\hat{G}_n$, $\hat{H}_N$, i.e. the empirical distribution functions of the X-sample, of the Y-sample, and of the pooled sample, respectively, and then, by means of (25) and a convenient kernel estimation to get the estimators of the densities $\bar{f}$, $\bar{g}$, etc. Different details and variations can be found e.g. in Behnen and Neuhaus (1983), Behnen, Neuhaus and Ruymgaart (1983), Hušková et al. (1984), Behnen and Hušková (1984); especially enlightening discussions are in Behnen (1984), and Neuhaus (1987).

A different and pioneering approach to the two-sample problem was analysed in the book by Behnen and Neuhaus (1989). They do suppose that F and G differ by a location (shift) parameter, but this shift parameter is not constant for all $x \in (-\infty, \infty)$. Instead, they suppose there is some shift function $D(x)$ reflecting different reactions in different parts of the population (e.g. it is plausible that the extreme parts of the population react in quite another way than the central part). Thus instead of using the exact location (shift) model $F(x) = G(x-\theta)$ for all x, they use the generalized shift model

$$F(x) = F\big(x - \theta D(x)\big) \quad \text{for all } x.$$

The shift function D is supposed to be bounded, $D \geq 0$, $D \neq 0$, and has a bounded derivative. Again, F and G are unknown, and the estimation of the optimal score function in these circumstances is investigated thoroughly in the book.

PROBLEMS AND COMPLEMENTS TO CHAPTER 8

Section 8.1

1. Let $U_N^{(1)} < \ldots < U_N^{(N)}$ be an ordered sample from the uniform distribution on $(0,1)$. Put $Z(i/(N+1)) = N^{\frac{1}{2}}(U_N^{(i)} - i/(N+1))$ and complete the definition of $Z(t)$, $\leq t \leq 1$, by assuming linearity within the intervals $((i-1)/(N+1), i/(N+1))$, and by assuming $Z(0) = Z(1) = 0$. Put $V_i = U_N^{(i)} - U_N^{(i-1)}$, with $U_N^{(0)} = 0$, $U_N^{(N+1)} = 1$, and let $V_N^{(1)} < \ldots < V_N^{(N+1)}$ denote the V_i's rearranged according to magnitude. Finally put $V_N^{(.)} = (V_N^{(1)}, \ldots, V_N^{(N+1)})$. Prove subsequently the following propositions:

(i) The vectors $(Z(t_1), \ldots, Z(t_k))$, $1 \leq t_1 < \ldots < t_k \leq 1$, are asymptotically jointly normal with zero means and variances-covariances $\sigma_i^2 =$

$t_i(1-t_i)$, $\sigma_{ij} = t_i(1-t_j)$, $1 \le i < j \le k$.

$$\text{(ii)} \quad P\Big(U_N^{(i)} = \sum_{j=1}^{i} v_N^{(r_j)},\ 1 \le i \le N \mid V_N^{(.)} = v_N^{(.)}\Big) = 1/N!, \ r \in R.$$

$$\text{(iii)} \quad \max_{1\le i\le N+1} \Big(V_N^{(i)} - \frac{1}{N+1}\Big)^2 \Big/ \sum_{j=1}^{N} \Big(V_N^{(j)} - \frac{1}{N+1}\Big)^2 \to 0$$

in probability.

$$\text{(iv)} \qquad \lim_{\delta\to 0} \liminf_{N\to\infty} P\big(\max_{|t-s|<\delta} |Z_N(t) - Z_N(s)| > \varepsilon\big) = 1.$$

[Hint: utilize (ii) and (iii) and Lemma 6.3.5.2.]

(v) The random processes $Z_N(t)$, $0 \le t \le 1$, converge in distribution in $C[0,1]$, to a Brownian bridge. [Hint: utilize (i), (iv) and Theorem 6.3.2.1.]

2. The method of seeking mutual densities of two normal distributions, differing in their expectations, is appropriate not only for processes but also for degenerate finite-dimensional distributions. Consider the k-dimensional distribution P with variances $1-\kappa_j^2$, $1 \le j \le k$, and covariances $-\kappa_j\kappa_g$, $1 \le j \ne g \le k$, where $\sum_{j=1}^{k} \kappa_j^2 = 1$, and with zero expectations. On the other hand, consider the distributions Q_b with the same variances-covariances, but with expectations $\mathsf{E}Z_j = b\Theta_j$, where $\sum_{j=1}^{k} \Theta_j^2 = 1$ and $\sum_{j=1}^{k} \Theta_j\kappa_j = 0$. Show that Q_b is absolutely continuous relative to P and that the Radon-Nikodym derivative equals

$$\frac{\mathrm{d}Q_b}{\mathrm{d}P} = \exp\Big(\sum_{j=1}^{k} \Theta_j z_j - \tfrac{1}{2}b^2\Big).$$

[Hint: Show that $\xi = b\sum_{j=1}^{k} \Theta_j Z_j$ satisfies $\mathsf{E}(Z_j\xi) = b\Theta_j$, $1 \le j \le k$.]

Section 8.2

3. Asymptotic efficiency for H_1. (We employ Definition 8.2.1 for one-sided tests and Definition 8.2.2 with $k=2$ for two-sided tests.)

Theorem: Consider testing H_1 against $q = \prod_{i=1}^{N} f_0(x_i - \Delta)$, $\Delta \ge 0$, with $I(f_0)\Delta^2 N \to b^2$, $0 < b^2 < \infty$, by a test based on $S_N^+ = \sum_{i=1}^{N} a_N(R_i^+)\,\mathrm{sign}\, X_i$, where the functions $a_N(1+[uN])$ converge in quadratic mean to a function $\varphi^+(u)$. Then the asymptotic efficiency of the S_N^+-test equals

$$e = \left[\int_0^1 \varphi^+(u)\varphi^+(u, f_0)\,\mathrm{d}u\right]^2 \cdot \left\{\int_0^1 [\varphi^+(u) - \bar{\varphi}^+]^2\,\mathrm{d}u \int_0^1 [\varphi^+(u, f_0)]^2\,\mathrm{d}u\right\}^{-1}.$$

The same result holds true for the $|S_N^+|$-test if $K = \{q_1, q_2\}$, with $q_t = \prod_{i=1}^{N} f_0(x_i + (-1)^t\Delta)$, $t = 1, 2$. [Utilize Theorem 7.2.5.1.]

4. Asymptotic efficiency for H_2. (We employ Definition 8.2.1 for one-sided tests, Definition 8.2.2 with $k = 2$ for two-sided tests.)

Theorem: Consider testing H_2 against $q_\Delta = \prod_{i=1}^{N} h_\Delta(x_i, y_i)$, with $h_\Delta(x, y) = f_0(x)g_0(y - \Delta m(x))$, where $I(f_0)I(g_0)\Delta^2 N \to b^2$, $0 < b^2 < \infty$, and $\mathsf{E}m(X) = 0$ and $\mathsf{E}m^2(X) < \infty$. Consider a test based on $S_N = \sum_{i=1}^{N} a_N(R_i)$ $b_N(Q_i)$, where the functions $a_N(1 + [uN])$ and $b_N + (1 + [uN])$ converge in quadratic mean to the functions $\varphi(u)$ and $\psi(u)$, respectively. Then the asymptotic efficiency of the S_N-test equals

$$e = \left[\int_0^1 \varphi(u)m(F_0^{-1}(u))\,\mathrm{d}u \int_0^1 \psi(u)\varphi(u, g_0)\,\mathrm{d}u\right]^2 \cdot\left\{I(g_0)\int_0^1 m^2(F_0^{-1}(u))\,\mathrm{d}u \int_0^1 [\varphi(u) - \bar{\varphi}]^2\,\mathrm{d}u \cdot \int_0^1 [\psi(u) - \bar{\psi}]^2\,\mathrm{d}u\right\}^{-1}.$$

The same result holds for the $|S_N - \mathsf{E}S_N|$-test, if $K = \{q_1, q_2\}$, with

$$q_t = \prod_{i=1}^{N} h_\Delta^t(x_i, y_i),\ h_\Delta^t(x, y) = f_0(x)g_0\big(y + (-1)^t\Delta m(x)\big),\ t = 1, 2.$$

[Hint: Assume the X_i's fixed and utilize Theorem 8.2.1.1 with $c_i = a_N(R_i)$ and $d_i = m(X_i)$, $1 \le i \le N$.]

5. Asymptotic efficiency for H_3. Here we employ Definition 8.2.2.

Theorem: Consider testing H_3 against q_Δ given by (7.3.1.9), under assumption (7.3.1.10). Further consider the test based on the statistic Q_n given by (7.3.1.11). Then, for $n \to \infty$, the asymptotic efficiency of the Q_n-test satisfies the inequality

$$e \ge \frac{1}{k-1}\Big[\sum_{j=1}^{k} a(j)a_k(i, f_0)\Big]^2\Big\{I(f_0)\sum_{j=1}^{k}\big(a(j) - \bar{a}\big)^2\Big\}^{-1}.$$

Here, no rank test is asymptotically most powerful, and the maximum possible efficiency of a rank test satisfies the inequality

$$\bar{e} \geq \frac{1}{k-1} \sum_{i=1}^{k} a_k^2(i, f_0)/I(f_0)$$

with $1 > \bar{e}$. If f_0 is normal, then the case of equality occurs in the two relations. If f_0 is not normal, then, presumably, the inequalities are sharp. [Utilize the end of Subsection 7.3.1.]

6. Consider contiguous location problems concerning H_0 and H_1 and for a normal density f_0. Then we obtain the following efficiencies for some common tests:

the Wilcoxon test (one-sample, two-samples)	$3/\pi \doteq 0.95$
the Kruskal-Wallis test	$3/\pi \doteq 0.95$
the median test, the sign test	$2/\pi \doteq 0.64$.

[See Pitman (1948), Andrews (1954), Noether (1958), Hájek (1956), Mood (1954), Chakravarti, Leone and Alanen (1962). Utilize Theorems 8.2.1.1 and 8.2.2.1, and Problem 3.]

7. Consider contiguous scale problems concerning H_0 and H_1 and for a normal density. Then we obtain the following efficiencies for some common tests:

the Ansari-Bradley test:	$6/\pi^2 \doteq 0.61$
the quartile test	$8[(2\pi)^{-\frac{1}{2}} k_{0.75} \exp(-\frac{1}{2}k_{0.75}^2)]^2 \doteq 0.37$
the Mood test	$15/2\pi^2 \doteq 0.76$

where $\Phi(k_{0.75}) = 0.75$. [Ansari and Bradley (1960), Mood (1954). Utilize Theorem 8.2.1.1 with H_0, $\varphi(u, f_0)$ and $I(f_0)$ replaced by H_1, $\varphi_1(u, f_0)$ and $I_1(f_0)$, respectively.]

8. For location problems concerning H_0 and the logistic density f_0 the efficiency of the Wilcoxon test equals 1, and that of the median test 0.75.

9. For scale problems concerning H_0 and the density $f_0 = \frac{1}{2}(1+|x|)^{-2}$ the efficiency of the Ansari-Bradley test equals 1, and that of the quartile test 0.75.

10. Prove that for location problems in H_3 with normal f_0 the asymptotic efficiency of the Friedman test equals $(3/\pi)(k/k+1)$. [Noether (1958). Utilize Problem 5.]

11. Consider testing H_2 against $q_\Delta = \prod_{i=1}^{N} h_\Delta(x_i, y_i)$, $h_\Delta(x, y) = (2\pi)^{-1}\cdot$

$\exp\{-\frac{1}{2}x^2 - \frac{1}{2}(y-\Delta x)^2\}$, by means of the statistics $S_N = \sum_{i=1}^{N} a_N(R_i)a_N(Q_i)$, where the function $a_N(1+[uN])$ converges in the quadratic mean to $\varphi(u)$. Then, under $N\Delta^2 \to b^2$, $0 < b^2 < \infty$, the asymptotic efficiency of the S_N-test equals

$$e = \left\{ \int_0^1 \varphi(u)\Phi^{-1}(u)\,\mathrm{d}u \left(\int_0^1 [\varphi(u) - \bar{\varphi}]^2\,\mathrm{d}u \right)^{-1} \right\}^2.$$

Thus the asymptotic efficiencies of the Spearman (Kendall) test, the quadrant test and the Fisher-Yates test equal $(3/\pi)^2$, $(2/\pi)^2$ and 1, respectively.

The same holds for

$$h_\Delta(x,y) = (2\pi\sigma_1\sigma_2)^{-1}(1-\rho^2)^{-\frac{1}{2}} \\ .\exp\left\{ -\frac{1}{2(1-\rho^2)}\left[\frac{(x-\mu)^2}{\sigma_1^2} - 2\rho\frac{(x-\mu)(y-\nu)}{\sigma_1\sigma_2} + \frac{(y-\nu)^2}{\sigma_2^2}\right]\right\},$$

with $\Delta = \rho/(1-\rho^2)$, and any fixed $(\mu, \nu, \sigma_1^2, \sigma_2^2)$, and for the $|S_N - \mathsf{E}S_N|$-test and for the alternative $K = \{q_1, q_2\}$ where q_t corresponds to

$$h_\Delta^t(x,y) = (2\pi)^{-1}\exp[-\tfrac{1}{2}x^2 - \tfrac{1}{2}(y + (-1)^t\Delta x)^2], \quad t = 1, 2.$$

12. Let f_0 be any density satisfying $\int_{-\infty}^{\infty} x^2 f_0(x)\,\mathrm{d}x < \infty$. Then, for contiguous location problems, the asymptotic efficiency of the Fisher-Yates-Terry-Hoeffding test is greater than or equal to the asymptotic efficiency of the t-test and equality occurs only for f_0 of normal type. [Chernoff and Savage (1958).] The paper by Mikulski (1963) shows that a test based on $S_{mn} = \sum_{i=1}^{m} a_n(R_i, f)$ uniformly surpasses the corresponding best parametric test only if f is of normal type. If f is of non-normal type, then the best parametric test corresponding to f has a greater efficiency for at least one f_0, if f_0 is true.

13. Let e_W be the asymptotic efficiency of the Wilcoxon test for a contiguous location problem with arbitrary density f_0, and let e_t be the efficiency of the t-test under the same situation.

Then $e_W \geq 0.864 e_t$. [Hodges and Lehmann (1956).] The same result holds for the one-sample Wilcoxon test and for the k-sample Kruskal-Wallis test.

14. Prove that

$$\int_0^1 \psi(\alpha, u)\,\mathrm{sign}(2u-1)\,\mathrm{d}u = 1 - 2\Phi(-(-2\log\alpha)^{\frac{1}{2}})/\alpha$$

with $\psi(\alpha, u)$ given by (7.4.5.3).

15. Let $e(\alpha, f_0)$ be the local asymptotic efficiency of the one-sided Kolmogorov-Smirnov test for location problems and for a density f_0, $I(f_0) < \infty$. Show that

$$\bar{e}(\alpha) = \sup_{f_0} e(\alpha, f_0) = 4\pi^2(-\log\alpha)\exp(k_{1-\alpha}^2)\int_0^1 [\psi(\alpha, u)]^2\,du,$$

and $\bar{e}(\alpha) \to 1$ for $\alpha \to 0$. Check by computations that $\bar{e}(0.05) \doteq 0.852$, $\bar{e}(0.01) \doteq 0.864$.

Chapter 9

Rank estimates and asymptotic linearity in regression parameters

In this chapter we will complement the main contents of the book by some important closely related topics. However, for brevity, we will present only the basic ideas, not developing them further, and omitting the details and proofs (with the two exceptions of Theorems 9.2.1.1 and 9.2.1.2 which seem to be of special importance).

The estimates based on rank statistics will be called briefly R-estimates.

9.1 R-ESTIMATES OF LOCATION AND REGRESSION

9.1.1 R-estimates of shift between two samples. First, on a simple model we will show the basic idea going back to Hodges and Lehmann (1963) and Sen (1963). As in (3.4.4.1), consider the density

$$q_\Delta(x_1, \ldots, x_m, x_{m+1}, \ldots, x_{m+n}) = \prod_{i=1}^{m} f(x_i - \Delta) \prod_{i=m+1}^{m+n} f(x_i), \qquad \Delta > 0 \tag{1}$$

where f is some density. This means that the first density generating the sample $X_1, \ldots, X_m$ is shifted to the right with respect to the second density generating the sample $X_{m+1}, \ldots, X_{m+n}$, usually denoted by $Y_1, \ldots, Y_n$. A natural idea for estimating Δ is then (after having observed the X_i's and Y_j's) to take $X_1 - \Delta, \ldots, X_m - \Delta$ as functions of Δ, and to estimate Δ so that this shifted sample is aligned as closely as possible to the sample $Y_1, \ldots, Y_n$.

The words 'aligned as closely as possible' might have different meanings, but Hodges and Lehmann (1963) proposed the following commonly accepted general method of solving this problem.

Consider a test statistic

$$h = h(X_1, \ldots, X_m; Y_1, \ldots, Y_n)$$

for the hypothesis $H_0 : \Delta = 0$ against the alternatives $\Delta > 0$. We will assume that

(A) $h(x_1 + a, \ldots, x_m + a; y_1, \ldots, y_n)$ is a non-decreasing function of a for all $\boldsymbol{x} = (x_1, \ldots, x_m)$ and $\boldsymbol{y} = (y_1, \ldots, y_n)$,

(B) when $\Delta = 0$, the distribution of $h(X_1, \ldots, X_m; Y_1, \ldots, Y_n)$ is symmetric about a fixed point μ for all continuous f.

We will use obvious conventions for relations concerning the vectors $\boldsymbol{x}$ and $\boldsymbol{y}$, and define

$$\begin{aligned} \Delta^* &= \sup\{\Delta : h(\boldsymbol{x} - \Delta; \boldsymbol{y}) > \mu\}, \\ \Delta^{**} &= \inf\{\Delta : h(\boldsymbol{x} - \Delta, \boldsymbol{y}) < \mu\}, \end{aligned} \tag{2}$$

and, for suitable functions h, the random variable

$$\widehat{\Delta} = (\Delta^* + \Delta^{**})/2 \tag{3}$$

is taken to be an estimate of the shift parameter Δ.

A suitable rank statistic for the above purpose is exactly that in (3.4.4.3) for testing the two-sample location problem, i.e.

$$h(X_1, \ldots, X_m; Y_1, \ldots, Y_n) = \sum_{i=1}^{m} a_N(R_i, f), \tag{4}$$

provided (A) and (B) hold true. As usual, here a_N are the scores, $N = m + n$, $R_1, R_2, \ldots, R_{m+n}$ are the ranks of $X_1, \ldots, X_m$, $Y_1, \ldots, Y_n$ in the pooled sample. Note that (A) is satisfied if the corresponding score function $\varphi(u, f)$ is non-decreasing, and (B) is satisfied if f is symmetric or if $m = n$.

Example 1. The estimate based on the two-sample Wilcoxon statistic. By (4.1.1.6), we can use

$$h = \sum_{i=1}^{m} R_i \tag{5}$$

and the above-described procedure for getting $\widehat{\Delta}$. However, for simplifying the computations, we may use the Mann-Whitney form, and then, symbolically,

$$\widehat{\Delta} = \operatorname{med}(\mathbf{X} - \mathbf{Y}). \tag{6}$$

(Cf. Problem 2 for Section 9.1.)

Example 2. The estimate based on the normal scores statistic. By (4.1.1.2), we can now employ

$$h = \sum_{i=1}^{m} a_{m+n}(R_i), \tag{7}$$

where $a_{m+n}(i) = \mathsf{E}V_{m+n}^{(i)}$, with $V_{m+n}^{(1)} < \ldots < V_{m+n}^{(m+n)}$ being an ordered sample from the standardized normal distribution, cf. (4.1.1.1). Of course, this procedure is more clumsy than in Example 1, since we must know the expectations of normal order statistics. (Cf. also Problem 3 for Section 9.1.) For both examples, Hodges and Lehmann (1963) describe certain 'hand aids' for facilitating the computations.

Hodges and Lehmann (1963) also investigated different properties of the estimate $\hat{\Delta}$: its translation equivariance (see Problem 1), symmetry of its distribution about Δ (implying unbiasedness of $\hat{\Delta}$), its median unbiasedness, and its asymptotic normality. However, there were some theoretical inconsistencies in Hodges and Lehmann (1963), but later on all this was streamlined by asymptotic linearity results of Jurečková, Koul, Jaeckel and Sen. In particular, the basic asymptotic properties of $\hat{\Delta}$ have been established in a unified manner by incorporating some asymptotic uniform linearity results to be presented in Section 9.2.

Further, if we have sequences of test statistics h_N and h'_N, and the corresponding estimates $\hat{\Delta}_N$ and $\hat{\Delta}'_N$, then under general conditions, the *asymptotic efficiency* of $\hat{\Delta}'_N$ relative to $\hat{\Delta}_N$ (=reciprocal ratio of asymptotic variance) equals the corresponding Pitman efficiency of the sequences of tests based on h'_N and h_N. The connection between the asymptotic efficiency of R-estimates and the Pitman efficiency of parallel rank tests has been thoroughly exploited for general scores by Sen (1966). Here also, it can be seen that the asymptotic linearity results in Section 9.2 provide an easy access.

It is interesting to note that the same alignment principle that generates the point R-estimates of the shift parameter Δ also provides *distribution-free confidence intervals* for Δ. Towards this end, suppose that $h(\boldsymbol{X}, \boldsymbol{Y})$ is a rank statistic, so that under the null hypothesis $\Delta = 0$ it is EDF. Hence, given m, n, corresponding to a prechosen α $(0 < \alpha < 1)$, we can determine α_{mn} $(\leq \alpha)$ and two numbers $h_{mn}^{(1)}$, $h_{mn}^{(2)}$ (independently of the underlying continuous distribution function F) such that

$$P_0(h_{mn}^{(1)} \leq h(\boldsymbol{X}, \boldsymbol{Y}) \leq h_{mn}^{(2)}) = 1 - \alpha_{mn} \geq 1 - \alpha;$$

if $h(.)$, under the null hypothesis, has a symmetric distribution around 0, we may even take $h_{mn}^{(1)} = -h_{mn}^{(2)}$. Let us then define

$$\begin{aligned}\hat{\Delta}_L &= \sup\{b : h(\boldsymbol{x} - b; \boldsymbol{y}) > h_{mn}^{(2)}\},\\ \hat{\Delta}_U &= \inf\{b : h(\boldsymbol{x} - b; \boldsymbol{y}) < h_{mn}^{(1)}\}.\end{aligned}$$

Then, similarly as in (2) and (3), we obtain that

$$P_\Delta(\hat{\Delta}_L \leq \Delta \leq \hat{\Delta}_U) = 1 - \alpha_{mn} \geq 1 - \alpha.$$

Thus, $(\hat{\Delta}_L, \hat{\Delta}_U)$ provides a distribution-free confidence interval for Δ (see Lehmann (1963), Sen (1963, 1966), and others). Like the point estimate too, this confidence interval is translation equivariant and robust. Moreover, the two-sample location model is a special case of the regression model to be considered later on. Hence, the asymptotic properties of this confidence interval can most conveniently be studied by incorporating some asymptotic linearity results on linear rank statistics (see Section 9.2).

Remark. Though we will digress now from the main topic of this section, we might note that also some estimates based on empirical distribution functions and on Kolmogorov-Smirnov statistics were proposed and studied, e.g. Boulanger (1983), Boulanger and van Eeden (1983), Johnson and Killeen (1985), Schuster and Narvarte (1973), Rao, Schuster and Littell (1975), etc.

9.1.2 R-estimates of shift of one sample. Now we start from the density (3.4.9.1), namely

$$q_\Delta(x_1, \ldots, x_N) = \prod_{i=1}^{N} f(x_i - \Delta), \quad \Delta \geq 0, \tag{1}$$

where f is some density symmetric about zero.

Hodges and Lehmann (1963) propose the following method for estimating Δ on the basis of a sample $X_1, \ldots, X_N$. Considerations similar to those for the two-sample problem in 9.1.1 suggest basing such an estimate on a test statistic $h = h(X_1, \ldots, X_N)$ for the hypothesis $\Delta = 0$ against $\Delta > 0$. We will assume that

(C) $h(x_1 + a, \ldots, x_N + a)$ is a non-decreasing function of a for each $\boldsymbol{x} = (x_1, \ldots, x_N)$,

(D) for $\Delta = 0$, the distribution of $h(X_1, \ldots, X_N)$ is symmetric about a fixed point μ for all continuous f symmetric about zero.

Now let

$$\Delta^* = \sup\{\Delta : h(\mathrm{x} - \Delta) > \mu\}, \quad \Delta^{**} = \inf\{\Delta : h(\boldsymbol{x} - \Delta) < \mu\}, \tag{2}$$

and our estimate of the shift Δ is

$$\widehat{\Delta} = (\Delta^* + \Delta^{**})/2. \tag{3}$$

As for rank statistics, we may use that in (3.4.9.2) for testing symmetry, namely

$$h_0(X_1, \ldots, X_N) = \sum_{i=1}^{N} a_N^+(R_i^+, f) \operatorname{sign} X_i. \tag{4}$$

However, we may also use the statistic

$$h(X_1, \ldots, X_m) = \sum_{X_i > 0} a_N^+(R_i^+, f), \tag{5}$$

(as, in fact, Hodges and Lehmann (1963) do), because by (4.5.1.1) the statistics (4) and (5) are equivalent. In both cases we need h_0 and h to satisfy (C) and (D). The quantities a_N^+ and R_i^+ are those in (3.4.3.6) and Subsection 3.1.3, respectively.

Example 1. The estimate based on the one-sample Wilcoxon statistic. Following (5), we can use for estimation of Δ

$$h = \sum_{X_i > 0} R_i^+. \tag{6}$$

However, Tukey (1949) found a simple convenient form of h (see Problem 4 for Section 9.1).

Example 2. The estimate based on the sign test. (5) becomes now

$$h = \sum_{X_i > 0} 1, \tag{7}$$

i.e. the number of positive X_i's, and it can be seen that $\widehat{\Delta} = \text{med}(X_1, \ldots, X_N)$.

Hodges and Lehmann (1963) then investigate also for the one-sample case the properties of $\widehat{\Delta}$ analogous to those for the two-sample case, mentioned in Subsection 9.1.1.

As for distribution-free confidence intervals for Δ, their construction is entirely analogous to that in Subsection 9.1.1.

9.1.3 *R*-estimates of regression parameters.

Consider a simple regression model

$$Y_i = \theta + \beta c_i + e_i, \quad i = 1, \ldots, n, \tag{1}$$

where the c_i are known regression constants, not all equal, θ and β stand for the intercept and slope parameters, and the errors e_i are independent identically distributed random variables with a continuous (unknown) distribution function F, defined on $(-\infty, \infty)$. Assume that F admits an absolutely continuous density f. Then the estimating equations for the maximum likelihood estimates of (θ, β) are

$$\sum_{i=1}^{n} \left(\frac{1}{c_i}\right) \{-f'(Y_i - \theta - \beta c_i)/f(Y_i - \theta - \beta c_i)\} = 0. \tag{2}$$

If f is a normal density, the maximum likelihood estimates $(\hat{\theta}, \hat{\beta})$ agree with the classical least squares estimates, namely,

$$\hat{\beta} = C_n^{-2} \sum_{i=1}^{n} Y_i(c_i - \bar{c}_n), \quad \hat{\theta} = \bar{Y}_n - \hat{\beta}\bar{c}_n, \tag{3}$$

where

$$\bar{Y}_n = \frac{1}{n} \sum_{i=1}^{n} Y_i, \quad \bar{c}_n = \sum_{i=1}^{n} c_i \quad \text{and} \quad C_n^2 = \sum_{i=1}^{n} (c_i - \bar{c}_n)^2. \tag{4}$$

When f departs from normality, the least squares estimates may become inefficient, and they may be quite sensitive to outliers or error contaminations (even to a smaller extent). On the other hand, for a non-normal f, the estimating equations for the maximum likelihood estimates may no longer be linear, and a closed algebraic solution may not exist. Also, if the true density f is not the same as an assumed one (say, g), then the maximum likelihood estimates based on g may no longer retain its (asymptotic) efficiency or other optimality properties. Or, in other words, it may not be very *robust* to plausible departures from the assumed model. Estimates based on appropriate rank statistics, incorporating the same alignment principle as in the preceding subsections, have excellent robustness prospects, and they are distribution-free in a true sense.

Keeping the Wilcoxon scores one- and two-sample estimates in mind, we illustrate the estimation of the slope β using the Kendall τ statistic (Sen (1968)). In Subsection 4.6.2, we considered the Kendall τ statistic for the bivariate problem. Here, for regression, we define its analogue

$$T_n = \binom{n}{2}^{-1} \sum_{1 \le i < j \le n} \text{sign}(c_i - c_j)\,\text{sign}(Y_i - Y_j). \tag{5}$$

Let us consider next the aligned observations $Y_i(b) = Y_i - bc_i$, $i = 1, \dots, n$, where b is real, and define

$$T_n(b) = \binom{n}{2}^{-1} \sum_{1 \le i < j \le n} \text{sign}(c_i - c_j)\,\text{sign}(Y_i(b) - Y_j(b)). \tag{6}$$

Without loss of generality, we may set $c_1 \ge c_2 \ge \dots \ge c_n$; $\bar{c}_n = 0$, and note that by simple steps,

$$\begin{aligned} T_n(b) &= \binom{n}{2}^{-1} \sum_{1 \le i < j \le n} \text{sign}(c_i - c_j)\,\text{sign}(Y_i - Y_j - b(c_i - c_j)) \qquad (7) \\ &= \binom{n}{2}^{-1} \sum_{S_n} \text{sign}\left(\frac{Y_i - Y_j}{c_i - c_j} - b\right), \quad b \in (-\infty, \infty), \end{aligned}$$

where

$$S_n = \{(i,j) : \; c_i > c_j, \; 1 \le i < j \le n\}. \tag{8}$$

Clearly then $T_n(b)$ is non-increasing in b and if we define $Z_{ij} = (Y_i - Y_j)/(c_i - c_j)$, $(i,j) \in S_n$, and denote their ordered values by $Z_{(1)} \le \ldots \le Z_{(N^*)}$ where N^* = cardinality of S_n $(n - 1 \le N^* \le \binom{n}{2})$, then $T_n(b) = N^*/\binom{n}{2}$ for $b < Z_{(1)}$, $-N^*/\binom{n}{2}$ for $b > Z_{(N^*)}$, and at each $Z_{(k)}$, it steps down by $-2/\binom{n}{2}$. Moreover, for $b = \beta$, $T_n(\beta)$ has the same distribution as $T_n(0)$ has under $H_0 : \beta = 0$; so $T_n(0)$ is EDF with $\mathsf{E}_0 T_n(0) = 0$. Therefore, we may define

$$\hat{\beta}_n = \tfrac{1}{2}(\hat{\beta}_{n,1} + \hat{\beta}_{n,2}), \tag{9}$$

where

$$\hat{\beta}_{n,1} = \sup\{b : \; T_n(b) > 0\}, \quad \hat{\beta}_{n,2} = \inf\{b : \; T_n(b) < 0\}. \tag{10}$$

Specifically, in the current case, we have

$$\begin{aligned} \hat{\beta}_n &= \text{median}\,\{Z_{ij}; \; (i,j) \in S_n\} \\ &= \begin{cases} Z_{(M)}, & \text{if } N^* = 2M - 1, \\ \frac{1}{2}(Z_{(M)} + Z_{(M+1)}), & \text{if } N^* = 2M. \end{cases} \end{aligned} \tag{11}$$

Having obtained $\hat{\beta}_n$, we may consider the residuals

$$\hat{Y}_i = Y_i - \hat{\beta}_n c_i, \quad i = 1, \ldots, n, \tag{12}$$

and use the Wilcoxon signed-rank statistic on these residuals (as in the preceding subsection) to estimate θ. Namely, we have here

$$\hat{\theta}_n = \text{median}\,\{\tfrac{1}{2}(\hat{Y}_i + \hat{Y}_j) : \; 1 \le i < j \le n\}. \tag{13}$$

Note that the residuals in (12) may no longer be independent or even identically distributed. Hence, $\hat{\theta}_n$ may not be strictly distribution-free, though $\hat{\beta}_n$ is so. Nevertheless, they share a common robustness property that the underlying distribution function F can be made to vary over a broad class of distribution functions. Since the distribution function of $T_n(0)$, under $\beta = 0$, does not depend on F, we can find a pair of values $C_{n\alpha}^{(1)}$ and $C_{n\alpha}^{(2)}$, and a number $\alpha_n : 0 < \alpha_n \le \alpha$, such that

$$P_0\{C_{n\alpha}^{(1)} \le T_n(0) \le C_{n\alpha}^{(2)}\} = 1 - \alpha_n \ge 1 - \alpha. \tag{14}$$

Then, as in the preceding subsections, we set

$$\begin{aligned} \hat{\beta}_{n,L} &= \sup\{b : \; T_n(b) > C_{n,\alpha}^{(2)}\}; \\ \hat{\beta}_{n,U} &= \inf\{b : \; T_n(b) < C_{n,\alpha}^{(1)}\}, \end{aligned} \tag{15}$$

and obtain a distribution-free confidence interval for β:

$$P_\beta\{\hat{\beta}_{n,L} \leq \beta \leq \hat{\beta}_{n,U}\} = 1 - \alpha_n \geq 1 - \alpha. \tag{16}$$

In the present case, $C_{n\alpha}^{(1)} = -C_{n\alpha}^{(2)}$ and $\alpha_n \to \alpha$ as $n \to \infty$. Asymptotic properties of the estimates, studied by Sen (1968), are based on the asymptotic linearity of $T_n(b)$ in b in a neighbourhood of β. Such linearity results in a more general setup will be considered in the next section. In order to motivate further such results, we consider here the Adichie (1967) R-estimate of (β, θ) based on general linear and signed-rank statistics.

Define the $Y_i(b)$ as in after (5), and let $R_i(b) = \sum_{j=1}^{n} I(Y_j(b) \leq Y_i(b))$ be the rank of $Y_i(b)$ within this set, for $i = 1, \ldots, n$. Consider scores $a_n(1) \leq \ldots \leq a_n(n)$ (without loss of generality) and define

$$\begin{aligned} L_n(b) &= \sum_{i=1}^{n}(c_i - \bar{c}_n)a_n(R_i(b)) \\ &= \sum_{1 \leq i < j \leq n} \frac{1}{2n}(c_i - c_j)[a_n(R_i(b)) - a_n(R_j(b))], \end{aligned} \tag{17}$$

b real, where without loss of generality, set $c_1 \geq c_2 \geq \ldots \geq c_n$, not all equal. Recall that two lines: $Y_i(b)$, $Y_j(b)$, b real, are either parallel (when $c_i = c_j$) or they intersect at a single point $Z_{ij} = (Y_i - Y_j)/(c_i - c_j)$ (when $c_i \neq c_j$). Therefore, as in after (8), we consider the N^* ordered points $Z_{(1)} \leq \ldots \leq Z_{(N^*)}$, and note that at any $Z_{(r)}$, only two lines intersect, so that for $b = Z_{(r)} - 0$ and $b = Z_r + 0$, only these two aligned ranks are interchanged, but the rest remain the same. Thus, as in Sen (1969), we conclude that at $b = Z_{(r)}$, $L_n(b)$ has a decrement that depends on the pair (i, j) through $(c_i - c_j)[a_n(s) - a_n(s+1)]$, where s is the rank for $R_i(b)$. As such, $L_n(b)$ assumes constant values in the open intervals $(Z_{(r-1)}, Z_{(r)})$, $r = 1, \ldots, N^*$, and steps down at each of the points $Z_{(1)} \leq \ldots \leq Z_{(N^*)}$. Therefore, we conclude that for monotone non-decreasing scores,

$$L_n(b) \text{ is non-increasing in } b: \; -\infty < b < \infty. \tag{18}$$

We also note that under β, $L_n(\beta)$ has the same distribution as $L_n(0)$, under $\beta = 0$, and the latter is EDF, with mean 0 and variance $C_n^2 A_n^2$, where $A_n^2 = (n-1)^{-1} \sum_{i=1}^{n} [a_n(i) - \bar{a}_n]^2$. This enables us to equate, as in (9)–(10), $L_n(b)$ to 0, and define $\hat{\beta}_n$, the R-estimate of β, based on the linear rank statistic $L_n(.)$. Exercise 5 for 9.1 is set to verify that $\hat{\beta}_n$ is regression-equivariant and translation-equivariant. Moreover, $\hat{\beta}_n$ is a robust and consistent estimate of β. Further, under $\beta = 0$, $L_n(0)$ is EDF, and hence, for a given α

$(0 < \alpha < 1)$ and $c_1, \ldots, c_n$, we can find an α_n $(\leq \alpha)$ and a pair L^*_{n1}, L^*_{n2}, such that

$$P_0\{L^*_{n1} \leq L_n(0) \leq L^*_{n2}\} = 1 - \alpha_n \geq 1 - \alpha, \tag{19}$$

and this does not depend on F (continuous). Hence, using (18), we may proceed as in (15)–(16), and construct a distribution-free confidence interval for β based on $L_n(.)$.

Having obtained $\hat{\beta}_n$, we may consider the residuals $\hat{Y}_i$ in (12), and, as in the preceding subsection, use a general signed-rank statistic to estimate θ (Adichie 1967). In passing, we may remark that whereas for the Kendall τ statistic $T_n(b)$, the Sen-Theil estimate in (11) and (15) can be expressed in terms of suitable sample quantiles of the divided differences Z_{ij}, $(i, j) \in S_n$, for a general $L_n(.)$, such a closed expression may not be tenable, and an iterative procedure is usually advocated; the estimate in (11) or (15) can be used as a preliminary one in this iteration process. The asymptotic properties of these estimates can most conveniently be studied with the aid of asymptotic linearity results presented in Section 9.2.

9.2 ASYMPTOTIC LINEARITY OF RANK STATISTICS IN REGRESSION PARAMETERS

In the preceding section we have observed that aligned rank statistics are non-increasing in the shift and regression parameters (under fairly general regularity conditions), and this laid down the foundation of the theory of R-estimation as dual to the theory of rank tests. The EDF property of rank test statistics (under the hypothesis of invariance) provides the basic clue for the formulation of estimating equations for R-estimates. However, these estimating equations are generally non-linear in nature, and hence algebraic solutions are often precluded. A basic property of aligned rank statistics that plays the most fundamental role in the asymptotic theory of R-estimates of location and regression parameters, unfolded mostly in the late 1960s (viz., Jurečková (1969), Koul (1969), Sen (1969), among others), is the *asymptotic uniform linearity of rank statistics in shift (or regression) parameters.* This fundamental result paved the way to the unification of general asymptotics for R-estimates as well as rank tests, and led to many other significant developments in asymptotic methods; we refer to Jurečková and Sen (1996) for a comprehensive treatise of this basic topic. In the current section, we present a brief outline of the theory of such linearity results, and indicate their impact on the theory of rank tests and estimates.

9.2.1 Rank statistics, one-dimensional regression. First, we will start with rank statistics (based on usual ranks from Subsection 3.1.1), and with the one-dimensional regression parameter Δ. As an exception, we will also present for this simple case the detailed proofs of the main theorems, in order to introduce the reader to methods pertinent to these topics, and because these results seem to be pioneering for much further work.

For any positive integer N let us consider:

(a) a random sample $(X_{N1}, X_{N2}, \ldots, X_{NN})$ of independent identically distributed random variables whose distribution function is F, density is f, having finite Fisher information,

(b) a real vector of regression constants $(c_{N1}, c_{N2}, \ldots, c_{NN})$ such that

$$\sum_{i=1}^{N}(c_{Ni} - \bar{c}_N)^2 > 0 \tag{1}$$

$$\max_{1\le i\le N}(c_{Ni} - \bar{c}_N)^2 \cdot \Big[\sum_{j=1}^{N}(c_{Nj} - \bar{c}_N)^2\Big]^{-1} \to 0 \text{ for } N \to \infty \tag{2}$$

where $\bar{c}_N = (1/N)\sum_{i=1}^{N} c_{Ni}$,

(c) a real vector $(d_{N1}, d_{N2}, \ldots, d_{NN})$ such that

$$\sum_{i=1}^{N}(d_{Ni} - \bar{d}_N)^2 \le M \quad \text{for all } N \tag{3}$$

with some constant $M > 0$, $\bar{d}_N = (1/N)\sum_{i=1}^{N} d_{Ni}$, and

$$\max_{1\le i\le N}(d_{Ni} - \bar{d}_N)^2 \to 0 \quad \text{for all } N \to \infty, \tag{4}$$

(d) a real parameter Δ,

(e) a vector of ranks $(R^{\Delta}_{N1}, R^{\Delta}_{N2}, \ldots, R^{\Delta}_{NN})$ corresponding to variables $X_{N1} + \Delta d_{N1}$, $X_{N2} + \Delta d_{N2}, \ldots, X_{NN} + \Delta d_{NN}$, i.e. R^{Δ}_{Ni} is equal to the number of $(X_{Nj} + \Delta d_{Nj})$'s which are $\le X_{Ni} + \Delta d_{Ni}$ provided all the mentioned variables are different.

Remark. For simplicity, we will omit indices N in X_{Ni}, c_{Ni}, d_{Ni} and R^{Δ}_{Ni} in the sequel, whenever possible.

(f) now let us consider the statistic

$$S_{\Delta N} = \sum_{i=1}^{N} c_i a_N(R^{\Delta}_i) \tag{5}$$

where $a_N(1), a_N(2), \ldots, a_N(N)$ are either the usual scores (cf. Subsection 3.4.3)

$$a_N(i) = \mathsf{E}\varphi(U^{(i)}), \quad i = 1, 2, \ldots, N, \tag{6}$$

$U^{(1)} < U^{(2)} < \ldots < U^{(N)}$ being an ordered sample from the uniform distribution on $[0, 1]$, or approximate scores

$$a_N(i) = \varphi\big(1/(N+1)\big), \tag{7}$$

where φ is some square integrable non-constant non-decreasing score generating function.

Remark. If $\Delta = 1$, we have the case of a usual general regression (cf. Subsections 7.2.4 and 7.2.1).

As for the finite-sample monotonicity of $S_{\Delta N}$ we have the following

Theorem 1 *Let N be fixed, and the assumptions* (a), (d), (e), (f) *be satisfied with real numbers $c_1, c_2, \ldots, c_N$ and $d_1, d_2, \ldots, d_N$ for which*

$$(c_i - c_j)(d_i - d_j) \geq 0 \quad [(c_i - c_j)(d_i - d_j) \leq 0] \tag{8}$$

for all $i, j = 1, 2, \ldots, N$. Then the statistic $S_{\Delta N}$ with the monotone scores $a_N(i)$, $i = 1, 2, \ldots, N$, is a non-decreasing [(non-increasing)] step-function of Δ a.s.

Proof. It suffices to consider the case

$$(c_i - c_j)(d_i - d_j) \geq 0 \quad \text{for } i, j = 1, 2, \ldots, N. \tag{9}$$

We may suppose without loss of generality that

$$d_1 \leq d_2 \leq \ldots \leq d_N. \tag{10}$$

Let us fix a vector $(x_1, x_2, \ldots, x_N)$ with different components, and let Δ_1, Δ_2 ($\Delta_1 < \Delta_2$) be two values of Δ such that $S_{\Delta_j N}$ is well-defined for $j = 1, 2$. If $k < \ell$ and $R_k^{\Delta_1} < R_\ell^{\Delta_1}$, clearly $x_k + \Delta_1 d_k < x_\ell + \Delta_1 d_\ell$, and this implies $x_k + \Delta_2 d_k < x_\ell + \Delta_2 d_\ell$ in view of (10), hence $R_k^{\Delta_2} < R_\ell^{\Delta_2}$. This means that the permutation $(R_1^{\Delta_2}, \ldots, R_N^{\Delta_2})$ is better ordered than $(R_1^{\Delta_1}, \ldots, R_N^{\Delta_1})$ in the sense of Lehmann's (1966) definition. By a slight generalization of Corollary 2 of Theorem 5 in Lehmann (1966), and taking into account (9) and (10), we get

$$\sum_{i=1}^{N} c_i a_N(R_i^{\Delta_1}) \leq \sum_{i=1}^{N} c_i a_N(R_i^{\Delta_2}),$$

hence $S_{\Delta N}$ is a non-decreasing function of Δ.

Nevertheless, it may happen that the set

$$C = \{\Delta;\ x_i + \Delta d_i = x_j + \Delta d_j \text{ for at least one pair } (i, j)\}$$

is non-empty and then $S_{\Delta N}$ is not defined at the points of C. However, on splitting the real line by the points of C, one can show that $S_{\Delta N}$ is constant on each part of this splitting. (The proof of this assertion is left to the reader as an exercise.) The proof of Theorem 1 is thus completed. □

The next Theorem 2, the basic theorem of this section, shows that $S_{\Delta N}$ is probabilistically and asymptotically a linear function of Δ. We will suppose that $S_{\Delta N}$ is well defined for all Δ, since its definition at the points of discontinuity may be completed so that $S_{\Delta N}$ is continuous either from the left or from the right.

Theorem 2 *Let the assumptions* (a)–(f) *be satisfied. If* $(c_i - c_j)(d_i - d_j) \geq 0$, $[(c_i - c_j)(d_i - d_j) \leq 0]$ *for* $i, j = 1, 2, \ldots, N$ *and for all* N, *then*

$$P\Big(\max_{|\Delta| \leq C} |S_{\Delta N} - S_{0N} - \Delta b_N| \geq \varepsilon (\mathsf{var}\, S_{0N})^{1/2}\Big) \to 0 \quad \text{as } N \to \infty \tag{11}$$

for any $\varepsilon > 0$ *and* $C > 0$. *Here* b_N *denotes*

$$b_N = \Big[\sum_{i=1}^{N} (c_i - \bar{c})(d_i - \bar{d})\Big] \cdot \int_0^1 \varphi(u)\varphi(u, f)\, du, \tag{12}$$

and $\varphi(u, f)$ *the usual score function* (2.2.4.3).

Proof. First, we will prove the theorem for the scores (6). We may suppose without loss of generality that

$$\sum_{i=1}^{N} c_i = 0, \quad \sum_{i=1}^{N} c_i^2 = 1 \quad \text{for all } N \geq 2. \tag{13}$$

We will now use the following metric on the space $\mathcal{M}$ of random variables: For $X, Y \in \mathcal{M}$ define

$$d(X, Y) = \inf\{\varepsilon > 0;\ P(|X - Y| \geq \varepsilon) < \varepsilon\}. \tag{14}$$

It is known that $d(.\,,.)$ is a metric on equivalence classes of $\mathcal{M}$ where $X \sim Y$ if and only if $X = Y$ a.e. Further,

$$d(X_n, X) \to 0$$

if and only if $X_n \to X$ in probability (see e.g. Loève (1955)).

The proof of Theorem 2 will be divided into several steps.

(i) For $k = 1, 2, \ldots$, and $i = 1, 2, \ldots, k$, consider the functions $\varphi^{(k)}(u)$ defined by

$$\varphi^{(k)}(u) = \varphi\big(i/(k+1)\big), \quad (i-1)/k \leq u < i/k. \tag{15}$$

Each function $\varphi^{(k)}$ is non-decreasing and bounded on the interval $(0,1)$. By Lemma 6.1.6.1 we have

$$\lim_{k\to\infty}\int_0^1 [\varphi^{(k)}(u)-\varphi(u)]^2\,\mathrm{d}u = 0. \tag{16}$$

Now consider the statistics

$$S_{\Delta N}^{(k)} = \sum_{i=1}^{N} c_i a_N^{(k)}(R_i^{\Delta}) \tag{17}$$

where the scores $a_N^{(k)}(i)$ are generated by $\varphi^{(k)}$ as the usual scores.

Lemma 1 *For each $\varepsilon > 0$ there is a positive integer k_0 such that for any integer $k > k_0$ there exists an index $N_0(k)$ such that*

$$d(S_{0N}, S_{0N}^{(k)}) < \varepsilon \quad \textit{for all } N > N_0(k). \tag{18}$$

Proof. The inequality (6.1.6.6) implies

$$\begin{aligned}\mathsf{E}(S_{0N}-S_{0N}^{(k)})^2 \;&\le\; 1/(N-1)\Big(\sum_{i=1}^{N} c_i^2\Big)\sum_{j=1}^{N}[a_N(j)-a_N^{(k)}(j)]^2 \qquad (19)\\ &=\; N/(N-1)\int_0^1 [\varphi_N(u)-\varphi_N^{(k)}(u)]^2\,\mathrm{d}u\end{aligned}$$

where $\varphi_N(u) = a_N(i)$, and $\varphi_N^{(k)}(u) = a_N^{(k)}(i)$, both for $(i-1)/N \le u < i/N$, $i = 1,2,\ldots,N$. Let us fix $\varepsilon > 0$. By (16) there exists k_0 such that for all $k > k_0$

$$\int_0^1 [\varphi^{(k)}(u)-\varphi(u)]^2\,\mathrm{d}u < \varepsilon^3/96. \tag{20}$$

On the other hand, Theorem 6.1.4.2 implies for every k the existence of $N_0(k)$ such that for all $N > N_0(k)$

$$\begin{aligned}\int_0^1 [\varphi_N(u)-\varphi(u)]^2\,\mathrm{d}u \;&<\; \varepsilon^3/192 \qquad (21)\\ \int_0^1 [\varphi_N^{(k)}(u)-\varphi^{(k)}(u)]^2\,\mathrm{d}u \;&<\; \varepsilon^3/192.\end{aligned}$$

Inequalities (19), (20), (21) together imply $\mathsf{E}(S_{0N}-S_{0N}^{(k)})^2 < (\varepsilon/2)^3$. (18) then follows from the inequality

$$P(|S_{0N}-S_{0N}^{(k)}| \ge \varepsilon/2) \le \mathsf{E}(S_{0N}-S_{0N}^{(k)})^2(\varepsilon/2)^{-2} < \varepsilon/2$$

holding for $k > k_0$ and $N > N_0(k)$. □

(ii) Considering the statistics

$$T_{0N}^{(k)} = \sum_{i=1}^{N} c_i \varphi^{(k)}(F(X_i)), \quad k = 1, 2, \ldots, \tag{22}$$

by the proof of Theorem 6.1.5.1 we get

$$\lim_{N\to\infty} d(S_{0N}^{(k)}, T_{0N}^{(k)}) = 0 \quad \text{for any } k = 1, 2, \ldots. \tag{23}$$

(iii) Further, we shall need

Lemma 2 *Let $T_{\Delta N}^{(k)}$ denote the statistic*

$$T_{\Delta N}^{(k)} = \sum_{i=1}^{N} c_i \varphi^{(k)}[F(X_i + \Delta(d_i - \bar{d}))]. \tag{24}$$

Then

$$\lim_{N\to\infty} d(T_{0N}^{(k)}, T_{\Delta N}^{(k)} - \mathsf{E}T_{\Delta N}^{(k)}) = 0 \tag{25}$$

for any Δ and any $k = 1, 2, \ldots$.

Proof. We have

$$\begin{aligned} &\mathsf{var}(T_{0N}^{(k)} - T_{\Delta N}^{(k)}) \\ &\leq \sum_{i=1}^{N} c_i^2 \int \{\varphi^{(k)}(F(x)) - \varphi^{(k)}[F(x + \Delta(d_i - \bar{d}))]\}^2 \, \mathrm{d}F(x). \end{aligned} \tag{26}$$

The convergence

$$F(x + \Delta(d_i - \bar{d})) \to F(x) \quad \text{for } N \to \infty \tag{27}$$

is uniform for all $x \in (-\infty, \infty)$ and $i = 1, 2, \ldots, N$ due to the continuity of F and (4). The set A of discontinuity points of $\varphi^{(k)}$ is at most countable, hence the convergence

$$\lim_{N\to\infty} \varphi^{(k)}[F(x + \Delta(d_i - \bar{d}))] = \varphi^{(k)}(F(x)) \tag{28}$$

holds uniformly for $i = 1, 2, \ldots, N$ a.e. with respect to F, since the exceptional set $B = F^{-1}(A)$ satisfies $\int_B \mathrm{d}F(x) = \int_A \mathrm{d}u = 0$. Hence we can use Lebesgue's theorem for the integrals in (26) which will converge uniformly to zero for $i = 1, 2, \ldots, N$. Chebyshev's inequality and $\mathsf{E}T_{0N}^{(k)} = 0$ finally imply the assertion of Lemma 2. □

(iv) Next, we shall prove the following

Lemma 3 *We have*

$$\lim_{N\to\infty} d(S_{\Delta N}^{(k)}, T_{\Delta N}^{(k)}) = 0 \tag{29}$$

for any Δ *and any* $k = 1, 2, \ldots$.

Proof. If P_N is the distribution with density $p_N = \prod_{i=1}^{N} f(x_i)$, and $Q_{\Delta N}$ the distribution with density $q_{\Delta N} = \prod_{i=1}^{N} f(x_i - \Delta(d_i - \bar{d}))$, then the densities $q_{\Delta N}$ are contiguous to the densities p_N, as follows from the remarks about (7.2.4.15).

Denote $S_{0N}^{(k)} = S_0^{(k)}(X_1, \ldots, X_N)$ and $T_{0N}^{(k)} = T_0^{(k)}(X_1, \ldots, X_N)$ for this proof. By (23),

$$P_N\big(|S_0^{(k)}(X_1, \ldots, X_N) - T_0^{(k)}(X_1, \ldots, X_N)| \geq \eta\big) \to 0 \quad \text{for } N \to \infty$$

for any $\eta > 0$. The contiguity now implies

$$Q_{\Delta N}\big(|S_0^{(k)}(X_1, \ldots, X_N) - T_0^{(k)}(X_1, \ldots, X_N)| \geq \eta\big) \to 0 \text{ for } N \to \infty,$$

which may be written as

$$\begin{aligned} P_N\big(|S_0^{(k)}&(X_1 + \Delta d_1, \ldots, X_N + \Delta d_N) \\ &- T_0^{(k)}(X_1 + \Delta(d_1 - \bar{d}), \ldots, X_N + \Delta(d_N - \bar{d}))| \geq \eta\big) \to 0, \end{aligned}$$

for the statistic $S_{0N}^{(k)}$, depending only on ranks, is invariant to the translation of the whole sample. The last relation may again be rewritten as

$$P\big(|S_{\Delta N}^{(k)} - T_{\Delta N}^{(k)}| \geq \eta\big) \to 0 \quad \text{for } N \to \infty \text{ and any } \eta > 0. \qquad \square$$

(v) Now, let us make Lemma 2 more precise.

Lemma 4 *There exists an integer* $k_1 > 0$ *such that for any* $k > k_1$ *and any* Δ

$$\lim_{N\to\infty} d(T_{0N}^{(k)}, T_{\Delta N}^{(k)} - \Delta b_N^{(k)}) = 0 \tag{30}$$

where

$$b_N^{(k)} = \Big[\sum_{i=1}^{N} c_i(d_i - \bar{d})\Big] \cdot \int_0^1 \varphi^{(k)}(u)\varphi(u, f)\, du. \tag{31}$$

Proof. The convergence (16) implies

$$\lim_{k\to\infty} \int_0^1 (\varphi^{(k)}(u) - \bar{\varphi}^{(k)})^2\, du = \int_0^1 (\varphi(u) - \bar{\varphi})^2\, du > 0,$$

the positiveness being a consequence of assumption (f) at the beginning of Subsection 9.2.1. It follows that there exists a k_1 such that $\int_0^1 (\varphi^{(k)}(u) - \bar{\varphi}^{(k)})^2\,du > 0$ for all $k > k_1$. By Theorem 7.2.4.1 and the remarks about 7.2.4.15, the statistics $T_{\Delta N}^{(k)}$ are for any Δ and any $k > k_1$ asymptotically normal $(\Delta b_N^{(k)}, \sigma^{(k)2})$, where

$$\sigma^{(k)2} = \sum_{i=1}^{N} c_i^2 \int_0^1 (\varphi^{(k)}(u) - \bar{\varphi}^{(k)})^2\,du = \int_0^1 (\varphi^{(k)}(u) - \bar{\varphi}^{(k)})^2\,du.$$

On the other hand, the statistics $T_{0N}^{(k)}$ are asymptotically normal $(0, \sigma^{(k)2})$ for all $k > k_1$, as follows by Theorem 6.1.5.1. Returning to Lemma 2, we see that

$$\lim_{N\to\infty} [\mathsf{E}T_{\Delta N}^{(k)} - \Delta b_N^{(k)}]^2 \cdot (\sigma^{(k)})^{-2} = 0 \quad \text{for } k > k_1$$

and Lemma 4 follows from Lemma 2 again. □

(vi) We will need further the following lemma on contiguous sequences. The phrase 'for almost all N' means 'with the possible exclusion of a finite number of N'.

Lemma 5 *Let P_N defined by densities p_N, and Q_N defined by densities q_N, be two sequences of probability measures. If the densities q_N are contiguous to p_N, then for every $\varepsilon > 0$ there is some $\delta > 0$ such that $Q_N(A_N) < \varepsilon$ is satisfied for almost all N for every sequence of sets A_N for which $P_N(A_N) < \delta$ for almost all N.*

Proof. By definition of contiguity, we have $Q_N(A_N) \to 0$ whenever $P_N(A_N) \to 0$ for $N \to \infty$. Suppose that it is possible, for some $\varepsilon_0 > 0$ and for any positive integer k, to find a sequence $\{A_{N,k}\}_{N=1}^{\infty}$ of sets such that

$$P_N(A_{N,k}) < 1/2^k \quad \text{for } N > N_0(k), \tag{32}$$

$$Q_N(A_{N,k}) \geq \varepsilon_0 \tag{33}$$

for infinitely many values of N. (33) implies that there exists $N_k^* > N_0(k)$ for $k = 1, 2, \ldots$ such that $Q_{N_k^*}(A_{N_k^*,k}) \geq \varepsilon_0$ and that the numbers N_k^*, $k = 1, 2, \ldots$, may be chosen so that $N_k^* < N_{k+1}^*$, $k = 1, 2, \ldots$. Let $\{B_N\}$ be the sequence of sets:

$$B_N = \begin{cases} A_{N_k^*,k} & \text{for } N = N_k^*,\ k = 1, 2, \ldots \\ \emptyset & \text{for other } N. \end{cases} \tag{34}$$

We see then that $P_N(B_N) \to 0$ for $N \to \infty$ but $Q_N(B_N) \not\to 0$, which contradicts the assumption of contiguity of q_N to p_N. □

Lemma 6 *For any fixed Δ and any $\varepsilon > 0$ there exists an integer $k_2 > 0$ such that for every integer $k > k_2$ there is an integer $N_2(k) > 0$ and for any $N > N_2(k)$ we have*

$$d(S_{\Delta N}, S_{\Delta N}^{(k)}) < \varepsilon. \tag{35}$$

Proof. Let P_N, $Q_{\Delta N}$ be the distributions with the densities p_N, $q_{\Delta N}$ which were defined in the proof of Lemma 3. By Lemma 5, contiguity of $q_{\Delta N}$ to p_N implies for every $\varepsilon > 0$ the existence of $\delta > 0$ such that $Q_{\Delta N}(A_N) < \varepsilon/2$ for almost all N for any sequence of sets for which $P_N(A_N) < \delta$ for almost all N. By Lemma 1, putting $\eta = \min(\varepsilon/2, \delta)$ in place of ε, there is an integer $k_2 > 0$ such that for every $k > k_2$ there exists $N_1(k)$ where for all $N > N_1(k)$ it holds that (using the notation of the proof of Lemma 3)

$$P_N\big(|S_0(X_1,\dots,X_N) - S_0^{(k)}(X_1,\dots,X_N)| \geq \varepsilon/2\big) < \delta.$$

Lemma 5 then establishes the existence of $N_2(k)$ for every $k > k_2$ such that

$$Q_{\Delta N}\big(|S_0(X_1,\dots,X_N) - S_0^{(k)}(X_1,\dots,X_N)| \geq \varepsilon/2\big) < \varepsilon/2$$

for all $N > N_2(k)$. This may be rewritten as

$$\begin{aligned} P_N\big(|S_0(X_1 + \Delta d_1, \dots, X_N + \Delta d_N) \\ -S_0^{(k)}(X_1 + \Delta d_1, \dots, X_N + \Delta d_N)| \geq \varepsilon/2\big) < \varepsilon/2, \end{aligned}$$

which means that (35) is true for $k > k_2$ and $N > N_2(k)$. □

(vii) Finally, we get a relation between S_{0N} and $S_{\Delta N}$.

Lemma 7 *For any fixed Δ and any $\varepsilon > 0$ there exists an integer $k^* > 0$, such that an index $N^*(k)$ may be found to any $k > k^*$ and for all $N > N^*(k)$ it holds that*

$$d(S_{0N}, S_{\Delta N} - \Delta b_N^{(k)}) < \varepsilon. \tag{36}$$

Proof. We have

$$\begin{aligned} &d(S_{0N}, S_{\Delta N} - \Delta b_N^{(k)}) \\ &\leq d(S_{0N}, S_{0N}^{(k)}) + d(S_{0N}^{(k)}, T_{0N}^{(k)}) + d(T_{0N}^{(k)}, T_{\Delta N}^{(k)} - \Delta b_N^{(k)}) \\ &+d(T_{\Delta N}^{(k)} - \Delta b_N^{(k)}, S_{\Delta N}^{(k)} - \Delta b_N^{(k)}) + d(S_{\Delta N}^{(k)} - \Delta b_N^{(k)}, S_{\Delta N} - \Delta b_N^{(k)}) \end{aligned} \tag{37}$$

for any $k, N = 1, 2, \dots$. The desired result then follows from Lemmas 1, 3, 4, 6 and from (23). □

Lemma 8 *For each* Δ *and* $\varepsilon > 0$ *we have*

$$\lim_{N\to\infty} P(|S_{\Delta N} - S_{0N} - \Delta b_N| \geq \varepsilon) = 0$$

where $b_N = [\sum_{i=1}^{N} c_i(d_i - \bar{d})] \int_0^1 \varphi(u)\varphi(u,f)\,du$.

Proof. The Schwarz inequality and (3) imply

$$\begin{aligned} |b_N^{(k)} - b_N| &= \Big|\Big[\sum_{i=1}^{N} c_i(d_i - \bar{d})\Big] \int_0^1 [\varphi^k(u) - \varphi(u)]\varphi(u,f)\,du\Big| \\ &\leq M^{1/2}\Big|\int_0^1 (\varphi^{(k)}(u) - \varphi(u))\varphi(u,f)\,du\Big|. \end{aligned}$$

The right-hand side of the inequality tends to zero for $k \to \infty$ by (16) and assumption (a), hence $\lim_{k\to\infty} b_n^{(k)} = b_N$ uniformly in N. The assertion of Lemma 8 now follows from Lemma 7. □

(viii) Consider now the scores (7). We use the notation

$$\begin{aligned} a_N(i) &= \mathsf{E}\varphi(U^{(i)}), & a_N^*(i) &= \varphi\big(i/(N+1)\big), \\ S_{\Delta N} &= \sum_{i=1}^{N} c_i a_N(R_i^{\Delta}), & S_{\Delta N}^* &= \sum_{i=1}^{N} c_i a_N^*(R_i^{\Delta}). \end{aligned}$$

By Lemma 6.1.6.1 and Theorem 6.1.6.1 we see that

$$\lim_{N\to\infty} P(|S_{0N} - S_{0N}^*| \geq \varepsilon) = 0 \quad \text{for any } \varepsilon > 0. \tag{38}$$

Therefore, in view of the contiguity of $q_{\Delta N}$ to p_N, it follows that

$$\lim_{N\to\infty} P(|S_{\Delta N} - S_{\Delta N}^*| \geq \varepsilon) = 0 \quad \text{for any } \varepsilon > 0. \tag{39}$$

Then (38), (39) and Lemma 8 imply

$$\lim_{N\to\infty} P(|S_{\Delta N}^* - S_{0N}^* - \Delta b_N| \geq \varepsilon) = 0 \quad \text{for any } \varepsilon > 0.$$

This shows that Lemma 8 holds true also for the scores (7). □

(ix) We now complete the proof of Theorem 2. Let C, ε, η be any positive numbers. Let us partition the interval $[-C, C]$ by $-C = \Delta_0 < \Delta_1 < \ldots < \Delta_r = C$ so that

$$\Big|(\Delta_i - \Delta_{i-1}) \int_0^1 \varphi(u)\varphi(u,f)\,du\Big| < \varepsilon/2 \cdot M^{-1/2} \tag{40}$$

for $i = 1, 2, \ldots, r$, where $M > 0$ is the constant satisfying $\sum_{i=1}^{N} (d_i - \bar{d})^2 \leq M$ for all N as in (3). Lemma 8 guarantees the existence of N_0 such that for any $N > N_0$

$$P(|S_{\Delta_i N} - S_{0N} - \Delta_i b_N| \geq \varepsilon/4) < \eta(r+1) \tag{41}$$

for $i = 1, 2, \ldots, r$.

Let Δ be any point of the interval $[-C, C]$. Then there exists h, $1 \leq h \leq r$, such that $\Delta_{h-1} \leq \Delta \leq \Delta_h$ and the following inequality holds true:

$$\begin{aligned} |S_{\Delta N} - S_{0N} - \Delta b_N| \leq\ & |S_{\Delta_h N} - S_{0N} - \Delta_h b_N| \\ & + |b_N(\Delta_h - \Delta_{h-1}| \\ & + |S_{\Delta_{h-1} N} - S_{0N} - \Delta_{h-1} b_N|. \end{aligned} \tag{42}$$

Now we prove (42). Suppose that c_i and d_i, $i = 1, 2, \ldots, N$, satisfy $(c_i - c_j)(d_i - d_j) \geq 0$, $i, j = 1, 2, \ldots, N$. Then the statistic $S_{\Delta N}$ is a non-decreasing function of Δ by Theorem 1. If $S_{\Delta N} - S_{0N} - \Delta b_N \geq 0$, then

$$\begin{aligned} |S_{\Delta N} - S_{0N} - \Delta b_N| \leq\ & S_{\Delta_h N} - S_{0N} - \Delta b_N \\ \leq\ & |S_{\Delta_h N} - S_{0N} - \Delta_h b_N| + |b_N(\Delta_h - \Delta_{h-1})| \\ & + |S_{\Delta_{h-1} N} - S_{0N} - \Delta_{h-1} b_N|. \end{aligned}$$

The case $S_{\Delta N} - S_{0N} - \Delta b_N < 0$ is handled quite analogously. We can come to the same conclusions if c_i and d_i satisfy $(c_i - c_j)(d_i - d_j) \leq 0$, $i, j = 1, 2, \ldots, n$, where the statistic S_Δ is a non-increasing function of Δ.

The inequality (42) implies

$$\max_{|\Delta| \leq C} |S_{\Delta N} - S_{0N} - \Delta b_N| \leq 2 \max_{0 \leq i \leq r} |S_{\Delta_i N} - S_{0N} - \Delta_i b_N| + \varepsilon/2. \tag{43}$$

It may be shown that $\max_{|\Delta| \leq C} |S_{\Delta N} - S_{0N} - \Delta b_N|$ is a random variable so that the probability that it is $\geq \varepsilon$ is well defined. Clearly (40) and (41) imply

$$\begin{aligned} & P(\max_{|\Delta| \leq C} |S_{\Delta N} - S_{0N} - \Delta b_N| \geq \varepsilon) \\ & \quad \leq \sum_{i=0}^{r} P(|S_{\Delta_i N} - S_{0N} - \Delta_i b_N)| \geq \varepsilon/4) < \eta \end{aligned} \tag{44}$$

for $N > N_0$.

By Theorem 6.1.6.1 the statistics S_{0N} are asymptotically normal $(0, \sigma^2)$ with

$$\sigma^2 = \sum_{i=1}^{N} c_i^2 \int_0^1 (\varphi(u) - \bar{\varphi})^2 \, du = \int_0^1 (\varphi(u) - \bar{\varphi})^2 \, du$$

or $\sigma^2 = \mathsf{var}\, S_{0N}$. Thus $\mathsf{var}\, S_{0N} \to \int_0^1 (\varphi(u) - \bar{\varphi})^2 \, du$ for $N \to \infty$, which, taking into account (44), implies that

$$\lim_{N\to\infty} P\big(\max_{|\Delta| \le C} |S_{\Delta N} - S_{0N} - \Delta b_N| \ge \varepsilon (\mathsf{var}\, S_{0N})^{1/2} \big) = 0.$$

The proof of Theorem 2 is complete. □

All results so far, including the proofs, in this Subsection 9.2.1 are due to Jurečková (1969).

Turning now to similar asymptotic linearity results but in *almost sure convergence*, we can note the following.

Under the assumptions displayed at the beginning of this subsection, and under some additional ones, mainly

(α) $f(x) > 0$ for all x, f is absolutely continuous and the integrals $\int_{-\infty}^{\infty} (f'(x-t))^2 / f(x) \, dx$ are bounded for $|t| \le \delta$ for some $\delta > 0$,

(β) $\max_{1 \le i \le N} |d_{Ni} - \bar{d}_N| = O(N^{-1/4-\eta})$ for some $\eta > 0$,

(γ) φ is non-decreasing and $\varphi'(u)$ is bounded on $[0, 1]$,

Jurečková (1973) proved for $S_{\Delta N}$ with the approximate scores (7) an analogue of Theorem 2, but where the convergence in probability is strengthened to the convergence with probability 1.

Another result of this kind, dealing with the almost sure convergence, may be found in Ghosh and Sen (1972).

As for the asymptotic linearity of rank statistics in the shift (location) parameter between *two samples*, it suffices to put in the above Theorems $d_{N1} = d_{N2} = \ldots = d_{Nm}$, $d_{Nm+1} = \ldots = d_{NN} = 0$, and $c_{N1} = \ldots = c_{Nm} = 1$, $c_{Nm+1} = \ldots = c_{NN} = 0$ (cf. Subsection 3.4.4).

9.2.2 Rank statistics, multidimensional regression. Let $Y_1, Y_2, \ldots, Y_N$ be independent random variables $\boldsymbol{\Delta} = (\Delta_1, \Delta_2, \ldots, \Delta_K)$ be a real vector of fixed regression parameters, $X_N = \{x_{ji}\}$ be a $K \times N$ design matrix with K-dimensional columns $\mathsf{x}^{(i)}$, $i = 1, \ldots, N$, and with N-dimensional rows $\mathsf{x}_{(j)}$, $j = 1, \ldots, K$. Further, let

$$\lim_{N\to\infty} (\mathsf{x}_{(\ell)} - \bar{x}_\ell)(\mathsf{x}_{(j)} - \bar{x}_j)' = \xi_{\ell,j}, \quad \ell, j = 1, \ldots, K,$$

where $\boldsymbol{\Xi} = \{\xi_{\ell j}\}_{\ell,j=1}^K = \{\xi^{(1)}, \ldots, \xi^{(K)}\}$ is a positive definite $K \times K$ matrix.

Suppose that Y_i has, under the null hypothesis $\boldsymbol{\Delta} = \boldsymbol{\Delta}^0$, the distribution function

$$F(y - \alpha - \boldsymbol{\Delta}^0 \mathsf{x}^{(i)}), \quad i = 1, 2, \ldots, N, \tag{1}$$

and consider the rank statistics

$$S_{Nj}(Y - \boldsymbol{\Delta}\mathsf{x}) = \sum_{i=1}^{N} (x_{ji} - \bar{x}_j) a_N(R_i^{\Delta}), \quad j = 1, 2, \ldots, K, \tag{2}$$

where R_i^{Δ} is the rank of the variable $Y_i - \mathbf{\Delta}x^{(i)}$ among $Y_1 - \mathbf{\Delta}x^{(1)}, \ldots,$ $Y_N - \mathbf{\Delta}x^{(N)}$.

We will now roughly describe a generalization of the linearity result in Theorem 9.2.1.2.

Let P_{Δ^0} denote the probability distribution with the density

$$p_{\Delta^0} = \prod_{i=1}^{N} f(y_i - \alpha - \mathbf{\Delta}^0 x^{(i)}); \tag{3}$$

let $\|\mathbf{\Delta} - \mathbf{\Delta}^0\| = [(\mathbf{\Delta} - \mathbf{\Delta}^0)(\mathbf{\Delta} - \mathbf{\Delta}^0)']^{1/2}$. Then, under certain conditions (partly identical to those in 9.2.1, partly special and somewhat complicated due to multidimensionality, namely certain ordering properties on the rows of X_N),

$$P_{\Delta^0}\{\max_{\|\mathbf{\Delta}-\mathbf{\Delta}^0\|\le C} |S_{Nj}(Y - \mathbf{\Delta}x) - S_{Nj}(Y - \mathbf{\Delta}^0 x) \\ + \gamma(\mathbf{\Delta} - \mathbf{\Delta}^0)\xi^{(j)}| \ge \varepsilon\} \to 0 \quad \text{as } N \to \infty, \tag{4}$$

holds for any $\varepsilon > 0$, $C > 0$, $\gamma = \int_0^1 \varphi(u)\varphi(u,f)\,du$, $j = 1, 2, \ldots, K$.

This result was proved by Jurečková (1971a). For an analogous result under similar but slightly different conditions on X_N, cf. Kraft and van Eeden (1972). Heiler and Willers (1988) proved related results removing Jurečková's ordering properties of X_n and replacing them by standard assumptions from the least squares theory.

9.2.3 Signed-rank statistics. The ranks R_i^+ and the signed-rank statistics were defined in 3.1.3, and investigated in 3.4.9 and in 4.5.1. They are used for basic densities symmetric about zero.

The results on the asymptotic linearity of signed-rank statistics in regression parameters, analogous to those in 9.2.1 and 9.2.2, were obtained by van Eeden (1972). Related results may be found e.g. in Jurečková (1971b), Kraft and van Eeden (1972), Sen (1980), and Tardif (1985).

9.3 RANK ESTIMATION OF REGRESSION PARAMETERS

9.3.1 Basic methods. First, a short remark on the simplest case of regression in which the random variables Y_i satisfy $Y_i = \alpha^0 + \beta^0 x_i + e_i$, where α^0 is the main additive effect, β^0 an unknown one-dimensional regression parameter, x_i the 'explaining variable', e_i the error term. Historically, a rank estimate of β^0 and α^0 was proposed by Adichie (1967). Bauer (1973) presented a simple computation technique for it.

However, we will put this simple case aside, and only refer to the literature. Instead, we will present here the main ideas for a more interesting case — that of multidimensional regression. We will now return to the basic setup and notation of Subsection 9.2.2.

Thus, we have independent random variables $Y_1, Y_2, \ldots, Y_N$, where Y_i has the distribution function (9.2.2.1). However, all pertaining expressions and methods in the sequel are invariant with respect to the main additive effect α, so they cannot be used for estimating α^0; thus we may and will assume at present $\alpha^0 = 0$. Our goal is to estimate $\mathbf{\Delta}^0 = (\Delta_1^0, \Delta_2^0, \ldots, \Delta_K^0)$. We might recall that

$$\delta_i(\mathbf{\Delta}) = Y_i - \mathbf{\Delta} x^{(i)}, \quad i = 1, 2, \ldots, N, \tag{1}$$

are the residuals of Y_i with respect to $\mathbf{\Delta}$. Hence, in other words, R_i^{Δ} is the rank of residual $\delta_i(\mathbf{\Delta})$ among all residuals $\delta_1(\mathbf{\Delta}), \ldots, \delta_N(\mathbf{\Delta})$.

The classical 'method of least squares' for estimating $\mathbf{\Delta}^0$ consists in minimizing the sum of squares of the residuals:

$$\sum_{i=1}^{N} \delta_i^2(\mathbf{\Delta}) = \sum_{i=1}^{N} [Y_i - \mathbf{\Delta} x^{(i)}]^2 = \min! \tag{2}$$

However, the estimates obtained by this method are very sensitive to outlying observations and fail for long-tailed basic distributions.

Therefore, trying to remove these disadvantages, we might use some rank methods, namely those based on $S_{Nj}(Y - \mathbf{\Delta} x)$ in (9.2.2.2). We may expect that such methods are more robust. Let us describe now their main ideas.

Under the null hypothesis $\mathbf{\Delta} = \mathbf{\Delta}^0$, it follows easily by Theorem 6.1.5.1 that the random vector

$$S_N(Y - \mathbf{\Delta}^0 x) = (S_{N1}(Y - \mathbf{\Delta}^0 x), \ldots, S_{NK}(Y - \mathbf{\Delta}^0 x)) \tag{3}$$

is asymptotically normal

$$\left(\mathbf{0}, \Xi \int_0^1 (\varphi(u) - \bar{\varphi})^2 \, du\right) \tag{4}$$

where Ξ is the $K \times K$ matrix defined just before (9.2.2.1). So it is natural to estimate $\mathbf{\Delta}^0$ by such $\mathbf{\Delta}$ for which $S_N(Y - \mathbf{\Delta} x)$ is 'as near to zero as possible'. But what does this phrase mean for a vector? In other words, we would like to solve the system of 'approximate equations'

$$S_{Nj}(Y - \mathbf{\Delta} x) \simeq 0 \quad \text{for } j = 1, \ldots, K \text{ simultaneously.} \tag{5}$$

The solutions of this problem are naturally again called R-estimates (of regression parameters). The interpretation of (5) and the phrase above it may be different, and we will now display three proposed methods for doing it.

Method 1. Jurečková (1971a) proposed to take for an estimate of $\mathbf{\Delta}^0$ any solution $\mathbf{\Delta}$ of the minimization problem

$$\sum_{j=1}^{K} |S_{Nj}(Y - \mathbf{\Delta}x)| = \min! \tag{6}$$

There may be more that one solution of (6). Jurečková (1971a) also proved under the assumptions in this chapter, that any such estimate is asymptotically normal

$$\left(\mathbf{\Delta}^0, \int_0^1 (\varphi(u) - \bar{\varphi})^2 \, du \cdot \left[\int_0^1 \varphi(u)\varphi(u,f)\, du\right]^{-2} \cdot \Xi^{-1}\right). \tag{7}$$

Method 2. Jaeckel (1972) started in principle from the method of least squares expressed by (2) but, in addition, he weighted the residuals by some rank scores. Namely, for getting the estimates of $\mathbf{\Delta}^0$, we need to minimize

$$\sum_{i=1}^{N} [Y_i - \mathbf{\Delta}x^{(i)}] a_N(R_i^{\Delta}) = \min! \tag{8}$$

It may be shown that these estimates yield also a kind of 'solutions' of (5). For a further generalization cf. McKean, and Hettmansperger (1976).

Method 3. Koul (1971) considered the ellipsoid

$$\{\mathbf{\Delta} : (S_N(Y - \mathbf{\Delta}x))'\Xi^{-1} S_N(Y - \mathbf{\Delta}x) \leq \chi^2_{\alpha,K}\} \tag{9}$$

with $\chi^2_{\alpha,K}$ being the α-critical value of the χ^2_K distribution. This ellipsoid is an asymptotic confidence region for $\mathbf{\Delta}^0$, and its centre of gravity is a point estimate of $\mathbf{\Delta}^0$.

All three of the above methods are asymptotically equivalent in the sense that the resulting R-estimates of $\mathbf{\Delta}^0$ have the same asymptotic distributions and efficiencies.

In investigations of all these R-estimates obtained by these three methods, the result (9.2.2.4) on the asymptotic linearity of rank statistics in regression parameters is an important tool. E.g. when proving the asymptotic normality (cf. (7)) of the R-estimate, one of the basic devices it to replace the original rank statistics by their 'probabilistic' and 'asymptotic' linear approximations as in (9.2.2.4) (cf. Jurečková 1971a). Or, when Koul (1971) proves the asymptotic boundedness of the ellipsoid (9), he uses the same linearity result.

Just before and after (9.2.2.4) it was remarked that Jurečková (1971a), who proposed Method 1 for estimating $\mathbf{\Delta}^0$, assumed certain ordering properties on the rows of X_N. These restrictions were removed by Heiler and Willers (1988).

Good surveys of methods for estimating $\mathbf{\Delta}^0$ were presented by Jurečková (1975) and (1984).

If we happen to know that $\alpha^0 \neq 0$, our aim might be to estimate it. To this end, we need to suppose that $f(y)$ is symmetric about zero, $\sum_{i=1}^{N} x_{ji} = 0$ for $j = 1, 2, \ldots, K$; $N = 2, 3, \ldots$, and we need first some estimate $\hat{\mathbf{\Delta}}$ of $\mathbf{\Delta}^0$. Then the procedure resembles essentially that of 9.1.2 of estimating the shift of one sample by means of signed-rank statistics. In some detail:

We start from the signed-rank statistics

$$S_N^+(Y - \alpha - \mathbf{\Delta}\mathsf{x}) = \sum_{i=1}^{N} a_N^+(R_i^+, f)\,\mathrm{sign}(Y_i - \alpha - \mathbf{\Delta}\mathsf{x}^{(i)}), \tag{10}$$

where $a_N^+(i, f)$ are the scores from (3.4.3.6) for testing symmetry (generated by a non-negative non-decreasing square-integrable function), and R_i^+ is the rank of $|Y_i - \alpha - \mathbf{\Delta}\mathsf{x}^{(i)}|$ among the variables $|Y_1 - \alpha - \mathbf{\Delta}\mathsf{x}^{(1)}|, \ldots, |Y_N - \alpha - \mathbf{\Delta}\mathsf{x}^{(N)}|$. Then the estimate of α is

$$\hat{\alpha}_N = (\alpha_N^* + \alpha_N^{**})/2 \tag{11}$$

where

$$\begin{aligned} \alpha_N^* &= \sup\{\alpha;\ S_N^+(Y - \alpha - \hat{\mathbf{\Delta}}\mathsf{x}) > 0\}, \\ \alpha_N^{**} &= \inf\{\alpha;\ S_N^+(Y - \alpha - \hat{\mathbf{\Delta}}\mathsf{x}) < 0\}. \end{aligned}$$

(For more details cf. Jurečková (1971a, 1971b).)

9.3.2 Linearized rank estimates. Here we will mention very briefly the idea of a useful technique for obtaining good and simple estimates of regression parameters. We need to start with some reasonably good consistent preliminary estimates (as e.g. those generated by the method of least squares or maximum likelihood estimates). Then we correct them with the help of asymptotic linearity of rank statistics (cf. Subsection 9.2.1 and 9.2.2), which gives rise to so-called 'linearized' rank estimates. Practically the same idea gives rise to the 'one-step versions' of the estimates: usually the estimates are solutions of some minimization problems (cf. Subsection 9.3.1). So, starting with a preliminary estimate, we apply then one step of the Gauss-Newton iterative method to the equation defining the desired estimate.

These methods yield, under mild conditions, very good results, and they are computationally more feasible.

For lack of space, we must stop here, giving no actual formulas. The idea is due to Kraft and van Eeden (1970, 1972) where one can find details and examples. For a brief introduction to the method, cf. also Jurečková (1975, 1984). There are many more investigations of these methods, e.g. McKean and Hettmansperger (1978), Boulanger (1983), Boulanger and van Eeden (1983), and Antille (1974).

PROBLEMS AND COMPLEMENTS TO CHAPTER 9

Section 9.1

1. Show that the estimate $\hat{\Delta}$ introduced in Subsection 9.1.1 is translation equivariant, namely $\hat{\Delta}(\boldsymbol{X}+a, \boldsymbol{Y}) = \hat{\Delta}(\boldsymbol{X}, \boldsymbol{Y}) + a$ for all real a. [Hodges and Lehmann (1963).]

2. In Example 1 in 9.1.1, we may use the Mann-Whitney form of the Wilcoxon statistic (cf. Problem 1 for Section 4.1, or the remark following the description of the Wilcoxon test). Prove that the estimate $\hat{\Delta}$ equals the median of the set of mn differences $X_i - Y_j$, $i = 1, \ldots, m$; $j = 1, \ldots, n$, or, symbolically, $\hat{\Delta} = \text{med}(\boldsymbol{X} - \boldsymbol{Y})$. [Hodges and Lehmann (1963).]

3. Generalization of Problem 2. Consider again the two-sample problem with the (separately ordered) order statistics $X^{(1)} < X^{(2)} < \ldots < X^{(m)}$ and $Y^{(1)} < Y^{(2)} < \ldots < Y^{(n)}$. Let the score function φ be skew-symmetric around 0 (so that the scores satisfy $a_N(k) = -a_N(N+1-k)$, $k = 1, 2, \ldots, N$). Then the estimate $\hat{\Delta}$ equals the 'weighted median' of the set of mn pairwise differences $\Delta_{ij} = X^{(i)} - Y^{(j)}$, $i = 1, \ldots, m$; $j = 1, \ldots, n$, where each Δ_{ij} is assigned the weight proportional to $a_N(i+j) - a_N(i+j-1)$. [Jaeckel (1972), Sec. 4, Th. 4 and Corollary 2.]

4. In Example 1 in 9.1.2, show that h defined by (6) may be expressed as $h =$ number of pairs (i, j) with $1 \leq i \leq j \leq N$ such that $X_i + X_j > 0$. The possible values of h are then the integers $0, 1, \ldots, N(N+1)/2$, and $\hat{\Delta}$ is the median of all the variables $(X_i + X_j)/2$ of this sort.

5. Consider the regression (9.1.3.1). Prove that the R-estimate $\hat{\beta}_n$ of β, defined after (9.1.3.18), is regression-equivariant and translation-equivariant.

Section 9.2

6. In the proof of Theorem 1 show that $S_{\Delta N}$ is constant on each part of the splitting generated by C. [Jurečková (1969).]

7. Theorem 2 is also valid if the score generating function φ is a difference of two monotone square integrable functions. [Jurečková (1969).]

Chapter 10

Miscellaneous topics in regression rank tests

10.1 ALIGNED RANK TESTS

The theory of rank tests, as presented in the preceding chapters, pertains to a null hypothesis of invariance (under appropriate groups of transformations that map the sample space onto itself), against suitable (parametric as well as nonparametric) alternatives. In a relatively more general setup, a null hypothesis, due to the presence of some nuisance parameter(s), may be composite, and in that way, lacking invariance, the natural appeal and simplicity of the classical rank tests may no longer be tenable. For example, in a randomized block design, due to the block-effects, that could be stochastic, ranks, ignoring the blocking, may no longer have the discrete uniform distribution (under the null hypothesis of no treatment effect). Elimination of these block effects by intra-block transformations may result in their dependence (within the blocks), and thereby may distort the EDF nature of the overall ranks based on such residuals. A second but more concrete example pertains to the linear model, represented as

$$\boldsymbol{Y} = \mathbf{X}_1\beta_1 + \mathbf{X}_2\beta_2 + e, \tag{12}$$

where $\boldsymbol{Y}$ and $\boldsymbol{e}$ are random n-vectors, $\mathbf{X}_1, \mathbf{X}_2$ are $n \times p_1$, $n \times p_2$ design matrices of known elements, and β_1, β_2 are unknown parametric vectors of dimensions p_1, p_2, respectively. Consider the *subhypothesis testing* problem :

$$H_0 : \beta_1 = \mathbf{0} \quad \text{against} \quad K : \beta_1 \neq \mathbf{0}, \tag{13}$$

where β_2 is treated as nuisance (so is the distribution function F for the errors). Clearly, under H_0, Y_i, the elements of $\boldsymbol{Y}$ may not be identically distributed. Therefore, if we rank the Y_i (among themselves), under H_0, the rank vector may not take each permutation of $(1, \ldots, n)$ with the same probability $(n!)^{-1}$, so that a test based on a linear rank statistic may no longer be EDF under H_0.

The *ranking after alignment* principle for the randomized block design, primarily due to Hodges and Lehmann (1962), provides an alternative avenue for interblock comparisons, and thereby enhancing the information contained in the rankings. This generally leads to a more efficient testing procedure, though often resulting in only being conditionally distribution-free (CDF). The theory of aligned rank tests for the subhypothesis testing problem has been systematically developed by Sen and Puri (1977), citing earlier references as well. These tests are generally only asymptotically distribution-free (ADF). We shall find it convenient to discuss these two types (i.e., CDF and ADF) of aligned rank tests separately in the following two subsections.

10.1.1 Aligned rank tests for two-way layouts. Consider a set of n blocks of p (≥ 2) plots each where p different treatments are applied (each once), and let X_{ij} be the response of the plot in the ith block, receiving the jth treatment, for $j = 1, \ldots, p$; $i = 1, \ldots, n$. We consider the following linear model (wherein we drop the assumption of normality of the errors):

$$X_{ij} = \mu + \beta_i + \tau_j + e_{ij}, \quad i = 1, \ldots, n; \; j = 1, \ldots, p, \tag{1}$$

where μ stands for the *mean effect*, β_i for the ith *block effect*, τ_j for the jth *treatment effect*, and the errors e_{ij} are assumed to be independent identically distributed random variables with a common, continuous distribution function F. Without loss of generality, we set $\tau'\mathbf{1} = 0$, and suppose that the distribution function F, though arbitrary, is centred at the origin. Thus, we have a semiparametric linear model with an arbitrary F. In this setup, we may even allow the block effects to be stochastic (as is usually the case in practice), so that our model would correspond to a *mixed-effects* model. Consider the null hypothesis

$$H_0 : \tau = \mathbf{0} \quad \text{against} \quad K : \tau \neq \mathbf{0}, \tag{2}$$

treating μ, β and F as nuisance parameters. The intra-block rank tests (viz., the Friedman test, see Subsection 4.7.1) are based on the within-block ranks (that do not depend on μ, β_i), and have been characterized as EDF under H_0. However, such intra-block rankings do not convey the information that can be acquired from the inter-block comparisons, especially when the block-effects are additive (in the sense that the error distribution function F remains the same from block to block). To illustrate this salient point, consider the simplest case of $p = 2$, where intra-block rankings reduce to the signs of the paired differences, so that a rank test is effectively a median or sign test. A more general signed-rank statistic can be formulated from the n paired differences arising out of the n blocks, and in general, such a signed-rank test would be more efficient than the sign test. Motivated by this simple case, Hodges and Lehmann (1962) advocated the following alignment procedure.

Let $\tilde{X}_i$ be a location-statistic based on the p observations $X_{i1}, \ldots, X_{ip}$ in the ith block, for $i = 1, \ldots, n$; this could be the mean, median, modified mean or some other translation-equivariant statistic that is symmetric in its p arguments. Define the aligned observations as

$$\hat{X}_{ij} = X_{ij} - \tilde{X}_i, \text{ for } j = 1, \ldots, p; \ i = 1, \ldots, n. \tag{3}$$

Note that under H_0, for each i, the aligned observations $\hat{X}_{i1}, \ldots, \hat{X}_{ip}$, though not independent, are exchangeable or interchangeable random variables, and for different i, these vectors are independent. Moreover these random variables are independent of the block effects, so that ranking ignoring the blocks makes sense. Let R_{ij} be the rank of $\hat{X}_{ij}$ among the $N = np$ aligned observations (covering all blocks), for $i = 1, \ldots, n$; $j = 1, \ldots, p$; these R_{ij} are therefore the numbers $1, \ldots, N$, permuted in some random manner. Let $\mathbf{R}_i' = (R_{i1}, \ldots, R_{ip})$, for $i = 1, \ldots, n$, and let

$$\mathbf{R} = (\mathbf{R}_1, \ldots, \mathbf{R}_n) \text{ of order } p \times n \tag{4}$$

be the *rank collection matrix*. Note that there are $(N!)$ possible realizations of $\mathbf{R}$, and the distribution of $\mathbf{R}$ over this set may generally depend on the unknown F (because of the stochastic dependence of the $\hat{X}_{ij}$ for a common i $(= 1, \ldots, n)$). For this reason, we incorporate the Chatterjee-Sen (1964) rank-permutation principle to obtain some conditionally distribution-free (CDF) tests based on the aligned rankings. Corresponding to a given $\mathbf{R}$, we generate a set $\mathcal{R}_N$ of $n!(p!)^n$ possible rank-collection matrices that can be obtained from $\mathbf{R}$ by permuting its n columns (among themselves) and by permuting within each column the set of p rankings among themselves. Using the within-block exchangeability of the aligned observations along with the independence (and hence interchangeability) of the n vectors of aligned observations, we claim that under H_0, the conditional distribution of $\mathbf{R}$ on $\mathcal{R}_N$ is discrete uniform with each possible element (matrix) of $\mathcal{R}_N$ having the same (conditional) probability $(n!(p!)^n)^{-1}$. We denote this (conditional) permutational probability law by $\mathcal{P}_N$. Aligned rank tests are based on this permutational probability law $\mathcal{P}_N$, and hence, they are CDF under H_0.

As in the case of several independent samples, we conceive of a set of scores $a_N(1), \ldots, a_N(N)$ which (without loss of generality) we take as monotone in $k(\leq N)$. Let then

$$\mathbf{T}_N = (T_{N1}, \ldots, T_{Np})'; \; T_{Nj} = n^{-1} \sum_{i=1}^{n} a_N(R_{ij}), \; j = 1, \ldots, p. \tag{5}$$

Let $\bar{a}_N = N^{-1} \sum_{i=1}^{N} a_N(i)$, and let

$$\bar{a}_{N,i} = p^{-1} \sum_{j=1}^{p} a_N(R_{ij}), \quad i = 1, \ldots, n \tag{6}$$

be the within-block average scores; these are all random in nature, and $n^{-1} \sum_{i=1}^{n} \bar{a}_{N,i} = \bar{a}_N$(non-random). We also define

$$\sigma^2_{\mathcal{P}_N} = \frac{1}{n(p-1)} \sum_{i=1}^{n} \sum_{j=1}^{p} [a_N(R_{ij}) - \bar{a}_{N,i}]^2. \tag{7}$$

Then it is easy to verify that $\mathsf{E}_{\mathcal{P}_N} \mathbf{T}_N = \bar{a}_N \mathbf{1}$, and

$$n \, \mathsf{var}_{\mathcal{P}_N}(\mathbf{T}_N) = \{\mathbf{I} - p^{-1}\mathbf{1}\mathbf{1}'\}\sigma^2_{\mathcal{P}_N}. \tag{8}$$

The covariance matrix in (8) and (10) is of rank $p-1$ (as $\mathbf{1}'\mathbf{T}_N = p(\bar{a}_N)$ is non-random), and $\mathbf{I} - p^{-1}\mathbf{1}\mathbf{1}'$ is an idempotent matrix; using the conventional quadratic form formulation, we consider the test statistic (Sen, 1968d):

$$\begin{aligned} \mathcal{L}_N &= (\mathbf{T}_N - \bar{a}_N\mathbf{1})'(\mathsf{var}_{\mathcal{P}_N}(\mathbf{T}_N))^-(\mathbf{T}_N - \bar{a}_N\mathbf{1}) \\ &= n\sigma^{-2}_{\mathcal{P}_N} \sum_{j=1}^{p} (T_{n,j} - \bar{a}_N)^2. \end{aligned} \tag{9}$$

The permutational distribution of $\mathcal{L}_N$, generated by the uniform discrete (conditional) distribution of $\mathbf{R}$ over the set R_N, provides us with a CDF test for $H_0 : \tau = \mathbf{0}$. Moreover, using the set of $(p!)^n$ within-block permutations of the elements of $\mathbf{R}_i$ (independently for different blocks), a central limit theorem works out well for $\mathbf{T}_N$. We therefore claim that under $\mathcal{P}_N$,

$$n^{1/2}(\mathbf{T}_N - \bar{a}_N\mathbf{1}) \text{ are asymptotically normal } (\mathbf{0}, \; \sigma^2_{\mathcal{P}_N}(\mathbf{I} - p^{-1}\mathbf{1}\mathbf{1}')), \tag{10}$$

and this asymptotic normality (in probability) in conjunction with the Cochran theorem yields that the asymptotic permutation distribution of $\mathcal{L}_N$ is central χ^2 with $p-1$ degrees of freedom, in probability. Therefore, under the null hypothesis, $\mathcal{L}_N$ has an asymptotically central χ^2 distribution with $p-1$ degrees of freedom. Sen (1968d) studied the asymptotic non-null distribution of $\mathcal{L}_N$, under local (Pitman-type) alternatives based on a multivariate Chernoff-Savage (1958) approach (without invoking contiguity). However, in line with the contiguity of alternative hypotheses presented in Chapters 7 and 8 it is easy to formulate such contiguous alternatives for this problem as follows.

We assume that the errors e_{ij} in (1) are independent identically distributed random variables with an absolutely continuous density $f(x)$ having finite Fisher information $I(f)$. Further, we consider local alternatives K_n of the type

$$K_n : \ \tau = n^{-1/2}\theta, \text{ for some } p\text{-variate } \theta. \tag{11}$$

For simplicity of presentation, we take the case of $\tilde{X}_i = \bar{X}_i = p^{-1} \sum_{j=1}^{p} X_{ij}$, so that under $\{K_n\}$,

$$\hat{X}_{ij} = \hat{e}_{ij} + n^{-1/2}\tau_j; \ \hat{e}_{ij} = e_{ij} - p^{-1} \sum_{k=1}^{p} e_{ik}, \tag{12}$$

for $j = 1, \ldots, p$; $i = 1, \ldots, n$. Further, note that $(\hat{e}_{i1}, \ldots, \hat{e}_{ip})$ are exchangeable random variables, independent for different i. Let then P_n and Q_n be the joint probability law for $(\hat{\boldsymbol{X}}_1, \ldots, \hat{\boldsymbol{X}}_n)$ under the null hypothesis H_0 and K_n respectively. Then it is easy to verify that under the assumed regularity conditions on the density f, $\{Q_n\}$ is contiguous to $\{P_n\}$ [see Problem 1]. This enables us to extend the asymptotic normality results [on $n^{1/2}(\mathbf{T}_N - \bar{a}_N \mathbf{1})$] from the null case (mentioned before) to that under $\{K_n\}$, resulting only in a shift of the centring (vector) that depends on θ, the score function and the density f [Problem 2]. Consequently, the asymptotic non-null distribution of $\mathcal{L}_N$, under $\{K_n\}$, comes out as a non-central χ^2 distribution with $p-1$ degrees of freedom and a non-centrality parameter $\Delta_{\mathcal{L}}$ that can be derived in the following manner.

We denote the marginal distribution function of $\hat{X}_{ij}$, under H_0, by $F_0(x)$; note that this is the same for all $i = 1, \ldots, n$; $j = 1, \ldots, p$, and the form of F_0 depends on the density f of the errors as well as the choice of the estimator $\tilde{X}_i$. Also, let $F_0^*(x, y)$ be the joint distribution function of $(\hat{X}_{i1}, \hat{X}_{i2})$, under H_0; this distribution function pertains to any pair within the same block. Further, let the scores $a_N(k)$ be generated by a score generating function $\varphi(u), u \in (0, 1)$, in the same manner as in the case of independent samples. Without loss of generality we assume that $\bar{\varphi} = \int_0^1 \varphi(u) \, \mathrm{d}u = 0$. Also, we assume that $\varphi(.)$ is square integrable (on $[(0, 1)]$), and denote

$$A_\varphi^2 = \int_0^1 \varphi^2(u) \, \mathrm{d}u = \int_{-\infty}^{\infty} \varphi^2(F_0(x)) \, \mathrm{d}F_0(x)),$$

$$\rho_\varphi = A_\varphi^{-2}\Big\{\int_{-\infty}^{\infty}\int_{-\infty}^{\infty}\varphi(F_0(x))\varphi(F_0(y))\,\mathrm{d}F_0^*(x,y)\Big\}. \tag{13}$$

Then, it is easy to verify that under H_0 (and hence, by contiguity, under $\{K_n\}$),

$$\sigma^2_{\mathcal{P}_N} \rightarrow A_\varphi^2(1-\rho_\varphi) \text{ in probability;} \tag{14}$$

we leave the derivation (Sen 1968d) as Problem 3. We also define the functional

$$\gamma(\varphi, F_0) = \gamma_\varphi^0 = \int_0^1 \varphi(u)\varphi(u, f_0)\,\mathrm{d}u, \tag{15}$$

where

$$\varphi(u, f_0) = -f_0'(F_0^{-1}(u))/f_0(F_0^{-1}(u)),\ u\ \in (0,1) \tag{16}$$

is the score function from (2.2.4.3) for the density f_0 corresponding to the distribution function F_0 of the aligned residual $\hat{e}_{11}$. Then by some standard computations (left as Problem 4), we obtain that under $\{K_n\}$, $\mathcal{L}_N$ has asymptotically a non-central χ^2 distribution with $p-1$ degrees of freedom and non-centrality parameter

$$\Delta_{\mathcal{L}} = \theta'\theta(\gamma_\varphi^0)^2/\{A_\varphi^2(1-\rho_\varphi)\}. \tag{17}$$

Side by side, we consider the classical ANOVA (variance-ratio) test statistic, denoted by $\mathcal{F}_n$. The asymptotic non-null distribution of $(p-1)\mathcal{F}_n$, under $\{K_n\}$, is easily shown (see Problem 5) to be non-central χ^2 with $p-1$ degrees of freedom and non-centrality parameter

$$\Delta_{\mathcal{F}} = \theta'\theta/\sigma_e^2, \tag{18}$$

where σ_e^2 stands for the variance of the error e_{11}. If we denote the variance of the distribution function F_0 (of the aligned errors) by $\sigma^2_{F_0}$ and note that $\sigma^2_{F_0} = \sigma_e^2(1-1/p)$, we may conclude from the above that the asymptotic relative efficiency (ARE) of the aligned rank test based on $\mathcal{L}_N$ with respect to the classical ANOVA test is given by

$$\mathrm{ARE}(\mathcal{L};\mathcal{F}) = \frac{p}{(p-1)(1-\rho_\varphi)}\{\sigma^2_{F_0}(\gamma_\varphi^0)^2/A_\varphi^2\}. \tag{19}$$

Using the interchangeability of the $\hat{e}_{ij}, j = 1,\ldots,p$, it follows easily (Sen 1968d) that

$$\rho_\varphi \geq -1/(p-1), \quad \text{for all } \varphi, \tag{20}$$

where the equality sign holds only when $\varphi(F(x)) \equiv x$, excepting on a set of null measure (see Problem 6). As such, we have

$$\mathrm{ARE}(\mathcal{L};\mathcal{F}) \geq \{(\gamma_\varphi^0\sigma_{F_0}/A_\varphi)\}^2, \tag{21}$$

where the right hand side stands for the ARE in the case of independent samples. If we carry out a similar analysis for intrablock rank tests, we arrive at a complementary picture where the ARE is bounded from above, and the upper bound is achieved only for large values of p. This shows that the alignment principle can enhance the efficacy of rank tests. We skip further details in this setup, and refer to Puri and Sen (1971, Ch. 7) for some related results.

10.1.2 Aligned rank statistics for subhypothesis testing. Let us incorporate an intercept parameter β_0, and extend the linear model in (10.1.1) as

$$\boldsymbol{Y} = \beta_0 \mathbf{1} + \mathbf{X}_1 \beta_1 + \mathbf{X}_2 \beta_2 + \boldsymbol{e}, \tag{1}$$

where (without loss of generality) we may assume that $\mathbf{1}'\mathbf{X}_j = \mathbf{0}$, $j = 1, 2$. We consider here the null and alternative hypotheses as

$$H_0 : \beta_1 = \mathbf{0} \quad \text{against} \quad H_{(1)} : \beta_1 \neq \mathbf{0}, \tag{2}$$

treating β_0, β_2 both as nuisance parameters. Note that under the above null hypothesis,

$$\boldsymbol{Y} = \beta_0 \mathbf{1} + \mathbf{X}_2 \beta_2 + \boldsymbol{e}, \quad \beta_0, \beta_2 \text{ nuisance}, \tag{3}$$

so that under H_0, the Y_i may not be identically distributed, and hence the permutation invariance structure discussed in earlier chapters may not hold here, and as a result, an EDF rank test may not generally exist. The basic idea is therefore to estimate the nuisance parameter β_2 (based on suitable rank statistics), form the aligned observations (residuals) by incorporating such an estimate, and then base a rank test on these aligned observations. Note that the ranks are invariant to shifts, so that β_0, though nuisance, does not affect the linear rank statistics on which a test is based. The alignment principle works out well in an asymptotic setup, and in this context, the asymptotic linearity of linear rank statistics (vectors) in the regression parameter(s), as presented in the preceding chapter, plays a key role; by construction, such aligned rank tests are usually ADF but not generally CDF or EDF. A similar problem crops up if we want to test for the null hypothesis $\beta_0 = 0$ against $\beta_0 \neq 0$, treating β as a nuisance parameter; we may need to use aligned signed rank statistics for this hypothesis testing problem. More generally, a null hypothesis may involve both β_0, β_1, treating β_2 as nuisance. We shall append some comments on these testing problems at the end of this subsection.

The model (1) (at this moment ignoring the term with β_0 for the above reasons) may be concisely rewritten as

$$\boldsymbol{Y} = \mathbf{X}\beta + \boldsymbol{e},$$

where the $(n \times (p_1 + p_2))$-matrix $\mathbf{X}$ is partitioned as $\mathbf{X} = (\mathbf{X}_1, \mathbf{X}_2)$, with

$\mathbf{X}_1$ ($\mathbf{X}_2$) being a $(n \times p_1)$-matrix ($(n \times p_2)$-matrix, respectively), and a $(p_1 + p_2)$-vector β partitioned as $\beta = (\beta_1, \beta_2)$, with a p_1-vector β_1, p_2-vector β_2, etc. similarly for other vectors. (In such a partitioning, the first part always corresponds to β_1 to be tested, the second part corresponds to a nuisance β_2.)

For every $(p_1 + p_2)$-vector $\mathbf{b}$, we define an aligned linear rank statistics vector as

$$\mathbf{L}_n(\mathbf{b}) = \sum_{i=1}^{n} \mathbf{x}_i a_n(R_{ni}(\mathbf{b})), \tag{4}$$

where the $\mathbf{x}_i'$ are the row vectors for $\mathbf{X}$, and later on we shall partition these as $\mathbf{x}_{i(1)}', \mathbf{x}_{i(2)}'$; also $R_{ni}(\mathbf{b})$ stands for the rank of $Y_i(\mathbf{b}) = Y_i - \mathbf{x}_i'\mathbf{b}$ among the n aligned observations $Y_k(\mathbf{b})$, $k = 1, \ldots, n$, for $i = 1, \ldots, n$. In view of the nuisance nature of β_2, we need to estimate the same under the null hypothesis model. For this purpose, we define

$$\mathbf{L}_{n(2)}(\mathbf{b}_2) = \sum_{i=1}^{n} \mathbf{x}_{i(2)} a_n(R_{ni}^*(\mathbf{b}_2)), \tag{5}$$

where $R_{ni}^*(\mathbf{b}_2)$ stands for the rank of $Y_i - \mathbf{x}_{i(2)}'\mathbf{b}_2$ among the n aligned observations $Y_k - \mathbf{x}_{i(2)}'\mathbf{b}_2$, $k = 1, \ldots, n$. Then as in Section 9.1 (R-estimation), we define a rank-estimate $\hat{\beta}_{2n}$ of β_2, under H_0, as

$$\hat{\beta}_{2n} = \text{ arg min}\{ \, ||\mathbf{L}_{n(2)}(\mathbf{b}_2)|| : \mathbf{b}_2 \; p_2\text{-variate}\}, \tag{6}$$

where $||.||$ can be taken as L_1-, L_2-, or even the sup-norm. Having obtained this estimate of the nuisance parameter, we consider the residuals (aligned observations):

$$\hat{Y}_{ni} = Y_i - \mathbf{x}_{i(2)}'\hat{\beta}_{2n}, \quad i = 1, \ldots, n. \tag{7}$$

Note that these aligned observations may no longer be independent or marginally identically distributed. Let us denote by $\hat{R}_{ni}$ the rank of $\hat{Y}_{ni}$ among the $\hat{Y}_{nk}$, $k = 1, \ldots, n$, and consider the vector of aligned linear rank statistics:

$$\hat{\mathbf{L}}_{n(1)} = \sum_{i=1}^{n} \mathbf{x}_{i(1)} a_n(\hat{R}_{ni}). \tag{8}$$

At this stage, let us bring some auxiliary results, mostly adapted from Sections 9.1 and 9.2, that are relevant to the subsequent developments here. First, based on the asymptotic linearity results for linear rank statistics, studied in Section 9.2, we consider the vector linear rank statistic $\mathbf{L}_n(\mathbf{b})$, take (without loss of generality) $\beta = \mathbf{0}$, and claim that for every finite, positive C,

$$\sup_{||\mathbf{b}|| \leq C} \{n^{-1/2} ||\mathbf{L}_n(n^{-1/2}\mathbf{b}) - \mathbf{L}_n(\mathbf{0}) + \gamma n^{-1/2}(\mathbf{X}'\mathbf{X})\mathbf{b}||\} \tag{9}$$
$$\to 0 \text{ in probability,}$$

where

$$\gamma = \gamma(\varphi, f) = \int_0^1 \varphi(u)\varphi(u, f)\, du, \tag{10}$$

and $\varphi(u, f) = -f'(F^{-1}(u))/f(F^{-1}(u))$, $u \in (0, 1)$, as in (2.2.4.3). Problem 7 is set for verification of this basic linearity result. Therefore, under H_0,

$$n^{1/2}(\hat{\beta}_{2n} - \beta_2) \text{ is asymptotically } p\text{-variate normal } (\mathbf{0}, (\gamma)^{-2} A_\varphi^2 \mathbf{\Lambda}_{22}^{-1}), \tag{11}$$

where $A_\varphi^2 = \int_0^1 (\varphi(u) - \bar{\varphi})^2\, du$, and we assume that for $\mathbf{C}_n = \mathbf{X}'\mathbf{X}$,

$$\lim_{n\to\infty} n^{-1}\mathbf{C}_n = \mathbf{\Lambda} \text{ exists, and is positive definite,} \tag{12}$$

and partition it as

$$\mathbf{\Lambda} = ((\mathbf{\Lambda}_{ij})),\ \mathbf{\Lambda}_{ij} \text{ is of order } p_i \times p_j,\ i, j = 1, 2. \tag{13}$$

Problem 8 is set to verify the above result. Secondly, based on the asymptotic linearity results in Section 9.2, we claim that under H_0,

$$\hat{\beta}_{2n} - \beta_2 = \gamma^{-1}(\mathbf{X}_2'\mathbf{X}_2)^{-1} \sum_{i=1}^n \mathbf{x}_{i(2)} \varphi(F(e_i)) + o_p(n^{-1/2}). \tag{14}$$

Actually this first-order representation result implies the asymptotic normality result in (11); see Problem 9. Thirdly, based on the same linearity results (along with the fact that by virtue of (11), $||\hat{\beta}_{2n} - \beta_2|| = O_p(n^{-1/2})$), we claim that under H_0,

$$n^{-1/2}(\hat{\mathbf{L}}_{n1} - \mathbf{L}_{n1}(\mathbf{0}) + \gamma n^{-1/2}\mathbf{C}_{n12}(\hat{\beta}_{2n} - \beta_2)) = o_p(1), \tag{15}$$

(see Problem 10), so that by (12) and (13),

$$n^{-1/2}\hat{\mathbf{L}}_{n1} = n^{-1/2}\mathbf{L}_{n1}(\mathbf{0}) - n^{-1/2}\mathbf{C}_{n12}\mathbf{C}_{n22}^{-1} \sum_{i=1}^n \mathbf{x}_{i(2)}\varphi(F(e_i)) + o_p(1). \tag{16}$$

(Problem 11). Further, by the projection results on linear rank statistics in the null case, we have under H_0,

$$n^{-1/2}\mathbf{L}_{n1}(\mathbf{0}) = n^{-1/2} \sum_{i=1}^n \mathbf{x}_{i(1)}\varphi(F(e_i)) + o_p(1), \tag{17}$$

(Problem 12), so that we obtain from the above that under H_0,

$$n^{-1/2}\hat{\mathbf{L}}_{n1} = n^{-1/2} \sum_{i=1}^n (\mathbf{x}_{i(1)} - \mathbf{C}_{n12}\mathbf{C}_{n22}^{-1}\mathbf{x}_{i(2)})\varphi(F(e_i)) + o_p(1). \tag{18}$$

Consequently, if we let

$$\mathbf{x}_{i(1:2)} = \mathbf{x}_{i(1)} - \mathbf{C}_{n12}\mathbf{C}_{n22}^{-1}\mathbf{x}_{i(2)},\ i = 1, \ldots, n, \tag{19}$$

and write

$$\mathbf{C}_{n11.2} = \mathbf{C}_{n11} - \mathbf{C}_{n12}\mathbf{C}_{n22}^{-1}\mathbf{C}_{n21} = \sum_{i=1}^{n} \mathbf{x}_{i(1:2)}\mathbf{x}'_{i(1:2)}, \tag{20}$$

then using (18), (19) and (20), we can construct an appropriate quadratic form in $\hat{\mathbf{L}}_{n1}$ as a test statistic for testing $H_0 : \beta_1 = \mathbf{0}$ against $H_1 : \beta_1 \neq \mathbf{0}$. As in Sen and Puri (1977), we consider the following aligned rank test statistic:

$$\hat{\mathcal{L}}_{n1} = A_n^{-2}\{\hat{\mathbf{L}}'_{n1}\mathbf{C}_{n11.2}^{-1}\hat{\mathbf{L}}_{n1}\}, \tag{21}$$

where $A_n^2 = (n-1)^{-1}\sum_{i=1}^{n}(a_n(i) - \bar{a}_n)^2$. It follows from the classical central limit theorem along with the Cochran theorem on normal quadratic forms that under H_0, $\hat{\mathcal{L}}_{n1}$ has an asymptotically central χ^2 distribution with p_1 degress of freedom. If H_0 does not hold, it can be easily shown (Problem 13) that for any fixed $\beta_1 \neq \mathbf{0}$, $n^{-1}\hat{\mathcal{L}}_n$ converges in probability to a positive constant (that depends on β_1 as well as $\mathbf{C}_{n11.2}$), so that $\hat{\mathcal{L}}_{n1}$ is $O_p(n)$; therefore as n increases indefinitely, it would exceed the critical level with probability converging to one. This establishes the consistency of the test based on $\hat{\mathcal{L}}_{n1}$. In this way, we have a class of ADF aligned rank tests for testing H_0 against H_1, and again within this class, we can choose an appropriate one to have other desirable properties against specific families of alternatives.

In order to study asymptotic power properties of such aligned rank tests, we consider appropriate local (contiguous) alternatives:

$$H_n : \beta_1 = \beta_{1n} = n^{-1/2}\lambda, \text{ for some fixed } p_1\text{-variate } \lambda, \tag{22}$$

and the density f is assumed to be absolutely continuous with finite Fisher information $I(f)$. Thus, under H_n, the Y_i are independent with densities $f(y - n^{-1/2}\lambda'\mathbf{x}_{i1} - \beta_2'\mathbf{x}_{i2})$, so that the likelihood ratio statistic can be expressed as

$$\prod_{i=1}^{n}\{f(Y_i - n^{-1/2}\lambda'\mathbf{x}_{i(1)} - \beta_2'\mathbf{x}_{i(2)})/f(Y_i - \beta_2'\mathbf{x}_{i(2)})\}. \tag{23}$$

Note that under H_0, the $Y_i - \beta_2'\mathbf{x}_{i(2)}$ are independent identically distributed random variables with the density $f(y)$, and hence, if we proceed as in the case of a simple regression model, treated in Chapter 7, we claim that $\{q_n\}$, the joint density of the $\{Y_i, i \leq n\}$ under $\{H_n\}$, is contiguous to $\{p_n\}$, the joint density under H_0; Problem 14 is set to verify this assertion.

By virtue of this contiguity and the representation in (18), we conclude that under $\{H_n\}$, $n^{-1/2}\hat{\mathbf{L}}_{n1}$ is asymptotically normal with mean $\gamma\mathbf{\Lambda}_{11.2}\lambda$ and covariance matrix $A_\varphi^2\mathbf{\Lambda}_{11.2}$, where γ and A_φ^2 are defined in (10) and after (11), and

$$\begin{aligned}\mathbf{\Lambda}_{11.2} &= \lim_{n\to\infty} n^{-1}\mathbf{C}_{n11.2} \\ &= \mathbf{\Lambda}_{11} - \mathbf{\Lambda}_{12}\mathbf{\Lambda}_{22}^{-1}\mathbf{\Lambda}_{21}.\end{aligned} \tag{24}$$

(See Problem 15). As a result, by the Cochran theorem, we claim that under $\{H_n\}$,

$$\begin{aligned}&\hat{\mathcal{L}}_{n1} \text{ has asymptotically the non-central } \chi^2 \text{ distribution} \\ &\text{with } p_1 \text{ degrees of freedom and non-centrality parameter } \Delta_{\mathcal{L}},\end{aligned} \tag{25}$$

where

$$\Delta_{\mathcal{L}} = \gamma^2 A_\varphi^{-2}(\lambda'\mathbf{\Lambda}_{11.2}^{-1}\lambda). \tag{26}$$

(See Problem 16.) Note that this asymptotic non-central distribution holds for all score functions $\varphi(.)$ that are square integrable inside $(0,1)$, when $I(f) < \infty$. Further the degrees of freedom is the same (p_1) for all such score functions, while the non-centrality parameter depends on the score function φ explicitly through $\gamma^2 A_\varphi^{-2}$. Since the tail probability of a non-central χ^2 distribution is a monotone non-decreasing function of its non-centrality parameter (when the degrees of freedom are held fixed), we can easily identify an asymptotically optimal test within this class of aligned rank tests as the one based on a score function φ_0 for which

$$\gamma(\varphi_0, f)/A_{\varphi_0} = \sup\{\gamma(\varphi, f)/A_\varphi : \varphi \in L_2(0,1)\}. \tag{27}$$

(See Problem 17.)

As always, letting $\varphi(u,f)$, $0 < u < 1$, be the score function defined in (2.2.4.3), and noting that $\int_0^1 \varphi^2(u,f)\,\mathrm{d}u = I(f)$, we may write

$$\begin{aligned}\gamma^2 A_\varphi^{-2} &= I(f)\rho^2(\varphi,\varphi_0); \\ \rho^2(\varphi,\varphi_0) &= \Big\{\int_0^1 \varphi(u)\varphi_0(u)\,\mathrm{d}u\Big\}^2 \cdot \|\varphi\|^{-2}\|\varphi_0\|^{-2}.\end{aligned} \tag{28}$$

As a result, we conclude that for the asymptotically optimal score function, we have

$$\varphi_0(u) \equiv \varphi_f(u,f), \quad u \in (0,1). \tag{29}$$

(See Problem 18.) Side by side, if we consider the usual likelihood ratio test for H_0 vs. H_1 (or its asymptotically equivalent form based on the classical Wald test statistic or the Rao scores test statistic), then it can be shown that for the sequence of alternative hypotheses $\{H_n\}$, we would have the non-central χ^2 distribution with p_1 degrees of freedom and non-centrality parameter

$$\Delta^* = I(f)(\lambda' \mathbf{\Lambda}_{11.2}^{-1} \lambda). \tag{30}$$

As a result we claim that an aligned rank test based on the score function φ_0 is asymptotically efficient for testing H_0 against $\{H_n\}$. This result is in agreement with the case of the simple regression model treated in earlier chapters. Therefore, for the subhypothesis testing problem, the alignment principle retains the basic properties of rank tests in an asymptotic setup. In this context, we may note that to carry out this aligned rank test, an essential ingredient is the estimation of the nuisance parameter β_2, and that has to be based on the same type of linear rank statistics so that the asymptotic linearity results can be properly incorporated to streamline the asymptotic distribution theory of such aligned rank test statistics. A rather discouraging aspect of R-estimation of regression parameters in a general linear model (such as the one considered here) is that if p_2 is greater than 1 then even for most simple score functions, such as the Wilcoxon scores, an exact or algebraic solution to the allied estimating equations may not exist, and therefore, iterative solutions are advocated. Such an iterative trial and error solution may be sometimes quite tedious (the same criticism may also be leveled against the maximum likelihood estimates when f does not belong to the exponential family of densities). In this respect, the tests based on regression rank scores estimates, as introduced in the next section, have a distinct advantage (Jurečková and Sen 1993): they do not entail the estimation of the nuisance parameters, and hence they are computationally more attractive. Moreover, based on a common score function, such regression rank procedures and aligned rank procedures are asymptotically equivalent under the null as well as local aternatives, and hence, they share the same asymptotic properties.

Let us briefly consider aligned rank tests for testing the null hypothesis of no intercept (i.e., β_0 is null), treating a part or all of the regression vector β as nuisance. Essentially, aligned rank tests involve the estimation of the nuisance parameters based on the same type of aligned linear rank statistics as used before, and then based on the residuals or aligned observations, use an appropriate signed rank statistic as a suitable test statistic. In this context too, asymptotic linearity of signed linear rank statistics in location and regression parameters plays the basic role, and the rest of the developments are similar to the case of linear rank statistics treated here. We omit these details, and refer to Puri and Sen (1985, Ch.7) where a more detailed treatment of aligned rank tests has been outlined.

10.2 REGRESSION RANK SCORES

Regression rank scores (RRS), introduced primarily by Gutenbrunner and Jurečková (1992), have in a short time become a focal point in robust statistical inference for linear models. They are the natural generalization of the Hájek (1965) rank scores, developed for the regression model (and discussed in our Subsections 4.3.3 and 6.3.6), to more general linear models, and as a result, they are in the main domain of the theory of rank tests. As a matter of fact, estimates based on RRS are closely related to the R-estimates in linear models (which were introduced and discussed in the preceding chapter). Tests (for composite hypotheses) in linear models based on RRS may also have certain advantages over others. The two concluding sections of Chapter 6 of Jurečková and Sen (1996) contain an up-to-date, unified treatment of various aspects of RRS. We therefore avoid the details to a certain extent.

Regression quantiles (RQ), developed by Koenker and Bassett (1978), are the precursors of RRS, and the duality of RRS and RQ, both based on a common *linear programming* methodology, is an extension of the duality of *order statistics* and vector of *ranks*, discussed in earlier chapters, from the location to linear models.

In linear models, estimates based on RRS have close proximity to the classical (R-)*estimates*, and some general (asymptotic) equivalence results in this direction are due to Jurečková and Sen (1993). This equivalence can be studied conveniently by incorporating uniform asymptotic linearity results (in the regression parameters) of both RRS and rank statistics. There are other related and more recent developments along this line, reported in Chapter 6 of Jurečková and Sen (1996), and we shall omit these details. In the context of *subhypothesis testing* in linear models (where a subset of the parameters is treated as *nuisance*), tests based on *linear RRS statistics* (LRRSS) compare favourably with aligned rank tests which we presented in the preceding section. Because of some *equivariance* (and *invariance*) properties of RRS, the LRRSS may have some computational advantages too; these properties have been studied in detail by Gutenbrunner et al. (1993). Studentizing scale statistics based on RRS have been considered by Jurečková and Sen (1994). A basic development in this context, due to Jurečková (1992a, b), is the *stochastic uniform asymptotic linearity* result (in the regression parameters) that is enjoyed by the LRRSS, as well as the linear rank statistics. These linearity results for linear rank statistics were discussed in Section 9.2, and parallel results for RRS statistics are systematically reviewed in Chapter 6 of Jurečková and Sen (1996) in diverse statistical inference problems. Sen (1996) considered RRS based inference in mixed-effects models. We present only an outline of these developments.

10.2.1 The duality of RQ and RRS. For observable random variables $Y_1, \ldots, Y_n$, consider the usual *fixed-effects* linear model:

$$Y_i = \beta' \mathbf{c}_i + e_i, \quad i = 1, \ldots, n; \tag{1}$$

where the unobserved e_i are independent and identically distributed random variables with a continuous (unspecified) distribution function F, defined on $(-\infty, +\infty)$, the $\mathbf{c}_i = (c_{i1}, \ldots, c_{ip})'$ are p-vectors, $p \geq 2$ of known regression constants (not all equal), and $\beta = (\beta_1, \ldots, \beta_p)'$ is a vector of unknown parameters. Note that in view of the unspecified nature of F, the scale parameter, if any, has been tacitly absorbed in F itself. Without loss of generality, we may set $c_{i1} = 1$, $i = 1, \ldots, n$, and let $\bar{c}_{nj} = n^{-1} \sum_{i=1}^{n} c_{ij} = 0$, $j = 2, \ldots, p$. Thus, β_1 is the intercept parameter, while the rest of β relates to regression parameter(s). We appraise the role of RRS in robust estimation of β, as well as testing for (sub-)hypotheses on β.

Koenker and Bassett (1978) extended the notion of *sample quantiles* to linear models, and termed them regression quantiles (RQ). For an α, $0 < \alpha < 1$, consider the norm

$$\rho_\alpha(x) = |x|\{\alpha I(x > 0) + (1 - \alpha) I(x < 0)\}, \quad -\infty < x < \infty, \tag{2}$$

and define the α-RQ of β as

$$\hat{\beta}_n(\alpha) = \text{argmin}\{\sum_{i=1}^{n} \rho_\alpha(Y_i - \mathbf{t}'\mathbf{c}_i) : \ p\text{-variate } \mathbf{t}. \tag{3}$$

They showed that $\hat{\beta}_n(\alpha)$ is a solution of the following linear programming problem:

$$\alpha \sum_{i=1}^{n} r_i^+ + (1 - \alpha) \sum_{i=1}^{n} r_i^- = \min! \tag{4}$$

subject to

$$\sum_{j=1}^{p} c_{ij} \hat{\beta}_j + r_i^+ - r_i^- = Y_i, \quad i = 1, \ldots, n; \tag{5}$$

$$p\text{-variate } \hat{\beta}; \ r_i^+ \geq 0, \ r_i^- \geq 0, \quad i = 1, \ldots, n;$$

where r_i^+ and r_i^- are respectively the positive and negative parts of the residuals $Y_i - \hat{\beta}'\mathbf{c}_i$, $i = 1, \ldots, n$; these are non-negative by definition. Koenker and d'Orey (1993) considered the computational aspects of RQ's. Dual linear programming underlies the RRS estimation of β.

For $\alpha, 0 < \alpha < 1$, the RR vector $\hat{\mathbf{a}}_n(\alpha) = (\hat{a}_{n1}(\alpha), \ldots, \hat{a}_{nn}(\alpha))'$ is defined as the solution to the linear programming problem:

$$\sum_{i=1}^{n} Y_i \hat{a}_{ni}(\alpha) = \max! \tag{6}$$

subject to

$$\sum_{i=1}^{n} c_{ij}\hat{a}_{ni}(\alpha) = (1-\alpha)\sum_{i=1}^{n} c_{ij}, \quad j = 1, \ldots, p, \tag{7}$$
$$\hat{a}_{ni}(\alpha) \in [0,1], \quad \text{for all } i = 1, \ldots, n.$$

For computational aspects, we refer to Gass (1985). RRS are then adapted to some score generating functions that yield some LRRSS.

10.2.2 RRS estimates and connection to location model. In the particular case of $p = 1$ (i.e., the location model), if we let, as in Subsection 6.3.6,

$$\begin{aligned} a_n^*(j,\alpha) \quad &= \quad 0, j - n\alpha, \text{ or } 1, \text{according as} \\ & \qquad j < n\alpha, \ j-1 \leq n\alpha \leq j, \text{ or } j > n\alpha + 1, \end{aligned} \tag{1}$$

then on letting R_i be the rank of Y_i among $Y_1, \ldots, Y_n$, for $i = 1, \ldots, n$, we have

$$\hat{a}_{ni}(\alpha) = a_n^*(R_i; \alpha), \quad i = 1, \ldots, n. \tag{2}$$

These scores $a_n^*(R_i, \alpha)$, $i = 1, \ldots, n$, were considered by Hájek (1965) in formulating a Kolmogorov-Smirnov type test against regression alternatives. This was the basic motivation in Gutenbrunner and Jurečková (1992) underlying their proposed RRS.

For the estimation of β the RRS are generally adopted to a *score generating function* to formulate suitable LRRS statistics. Let $\varphi(u)$, $0 < u < 1$, be a non-decreasing and square integrable score generating function, and defining the RR vector $\hat{\mathbf{a}}_n(\alpha)$ as before, we compute the scores $\hat{\mathbf{b}}_n = (\hat{b}_{n1}, \ldots, \hat{b}_{nn})'$ by letting

$$\hat{b}_{ni} = -\int_0^1 \varphi(\alpha)\, \mathrm{d}\hat{a}_{ni}(\alpha), \quad i = 1, \ldots, n. \tag{3}$$

Then a LRRS statistic can be defined as

$$\hat{L}_n(\varphi) = \sum_{i=1}^{n} d_{ni}\hat{b}_{ni}, \tag{4}$$

where the coefficients d_{ni} can be chosen on the basis of the regression vectors $\mathbf{c}_i$ that appear in the original linear model in (1). In the above setup, without loss of generality, we may take $\bar{\varphi} = \int_0^1 \varphi(u)\,du = 0$. Then for the estimation of β, consider a *dispersion measure*

$$D_n(\mathbf{t}) = \sum_{i=1}^{n} (Y_i - \mathbf{t}'\mathbf{c}_i)[\hat{b}_{ni}(\mathbf{Y} - \mathbf{C}\mathbf{t})], \quad p\text{-variate } \mathbf{t}, \tag{5}$$

where the $\hat{b}_{ni}(\mathbf{Y} - \mathbf{Ct})$ refer to the RRS based on the aligned observations $\mathbf{Y} - \mathbf{Ct} = (Y_1 - \mathbf{c}_1'\mathbf{t}, \ldots, Y_n - \mathbf{c}_n'\mathbf{t})'$. Incorporating the equivariance properties of the RRS, one may then consider the following RRS estimator:

$$\tilde{\beta}_n = \text{argmin}\{D_n(\mathbf{t}) : \ p\text{-variate } \mathbf{t}\}. \tag{6}$$

A similar procedure works out for any subset estimation, i.e., letting $\beta = (\beta_1, \beta_{(1)}', \beta_{(2)}')'$, for estimating either $\beta_{(j)}$, $j = 1, 2$, treating the other components as nuisance parameters.

For notational simplicity, we write

$$Y_i = \beta_0 + \beta'\mathbf{c}_i + e_i, \quad i = 1, \ldots, n, \tag{7}$$

where the e_i are independent identically distributed random variables with a continuous distribution function F, and as before, we let $\sum_{i=1}^{n} \mathbf{c}_i = \mathbf{0}$. Let then

$$\mathbf{V}_n = \sum_{i=1}^{n} \mathbf{c}_i\mathbf{c}_i' \tag{8}$$

and assume that $\mathbf{V}_n$ is positive definite. Then under some additional regularity conditions (viz., Theorem 6.7.4 of Jurečková and Sen, 1996), the following (first-order) asymptotic representation holds for $\tilde{\beta}_n$, the regression rank scores estimate of β as defined above:

$$(\tilde{\beta}_n - \beta) = \gamma^{-1}\mathbf{V}_n^{-1}\sum_{i=1}^{n} \mathbf{c}_i\varphi(F(e_i)) + o_p(n^{-1/2}), \tag{9}$$

where $\gamma = -\int_0^1 \varphi(u)\,\mathrm{d}f(F^{-1}(u))$. Whenever the density f is absolutely continuous with finite Fisher information, we have a finite γ, and we can express it as

$$\gamma = \int_0^1 \varphi(u)\varphi(u, f)\,\mathrm{d}u; \quad \varphi(u, f) \text{ defined by (2.2.4.3).} \tag{10}$$

A similar representation holds for RRS estimates for any subset of β.

The classical R-estimation theory outlined in Sections 9.2 and 9.3, works out well for this linear model, and if that involves the same score generating function $\varphi(.)$, then it follows from the discussions in Sections 9.2 and 9.3 that the same asymptotic representation holds for the classical R-estimates of β. This isomorphism was explored by Jurečková and Sen (1993); therefore under additional regularity conditions on the underlying density and score functions, the RRS estimate and R-estimate are asymptotically stochastically equivalent, and hence they share all the asymptotic properties (robustness, consistency, asymptotic multinormality, etc.) that are presented in detail in Chapter 6 in Jurečková and Sen (1996). Because of this asymptotic equivalence, tests based on the LRRS statistics also share the same asymptotic properties with the classical aligned rank tests based on linear rank statistics.

The fixed-effects model can also be extended to an ANOCOVA model:

$$Y_i = \beta' \mathbf{c}_i + \gamma' \mathbf{z}_i + e_i, \quad i = 1, \ldots, n, \tag{11}$$

where $\mathbf{Z}_i$ are concomitant variates, assumed to be independent identically distributed random variables. It is possible that the linearity of fixed-effects component holds, but the linearity of regression on concomitant variates may be doubtful. For this reason, Sen (1996) considered a partially linear model wherein the regression on the covariates is allowed to be nonparametric but the linearity of the fixed-effect part is assumed to be true. The RRS estimation theory has been extended to such mixed models, and the relative advantages over the fixed-effects model analysis are appraised.

10.3 RANK VERSUS OTHER ROBUST PROCEDURES

Throughout the text emphasis has been laid on rank tests because of their basic properties of invariance (under suitable groups of transformations that map the sample space onto itself), (often exact) distribution-freeness, and of course, good local as well as asymptotic properties that include robustness in a global sense. Apart from some of the situations (like the bivariate independence problem), all the rank tests and related estimates presented in this monograph relate to the so-called semiparametric (location/scale/regression) models wherein the basic normality or other parametric forms of the underlying densities have been dispensed with, and a nonparametric formulation has been advocated. The past thirty years have witnessed a phenomenal growth of research literature on semiparametric statistical inference procedures that include rank procedures as well as notable members. On the other hand, in a completely nonparametric setup too, statistical procedures based on the Hoeffding (1948a) U-statistics, von Mises (1947) functionals, and more generally, statistical functionals that are differentiable in a broader sense have cropped up in various contexts. A detailed treatment of U-statistics and its various generalizations, with due emphasis on their role in nonparametrics, can be found in Sen (1981) or Serfling (1980). Such statistics have certain desirable and (nonparametric) optimality properties. However, they need not be very robust for semiparametric (or parametric) models. For example, if we are interested in estimating $\theta(F) = \sigma^2(F)$, the variance of a distribution (F that is arbitrary but assumed to have a finite second moment), an optimal nonparametric estimate is the sample variance (that is a U-statistic):

$$S_n^2 = (n-1)^{-1} \sum_{i=1}^{n} (X_i - \bar{X}_n)^2, \tag{1}$$

where $X_1, \ldots, X_n$ stand for the sample observations and the sample mean $\bar{X}_n = n^{-1} \sum_{i=1}^{n} X_i$ is also another U-statistic that is a nonparametric optimal estimate of the mean of the distribution function F. If we look into these estimates from a parametric point of view where, for example, we let F be a normal distribution, $\bar{X}_n$ and S_n^2 are optimal estimates, but are known to be non-robust against any possible departure of F from normality; the extent of the non-robustness is even likely to be more for the sample variance. For this reason, a parametric test, such as the Student t-test, based on these U-statistics, may not be that robust against plausible departures from the assumed model. On the other hand, the Wilcoxon signed-rank test (for location, one-sample model) has been shown to be based on a U-statistic, and it is globally robust in the sense that the underlying distribution function F is allowed to be arbitrary (but continuous and symmetric); nevertheless, when F is normal, the Wilcoxon test is not optimal, even asymptotically, though it is about 95% (asymptotically) efficient for normal F. Such robustness considerations are more pertinent in the context of (point as well as interval) estimation of location, scale or regression parameters, and we would incorporate the R-estimates, introduced in Chapter 9, in this setup as well.

Robustness of statistical inference procedures, particularly estimates and test statistics, has been an area of intense research activity for the past thirty years. Huber (1964) introduced robust versions of maximum likelihood estimates of location, termed M-estimates, and provided a theoretical foundation of their basic properties. Basically, he had in mind the effect of a small or local departure from the model-based regularity assumptions on the performance characteristics of estimates, and he showed that in this respect his proposed M-estimates perform better; this has been termed an infinitesimal departure model, and the gross-error or outlier models pertain to this scheme. M-estimates have also been extended to the regression models as well as to other models that are more complex than the simple location model. From a local robustness perspective, such M-estimates, or tests based on such M-statistics, perform quite well. The recent monographs by Rieder (1994), and Jurečková and Sen (1996) are good sources to fathom such details. In this context, it is worth mentioning the following aspects of M-statistics based inference procedures, as contrasted with rank tests and R-estimates.

(1) *Scale-equivariance and EDF property.* M-estimates and M-statistics, in general, are not scale-equivariant (whereas rank statistics and R-estimates are). Therefore, even if we contemplate a local (contaminated) model:

$$F(x) = (1-\epsilon)F_0(x) + \epsilon H(x), \quad -\infty < x < \infty,\ \epsilon > 0, \tag{2}$$

where $H(x)$ has a much heavier tail than F_0, and even if we assume that both F_0 and H have a common location parameter, say θ, the formulation of an M-estimate of θ deems the knowledge of the scale parameter of F_0; similarly, the distribution of an M-statistic may depend generally on the scale parameter associated with F_0. In some cases, one advocates the use of studentization or some studentized M-statistics to eliminate this problem due to an unknown scale parameter. But such a process of studentization may make small sample prospects rather bleak and may also affect the robustness picture to a certain extent. On the other hand, rank tests for θ are scale-invariant (as the ranks or signed-ranks are unaffected by any scalar mutiplication of the observations), and R-estimates of location are thereby scale-equivariant too. Therefore, whereas a rank test for location is EDF, an M-test is generally only ADF. For the confidence interval problem, (for θ) there is greater concern. A confidence interval based on a rank statistic (and alignment principle) is distribution-free in the sense that it does not depend on the form of the distribution function F, whereas one needs to estimate the asymptotic variance of an M-statistic and incorporate its asymptotic normality to formulate such a confidence set (that may only be ADF). Therefore, in judging the robustness aspects of estimates, it is important to pay due attention to the scale-equivariance property, and rank procedures fair better in this respect.

Let us compare the pictures in the case of a simple linear model, such as the one in Subsection 10.1.2. In this case, for F other than a normal distribution function, there may not be a closed algebraic expression for the maximum likelihood estimates, M-estimates or R-estimates of the regression parameter (vector) β; all of them are to be obtained by iterative methods, and hence there might not be any strong computational incentive in favour of one over the others. Therefore, the main point of discrimination would be their scale-equivariance and distribution-freeness properties. In this respect, the picture is the same as in the location model, and again that puts rank procedures in a better global robustness perspective. There is, however, a point of discrimination between the classical aligned rank tests and tests based on regression rank scores (as presented in Section 10.2). As discussed in Subsection 10.1.2., an aligned rank test rests on the (rank-)estimation of the nuisance parameters in the formulation of the aligned observations. Since such estimates may not have a closed algebraic expression, and therefore may need to be obtained by an (iterative) trial and error solution, from a computational point of view, they are not very attractive. On the other hand, by virtue of their regression-invariance property, the regression rank scores do not explicitly involve the estimation of the nuisance parameters, and hence, might be computationally less laborious than the aligned rank procedures based on equivalent score generating functions. However, it might also have some bearing on the robustness aspects of the two asymptotically equivalent procedures.

(2) *Influence curves.* The basic motivation for incorporating M-stat-

istics in the process of drawing statistical conclusions stems from the desire to deemphasize the impact of outliers, or gross-errors, and this is largely achieved through the use of suitable influence curves that are bounded; the very formulation of the Huber (1964) M-estimate corresponds to an influence curve that agrees with the normal score in the central part, and is flat beyond this range. Hampel et al. (1986) have provided a detailed account of the role of influence curves in robust statistical inference for linear and location models; they have advocated the use of bounded influence curves and shown that such procedures have reasonable efficiency properties and they are robust too. On the contrary, for rank statistics, the influence curves need not be bounded. For example, if we use the normal scores linear or signed linear rank statistics for the regression or location models, the corresponding influence curves are not bounded. If we look into the LMPR test statistics for the various hypotheses, we will observe that whenever the score function $\varphi(u, f) = -f'(F^{-1}(u))/f(F^{-1}(u))$, $0 < u < 1$, is not bounded almost surely, these rank statistics would not have bounded influence functions. For Laplace or logistic densities, the score function is bounded, but for many others, including the normal, this is not the case. For Wilcoxon scores statistics, the influence curves are of course bounded. Thus, consideration of bounded influence curves will exclude many important rank tests and estimates that are globally robust and efficient too. In view of the basic difference between local and global robustness properties for M- and R-statistics, as explained before, the role of influence curves may not be isomorphic in these two cases. Nor should we be specially concerned about bounded influence curves in a global robustness perspective, as is the case with rank procedures.

(3) *L-functionals.* The use of best linear unbiased estimates (BLUE) of location and scale parameters (based on the sample order statistics) for drawing statistical conclusions goes back to Lloyd (1952), and a vast literature pertains to this domain; we refer to David (1980) for an excellent introduction to the subject matter. Basically, the BLUE's are of parametric flavour, as the optimal weights depend on the underlying density. Moreover, exact computations involved in this scheme can be very cumbersome, specially when the sample size is not small; for small sample sizes extensive tabulations have been made for these coefficients. For large samples, (asymptotically) ABLUE's have also been considered, and their scale-equivariance, robustness and other properties (including asymptotic normality) have been thoroughly studied. In the course of these studies, the passage from parametrics to nonparametrics has been fortified, and as of now, L-functionals or L-statistics are quite popularly incorporated in nonparametric statistical inference. Like the rank procedures (but unlike the M-procedures), these L-statistics are usable in various censored cases, and they are scale-equivariant too. However, unlike the rank tests, a test based on L-estimates may not be genuinely distribution-free, even for test-

ing the most simple hypotheses of invariance. There are certain asymptotic equivalence results on R- and L-statistics, eploited in detail by Jurečková and Sen (1996), that put these L-statistics in a quite favourable stand in nonparametrics as well. But for general linear models, L-statistics may have generally some less desirable properties, and looking at the EDF perspectives, we may conclude that rank procedures have greater appeal for the finite sample size models.

(4) *Generalized linear models.* There has been a spurt of activities in characterizing some models that are essentially semiparametric ones as transformed linear models or the so-called generalized linear models (GLM's); we refer to McCullagh and Nelder (1989) for an excellent treatise of this subject matter. Although the GLM's have a distinct parametric flavour, they are the precursors of the so-called semiparametric models, and include the celebrated Cox (1972) proportional hazards model as a bona fide member of this class. For this last model, a partial likelihood principle leads to a simple test statistic that is a direct extension of the well-known log-rank statistic. In that way, rank tests are relevant to GLM and semiparametric models as well. However, in general, in GLM's the associated estimating equations generally lead to more complex estimates that are not reducible to R-estimates, generally may not have closed algebraic expressions, and may not be very robust even in a local sense. Hence, the success of R-estimates in semiparametric models may have to be judged on a case by case basis.

(5) *Non-linear models.* Rank tests and derived estimates remain pertinent to such models as long as the basic invariance structure can be retained for such models. The rank tests are suitable when the null hypothesis is simple in the sense that under such a hypothesis, ranks are maximal invariants in a well-defined sense. This was the case with randomized block designs where it was shown that the intra-block ranking works out well even when the block and/or treatment effects are not necessarily additive, but the null hypothesis pertains to a stochastic interchangeability of the within block observations. This feature may not be generally true for the estimation problem, as there the parameters may be functional ones, and hence would require a somewhat different formulation. This is the case with nonparametric regression where the regression function of one variable on the other(s) may not be linear or even non-linear of a specific nature. In such a case, one generally assumes a smooth regression function, and incorporates either a kernel method or a nearest neighbour method of estimating the function using appropriate conditional functionals. Rank procedures, though relevant in this setup too, may need to be appraised properly for their suitability from robustness as well as efficiency perspectives.

Based on all the remarks made above, we are in a comfortable situation to advocate the general use of rank tests and derived estimates in a variety of models of general statistical interest. They compare very favourably with alternative ones that are common in nonparametrics and semiparametrics.

PROBLEMS AND COMPLEMENTS TO CHAPTER 10

Section 10.1.

1. Define the aligned observation vectors $\hat{\boldsymbol{X}}_i$, $i = 1, \ldots, n$, as in Subsection 10.1.1, and denote their joint distributions under H_0 and K_n by P_n and Q_n respectively. Use LeCam's lemmas to verify that $\{Q_n\}$ is contiguous to $\{P_n\}$.

2. Use the contiguity result in the preceding problem along with LeCam's third lemma, and extend the joint asymptotic normality of the aligned rank statistics to contiguous alternatives in K_n.

3. Derive the convergence in formula (10.1.1.14).

4. Show that $\mathcal{L}_N$, defined by (10.1.1.9), has asymptotically, under K_n, a non-central χ^2 distribution with $p-1$ degrees of freedom and non-centrality parameter (10.1.1.17).

5. Prove that the $(p-1)$-multiple of classical ANOVA (variance-ratio) test statistic has asymptotically, under K_n, a non-central χ^2 distribution with $p-1$ degrees of freedom and non-centrality parameter (10.1.1.18).

6. Prove the inequality (10.1.1.20), and that the equality sign holds in it only when $\varphi(F(x)) \equiv x$, excepting on a set of null measure.

7. Prove the asymptotic linearity result (10.1.2.9).

8. Show that $\hat{\beta}_{2n}$ is asymptotically normal, as stated in (10.1.2.11).

9. Show that the representation (10.1.2.14) holds, and that it implies the asymptotic normality in (10.1.2.11).

10. Prove the asymptotic result (10.1.2.15).

11. Prove the asymptotic result in relation (10.1.2.16).

12. Using the projection method, show (10.1.2.17).

13. Consider the setup of Subsection 10.1.2, and testing $H_0 : \beta_1 = \mathbf{0}$ against $H_1 : \beta_1 \neq \mathbf{0}$ by means of the statistic $\hat{\mathcal{L}}_{n1}$ given by (10.1.2.21). If H_0 does not hold, i.e. $\beta_1 \neq \mathbf{0}$, show that $n^{-1}\hat{\mathcal{L}}_n$ converges in probability to a positive constant, so that $\hat{\mathcal{L}}_{n1}$ is $O_p(n)$.

14. Verify that $\{q_n\}$, the joint density of $\{Y_i;\, 1 \leq i \leq n\}$ under contiguous alternatives $\{H_n\}$ given by (10.1.2.22), is contiguous to $\{p_n\}$, the joint density under H_0.

15. Show that, under $\{H_n\}$, $n^{-1/2}\hat{\mathbf{L}}_{n1}$ is asymptotically normal with parameters given before and in (10.1.2.24).

16. Prove that, under $\{H_n\}$, the statistic $\hat{\mathcal{L}}_{n1}$ has asymptotically the non-central χ^2 distribution as given in (10.1.2.25) and (10.1.2.26).

17. Having still the setup of Subsection 10.1.2, prove that the asymptotically optimal aligned rank test uses the score function φ_0 given by (10.1.2.27).

18. (Continuation) Verify the equality (10.1.2.29), i.e. $\varphi_0(u) \equiv \varphi(u, f)$, $0 \leq u \leq 1$.

Bibliography

Abbreviations:

AMS = *Annals of Mathematical Statistics*
AS = *Annals of Statistics*
JASA = *Journal of the American Statistical Association*
JRSS B = *Journal of the Royal Statistical Society, Series B*

Adichie, J.N. (1967). Estimates of regression parameters based on rank tests. *AMS* **38**, 894–904.

Adichie, J. N. (1978). Rank tests of sub-hypotheses in the general linear regression. *AS* **6**, 1012–1026.

Albers, W. (1974). *Asymptotic Expansions and the Deficiency Concept in Statistics.* Mathematical Centre Tracts **58**, Math. Centrum, Amsterdam.

Albers, W. (1975). Efficiency and deficiency considerations in the symmetry problem. *Stat. Neerlandica* **29**, 81–92.

Albers, W. (1978). A note on the Edgeworth expansion for the Kendall rank correlation coefficient. *AS* **6**, 923–925.

Albers, W. (1979). Asymptotic deficiencies of one-sample rank tests under restricted adaptation. *AS* **7**, 944–954.

Albers, W., Bickel, P.J. and van Zwet, W.R. (1976). Asymptotic expansions for the power of distribution free tests in the one-sample problem. *AS* **4**, 108–156.

Anděl, J. (1967). Local asymptotic power and efficiency of tests of Kolmogorov-Smirnov type. *AMS* **38**, 1705–1725.

Andersen, P. K., Borgan, O., Gill, R.D. and Keiding, N. (1993). *Statistical Models Based on Counting Processes.* Springer-Verlag, New York.

Anderson, T. W. (1960). A modification of the sequential probability ratio test to reduce the sample size. *AMS* **31**, 165–197.

Andrews, F.C. (1954). Asymptotic behaviour of some rank tests for analysis of variance. *AMS* **25**, 724–736.

Ansari, A.R. and Bradley, R.A. (1960). Rank-sum tests for dispersions. *AMS* **31**, 1174–1189.

Antille, A. (1974). A linearized version of the Hodges-Lehmann estimator. *AS* **2**, 1308–1313.

Bahadur, R.R. (1960). Stochastic comparison of tests. *AMS* **31**, 276–295.

Bahadur, R.R. (1967). Rates of convergence of estimates and test statistics. *AMS* **38**, 303–324.

Bahadur, R.R. (1971). *Some Limit Theorems in Statistics.* SIAM, Philadelphia.

Bartlett, N.S. and Govindarajulu, Z. (1968). Some distribution-free statistics and their application to the selection problem. *Ann. Inst. Statist. Math.* **20**, 79–97.

Bauer, D.F. (1973). On some nonparametric estimates for regression parameters. *Commun. Statist.* **2**, 225–234.

Behnen, K. (1971). Asymptotic optimality and ARE of certain rank-order tests under contiguity. *AMS* **42**, 325–329.

Behnen, K. (1983). Adaptive rank tests. *Acta Univ. Carolinae – Math. Phys.* **24**, 13–22.

Behnen, K. (1984). Adaptation of rank statistics by rank estimation of scores. *Proc. 3rd Prague Symp. Asympt. Statist.*, 1983. Ed. P. Mandl, M. Hušková. Elsevier Sci. Publ., Amsterdam 31–44.

Behnen, K. and Hušková, M. (1984). A simple algorithm for the adaptation of scores and power behavior of the corresponding rank test. *Commun. in Statist. – Theory Meth.* **13**, 305–325.

Behnen, K. and Neuhaus, G. (1975). A central limit theorem under contiguous alternatives. *AS* **3**, 1349–1353.

Behnen, K. and Neuhaus, G. (1983). Galton's test as a linear rank test with estimated scores and its local asymptotic efficiency. *AS* **11**, 588–599.

Behnen, K. and Neuhaus, G. (1989). *Rank Tests with Estimated Scores and Their Application.* Teubner, Stuttgart.

Behnen, K., Neuhaus, G. and Ruymgaart, F. (1983). Two-sample rank estimators of optimal nonparametric score-functions and corresponding adaptive rank statistics. *AS* **11**, 1175–1189.

Bennett, B. M. (1962). On multivariate sign tests. *JRSS B* **24**, 159–161.

Bennett, B. M. (1964). A bivariate signed rank test. *JRSS B* **26**, 457–461.

Beran, R. (1974). Asymptotically efficient adaptive rank estimates in location models. *AS* **2**, 63–74.

Beran, R. (1978). On efficient and robust adaptive estimator of location. *AS* **6**, 292–313.

Bickel, P.J. (1974). Edgeworth expansions in nonparametric statistics. *AS* **2**, 1–20.

Bickel, P.J., Klaassen, C.A.J., Ritov, Y. and Wellner, J.A. (1993). *Efficient and Adaptive Estimation For Semiparametric Models.* Johns Hopkins Univ. Press, Baltimore.

Bickel, P.J. and van Zwet, W.R. (1978). Asymptotic expansions for the power of distribution free tests in the two-sample problem. *AS* **6**, 937–1004.

Billingsley, P. (1968). *Convergence of Probability Measures.* Wiley, New York.

Birnbaum, Z.W. and Hall, R.A. (1960). Small sample distribution for multi-sample statistics of the Smirnov type, *AMS* **31**, 710–720.

Birnbaum, Z.W. and Klose, O.M. (1957). Bounds for the variance of the Mann-Whitney statistic. *AMS* **28**, 933–945.

Blackman, J. (1956). An extension of the Kolmogorov distribution. *AMS* **27**, 513–520. Correction: *AMS* **29** (1958), 318–324.

Blomqvist, N. (1950). On a measure of dependence between two random variables. *AMS* **21**, 593–600.

Blum, J.R., Kiefer, J. and Rosenblatt, M. (1961). Distribution free tests of independence based on the sample distribution function. *AMS* **32**, 485–498.

Blyth, C.R. (1958). Note on relative efficiency of tests. *AMS* **29**, 898–903.

Borovkov, A.A. (1962). On the problem of two samples. (Russian.) *Izvestia AN SSSR* **26**, 605–624.

Boulanger, A. (1983). Estimators of shift based on statistics of the Kolmogorov-Smirnov type. *Canad. J. Statist.* **11**, 271–284.

Boulanger, A. and van Eeden, C. (1983). Estimators of location based on Kolmogorov-Smirnov type statistics. *Canad. J. Statist.* **11**, 293–303.

Brown, G.W. and Mood. A.M. (1950). On median tests for linear hypotheses. *Proc. 2nd Berkeley Symposium*, Univ. Calif. Press, Los Angeles, Calif. 159–166.

Bukač, J. (1975). Critical values of the sign test, Algorithm AS 85. *Applied Statistics (JRSS Series C)*, **24**, 265–267.

Bukač, J. (1995). Formulas for critical values of Wilcoxon paired test and for Kendall τ. (Czech.) *Inform. Bull. Czech Statist. Soc.* **6**, No. 4 (December 1995), 32–35.

Burr, E.J. (1963). Distribution of the two-sample Cramér-von Mises criterion for small equal samples. *AMS* **34**, 95–101.

Büringer, H., Martin, H. and Schriever, K.H. (1980). *Nonparametric Sequential Selection Procedures.* Birkhäuser Verlag, Boston.

Butler, C. C. (1969). A test for symmetry using the sample distribution function. *AMS* **40**, 2209–2210.

Capon, J. (1961). Asymptotic efficiency of certain locally most powerful rank tests. *AMS* **32**, 88–100.

Carvalho, P.E.O. (1959). On the distribution of the Kolmogorov-Smirnov D-statistic. *AMS* **30**, 173–176.

Chakravarti, I.M., Leone, F.C. and Alanen, J.D. (1962). Asymptotic relative efficiency of Mood's and Massey's two sample tests against some parametric alternatives. *AMS* **33**, 1375–1383.

Chang, L.C. and Fisz, M. (1957). Exact distributions of the maximal values of some functions of empirical distribution functions. *Science Record, New Ser.* **1**, 341–346.

Chatterjee, S. K. (1966a). A bivariate sign-test for location. *AMS* **37**, 1771–1782.

Chatterjee, S. K. (1966b). A multisample nonparametric scale test based on U-statistics. *Calcutta Statist. Assoc. Bull.* **15**, 109–119.

Chatterjee, S. K. and Sen, P.K. (1964). Nonparametric tests for the bivariate two-sample location problem. *Calcutta Statist. Assoc. Bull.* **13**, 18–58.

Chatterjee, S. K. and Sen, P.K. (1965). Some nonparametric tests for the bivariate two-sample association problem. *Calcutta Statist. Assoc. Bull.* **14**, 14–34.

Chatterjee, S. K. and Sen, P.K. (1966). Nonparametric tests for the multivariate multisample location problem. *Essays in Probability and Statistics in Memory of S. N. Roy* (Ed. R. C. Bose et al.), Univ. of N. Carolina Press, Chapel Hill, pp. 197–228.

Chatterjee, S. K. and Sen, P.K. (1973a). Nonparametric testing under progressive censoring. *Calcutta Statist. Assoc. Bull.* **22**, 13–50.

Chatterjee, S. K. and Sen, P.K. (1973b). On Kolmogorov-Smirnov type tests for symmetry. *Ann. Inst. Statist. Math.* **25**, 288–300.

Chaudhuri, P. (1992). Generalized regression quantiles: Forming a useful toolkit for robust linear regression. *L-1-Statistical Analysis and Related Methods.* (ed. Y. Dodge), North Holland, Amsterdam, pp. 169–185.

Chernoff, H. and Savage, I.R. (1958). Asymptotic normality and efficiency of certain non-parametric test statistics. *AMS* **29**, 972–994.

Ciesielska, L. and Ledwina, T. (1983). On locally most powerful rank tests of independence. *Zastosowania Matematyki* **18**, 61–66.

Collatz, L. (1964). *Funktionalanalysis und numerische Mathematik.* Springer-Verlag, Heidelberg.

Cox, D.R. (1972). Regression models and life-tables. *JRSS B* **34**, 187–220.

Cox, D.R. (1975). Partial likelihood. *Biometrika* **62**, 269–276.

Cramér, H. (1928). On the composition of elementary errors. *Skand. Aktuarietid.* **11**, 13–74, 141–180.

Cramér, H. (1945). *Mathematical Methods of Statistics.* Almqvist & Wiksells, Uppsala. Princeton 1946.

Csáki, E. (1959). On two modifications of the Wilcoxon statistic (Hungarian). *Mag. Tud. Akad. Mat. Kutató Int. Közleményei* **4**, 313–319.

Csörgö, M. and Révész, P. (1981). *Strong Approximations in Probability and Statistics.* Akadémiai Kiadó, Budapest.

van Dantzig, D. (1951). On the consistency and the power of Wilcoxon's two sample test. *Indagationes math.* **13** (*Proc. Kon. Nederl. Akad. Wet.* **54**), 1–8.

Darling, D.A. (1960). Sur les théorèmes de Kolmogorov-Smirnov. *Teoriya veroyatnostey* **5**, 393–398.

David, H.A. (1980). *Order Statistics*, 2nd ed. Wiley, New York.

David, H.T. (1958). A three-sample Kolmogorov-Smirnov test. *AMS* **29**, 842–851.

David, S.T., Kendall, M.G. and Stuart, A. (1951). Some questions of distribution in the theory of rank correlation. *Biometrika* **38**, 131–140.

Dixon, W.J. (1953). Power functions of the sign test and power efficiency for normal alternatives. *AMS* **24**, 467–473.

Dixon, W.J. (1954). Power under normality of several nonparametric tests. *AMS* **25**, 610–614.

Dixon, W.J. and Mood, A.M. (1946). The statistical sign test. *JASA* **41**, 557–566.

Does, R.J.M.M. (1983). An Edgeworth expansion for simple linear rank statistics under the null hypothesis. *AS* **11**, 607–624.

Does, R.J.M.M. (1984). Asymptotic expansions for simple linear rank statistics under contiguous alternatives. *Proc. 3rd Prague Symp. Asympt. Statistics* 1983. Ed. P. Mandl, M. Hušková. Elsevier Sci. Publ., Amsterdam, 221–230.

Doob, J.L. (1949). Heuristic approach to the Kolmogorov-Smirnov theorems. *AMS* **20**, 393–403.

Doob, J.L. (1953). *Stochastic Processes.* J. Wiley, New York.

Doornbos, R. and Prins, H.J. (1958). On slippage tests. I, II, III. *Indagationes math.* **20** (*Proc. Kon. Nederl. Akad. Wet.* **61**), 38–46, 47–55, 438–447.

Drion, E.F. (1952). Some distribution-free tests for the difference between two empirical cumulative distribution functions. *AMS* **23**, 563–574.

Dunn, O.J. (1964). Multiple comparisons using rank sums. *Technometrics* **6**, 241–252.

Dupač, V. and Hájek, J. (1969). Asymptotic normality of simple linear rank statistics under alternatives II. *AMS* **40**, 1992–2017.

Dwass, M. (1956). The large-sample power of rank order tests in the two-sample problem. *AMS* **27**, 352–374.

Dwass, M. (1957). On the distribution of ranks and of certain rank order statistics. *AMS* **28**, 424–431.

Dwass, M. (1960). Some k sample rank-order tests. *Contr. to Prob. and Statistics, Essays in Honor of H. Hotelling.* Stanford Univ. Press, Stanford, Calif., 198–202.

van Eeden, C. (1963). The relation between Pitman's asymptotic relative efficiency of two tests and the correlation coefficient between their test statistics. *AMS* **34**, 1442–1451.

van Eeden, C. (1972). An analogue, for signed rank statistics, of Jurečková's asymptotic linearity theorem for rank statistics. *AMS* **43**, 791–802.

Elandt, R. (1957). A non-parametric test of tendency. *Bull. Acad. Polon. Sci. cl. V, ser. sci. biol.* **5**, 187–190.

Elandt, R. (1958). On certain interaction tests in serial experiments. The problem of stratification. (Polish.) *Zastosowania matematyki* **3**, 8–45.

Elandt, R. (1962). Exact and approximate power function of the non-parametric test of tendency. *AMS* **33**, 471–481.

Farlie, D.J.G. (1961). The asymptotic efficiency of Daniels's generalized correlation coefficients. *JRSS B* **23**, 128–142.

Feller, W. (1950). *An Introduction to Probability Theory and its Applications.* Wiley, New York.

Fellingham, S.A. and Stoker, D.J. (1964). An approximation for the exact distribution of the Wilcoxon test for symmetry. *JASA* **59**, 899–905.

Fieller, E.C., Hartley, H.O. and Pearson, E.S. (1957). Tests for rank correlation coefficients I. *Biometrika* **44**, 470–481.

Fieller, E.C. and Pearson, E.S. (1961). Tests for rank correlation coefficients II. *Biometrika* **48**, 29–40.

Fisher, R.A. and Yates, F. (1938). *Statistical Tables for Biological, Agricultural and Medical Research.* Oliver & Boyd, Edinburgh-London. (1st edition 1938, 5th edition 1957.)

Fisz, M. (1957). A limit theorem for empirical distribution functions. *Bull. Acad. Polon. Sci. Cl. III*, vol. **5**, 695–698.

Fisz, M. (1960). Some non-parametric tests for the k-sample problem. *Colloquium math.* **7**, 289–296.

Fix, E. and Hodges, J.L. Jr. (1955). Significance probabilities of the Wilcoxon test. *AMS* **26**, 301–312.

Fraser, D.A.S. (1957a). *Non-parametric Methods in Statistics.* J. Wiley, New York.

Fraser, D.A.S. (1957b). Most powerful rank-type tests. *AMS* **28**, 1040–1043.

Friedman, M. (1937). The use of ranks to avoid the assumption of normality implicit in the analysis of variance. *JASA* **32**, 675–701.

Friedman, M. (1940). A comparison of alternative tests of significance for the problem of m rankings. *AMS* **11**, 86–92.

Gass, S.I. (1985). *Linear Programming.* Encyclopedia of Statistical Sciences, **5**, Wiley, New York, 35–51.

Ghosh, M. (1973a). On a class of asymptotically optimal nonparametric tests for grouped data I. *Ann. Inst. Statist. Math.* **25**, 91–108.

Ghosh, M. (1973b). On a class of asymptotically optimal nonparametric tests for grouped data II. *Ann. Inst. Statist. Math.* **25**, 109–122.

Ghosh, M. and Sen, P.K. (1971). On a class of rank order tests for regression with partially informed stochastic predictors. *AMS* **42**, 650–661.

Ghosh, M. and Sen, P.K. (1972). On bounded length confidence interval for the regression coefficient based on a class of rank statistics. *Sankhya Ser. A* **34**, 33–52.

Gibbons, J.D. (1964a). On the power of two-sample rank tests on the equality of two distribution functions. *JRSS B* **26**, 293–304.

Gibbons, J.D. (1964b). A proposed two-sample rank test: the psi test and its properties. *JRSS B* **26**, 305–312.

Gibbons, J.D., Olkin, I. and Sobel, M. (1977). *Selecting and Ordering Populations: A New Statistical Methodology.* Wiley, New York.

Gihman, I.I. (1957). On a certain non-parametric test of homogeneity of k-samples. (Russian.) *Teoriya veroyatnostey* **2**, 380–384.

Gihman, I.I., Gnedenko, B.V. and Smirnov, N.V. (1956). Non-parametric methods in statistics. (Russian.) *Trudy 3-go vsesoyuz. mat. syezda*, vol. **3**, 320–334. AN SSSR 1958.

Glasser, G.J. and Winter, R.F. (1961). Critical values of the coefficient of rank correlation for testing the hypothesis of independence. *Biometrika* **48**, 444–448.

Gnedenko, B.V. (1954). Tests of homogeneity of probability distributions in two independent samples. (Russian.) *Math. Nachrichten* **12**, 29–66.

Gnedenko, B.V. and Korolyuk, V.S. (1951). On the maximal deviation of two empirical distributions. (Russian.) *Doklady AN SSSR* **80**, 525–528.

Govindarajulu, Z., LeCam, L. and Raghavachari, M. (1967). Generalizations of theorems of Chernoff and Savage on the asymptotic normality of test statistics. Proc. 5th Berkeley Symp. 1, Univ. Calif. Press, Los Angeles, Calif., pp. 609–638.

Gupta, S.S. and Huang, D.-Y. (1981). *Multiple Decision Theory: Recent Developments.* Springer, Berlin.

Gupta, S.S. and Panchapakesan, S. (1979). *Multiple Decision Procedures: Theory and Methodology of Selection and Ranking Populations.* Wiley, New York.

Gutenbrunner, C. (1986). Zur Asymptotic von Regression Quantile Prozessen und daraus abgeleiten Statistiken. Doctoral Dissertation, Universität Freiburg, Germany.

Gutenbrunner, C. and Jurečková, J. (1992). Regression rank scores and regression quantiles. *AS* **20**, 305–330.

Gutenbrunner, C., Jurečková, J., Koenker, R. and Portnoy, S. (1993). Tests of linear hypotheses based on regression rank scores. *J. Nonparam. Statist.* **2**, 307–331.

Haga, T. (1959/60). A two-sample rank test on location. *Ann. Inst. Stat. Math.* **11**, 211–219.

Hájek, J. (1955). Some rank distributions and their use. (Czech.) *Časopis pro pěstování matematiky* **80**, 17–31.

Hájek, J. (1956). Asymptotic efficiency of a certain sequence of tests. (Russian.) *Czech. Math. J.* **6(81)**, 26–29.

Hájek, J. (1960a). Limiting distributions in simple random sampling from a finite population. *Publ. Math. Inst. Hung. Acad. Sci.* vol. **5**, ser. A, fasc. 3, 361–374.

Hájek, J. (1960b). On a simple linear model in Gaussian processes. *Trans. 2nd Prague Conf. Inf. Theory, Stat. Dec. Functions, Random Proc.* Publishing House of the Czechoslowak Academy of Sciencies, Prague, 185–197.

Hájek, J. (1961). Some extensions of the Wald-Wolfowitz-Noether theorem. *AMS* **32**, 506–523.

Hájek, J. (1962). Asymptotically most powerful rank-order tests. *AMS* **33**, 1124–1147.

Hájek, J. (1965). Extension of the Kolmogorov-Smirnov test to regression alternatives. *Proc. Bernoulli-Bayes-Laplace Seminar*, Berkeley, 1963. Ed. LeCam, Univ. of California Press, pp. 45–60.

Hájek, J. (1968). Asymptotic normality of simple linear rank statistics under alternatives. *AMS* **39**, 325–346.

Hájek, J. (1970). Miscellaneous problems of rank test theory. In: *Nonparametric Techniques in Statistical Inference.* Ed. M.L. Puri, Cambridge Univ. Press, pp. 3–17.

Hájek, J. (1974). Asymptotic sufficiency of the vector of ranks in the Bahadur sense. *AS* **2**, 75–83.

Hájek, J. and Šidák, Z. (1967). *Theory of Rank Tests.* First Edition. Academia, Prague & Academic Press, New York.

Hampel, F.R., Ronchetti, E., Rousseeuw, P. and Stahel, W. (1986). *Robust Statistics - The Approach Based on Influence Functions.* Wiley, New York.

Hart, J.F. (1968). *Computer Approximations.* Wiley, New York.

Heiler, S. and Willers, R. (1988). Asymptotic normality of R-estimates in the linear model. *Statistics* **19**, 173–184.

Hemelrijk, J. (1952). A theorem on the sign test when ties are present. *Indagationes math.* **14** (Proc. Kon. Nederl. Akad. Wet. **55**), 322–326.

Hemelrijk, J. (1960). Experimental comparison of Student's and Wilcoxon's two-sample tests. *Symp. Quant. Methods in Pharmacology*, Leyden, pp. 118–134. North-Holland, Amsterdam; Interscience, New York 1961.

Hoadley, A.B. (1965). The theory of large deviations with statistical applications. Ph.D. Dissertation, Univ. of California at Berkeley.

Hodges, J.L. Jr. and Lehmann, E.L. (1956). The efficiency of some nonparametric competitors of the t-test. *AMS* **27**, 324–335.

Hodges, J.L. Jr. and Lehmann, E.L. (1962). Rank methods for combination of independent experiments in analysis of variance. *AMS* **33**, 482–497.

Hodges, J.L. Jr. and Lehmann, E.L. (1963). Estimates of location based on rank tests. *AMS* **34**, 598–611.

Hodges, J.L. Jr. and Lehmann, E.L. (1970). Deficiency. *AMS* **41**, 783–801.

Hoeffding, W. (1948a). A class of statistics with asymptotically normal distribution. *AMS* **19**, 293–325.

Hoeffding, W. (1948b). A non-parametric test of independence. *AMS* **19**, 546–557.

Hoeffding, W. (1950). "Optimum" non-parametric tests. *Proc. 2nd Berkeley Symposium.* Univ. Calif. Press, Los Angeles, Calif., 83–92.

Hoeffding, W. (1973). On the centering of a simple linear rank statistic. *AS* **1**, 54–66.

Hoeffding, W. and Rosenblatt, J.R. (1955). The efficiency of tests. *AMS* **26**, 52–63.

Hogg, R.V. (1974). Adaptive robust procedures: A partial review and some suggestions for future applications and theory. *JASA* **69**, 909–923.

Hogg, R.V., Fisher, D.M. and Randles, R.H. (1975). A two-sample adaptive distribution-free test. *JASA* **70**, 656–661.

Hogg, R.V. and Lenth, R.V. (1984). A review of some adaptive statistical techniques. *Commun. in Statist. – Theory Meth.* **13**, 1551–1579.

Huber, P. J. (1964). Robust estimation of a location parameter. *AMS* **35**, 73–101.

Huber, P. J. (1981). *Robust Statistics.* Wiley, New York.

Hušková, M. (1969/70). Asymptotic distribution of simple linear rank statistics for testing symmetry. *Zeit. Wahrsch. verw. Geb.* **14**, 308–322.

Hušková, M. (1983). Adaptive procedures. *Acta Univ. Carolinae – Math. Phys.* **24**, 41–48.

Hušková, M. (1984). Adaptive methods. In: P.R. Krishnaiah, P.K. Sen, eds., *Handbook of Statistics*, vol. 4. Elsevier Science Publishers, Amsterdam, pp. 347–358.

Hušková, M., Jurečková, J., Behnen, K. and Neuhaus, G. (1984). Two-sample adaptive linear rank tests and their Bahadur efficiencies. *Proc. 3rd Prague Symp. Asympt. Statist.*, 1983. Ed. P. Mandl, M. Hušková. Elsevier Sci. Publ., Amsterdam, 103–117.

Hušková, M. and Sen, P.K. (1985). On sequentially adaptive asymptotically efficient rank statistics. *Sequen. Anal.* **4**, 225–251.

Hušková, M. and Sen, P.K. (1986). On sequentially adaptive signed rank statistics. *Sequen. Analysis* **5**, 237–251.

Hüsler, J. (1987). On the two-sample adaptive distribution-free test. *Commun. in Statist. – Simula.* **16**, 55–68.

Ibragimov, J.A. (1956). On the composition of unimodal distributions. (Russian.) *Teoriya veroyatnostey* **1**, 283–288.

Isserlis, L. (1931). On the moments distributions of moments in the case of samples drawn from a limited universe. *Proc. Roy. Soc. A*, **132**, 586.

Jaeckel, L.A. (1972). Estimating regression coefficients by minimizing the dispersion of the residuals. *AMS* **43**, 1449–1458.

Janssen, A. and Mason, D.M. (1990). *Non-Standard Rank Tests.* Springer, New York.

Johnson, B.M. and Killeen, T.J. (1985). Kolmogorov-Smirnov confidence sets: the case of more than one sample. *Commun. Statist. – Theory Meth.* **14**, 2873–2885.

Jones, D.H. (1979). An efficient adaptive distribution-free test for location. *JASA* **74**, 822–828.

Jurečková, J. (1969). Asymptotic linearity of a rank statistic in regression parameter. *AMS* **40**, 1889–1900.

Jurečková, J. (1971a). Nonparametric estimate of regression coefficients. *AMS* **42**, 1328–1338.

Jurečková, J. (1971b). Asymptotic independence of rank test statistic for testing symmetry on regression. *Sankhya Ser. A* **33**, 1–18.

Jurečková, J. (1973). Almost sure uniform asymptotic linearity of rank statistics in regression parameter. *Trans. 6th Prague Conf. on Inf. Theory, Random Proc. and Statist. Dec. Functions*, Prague 1971. Academia, Prague, pp. 305–313.

Jurečková, J. (1975). Nonparametric estimation and testing linear hypotheses in the linear regression model. *Math. Operationsforsch. u. Statistik* **6**, 269–283.

Jurečková, J. (1984). *M*-, *L*- and *R*-estimators. In: P.R. Krishnaiah, P.K. Sen, eds., *Handbook of Statistics*, vol. 4. Elsevier Science Publishers, Amsterdam, pp. 463–485.

Jurečková, J. (1992a). Uniform asymptotic linearity of regression rank scores process. *Nonparametric Statistics and Related Topics* (ed. A.K.M.E. Saleh), North Holland, Amsterdam, pp. 217–228.

Jurečková, J. (1992b). Estimation in a linear model based on regression rank scores. *J. Nonparamet. Statist.* **1**, 197–203.

Jurečková, J. (1995). Regression rank scores: Asymptotic linearity and RR-estimators. *Proc. MODA'4* (eds. C.P. Kitsos and W.G. Müller), Physica Verlag, Heidelberg, pp. 193–203.

Jurečková and J., Sen, P.K. (1993). Asymptotic equivalence of regression rank scores estimators and R-estimators in linear models. *Statistics and Probability: A R. R. Bahadur Festschrift* (eds. J.K. Ghosh et al.) Wiley Eastern, New Delhi, pp. 279–292.

Jurečková, J. and Sen, P.K. (1994). Regression rank scores statistics and studentization in the linear model. *Proc. 5th Prague Confer. Asymptotic Statistics* (eds. M. Hušková and P. Mandl), Physica-Verlag, Vienna, pp. 111–121.

Jurečková, J. and Sen, P.K. (1996). *Robust Statistical Procedures: Asymptotics and Interrelations.* Wiley, New York.

Kamat, A.R. (1956). A two-sample distribution-free test. *Biometrika* **43**, 377–387.

Kemperman, J.H.B. (1959). Asymptotic expansions for the Smirnov test and for the range of cumulative sums. *AMS* **30**, 448–462.

Kendall, M.G. (1938). A new measure of rank correlation. *Biometrika* **30**, 81–93.

Kendall, M.G. (1948). *Rank Correlation Methods.* Griffin & Co, London. (1st edition 1948, 3rd edition 1962.)

Kendall, M.G., Kendall, S.F.H. and Babington Smith, B. (1938). The distribution of Spearman's coefficient of rank correlation in a universe in which all rankings occur an equal number of times. *Biometrika* **30**, 251–273.

Kendall, M.G. and Babington Smith, B. (1939). The problem of m rankings. *AMS* **10**, 275–287.

Kiefer, J. (1959). K-sample analogues of the Kolmogorov-Smirnov and Cramér--von Mises tests. *AMS* **30**, 420–447.

Klotz, J. (1962). Nonparametric tests for scale. *AMS* **33**, 498–512.

Klotz, J. (1963). Small sample power and efficiency for the one sample Wilcoxon and normal scores tests. *AMS* **34**, 624–632.

Klotz, J. (1964). On the normal scores two-sample rank test. *JASA* **59**, 652–664.

Klotz, J. (1967). Asymptotic efficiency of the two-sample Kolmogorov-Smirnov test. *JASA* **62**, 932–938.

Koenker, R. and Bassett, G. (1978). Regression quantiles. *Econometrika* **46**, 33–50.

Koenker, R. and d'Orey, V. (1993). A remark on computing regression quantiles. *Appl. Statist.* **36**, 383–393.

Kolmogorov, A.N. (1933). Sulla determinazione empirica di una legge di distribuzione. *Giorn. dell'Istituto Ital. Degli Attuari* **4**, 83–91.

Konijn, H.S. (1956). On the power of certain tests for independence in bivariate populations. *AMS* **27**, 300–323. Correction: *AMS* **29** (1958), 935.

Korolyuk, V.S. (1955a). On the deviation of empirical distributions for the case of two independent samples. (Russian.) *Izvestiya AN SSSR* **19**, 81–96.

Korolyuk, V.S. (1955b). Asymptotic expansions for goodness-of-fit statistics of A.N. Kolmogorov and N.V. Smirnov. (Russian.) *Izvestiya AN SSSR* **19**, 103–124.

Koul, H.L. (1969). Asymptotic behaviour of Wilcoxon type confidence regions in multiple linear regression. *AMS* **40**, 1950–1979.

Koul, H.L. (1971). Asymptotic behavior of a class of confidence regions based on ranks in regression. *AMS* **42**, 466–476.

Kraft, C.H. and van Eeden, C. (1970). Efficient linearized estimates based on ranks. In: M.L. Puri, ed., *Nonparametric Techniques in Statistical Inference.* Cambridge Univ. Press, pp. 267–273.

Kraft, C.H. and van Eeden, C. (1972). Linearized rank estimates and signed-rank estimates for the general linear hypothesis. *AMS* **43**, 42–57.

Kremer, E. (1983). Bahadur efficiency of linear rank tests – a survey. *Acta Univ. Carolinae – Math et Physica* vol. **24**, 61–75.

Kruskal, W.H. (1952). A nonparametric test for the several sample problem. *AMS* **23**, 525–540.

Kruskal, W.H. (1958). Ordinal measures of association. *JASA* **53**, 814–861.

Kruskal, W.H. and Wallis, W.A. (1952). Use of ranks in one-criterion variance analysis. *JASA* **47**, 583–621. Errata: *JASA* **48** (1953), 910.

van der Laan (1964). Exact power of some rank tests. Presented at the I.M.S. Conference in Bern in 1964.

LeCam, L. (1960). Locally asymptotically normal families of distributions. Univ. of Calif. Publ. in Stat. 3, 37–98.

LeCam, L. and Yang, G.L. (1990). *Asymptotics in Statistics: Some Basic Concepts.* Springer, New York.

Lehmann, E.L. (1951). Consistency and unbiasedness of certain nonparametric tests. *AMS* **22**, 165–179.

Lehmann, E.L. (1953). The power of rank tests. *AMS* **24**, 23–43.

Lehmann, E.L. (1959). *Testing Statistical Hypotheses.* J. Wiley, New York.

Lehmann, E.L. (1963a). A class of selection procedures based on ranks. *Math. Annalen* **150**, 268–275.

Lehmann, E.L. (1963b). Nonparametric confidence intervals for a shift parameter. *AMS* **34**, 1507–1512.

Lehmann, E.L. (1966). Some concepts of dependence. *AMS* **37**, 1137–1153.

Lehmann, E.L. and Stein, C. (1949). On the theory of some non-parametric hypotheses. *AMS* **20**, 28–45.

Lloyd, E.H. (1952). Least squares estimation of location and scale parameters using order statistics. *Biometrika* **34**, 41–67.

Loève, M. (1955). *Probability Theory.* Van Nostrand, Princeton. (1st edition 1955, 2nd edition 1960.)

Lukaszewicz, J. and Sadowski, W. (1960/61). On comparing several populations with a control population. *Zastosowania matematyki* **5**, 309–320.

Mann, H.B. and Whitney, D.R. (1947). On a test of whether one of two random variables is stochastically larger than the other. *AMS* **18**, 50–60.

Massey, F.J. Jr. (1951). The distribution of the maximum deviation between two sample cumulative step functions. *AMS* **22**, 125–128.

McCullagh, P. and Nelder, J.A. (1989). *Generalized Linear Models*, 2nd ed. Chapman & Hall, London.

McKean, J.W. and Hettmansperger, T.P. (1976). Tests of hypotheses based on ranks in the general linear model. *Commun. Statist.* **A5**, 693–709.

McKean, J.W. and Hettmansperger, T.P. (1978). A robust analysis of the general linear model based on one-step R-estimates. *Biometrika* **65**, 571–579.

McLeish, D.L. (1974). Dependent central limit theorems and invariance principles. *Ann. Probability* **2**, 620–628.

Mikulski, P.W. (1963). On the efficiency of optimal nonparametric procedures in the two sample case. *AMS* **34**, 22–32.

von Mises, R. (1931). *Wahrscheinlichkeitsrechnung.* Springer, Leipzig-Wien.

von Mises, R. (1947). On the asymptotic distribution of differentiable statistical fucntions. *AMS* **18**, 309–348.

Mohanty, G. (1979). *Lattice Path Counting and Applications.* Academic Press, Springer, New York.

Mood, A.M. (1950). *Introduction to the Theory of Statistics.* McGraw-Hill, New York.

Mood, A.M. (1954). On the asymptotic efficiency of certain nonparametric two--sample tests. *AMS* **25**, 514–522.

Moses, L.E. (1963). Rank tests of dispersion. *AMS* **34**, 973–983.

Mosteller, F. (1948). A k-sample slippage test for an extreme population. *AMS* **19**, 58–65.

Motoo, M. (1957). On the Hoeffding's combinatorial central limit theorem. *Ann. Inst. Stat. Math.* **8**, 145–154.

Natanson, I.P. (1957). *Theory of Functions of a Real Variable.* (Russian.) 2nd edition, Gos. izd. tech. teor. lit., Moscow.

Neuhaus, G. (1987). Local asymptotics for linear rank statistics with estimated score functions. *AS* **15**, 491–512. Addendum: *AS* **16** (1988), 1342–1343.

Nikitin, Ya. (1995). *Asymptotic Efficiency of Nonparametric Tests.* Cambridge Univ. Press.

Noether, G.E. (1955). On a theorem of Pitman. *AMS* **26**, 64–68.

Noether, G.E. (1958). The efficiency of some distribution-free tests. *Stat. Neerlandica* **12**, 63–73.

Noether, G.E. (1963). Note on the Kolmogorov statistic in the discrete case. *Metrika* **7**, 115–116.

Olds, E.G. (1938). Distributions of sums of squares of rank differences for small numbers of individuals. *AMS* **9**, 133–148.

Oosterhoff, J. and van Zwet, W.R. (1979). A note on contiguity and Hellinger distance. *Contributions to Statistics*, Jaroslav Hájek Memorial Volume (ed. J. Jurečková), Academia, Prague, 157–166.

Pearson, E.S. and Snow, B.A.S. (1962). Tests for rank correlation coefficients III. Distribution of the transformed Kendall coefficient. *Biometrika* **49**, 185–191.

Pfanzagl, J. (1979). First order efficiency implies second order efficiency. *Contributions to Statistics*, J. Hájek Memorial Volume, (ed. J. Jurečková), Academia, Prague, 167–196.

Pitman, E.J.G. (1937/38). Significance tests which may be applied to samples from any populations. I. *Suppl. JRSS* **4** (1937), 119–130. II. The correlation coefficient test. *Suppl. JRSS* **4** (1937), 225–232. III. The analysis of variance test. *Biometrika* **29** (1938), 322–335.

Pitman, E.J.G. (1948). *Notes on Nonparametric Statistical Inference.* Columbia Univ., New York (mimeographed).

Prášková, Z. (1976). Asymptotic expansion and a local limit theorem for the signed Wilcoxon statistic. *Comment. Math. Univ. Carolinae* **17**, 335–343.

Puri, M.L. and Sen, P.K. (1971). *Nonparametric Methods in Multivariate Analysis.* Wiley, New York.

Puri, M.L. and Sen, P.K. (1985). *Nonparametric Methods in General Linear Models.* Wiley, New York.

Pyke, R. and Shorack, G. R. (1968a). Weak convergence of two-sample empirical process and a new approach to Chernoff-Savage theorems. *AMS* **39**, 755–771.

Pyke, R. and Shorack, G. R. (1968b). Weak convergence and a Chernoff-Savage theorem for random sample sizes. *AMS* **39**, 1675–1685.

Randles, R.H. and Hogg, R.V. (1973). Adaptive distribution-free tests. *Commun. in Statist.* **2**, 337–356.

Rao, P.V., Schuster, E.F. and Littell, R.C. (1975). Estimation of shift and center of symmetry based on Kolmogorov-Smirnov statistics. *AS* **3**, 862–873.

Reimann, J. and Vincze, I. (1960). On the comparison of two samples with slightly different sizes. *Magyar Tud. Akad. Matem. Kutató Int. Közleményei* **5**, 293–309.

Rényi, A. (1953a). On the theory of order statistics. *Acta math. Acad. sci. hung.* **4**, 191–231.

Rényi, A. (1953b). Neue Kriterien zum Vergleich zweier Stichproben. (Hungarian.) *Magyar Tud. Akad. Alkalmazott Matem. Int. Közleményei* **2**, 243–265.

Rieder, H. (1994). *Robust Asymptotic Statistics.* Springer, Heidelberg.

Rijkoort, P.J. and Wise, M.E. (1953). Simple approximations and nomograms for two ranking tests. *Indagationes math.* **15** (*Proc. Kon. Nederl. Akad. Wet.* **56**), 294–302.

Rizvi, M.H. and Woodworth, G.G. (1970). On selection procedures based on ranks: counterexamples concerning least favorable configurations. *AMS* **41**, 1942–1951.

Ruberg, S.J. (1986). A continuously adaptive nonparametric two-sample test. *Commun. in Statist. – Theory Meth.* **15**, 2899–2920.

Ruymgaart, F.H. (1974). Asymptotic normality of nonparametric tests for independence. *AS* **2**, 892–910.

Ruymgaart, F.H., Shorack, G.R. and van Zwet, W.R. (1972). Asymptotic normality of nonparametric tests for independence. *AMS* **43**, 1122–1135.

Sadowski, W. (1960/61). A non-parametric test of comparing dispersions. *Zastosowania matematyki* **5**, 299–308.

Sarkadi, K. (1961). On Galton's rank order test. *Magyar Tud. Akad. Matem. Kutató Int. Közleményei* **6**, 127–131.

Savage, I.R. (1956). Contributions to the theory of rank order statistics – the two-sample case. *AMS* **27**, 590–615.

Savage, I.R. (1962). *Bibliography of Nonparametric Statistics.* Harvard Univ. Press, Cambridge, Mass.

Scheffé, H. (1947). A useful convergence theorem for probability distributions. *AMS* **18**, 434–438.

Schneller, W. (1989). Edgeworth expansions for linear rank statistics. *AS* **17**, 1103–1123.

Schuster, E.F. and Narvarte, J.A. (1973). A new nonparametric estimator of the center of a symmetric distribution. *AS* **1**, 1096–1104.

Sen, P.K. (1963). On the estimation of relative potency in dilution (-direct) assays by distribution-free methods. *Biometrics* **19**, 532–552.

Sen, P.K. (1964). On some properties of the rank-weighted means. *J. Indian Society of Agricultural Statistics* **16**, 51–61.

Sen, P.K. (1966). On a distribution-free method of estimating asymptotic efficiency of a class of nonparametric tests. *AMS* **37**, 1759–1770.

Sen, P.K. (1967). Asymptotically most powerful rank order tests for grouped data. *AMS* **38**, 1229–1239.

Sen, P.K. (1968a). Estimates of regression coefficients based on Kendall's tau. *JASA* **63**, 1379–1389.

Sen, P.K. (1968b). On a class of aligned rank order tests in two-way layouts. *AMS* **39**, 1115–1124.

Sen, P.K. (1968c). On a further robustness property of the test and estimator based on Wilcoxon's signed rank statistic. *AMS* **39**, 282–285.

Sen, P.K. (1968d). Robustness of some nonparametric procedures in linear models. *AMS* **39**, 1913–1922.

Sen, P.K. (1968e). Asymptotically efficient tests by the method of n rankings. *JRSS B* **30**, 312–317.

Sen, P.K. (1969). On a class of rank order tests for the parallelism of several regression lines. *AMS* **40**, 1668–1683.

Sen, P.K. (1970). On some convergence properties of one-sample rank order statistics. *AMS* **41**, 2137–2139.

Sen, P.K. (1972). Finite population sampling and weak convergence to a Brownian bridge. *Sankhya, Ser A* **34**, 85–90.

Sen, P.K. (1976). A two-dimensional functional permutational central limit theorem for linear rank statistics. *Ann. Probability* **4**, 474–479.

Sen, P.K. (1979). Weak convergence of some quantile processes arising in progressively censored tests. *AS* **7**, 414–431.

Sen, P.K. (1980). On almost sure linearity theorems for signed rank order statistics. *AS* **8**, 313–321.

Sen, P.K. (1981). *Sequential Nonparametrics: Invariance Principles and Statistical Inference.* Wiley, New York.

Sen, P.K. (1983). On permutational central limit theorems for general multivariate linear rank statistics. *Sankhya, Ser. A* **45**, 141–149.

Sen, P.K. (1991). Nonparametrics: Retrospectives and perspectives (with discussion). *J. Nonparam. Statist.* **1**, 1–68.

Sen, P.K. (1995). Statistical functionals, Hadamard differentiability and martingales. *A Festschrift for Professor J. Medhi* (ed. A.C. Borthakur and H. Chowdhury) Wiley Eastern, New Delhi, pp. 29–47.

Sen, P.K. (1996). Regression rank scores estimation in ANOCOVA. *AS* **24**, 1586–1602.

Sen, P.K. and Ghosh, M. (1971). On bounded length sequential confidence intervals based on one-sample rank order statistics. *AMS* **42**, 189–203.

Sen, P.K. and Ghosh, M. (1972). On strong convergence of regression rank statistics. *Sankhya, Ser A* **34**, 335–348.

Sen, P.K. and Ghosh, M. (1973). A law of iterated logarithm for one-sample rank order statistics and an application. *AS* **1**, 568–576.

Sen, P.K. and Ghosh, M. (1974). Sequential rank tests for location. *AS* **2**, 540–552.

Sen, P.K. and Krishnaiah, P.R. (1984). Selected tables for nonparametric statistics. In: P.R. Krishnaiah, P.K. Sen, eds., *Handbook of Statistics*, vol. 4. Elsevier Science Publishers, Amsterdam, pp. 937–958.

Sen, P.K. and Puri, M.L. (1967). On the theory of rank order tests for location in the multivariate one sample problem. *AMS* **38**, 1216–1228.

Sen, P.K. and Puri, M.L. (1977). Asymptotically distribution-free aligned rank order tests for composite hypotheses for general multivariate linear models. *Zeit. Wahrsch. verw. Geb.* **39**, 175–186.

Sen, P.K. and Singer, J.M. (1993). *Large Sample Methods in Statistics: An Introduction with Applications.* Chapman and Hall, New York.

Serfling, R.J. (1980). *Approximation Theorems of Mathematical Statistics.* Wiley, New York.

Shirahata, S. (1974). Locally most powerful rank tests for independence. *Bull. Math. Statist.* **16**, 11–21.

Siegel, S. (1956). *Nonparametric Statistics for the Behavioral Sciences.* McGraw-Hill, New York.

Siegel, S. and Tukey, J.W. (1960). A nonparametric sum of ranks procedure for relative spread in unpaired samples. *JASA* **55**, 429–445. Correction: *JASA* **56** (1961), 1005.

Singh, K. (1984). Asymptotic comparison of tests – a review. In: P.R. Krishnaiah, P.K. Sen, eds., *Handbook of Statistics*, vol. 4. Elsevier Science Publishers, Amsterdam, pp. 173–184.

Skorokhod, A.V. (1956). Limit theorems for stochastic processes. *Theory Prob. Appl.* **1**, 261–290.

Smirnov, N.V. (1939). Estimate of deviation between empirical distribution functions in two independent samples. (Russian.) *Bulletin Moscow univ.* **2**, No. 2, 3–16.

Steck, G.P. (1969). The Smirnov two-sample tests as rank tests. *AMS* **40**, 1449–1466.

Steel, R.G.D. (1959a). A multiple comparison rank sum test: treatments versus control. *Biometrics* **15**, 560–572.

Steel, R.G.D. (1959b). A multiple comparison sign test: treatments versus control. *JASA* **54**, 767–775.

Steel, R.G.D. (1960). A rank sum test for comparing all pairs of treatments. *Technometrics* **2**, 192–207.

Steel, R.G.D. (1961). Some rank sum multiple comparisons test. *Biometrics* **17**, 539–552.

Stone, M. (1967). Extreme tail probabilities for the null distribution of the two-sample Wilcoxon statistic. *Biometrika* **54**, 629–640.

Stone, M. (1968). Extreme tail probabilities for sampling without replacement and exact Bahadur efficiency of the two-sample normal scores test. *Biometrika* **55**, 371–375.

Strassen, V. (1964). An invariance principle for the law of iterated logarithm. *Zeitschr. Wahrscheim. verw. Gebiete* **3**, 211–226.

Stuart, A. (1954). The asymptotic relative efficiencies of tests and the derivatives of their power functions. *Skand. Aktuarietids.* **37**, 163–169.

Stuart, A. (1958). The measurement of estimation and test efficiency. *Bull. Int. Statist. Inst.* **36**, No. 3, 79–86.

Sukhatme, B.V. (1957). On certain two-sample non-parametric tests for variances. *AMS* **28**, 188–194.

Sukhatme, B.V. (1958a). Testing the hypothesis that two populations differ only in location. *AMS* **29**, 60–78.

Sukhatme, B.V. (1958b). A two-sample distribution-free test for comparing variances. *Biometrika* **45**, 544–549.

Sukhatme, B.V. (1960). Power of some two-sample non-parametric tests. *Biometrika* **47**, 355–362.

Sundrum, R.M. (1954). On Lehmann's two-sample test. *AMS* **25**, 139–145.

Tardif, S. (1985). Sur la linéarité asymptotique du premier ordre de statistiques de rang signé. *Canad. J. Statist.* **13**, 303–310.

Teichroew, D. (1955). Empirical power functions for nonparametric two-sample tests for small samples. *AMS* **26**, 340–344.

Terry, M.E. (1952). Some rank order tests which are most powerful against specific parametric alternatives. *AMS* **23**, 346–366.

Thompson, R., Govindarajulu, Z. and Doksum, K. (1967). Distribution and power of the absolute normal scores test. *JASA* **62**, 966–975.

Tukey, J.W. (1949). The simplest signed-rank tests. Mem. Report 17, Statistical Research Group, Princeton Univ.

Vincze, I. (1957). Einige zweidimensionale Verteilungs- und Grenzverteilungssätze in der Theorie der geordneten Stichproben I. *Magyar Tud. Akad. Mat. Kutató Int. Közleményei* **2**, 183–209.

Vincze, I. (1959). On some joint distributions and joint limiting distributions in the theory of order statistics II. *Magyar Tud. Akad. Mat. Kutató Int. Közleményei* **4**, 29–47.

Vincze, I. (1960). On two-sample tests based on order statistics. *Proc. 4th Berkeley Symposium*, Univ. Calif. Press, Los Angeles, Calif., vol. I, 695–705.

Vincze, I. (1963). A generating function in the theory of order statistics. *Publicationes Math.* **10**, 82–87.

van der Waerden, B.L. (1952/53). Order tests for the two-sample problem and their power. I, II, III. *Indagationes math.* **14** (*Proc. Kon. Nederl. Akad.*

Wet **55**), 453–458; *Indag.* **15** (*Proc.* **56**), 303–310, 311–316; Correction: *Indag.* **15** (Proc. **56**), 80.

van der Waerden, B.L. (1953). Ein neuer Test für das Problem der zwei Stichproben. *Math. Annalen* **126**, 93–107.

van der Waerden, B.L. (1956). The computation of the X-distribution. *Proc. 3rd Berkeley Symposium*, Univ. of Calif. Press, Los Angeles, Calif., vol. I, 207–208.

van der Waerden, B.L. and Nievergelt, E. (1956). *Tafeln zum Vergleich zweier Stichproben mittels X-test und Zeichentest.* Springer, Berlin-Göttingen-Heidelberg.

Wald, A. and Wolfowitz, J. (1940). On a test whether two samples are from the same population. *AMS* **11**, 147–162.

Wallace, D.L. (1959). Simplified beta-approximations to the Kruskal-Wallis H-test. *JASA* **54**, 225–230.

Walsh, J.E. (1946). On the power function of the sign test for slippage of means. *AMS* **17**, 358–362.

Walsh, J.E. (1962). *Handbook of Nonparametric Statistics.* Van Nostrand, Princeton.

Walsh, J.E. (1963). Bounded probability properties of Kolmogorov Smirnov and similar statistics for discrete data. *Ann. Inst. Stat. Math.* **15**, 153–158.

Wegner, L.H. (1956). Properties of some two-sample tests based on a particular measure of discrepancy. *AMS* **27**, 1006–1016.

Westenberg, J. (1948). Significance test for median and interquartile range in samples from continuous populations of any form. *Proc. Kon. Nederl. Akad. Wet.* **51**, 252–261.

Wieand, H.S. (1976). A condition under which the Pitman and Bahadur approaches to efficiency coincide. *AS* **4**, 1003–1011.

Wilcoxon, F. (1945). Individual comparisons by ranking methods. *Biometrics Bull.* **1**, 80–83.

Witting, H. (1960). A generalized Pitman efficiency for non-parametric test. *AMS* **31**, 405–414.

Woodworth, G.G. (1970). Large deviations and Bahadur efficiency of linear rank statistics. *AMS* **41**, 251–283.

Zajta, A. (1960). On the Lehmann test. (Hungarian.) *Magyar Tud. Akad. Mat. Kutató Int. Közleményei* **5**, 447–459.

van Zwet, W.R. (1982). On the Edgeworth expansion for the simple linear rank statistic. *Colloquia Math. Soc. J. Bolyai*, **32**, Budapest 1980. *Nonparametric Statistical Inference*, vol. 2, 889–909. North-Holland, Amsterdam.

van Zwet, W.R. (1984). A Berry-Esséen bound for symmetric statistics. *Zeitschr. Wahrscheinlich. verw. gebiete* **66**, 425–440.

Subject index

Author index

Index of mathematical symbols

A symbol is usually followed by a brief (and rather rough) indication of its meaning, intended to help the reader to recall it, and then by the number of the page where its definition is given.

PROBABILITY AND MATHEMATICAL STATISTICS

Thomas Ferguson, *Mathematical Statistics: A Decision Theoretic Approach*
Howard Tucker, *A Graduate Course in Probability*
K. R. Parthasarathy, *Probability Measures on Metric Spaces*
P. Révész, *The Laws of Large Numbers*
H. P. McKean, Jr., *Stochastic Integrals*
B. V. Gnedenko, Yu. K. Belyayev and A. D. Solovyev, *Mathematical Methods of Reliability Theory*
Demetrios A. Kappos, *Probability Algebras and Stochastic Spaces*
Ivan N. Pesin, *Classical and Modern Integration Theories*
S. Vajda, *Probabilistic Programming*
Robert B. Ash, *Real Analysis and Probability*
V. V. Fedorov, *Theory of Optimal Experiments*
K. V. Mardia, *Statistics of Directional Data*
H. Dym and H. P. McKean, *Fourier Series and Integrals*
Tatsuo Kawata, *Fourier Analysis in Probability Theory*
Fritz Oberhettinger, *Fourier Transforms of Distributions and Their Inverses: A Collection of Tables*
Paul Erdös and Joel Spencer, *Probabilistic Methods of Statistical Quality Control*
K. Sarkadi and I. Vincze, *Mathematical Methods of Statistical Quality Control*
Michael R. Anderberg, *Cluster Analysis for Applications*
W. Hengartner and R. Theodorescu, *Concentration Functions*
Kai Lai Chung, *A Course in Probability Theory, Second Edition*
L. H. Koopmans, *The Spectral Analysis of Time Series*
L. E. Maistrov, *Probability Theory: A Historical Sketch*
William F. Stout, *Almost Sure Convergence*
E. J. McShane, *Stochastic Calculus and Stochastic Models*
Robert B. Ash and Melvin F. Gardner, *Topics in Stochastic Processes*
Avner Friedman, *Stochastic Differential Equations and Applications, Volume 1, Volume 2*
Roger Cuppens, *Decomposition of Multivariate Probabilities*
Eugene Lukacs, *Stochastic Convergence, Second Edition*
H. Dym and H. P. McKean, *Gaussian Processes, Function Theory, and the Inverse Spectral Problem*
N. C. Giri, *Multivariate Statistical Inference*
Lloyd Fisher and John McDonald, *Fixed Effects Analysis of Variance*
Sidney C. Port and Charles J. Stone, *Brownian Motion and Classical Potential Theory*
Konrad Jacobs, *Measure and Integral*

K. V. Mardia, J. T. Kent and J. M. Biddy, *Multivariate Analysis*
Sri Gopal Mohanty, *Lattice Path Counting and Applications*
Y. L. Tong, *Probability Inequalities in Multivariate Distributions*
Michel Metivier and J. Pellaumail, *Stochastic Integration*
M. B. Priestly, *Spectral Analysis and Time Series*
Ishwar V. Basawa and B. L. S. Prakasa Rao, *Statistical Inference for Stochastic Processes*
M. Csörgö and P. Révész, *Strong Approximations in Probability and Statistics*
Sheldon Ross, *Introduction to Probability Models, Second Edition*
P. Hall and C. C. Heyde, *Martingale Limit Theory and Its Applications*
Imre Csiszár and János Körner, *Information Theory: Coding Theorems for Discrete Memoryless Systems*
A. Hald, *Statistical Theory of Sampling Inspection by Attributes*
H. Bauer, *Probability Theory and Elements of Measure Theory*
M. M. Rao, *Foundations of Stochastic Analysis*
Jean-Rene Barra, *Mathematical Basis of Statistics*
Harald Bergström, *Weak Convergence of Measures*
Sheldon Ross, *Introduction to Stochastic Dynamic Programming*
B. L. S. Prakasa Rao, *Nonparametric Functional Estimation*
M. M. Rao, *Probability Theory with Applications*
A. T. Bharucha-Reid and M. Sambandham, *Random Polynomials*
Sudhakar Dharmadhikari and Kumar Joag-dev, *Unimodality, Convexity, and Applications*
Stanley P. Gudder, *Quantum Probability*
B. Ramachandran and Ka-Sing Lau, *Functional Equations in Probability Theory*
B. L. S. Prakasa Rao, *Identifiability in Stochastic Models: Characterization of Probability Distributions*
Moshe Shaked and J. George Shanthikumar, *Stochastic Orders and Their Applications*
Barry K. Moser, *Linear Models: A Mean Model Approach*
Jaroslav Hájek, Zbyněk Šidák and Pranab K. Sen, *Theory of Rank Tests, Second Edition*